REGIONAL CLIMATES OF THE BRITISH ISLES

In recent yea[illegible] global climate chang[illegible] g have commanded unprecedented public intere[illegible] This book presents a [illegible] the climate of the British Isles. A regional appr[illegible]h provides a framewo[illegible] tween geography and meteorology, together wit[illegible]e resulting variety [illegible] he British Isles. The final chapter shifts the fo[illegible] from climatic variab[illegible] r time, evaluating the causes and regional imp[illegible]ions of climatic cha[illegible]

In Part One[illegible]ontributors examine [illegible] ates evolve from the interplay of meteorologic[illegible]onditions and the g[illegible] provides a concise introduction to mid-latitude [illegible]nates and is illustrat[illegible] xamples.

Part Two is a[illegible]-to-date survey of th[illegible] gions of the British Isles. The aim of each chapter [illegible]o assess the climatic distinctiveness of each region together with an explanation of local weather cont[illegible]. Each chapter includes a comprehensive summary of climatic data using the 1961–90 climatic aver[illegible]s.

Part Three r[illegible]s the spatial dimension of Part Two to climatic change. Following a review of the causes and history [illegible]matic change, evidence for future climatic change is compared with present day climatic variations a[illegible]eir consequences. Discussion of the implications of climatic change in this regional framework a[illegible]wledges increasing evidence for regionally diverse responses to the greenhouse effect.

This study [illegible]s regional climates in the context of a north-west/south-east climatic gradient across the British Isles [illegible]which the more changeable, dynamic and wetter climate of north-west Britain is distinguished fro[illegible]t of the south-east. Evidence for a recent strengthening of this contrast is compared with similar resul[illegible] rom the most recent climate models.

Dennis Wh[illegible]ler is a Reader in Geography, University of Sunderland; **Julian Mayes** is a Senior Lecturer in the Departm[illegible]nt of Environmental and Geographical Studies, Roehampton Institute London.

REGIONAL CLIMATES OF THE BRITISH ISLES

edited by

Dennis Wheeler and Julian Mayes

London and New York

First published 1997
by Routledge
11 New Fetter Lane, London EC4P 4EE
29 West 35th Street, New York, NY 10001

Typeset in Garamond by J&L Composition Ltd, Filey, North Yorkshire

Printed and bound in Great Britain by Butler & Tanner Ltd, Frome, Somerset

British Library Cataloguing in Publication Data
A catalogue record for this book is available from the British Library

Library of Congress Cataloging-in-Publication Data
A catalogue record for this book has been requested

ISBN 0–415–13930–9 (hbk)
0–415–13931–7 (pbk)

CONTENTS

PART 2: THE CHARACTER OF REGIONAL CLIMATES ACROSS THE BRITISH ISLES

FIGURES

INTRODUCTION

CHAPTER 1 THE ANATOMY OF REGIONAL CLIMATES IN THE BRITISH ISLES

CHAPTER 2 SOUTH-WEST ENGLAND AND THE CHANNEL ISLANDS

CHAPTER 5 THE MIDLANDS

CHAPTER 6 WALES

CHAPTER 7 NORTH-EAST ENGLAND AND YORKSHIRE

CHAPTER 8 NORTH-WEST ENGLAND AND THE ISLE OF MAN

CHAPTER 9 CENTRAL AND SOUTHERN SCOTLAND

CHAPTER 10 HIGHLAND AND ISLAND SCOTLAND

CHAPTER 11 IRELAND

CHAPTER 12 REGIONAL PERSPECTIVES ON CLIMATE VARIABILITY AND CHANGE

TABLES

CHAPTER 1 THE ANATOMY OF REGIONAL CLIMATES IN THE BRITISH ISLES

CHAPTER 2 SOUTH-WEST ENGLAND AND THE CHANNEL ISLANDS

CHAPTER 3 SOUTH-EAST ENGLAND

CHAPTER 8 NORTH-WEST ENGLAND AND THE ISLE OF MAN

CHAPTER 9 CENTRAL AND SOUTHERN SCOTLAND

CHAPTER 10 HIGHLAND AND ISLAND SCOTLAND

CHAPTER 11 IRELAND

CHAPTER 12 REGIONAL PERSPECTIVES ON CLIMATIC VARIABILITY AND CHANGE

CONTRIBUTORS

Brian Giles	School of Geography, University of Birmingham, PO Box 363, Birmingham B15 2TT
John Harrison	School of Environmental Science, University of Stirling, Stirling FK9 4LA
John Kings	School of Geography, University of Birmingham, PO Box 363, Birmingham B15 2TT
Julian Mayes	Department of Environmental and Geographical Studies, Roehampton Institute London, Wimbledon Parkside, London SW19 5NN
Allen Perry	Department of Geography, University of Wales at Swansea, Singleton Park, Swansea SA2 8PP
Marjory Roy	Department of Meteorology, University of Edinburgh, James Clerk Maxwell Building, Mayfield Road, Edinburgh EH9 3JZ
Graham Sumner	Department of Geography, St David's University College, Lampeter, Ceredigion SA48 7ED
Geoffrey Sutton	Norwich Weather Centre, Rouen House, Rouen Road, Norwich NR1 1RB
John Sweeney	Department of Geography, St Patrick's College, Maynooth, Co. Kildare, Republic of Ireland
Lance Tufnell	Department of Geography, University of Huddersfield, Queensgate, Huddersfield HD1 3DH
Dennis Wheeler	Department of Geography, School of the Environment, University of Sunderland, Sunderland SR1 3SD

PREFACE

This book has its origins in the Geography Departments of Hull University and of the University College of Swansea where the two editors received their undergraduate training and a lasting interest in the science, sometimes art, of climatology. This interest was sustained through the activities of the Royal Meteorological Society and the Association of British Climatologists, both of which provided an invaluable forum for the exchange of ideas and enthusiasms of British and Irish climatologists. It was in this fertile and amiable academic soil that the seeds of the idea for a thoroughly regional approach to climatic studies were sown and took root. The editors believe that the regional approach is appropriate at a time when regional anomalies in weather and climate have attracted extensive attention, and in the following chapters we hope to illustrate and account for patterns of regional meteorological diversity. This is by no means the first attempt to inform and engage a relatively wide audience on the weather of our islands. If this new text shows evidence of having inherited any of the characteristics and vigour of Hubert Lamb and C.E.P. Brook's *The English Climate*, of Tony Chandler and Stanley Gregory's *The Climate of the British Isles* and, perhaps most especially, of Gordon Manley's *Climate and the British Scene*, then it can be considered a healthy specimen indeed.

So many people have contributed, perhaps unwittingly, to this book that it is invidious to single out any in particular. But special thanks must go to Sarah Lloyd and Matthew Smith of Routledge for their patience and forbearance in waiting for the text over what was an unexpectedly long gestation period. Our colleagues and friends in the departments at Sunderland University and Roehampton Institute London, and in particular Mary Mackenzie, also deserve recognition for the support they have given. Thanks also go to the numerous cartographers in the Departments at Maynooth College, the University of Stirling, the University of Sunderland, the University of Wales (Lampeter) and the University of Wales (Swansea) who with their usual enthusiasm and skill produced many of the text illustrations. A large proportion of the statistical data were abstracted from the *Monthly Weather Reports* published by the Meteorological Office. Met Eireann provided the data used in Chapter 11. Our thanks go also to the contributing authors who so unwittingly offered themselves to be pestered and imposed upon when agreeing to provide the chapters at the heart of the book. Finally, we thank our families for their help and patience in a project which consumed so much of our time and attention.

ACKNOWLEDGEMENTS

The editors and authors are grateful to the following for permission to reproduce copyright material and in many cases for supplying the items in question:

The University of Dundee (NERC Satellite Unit) for Figures 3.5, 4.10, 6.12, 6.14, 7.2, 7.3, 7.5, 7.11 and 8.8; the Louth Naturalists, Antiquarian and Literary Society for Figure 4.8; the Institute of Hydrology for Figure 6.7; the *South Wales Evening Post* for Figures 6.15 and 6.16; the Friends of the Lake District for Figure 8.3; Derbyshire Times Newspapers Ltd for Figure 8.6; the *Irish Times* for Figure 11.11; the Geography Department of the University of Durham for Figure 12.10; the Royal Meteorological Society for permission to make direct quotation from *Weather* and also for Figure 12.20; HarperCollins Publishers for permission to quote from *Climate and the British Scene*; Studio Editions Ltd for permission to quote from *Weather Lore*; Terence Dalton Publishers for permission to quote from *Weather Patterns of East Anglia*; the Chartered Institution of Water and Environmental Management for permission to reproduce Figures 12.13, 12.15, 12.18 and 12.25. Individual acknowledgements are made to Mr Ian Currie for Figure 3.8; Mr Barry Horton for Figure 2.4; Mr Elwyn Jones and Mr Iolo Jones for Figure 6.7; Dr Lance Tufnell for Figures 8.2, 8.9 and 8.10: Mr J. McArthur for Figure 9.9; Mr T. Emerson and the Kingston Heritage Centre for Figure 12.11; Mr Terry Marsh and the Institute of Hydrology for Figure 12.22. The data used to prepare Figure 8.12 were kindly provided by the Atmospheric Research and Information Centre (Manchester). Dr Mike Hulme, Dr David Viner and the Climate LINK project of the Climatic Research Unit of the University of East Anglia kindly supplied information, data and permission to reproduce Figure 12.26. Mr Bill Lindsay's loan of historical material from Galloway and Carrick is much appreciated. Data for Keele University Weather Station were kindly provided by Mr Michael Edge. The editors would also like to take this opportunity to thank the following individuals for information and helpful discussion: Mr Jim Bacon of Anglia Television, Mr Ian Currie, Mr Bob Johnson of Tyne Tees Television, Mr Elwyn Jones, Mr Alistair Ozanne, Mr Clive Burlton of the Norwich Weather Centre, Mr John Powell and the staff of the National Meteorological Library at the Meteorological Office, Bracknell. A large volume of data was collated and processed by the 1996 graduate class in the MSc Applied Meteorology and Climatology at Birmingham University and by Ms Helen Lynch-Smith of Sunderland University's School of the Environment.

INTRODUCTION

In 1952 Gordon Manley's *Climate and the British Scene* was published and was followed in 1976 by Chandler and Gregory's *The Climate of the British Isles*. Both are still valuable texts but twenty years have elapsed since the latter's publication and in that time much has happened to sharpen our climatic awareness and curiosity. The aim of this book is to provide an up-to-date analysis of the climate of the British Isles in a regional format. It has in the past been customary to approach the topic thematically by analysing the geographical variation of each climatic element in turn. Yet climate and weather are not a disaggregated set of components and are experienced as a combination whose character is something more than the sum of its parts. Even Chandler and Gregory's essentially systematic approach concludes with a chapter on regional climates which includes the following observation: 'it is this set of complex interactions that produces what we understand and appreciate as the overall climate of any particular area, and to which we as individuals respond in our daily lives' (Chandler and Gregory 1976). Gordon Manley (1952) similarly observed, 'climate is however apprehended as a whole and through several senses. Let the reader therefore recall not merely the meteorological situation, but all the feelings and associations of the landscapes at various seasons.' The regional framework adopted here provides a fuller acknowledgement of this gestalt quality and allows us to examine and to interpret climate as it is experienced in different parts of Britain and Ireland. This regional approach provides also an opportunity to illustrate how the changing character of climate may be experienced across the British Isles.

Although this book aims to provide a climatic account of the regions of the British Isles, the authors do not contend that distinctive regional climates can be objectively identified. On the contrary, we proceed from the view that there is a continuum of climate variation across our islands within which many boundaries might be drawn depending upon the purpose of the study. Although the regions adopted in the following chapters share some common climatic themes there are also important differences between them. The existence of climatic contrasts across the British Isles has long been recognised, and no less a geographer than Sir Halford Mackinder (1902) stressed this diversity by stating, 'climate is average weather, and although the weather of Britain is everywhere changeable, yet the average result of the changes is not the same for all points'.

The text is presented in three parts. Part One, The Anatomy of Regional Climates in the British Isles, provides an explanatory framework for the variation of weather in different areas of the British Isles. After placing the subject of the book in the context of global meteorology, it shows how particular air masses are modified by topography, thereby creating the diversity of daily weather across the British Isles. A further aim of this section is to explain meteorological processes that are common to all regions. A glossary of terms is included at the end of the book to elaborate and clarify some of the themes.

Part Two, consisting of Chapters 2 to 11, contains analyses of the regional climates of the British Isles written by local experts. These contributions highlight both the distinctive climatic features of each region and the range of often little-known intra-regional variations of weather and climate that combine to shape climate as we experience it. Significant climatic events of recent decades provide case studies which illustrate the general principles introduced in Part One.

Part Three (Chapter 12) focuses on regional perspectives on climatic variability and change, and assesses the factors that govern climate change and how regional climates and the contrasts between them may alter in response to changing global themes. Discussion of the causes of climatic change is followed by analysis of the recent climatic variation. The implications of global warming may be expressed in contrasting ways in different regions. Computer-generated climate models that are able to produce regionally distinct scenarios provide a basis for this discussion in which a knowledge of the regional implications of particular circulation regimes allows us to compile regionally distinctive climatic scenarios. This process forms an important part of the downscaling of scenarios from the global to the regional level. Most importantly, the regional expression of apparently straightforward climate changes may be more complex and diverse than initial scrutiny of the overall models might suggest.

THE REGION IN CLIMATOLOGY

Regions are a familiar concept; many people identify themselves with their 'home' region and they occupy an important place in our collective consciousness. Although regional studies no longer claim the attention they were awarded by previous generations of geographers they remain a useful vehicle for conveying the sense of a spatially variable phenomenon such as weather and climate. We must of course be assured that variation takes place at a scale that can be registered between the regions. Problems arise, however, when we attempt to define them. As Minshull (1967) so clearly observed, 'the concept of region floats away when one tries to grasp it, and disappears when one looks directly at it and tries to focus'. This timely warning notwithstanding, the task confronting the authors is not that of divining natural climatic regions from the spectrum of variation that we observe; rather it is the better comprehension of a phenomenon that reveals a measurable degree of spatial contrast. Even Hettner (1908) was moved to observe, 'one cannot speak of true and false regional divisions, but only of purposeful or non-purposeful'. In this instance the purpose is the understanding of climatic variation across the British Isles.

The history of climatology is strewn with attempts to provide an objective spatial framework for study. In part this results from the statistical nature of climatological data, which allows for the ready identification of thresholds, the latter often defined in terms of botanical responses. Isothermal limits for numbers of frost-free days, critical states in the precipitation–evaporation balance and mean monthly temperatures have all been used for these purposes. Before we turn to consider the integrity of the regions here adopted it is therefore valuable to review these efforts.

Since the earliest times, when the Greeks attempted to distinguish between the torrid, frigid and temperate zones, schemes have been proposed whereby a sense of order can be imposed upon the spatially continuous variation in climate. Of these, the best-known is that of the German scientist Waldimir Köppen, whose first system was published in 1918 (see also Köppen 1936) and is based on five major climatic types: A – tropical moist climates; B – dry climates; C – moist mid-latitude climates with mild winters; D – moist mid-latitude climates with severe winters; and E – polar climates. Each of these classes can be subdivided at two lower levels according to the character and degree of seasonality of both the temperature and the precipitation regimes. The other notable contribution has been made by the American climatologist C. Warren Thornthwaite (1933), whose system was again based on precipitation and temperature.

Thornthwaite's system lays great emphasis on the response of natural vegetation to dominant climate and uses a precipitation–evaporation index to assist in defining five major humidity classes: rain forest, forest, grassland, steppe and desert.

While such global classifications possess an indisputable pedagogic value their contribution to defining climatic regions is questionable. Their reliance on precipitation and temperature is a limiting factor. The phenomena of sunshine, wind speed and humidity are largely overlooked, and such systems are generally insensitive to important variations in the balance of these climatic elements. Hence, for example, much of the British Isles would fall within Köppen's moist zone with mild winters, no pronounced dry season and long cool summers (all months below 22°C but at least four months above 10°C), and the system would fail to distinguish further important regional variations commonly experienced in the British Isles. Hartshorne (1949) has argued cogently that that such systems are little more than a convenient means by which type-combinations can be identified rather than climatic regions defined. That the same combinations can be found in several disparate locations tends to support this view.

Most current schemes were designed for a geographical scale greater than that embraced by the British Isles. As a result, these and other attempts such as those by Trewartha (1954) and Troll and Paffen (1965) are able to distinguish at best only the climatic regimes of highest ground of Scotland from a more general picture. The closest approach, at least in terms of the number and size of regions, to those adopted here is that of Thran and Broekhuizen (1965). Scotland is treated as a single unit and England and Wales are divided into four regions and Ireland into a further three.

Much finer spatial resolution is provided by Gregory (1976), who suggests a classificatory framework based on temperature (through length of growing season), rainfall and rainfall seasonality. Of the thirty-six possible groups only fifteen are identified. The resulting pattern of climate units sees much of eastern and central Britain and Ireland falling within just two categories but a more variable picture over western and upland areas (Figure i). White and Perry (1989) provide a useful assessment of climatic classification in the British Isles and show how initially intricate schemes may be simplified by statistical procedures.

On the other hand, attempts have been made to account for and describe the climate of regions defined in non-climatic terms. These have usually succeeded with no apparent loss of functionality. A good example is that of climate chapters in the county volumes of Stamp's Land Utilization Survey series or the climatic essays that accompany the British Association for the Advancement of Science handbooks. More inclusive are the two schemes adopted by the UK Meteorological Office. One of them provides the basis of its series of Climatological Memoranda published in the 1980s. Scotland and England are presented in thirteen volumes each, Wales in a further three and Northern Ireland in one. Meanwhile a coarser 'district' analysis based on eight regions is featured in the *Monthly Weather Report*. Figure i summarises some of these regional schemes. More recently, Gregory *et al.* (1991) produced a statistically coherent regional subdivision of Great Britain for the purposes of precipitation analysis. Even they, however, stressed that 'there is no uniquely correct way of partitioning the country'. This observation assumes yet greater significance if attention broadens to include all climatic elements.

The approach taken here lies between the two UK Meteorological Office frameworks and offers ten regions. These regions will be quickly recognised by the general public and by those with a closer interest in everyday matters such as politicians, planners and academics. It is not anticipated that the scheme will meet with universal approval as no such acceptable regional subdivision exists.

No special names are attached to the regions used in this study (Figure ii), all of which could be approximately identified on a map by anyone familiar with the geography of the British Isles. The precise boundaries will, however, provide ample scope for discussion and debate. For the most part the English regions, perhaps the least easy to define, consist

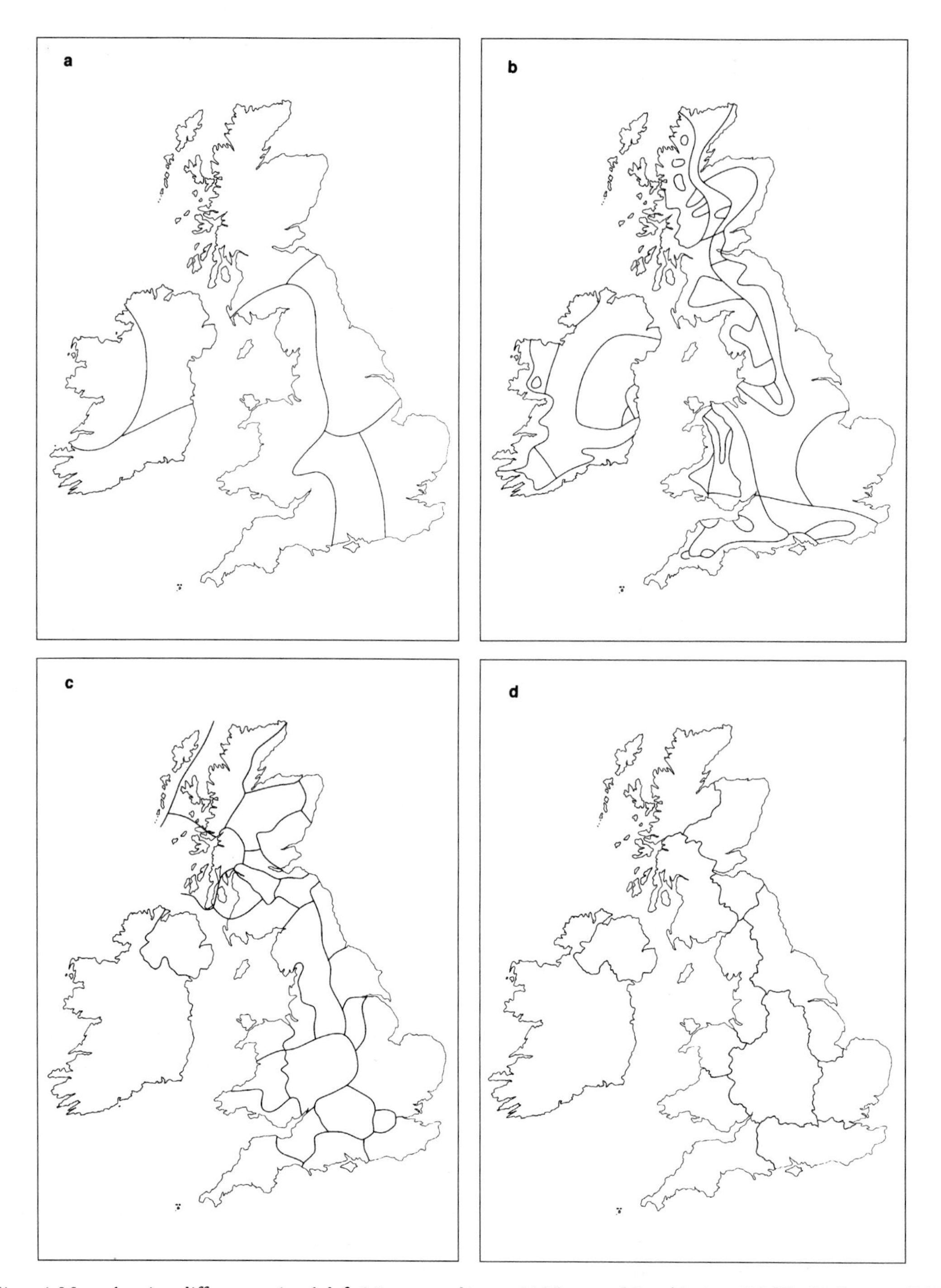

Figure i Maps showing different regional definitions according to (a) Thran and Broekhuizen (1965), (b) Gregory (1976), and for (c) Meteorological Office regions and (d) Meteorological Office districts

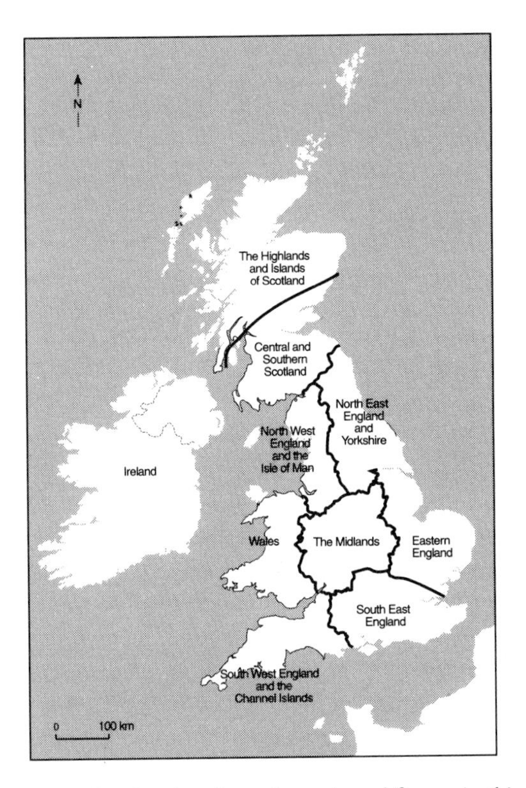

Figure ii Map showing the regions adopted for use in this book

largely of groupings of the climatically uninspiring units of counties. Wales is taken as a single entity, the boundary being that of its political administration. South and Central Scotland has been distinguished from England by the same means. The division between the latter and Highland Scotland could not be achieved using administrative boundaries, however, and the Strathclyde region in particular embraced too much of both to be accepted as a single unit within either. The editors have, with historical, geographical and cultural precedents, taken the 'Highland Line' as the divide. For similar reasons the natural geographic divide of the Pennines watershed is adopted as the boundary between north-west and north-east England. This latter division exemplified the problem of important sites which might lie on a boundary, hence data from Great Dun Fell are cited both in the chapter dealing with north-west England and that dealing with north-east England. Ireland has been taken as a single unit.

Climatic variation within these regions may be as great as that between them, and the often subtle, but sometimes coarse, intra-regional contrasts are portrayed and explained within each chapter. The regions embrace geographic dimensions mid-way between the synoptic-scale systems such as depressions and anticyclones and the scale at which local variations are observed, for example from urban to rural settings, from hilltop to valley floor, or coastal setting to inland sites. We expect them thus to reveal influences of both scales of activity. We also see at the regional scale contrasts in important landscape elements such as soils, drainage, natural and cultivated vegetation and water resources. Hence our perception and memories of the Scottish Highlands, of the South Downs, Dartmoor and the East Anglian Heights are as much determined by the consequences of the climatic contrasts such as natural vegetation, cloudiness and propensity to rainfall as they are by topography. In this way regional climates make a tangible contribution to the kaleidoscope of landscapes that characterise the British Isles.

A NOTE ON STATISTICS AND UNITS OF MEASUREMENT

Whilst the following chapters make reference to climatic events over the past century or more we have generally endeavoured to use the most recent (1961–90) statistics. Clearly, this is not possible in all cases, and where averages do not meet this criterion this is indicated to be the case. Similarly, the majority of sites selected for the tables and summaries are part of the UK Meteorological Office climatological network. The use of the most recent standard period accords with our intention to provide an up-to-date picture of the climate. At a time when we are increasingly preoccupied with climate change through time, it also allows us to draw some preliminary conclusions based on comparisons with averages from earlier standard periods for our most well-established stations.

The source of most of the data has been the *Monthly Weather Reports* published by the UK Meteorological Office. The various means, averages and totals have, whenever possible, been derived from the raw data in these publications. In a very few instances the record for individual stations might not be complete. Depending upon the method used to overcome the problem of missing data, the means cited here may differ slightly from those appearing in other publications.

Where appropriate, SI units are employed. There are, however, important exceptions to this where other units enjoy continued or widespread usage. For example, the knot (1 nautical mile per hour) remains the most commonly employed unit for the measurement and presentation of wind speed data. For similar reasons millibars (mb) are preferred to hectapascals (hPa). Because of the nature of the Celsius scale, with its arbitrary definition of the zero point, a distinction can be drawn between observed temperature values, which are denoted by °C, and temperature ranges or differences, which are described by 'degrees C'. When wind directions are being described the directions are written in full, for example 'south-westerly winds', but initial (capital) letters are used when making reference to the Lamb airflow categories – when referring to a W airflow, for example.

In Part 1, lettering in *italic* denotes terms defined in the Glossary (pp. 332–4).

REFERENCES

Chandler, T.J. and Gregory, S. (1976) 'Introduction' in T.J. Chandler and S. Gregory (eds) *The Climate of the British Isles*, London: Longman.

Gregory, J.M., Jones, P.D. and Wigley, T.M.L. (1991) 'Precipitation in Britain: an analysis of area-average data updated to 1989', *Int. J. Climatol.*, 11: 331–345.

Gregory, S. (1976) 'Regional climates', in T.J. Chandler and S. Gregory (eds) *The Climate of the British Isles*, London: Longman.

Hartshorne, R. (1949) *The Nature of Geography*, 2nd edn, Lancaster, PA: Association of American Geographers.

Hettner, A. (1908) 'Die geographische Einteilung der Erdoberfläche', *Geogr. Ztgchr*, 14: 1–13, 94–110, 137–150.

Köppen, W. (1936) 'Das geographische System der Klimate', in W. Köppen and R. Geiger (eds) *Handbuch der Klimatologie*, Berlin: Bornträger.

Manley, G. (1952) *Climate and the British Scene*, London: Collins.

Minshull, R. (1967) *Regional Geography*, London: Hutchinson.

Thornthwaite, C.W. (1933) 'The climates of the Earth', *Geogrl. Rev.*, 23: 433–440.

Thran, P. and Broekhuizen, S. (1965) *Agro-climatic Atlas of Europe*, Wageningen: PUDOC, Centre for Agricultural Publications and Documentation.

Trewartha, G.T. (1954) *An Introduction to Climate*, 3rd edn, New York: McGraw-Hill.

Troll, C. and Paffen, K.H. (1965) 'Seasonal climates of the Earth', in E. Rodenwaldt and H.J. Jusatz (eds) *World Maps of Climatology*, 2nd edn, Berlin: Springer.

White, E. and Perry, A.H. (1989) 'Classification of the climate of England and Wales based on agroclimatic data', *Int. J. Climatol.*, 9: 271–291.

PART 1

THE ANATOMY OF REGIONAL CLIMATES IN THE BRITISH ISLES

1

THE ANATOMY OF REGIONAL CLIMATES IN THE BRITISH ISLES

Julian Mayes and Dennis Wheeler

INTRODUCTION

Lamb (1964) has observed, 'Just as it is true that English history cannot be understood apart from world history, so to understand the English climate we must first learn what generates the various climates of the world'. Although Oke's (1992) scheme would place the British regions at the 'local' and 'meso' scales – negligible by global standards – Lamb's view retains its relevance, and a review of large-scale climatic processes provides the best foundation for the regional studies that follow.

The raw material out of which our weather is formed – the atmospheric gases, water vapour, heat and energy – are largely derived from elsewhere. Altered to a greater or lesser degree by its encounter with the British Isles, this air passes in turn onwards to other areas. In this sense the passage of *air masses* places the British Isles at a veritable crossroads in global-scale atmospheric movements with the polar latitudes to the north, the subtropics to the south; the vastness of the Atlantic to the west and of the Eurasian landmass to the east. Yet climatic forces operate over a wide range of space and time scales (Figure 1.1). Only by appreciating that range and the delicate balance of forces that direct the air masses and associated pressure weather systems can we comprehend the character of the climate of this part of the northern hemisphere and its regional expressions.

THE THREE-DIMENSIONAL ATMOSPHERE

The widespread use of weather maps encourages us to view the atmosphere in only two dimensions. Yet the conditions and changes in the vertical dimension are also important. Temperature and pressure changes are generally far more marked than in the horizontal plane (Figure 1.2). Within the troposphere, temperature changes with height are in the order of 5 to 10 degrees C in 1 km. This contrasts with average horizontal variation close to the Earth's surface between the tropics and the poles of the order of 50 degrees C in 10,000 km. Even across well-defined *warm* and *cold fronts*, horizontal temperature gradients rarely exceed 5 degrees C in 1 km. The behaviour of the upper *troposphere* also plays a major role in determining the movements and development of large-scale weather systems. Most importantly, the relief of some regions, which may expose areas to the inhospitable nature of conditions at higher levels, is one of the factors helping to account for inter-regional variability. Those who have followed the popular tourist path from Fort William to the Ben Nevis summit will readily testify to the rapidity with which conditions can change with altitude.

The Earth's atmosphere can be identified up to an altitude of approximately 120 km. At a height of just over 5 km air pressure is 500 mb, about half that at sea level, while at 50 km pressure is only

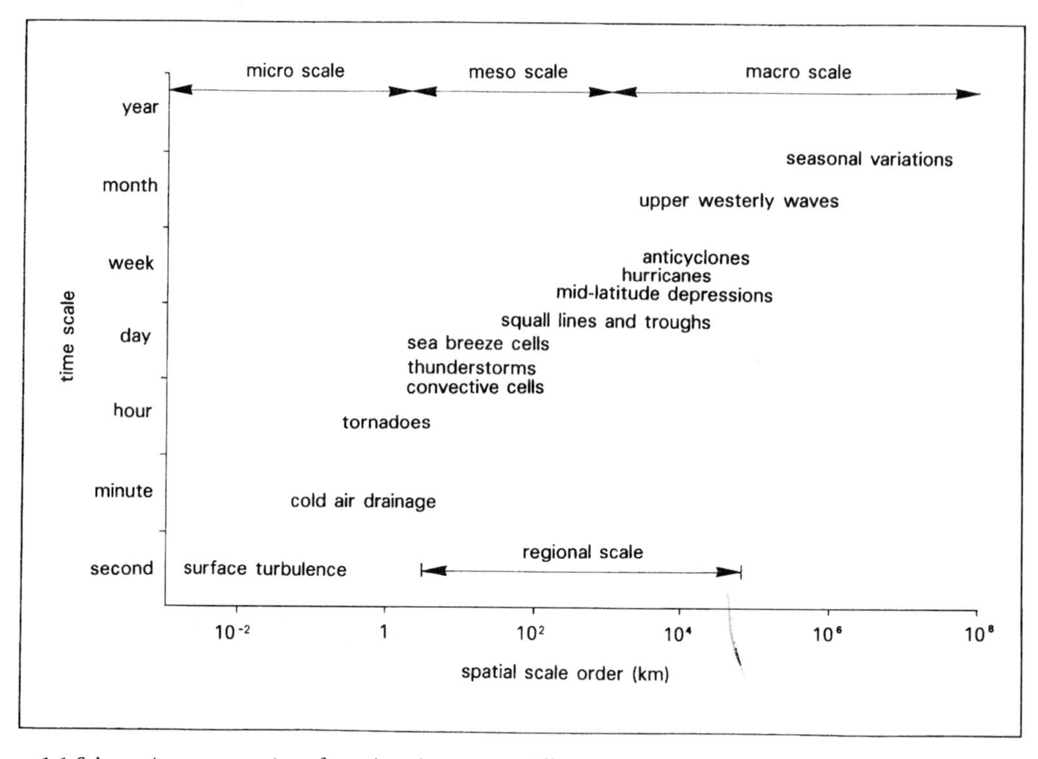

Figure 1.1 Schematic representation of weather elements at different time and space scales (after Orlanski, 1975)

about 1 mb (Figure 1.2). Air pressure and density change steadily, though non-linearly, with altitude. The temperature profile, however, is irregular and provides the means of demarcating the various levels. Most climatologists are concerned only with the behaviour of the two lowest levels, the troposphere and the *stratosphere*. The boundary between the two is the *tropopause* – a permanent *temperature inversion* formed by the changing temperature profile from a negative gradient (fall of temperature with height) in the troposphere to a constant or positive gradient in the stratosphere.

The troposphere, which varies in depth from 15 km at the equator to 9 km at the poles, contains 75 per cent of the atmosphere's mass and virtually all its water vapour. Because of the partial barrier to vertical motion provided by the tropopause, the climatic consequences of the interactions between the surface and the atmosphere are largely confined to this lowest and most dynamic of the atmosphere's layers. This dynamism is partly the result of the energy and moisture the troposphere receives from the Earth's surface, whose character therefore does much to determine local, regional and global climates.

THE ATMOSPHERE'S ENERGY AND MOISTURE SOURCES

The temperature profile of the troposphere reflects the significance and the sensitivity of the connections between the atmosphere and the Earth's surface. The temperature profile of the troposphere is well understood (Barry and Chorley 1992; Lamb 1972) and stems from the capacity of the atmosphere to allow short-wave solar radiation to pass through it with relatively little effect on its temperature. The average incident radiation at the upper atmosphere is 1,360 watts/m^2, of which only about 16 per cent is absorbed directly by the

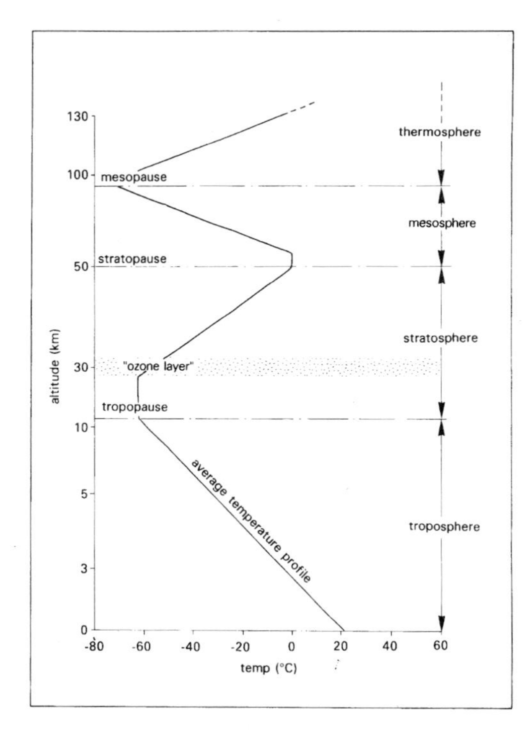

Figure 1.2 The atmosphere's vertical structure

atmosphere's constituent gases and particles of solid matter such as volcanic dust. A further 20 per cent is reflected from the upper surfaces of clouds and, globally, only about half of the upper atmosphere's incident radiation reaches the surface. It is at the surface that the solar radiation, mainly in the form of light and short-wave infra-red radiation in the waveband of 0.3 to 1.4 μm, is converted into heat energy in its greatest quantities, thereby raising surface temperatures to levels dictated by such properties as its *albedo* and *specific heat*. Specific heat (the heat energy required to raise the temperature of 1 gram of matter by 1 degree C) differs most notably between land and sea bodies, the former being only about one-tenth of the latter's 1.0 cal/g. The transparency of the seas allows radiation to penetrate for up to 30 m beneath the surface, whereas land surfaces are opaque in this respect. As a result, the heat energy capacity of the oceans is far greater than that of land surfaces. Albedo is highly variable (Table 1.1) and determines the

Table 1.1 Approximate albedos of different surfaces

Surface properties	*Albedo*
Dark and wet soils	0.05
Light and dry soils	0.40
Desert surfaces	0.20–0.45
Grass	0.16–0.26
Deciduous forest	0.15–0.18
Coniferous forest	0.09–0.15
Water (low incident angle)	0.10–1.00
Water (high incident angle)	0.03–0.10
Fresh snow	0.90–0.95
Melting snow	0.40–0.60

Source: After Oke (1992)

quantities of incoming radiation that are reflected and hence make no contribution to raising surface temperatures.

All warmed surfaces behave as radiating bodies emitting in the long-wave band between 4 and 30 μm. The waveband of this terrestrial radiation is greater than that of the Sun's radiation because of the Earth's lower surface temperature. It is within sections of this long waveband that radiant energy is intercepted and converted into heat by radiatively active gases such as carbon dioxide, water vapour and methane. These, among other materials, constitute the family of *greenhouse gases*. There is also much absorption and reradiation within the atmosphere, and only 5 per cent of terrestrial radiation passes immediately out of the atmosphere, much of it through the *radiation window* in the waveband 8 to 11 μm. It has been estimated that absorption by greenhouse gases increases the atmosphere's overall temperature by as much as 33 degrees C above that which would otherwise prevail.

Radiation is not the only means by which energy is transferred from the Earth's surface into the atmosphere; the processes of conduction and convection need also to be considered. Oke (1992) argues that conduction is important only over the scale of the few millimetres of air immediately above the surface and that the combination of free and forced (mixed) convection is the more important means by which the transfer occurs. Heat energy exists in two forms:

sensible heat and *latent heat*. Of the two the latter is several times more important but both provide much of the energy needed to drive the world's weather systems and major circulations.

The heating of the Earth's surface and consequent warming of the lower troposphere creates an *environmental lapse rate* with an average decrease of 0.65 degree C for every 100 m of altitude gained. Importantly, lapse rates over the British Isles are some of the steepest to be found anywhere in the world, signifying thereby the possibility of rapid climatic variation with altitude. They are nonetheless subject to variation according to the dominant air mass and the changes to which it may have been subjected. Fluctuations also occur through the seasons, though, as Taylor (1976) has pointed out, these are too irregular for us to discern an unambiguous seasonal rhythm.

The temperature response to energy fluxes across the atmosphere–Earth surface interface is spatially variable over short distances and depends on the surface properties such as vegetation, soil moisture, snow cover and extent of built-up area and their associated albedos and specific heats. The surface of dry soils, for example, will gain and lose heat energy rapidly and be more frost-prone as a result. Snow is also an efficient radiator, and many of the lowest temperatures found in the British Isles are recorded on calm, clear nights with extensive snow cover. Damp soils are less subject to rapid or severe heat loss. Also, the picture for urban areas is complicated by the presence of artificial heat sources giving rise to the urban heat-island effect. Significant temperature differences brought about by these processes are sustainable, however, only when winds are light; marked temperature gradients are soon dispersed by turbulent mixing when wind speeds exceed 10–12 knots. Depending on the frequency of settled weather these elements may nevertheless influence the long-term climatic record. Though local in character, these variations might also assume regional proportions where they dominate over wide areas. The extensive and prolonged snow cover of some of the highest Scottish hills, the seasonally dry soils of East Anglia, the damper soils of the Midlands plains or the suburban spread of London (Chandler 1965) come to mind as regional-scale expressions of these elements.

The Earth's surface, in particular the oceans, is also the only source of the other important atmospheric component: water vapour. Because the latent heat of vaporisation of water (the quantity of energy needed to evaporate a given mass of water) is very great at 595 cal/g, heat energy is used in large quantities in the process of evaporation. Water vapour also finds its way into the atmosphere from vegetated surfaces by the process of transpiration. In areas such as tropical rain forests the quantities of water vapour derived from transpiration are far from negligible but in the British Isles its role is less significant, and most water vapour has an oceanic source. The volumes of water exchanged by evaporation and precipitation across the major geographic units are summarised in Table 1.2. However, this table takes a static view of a dynamic system in which the overall global balance is achieved by atmospheric circulations transferring water vapour from oceans to land surfaces and by rivers returning water to the oceans.

The atmosphere can also lose heat and moisture to the ground. Atmospheric heat loss by conduction is limited but precipitation is altogether more important, and rain, snow and dew are a self-evident

Table 1.2 Ocean and terrestrial water balances

Location	*Evaporation*	*Precipitation*
Atlantic Ocean	124	89
Indian Ocean	132	117
Pacific Ocean	132	133
All oceans	126	114
Africa	43	69
Asia	31	60
Australia	42	47
Europe	39	64
North America	32	66
South America	70	163
All land areas	42	73

Sources: Bodyko (1971); Strahler and Strahler (1973)
Note: The units of measurement are centimetre depths of water per year averaged over each of the areas. Balances are made up by movement of water between oceans and land surface runoff

manifestation of the return of water to the surface. We must imagine therefore a constant exchange between atmosphere and surface of these two important components of heat energy and water. These factors alone lend special significance to ground-level characteristics insofar as they contribute to the character of the lower atmosphere. Such considerations form the basis of the concept of air masses but their expression can also be sought at the smaller, regional scale. Before returning to examine these issues in a little more detail we must turn our attention to large-scale, indeed global-scale, activities.

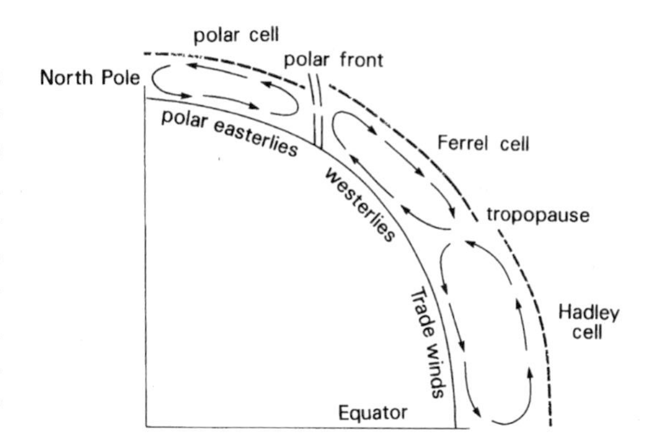

Figure 1.3 The Earth's simplified meridional circulations in vertical section

GLOBAL-SCALE CIRCULATIONS

Large-scale latitudinal variations in the solar energy receipts have important effects for they sustain the great global circulations that convey heat energy from areas of excess (of incoming over outgoing radiant energy) in the lower latitudes to areas of deficit in the higher latitudes (Barry and Chorley 1992). The fluctuating intensity of these circulations is governed by seasonality and by geographical factors such as the distribution of major land and water bodies (Lamb 1972). It is the weather of the British Isles, imported from elsewhere and modified by contact with the ocean and land surfaces, that brings with it not only moisture but the heat energy that ensures the very survival of the British and Irish peoples. Were we to rely exclusively on the purely local receipts of solar radiation to determine temperature levels we would find ourselves in a climate substantially cooler than that which currently prevails. In this important task the atmospheric circulations, with their burdens of sensible and latent heat, are assisted by those of the oceans, which, while not so rapid, convey through the high specific heat of water large quantities of sensible heat energy. In this respect the role of the North Atlantic Drift is critical for the climate over the British Isles and raises temperatures to 11 degrees Celsius above the global latitudinal average.

The air currents responsible for the redistribution of heat and energy were once thought of as a system comprising three cells of vertical air currents (Figure 1.3). Heating at the equator encourages uplift leading to the development of the *Hadley cell*, while cooling at the poles ensures a cold circulation at the highest latitudes; between the two is the *Ferrel cell*. However, this simple model fails to explain many aspects of the mid-latitude weather and climate, and scientists now view the Ferrel cell in particular as a much more dynamic zone in which the circulations around the constantly evolving cyclones and anticyclones promote the north-to-south exchange of air and heat energy at all tropospheric levels. The importance of this short-term view, as opposed to attempts to explain behaviour on some long-term average state, is a theme that we will return to when we examine the influence of airflow types on the weather of the regions.

The pattern of horizontal air movement is summarised in Figure 1.4. Allowing for the disruption to this idealised scheme caused by the continental landmasses and the perturbations of the mid-latitude weather systems, the overall picture is tolerably accurate. Air does not, it can be seen, move directly from high- to low-pressure areas. The effect of the Earth's rotation and the need to sustain mechanical equilibrium of the atmospheric system introduces the *Coriolis force* to the dynamics of airflow. As a result, the principal air currents are deflected to provide a system with easterly and westerly (zonal) components and are superimposed on the essentially north- and

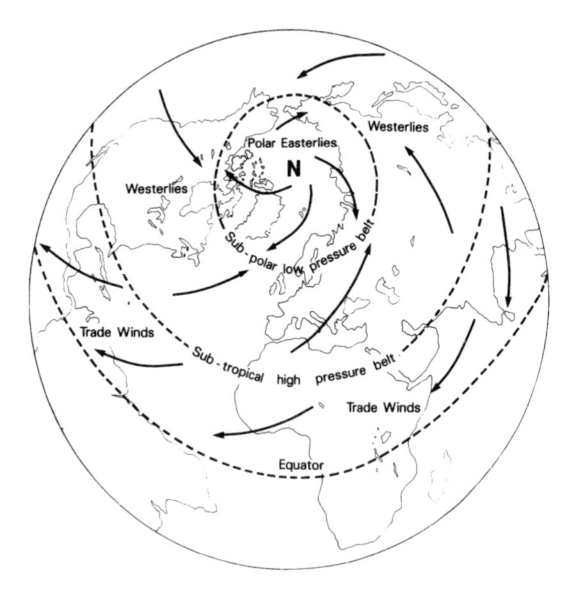

Figure 1.4 The generalised sea level wind circulations and pressure zones

southwards-directed (meridional) elements outlined in the preceding paragraph. The British Isles lie on the northerly margins of the mid-latitude westerly circulations and for that reason a large proportion of our weather comes from that direction. Hence, because of interactions with the land surface and often abrupt relief over which the airstreams pass, the west-to-east climatic gradient is more marked than the north-to-south gradient.

UPPER TROPOSPHERIC CIRCULATIONS

The principal motion of weather systems and air masses is governed by events in the middle and upper troposphere from approximately 3,000 m upwards. These are the altitudes of the upper westerly circulation. At ground level this circulation, which dominates between the subtropical and the polar latitudes, is represented by the westerlies shown in Figure 1.4. In the mid-latitudes the westerlies dominate through the depth of the troposphere but further north they override the polar easterlies, which prevail only over the lowest 2,000–3,000 m of the troposphere. The upper westerly airflow is, however, far from geometrically regular and undergoes varying degrees of distortion.

Upper-air circulations, though distant from the Earth's surface, are nonetheless responsive to its changing character. Lamb (1972) has shown that physiographic features such as the Rocky Mountains and the Tibetan Plateau extend far enough into the troposphere to cause distortion in the circumpolar vortex of winds. Responses are also induced by extensive areas of warm or of cool surface conditions, which will cause respectively northwards and southwards deflections of the upper airstreams. The seasonal fluctuations of temperature on the Asian landmass and the northwards intrusion of warm water in the North Atlantic Drift are important in this respect.

By these means the upper westerly airstreams are encouraged to meander northwards and southwards across the latitudes in the form of *Rossby waves*. The degree of meandering varies over a time scale of weeks and months and is described by the *zonal index*: the upper westerlies are said to have a high zonal index when the flow is confined to a narrow latitudinal band, but a low zonal index when a great degree of meandering develops. The significance of these variations lies in the fact that they do much to influence the movement and development of high- and low-pressure weather systems. The generally eastwards motion of depressions is governed by the ever-changing character of the upper circulations and in particular by the *jet streams* that mark the discontinuity in the tropopause where the warm tropical and cold polar air columns are juxtaposed along the polar front.

Most frequently the centres of depressions pass over or to the north of the British Isles, thus making 'westerly' or 'cyclonic' weather commonplace. A preference for more southerly routes in winter and more northerly routes in summer reflects seasonal changes in the global energy balance and latitudinal shifts of the major climatic zones. Because of the generally steeper thermal gradient across the latitudes in winter, depressions are often more well-developed at that season. There are, however, occasions when *meridional air flow* in the upper westerlies may direct weather systems on unusual routes. Where the upper airflow is northwards, drawing warm air into higher latitudes, an upper ridge will develop, the peak of which is a favoured location for high-pressure systems. Upper troughs also occur in which cool air is drawn southwards. Meridional airflow is important because it often leads to some of the British Isles' most anomalous weather patterns. One example is the *blocking* situation in which unusually persistent anticyclones in the mid-latitudes hinder the eastwards movement of depressions. As a result the jet stream might bifurcate, being unable to penetrate the anticyclone, and deflect rain-bearing depressions and fronts to higher or lower latitudes with dramatic consequences, as the drought summers of 1975, 1976 and, most recently, 1995 have all too clearly demonstrated. Within deep troughs the counterpart to the blocking high – the *cut-off low* – might develop. Even these anomalous conditions are rarely of sufficient range to eliminate regional contrasts within the British Isles. Though it is a fact often overlooked in southern Britain, the great droughts of 1975 and 1976 did not embrace areas such as western and northern Scotland or parts of western Ireland. Conversely, the period April to August 1984 saw less than 50 per cent of average rainfall in large parts of Highland Scotland, whilst south-east England was unusually wet. These variations indicate how the regional distribution of rainfall is sensitive to the longitude of any trough in the upper atmosphere when the weather pattern is blocked. A slow-moving trough centred, for example, to the west of the British Isles will 'anchor' low-pressure areas and fronts to western areas, with eastward progress being discouraged by the presence of a blocking area of high pressure. This can be especially common in late winter and early spring, when a blocking high may be entrenched over Scandinavia or Russia owing to the low temperatures there; such a situation arose in the winter of 1995/96. In these cases blocking can help to concentrate rainfall over western fringes of the British Isles. If a trough is centred over the North Sea, the rainfall distribution is likely to be reversed.

SYNOPTIC-SCALE CLIMATIC ELEMENTS

Other factors also cause frequent changes in our weather; were it otherwise, the westerlies would prevail without interruption, yet clearly this is not the case. The British Isles are situated close to the important boundary between the cold polar easterlies and warmer westerlies. Slight movements of this boundary will therefore introduce air from quite different source regions, either polar or tropical, to the British Isles. More importantly, it is at this boundary, known as the polar front, that *cyclogenesis* takes place. The movement of these cyclones and their attendant fronts, constituent polar and tropical air masses and circulations does much to determine the weather and climate of the mid-latitudes generally. They are the principal source of precipitation in the British Isles and are also associated with the steepest atmospheric pressure gradients and strongest winds. The warm and cold fronts are particularly important as zones where precipitation processes are most active. Within the warm sector of tropical air, where humidities are usually highest, the so-called *conveyor belt* motion of air is important and encourages the large-scale vertical movement of air.

Despite the significance of the mid-latitude depression – and well over one hundred pass over or close to our shores each year – the British Isles are not denied the direct influence of either the polar or the Azores anticyclonic systems and their extensions. We need, however, to recognise that the ever-changing character of the prevailing weather and the circulations within the major weather systems will provide regions with contrasting momentary degrees of exposure and protection, with consequent differences in the geography of rainfall, temperature and sunshine. Sawyer (1956) has made an exhaustive study of the manner in which regional rainfall patterns vary according to the movement of depressions across the British Isles.

It must also be remembered that the land and sea surfaces over which the weather systems pass will themselves experience temperature changes from season to season which in turn influence air masses passing over them. It would not be expected, for example, that tropical maritime air would respond to winter sea surface conditions in the same way as it would to summer states. Our instantaneous weather is partly accounted for by such short-term synoptic conditions and seasonal variations. Long-term climate averages reflect the frequency with which these different states and balances occur. Complex and variable though these might be, it is possible to tease from them synoptic schemes with consistent characteristics and regional patterns of response. This study of synoptic, weather or airflow types will be considered more carefully in a later section but it forms a vital component in any attempt to explain the climatic character of these islands and their constituent regions.

LOCAL AIR CIRCULATIONS

Microclimatological studies (Geiger 1959; Oke 1992) have demonstrated the significance of small-scale climatological processes. Differences in soils and vegetation or the character of built-up areas produce measurable effects on the climate of the so-called boundary layer. During spells of settled anticyclonic weather such local contrasts are most evident. A critical factor is the accumulation of small bodies of air with different temperatures, able thereby to produce corresponding contrasts in air pressure leading to small scale, subsynoptic airflows. In more turbulent conditions these small-scale differences have no opportunity to influence local weather.

The most ubiquitous of these local winds are coastal *sea breezes* (Simpson 1994). These circulations may extend up to 50 km inland, and may thus hold temporary sway over a large proportion of the land surface. The determining element in sea breezes is the contrasting thermal properties of land and sea surfaces. Intense summer radiation levels raise daytime land surface temperatures above those of the thermally more conservative seas and oceans. The consequent temperature and pressure contrasts in the air columns over land and sea initiate the sea breeze circulation (Figure 1.5) as the column of warm air over the land expands upwards to create a high-level pressure gradient to take air towards the sea. The reverse gradient prevails at ground level. These breezes rarely extend more than 1,000 m into the lower atmosphere but can penetrate far inland as the warm days progress, particularly in areas such as East Anglia where there are few physical obstacles to hinder them.

The principal result of sea breeze activity is to extend inland the sea's cooling influence. Because sea surface temperatures are lowest along the eastern coasts of England and Scotland this is where they are most effective, and summer maxima can be significantly reduced both day by day and in the long-term record. The breezes themselves are generally light and rarely sustain speeds in excess of 10 knots. The column of unstable air in the circulation's ascending limb may, however, create a local convergence zone or sea breeze front marked by a line of cloud with the possibility of showers. The reverse temperature and pressure gradient may become established on calm nights as land surfaces cool rapidly, giving rise to land breezes. These tend to be much less vigorous and extensive. Because of the conditions required for the formation of land and sea breezes they are almost exclusively a spring and summer phenomenon and are particularly active in early summer when sea surface temperatures have yet to make a full recovery from their late winter minima while the land surface is warming more rapidly in response to the lengthening day and more powerful Sun.

In areas of contrasting relief, hillsides will experience different degrees of heating depending upon their slope and orientation with respect to the incident solar radiation. On exposed hillsides air is warmed through conduction and small-scale mixing. The resulting decrease in density encourages buoyancy, with air moving upslope as an *anabatic*

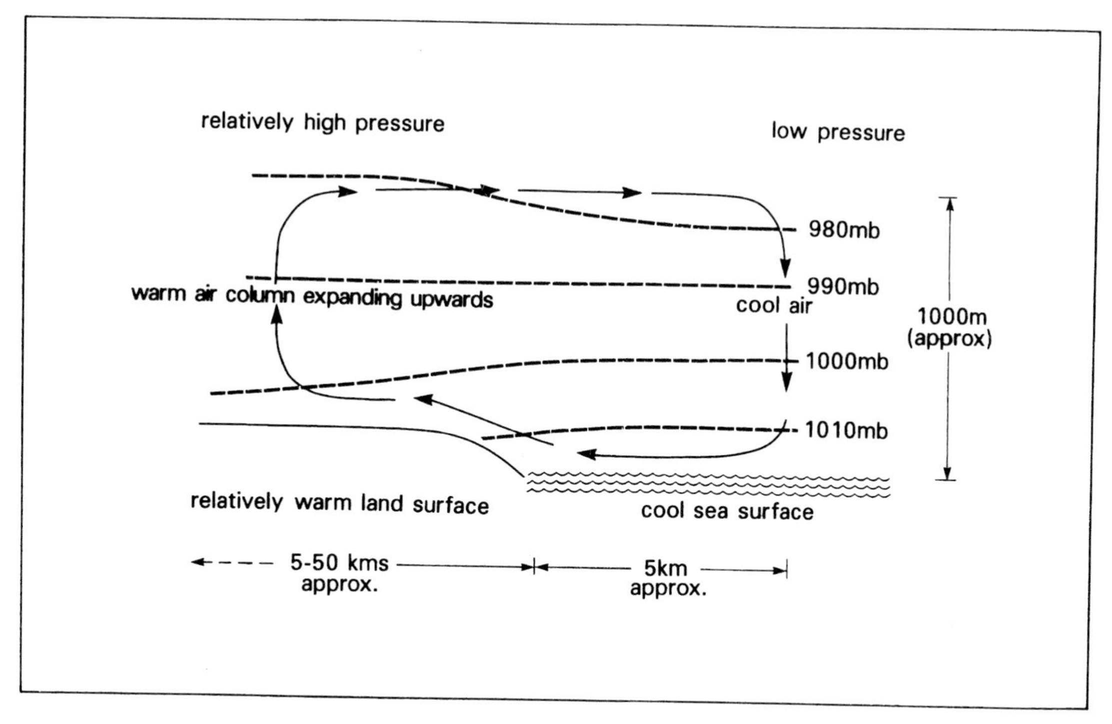

Figure 1.5 Section through a generalised sea breeze circulation

wind (Figure 1.6a). At night-time, radiant heat loss from land surfaces is no longer offset by solar radiation, and temperature of the land surface and the immediate atmosphere will fall. As the air cools it becomes more dense and tends to drift downslope as a *katabatic* airflow (Figure 1.6b). This creates nocturnal temperature inversions (though they may occasionally persist through the following day) that contain the coldest air close to ground level which, being denser than the overlying warmer layers, prevents vertical mixing. The valley floors then become *frost hollows* with colder conditions than those found on adjacent valley slopes and hilltops. Steeper slopes, especially if covered by snow (above which temperature falls can be most marked), encourage this activity yet further. Some locations, such as the Rickmansworth and Houghall frost hollows in Hertfordshire and County Durham respectively, are notorious, with air frosts possible even in summer and staggeringly low temperatures in winter under optimum conditions. This latter point is important because these local and light winds, perhaps of no more than 1 or 2 knots, depend very much upon the absence of pressure gradient winds which would promote sufficient turbulence to mix the lower air.

Rapid ground-level cooling can also, at any season, lead to night-time fogs when the air temperatures are lowered beyond the *dew point*. Such *radiation fogs* develop beneath local temperature inversions but are usually dispersed by the Sun's heating as the following day progresses. They may, however, persist when deep and well-developed. Any extensive low-lying area, such as the lower Thames valley, the damp Midlands plains of England or the Vale of York, is notably subject to these conditions.

Regional warming may result from the flow of air over hill and mountain barriers even of the order of 1,000 m which can induce the föhn effect. This results from moist air rising over windward slopes

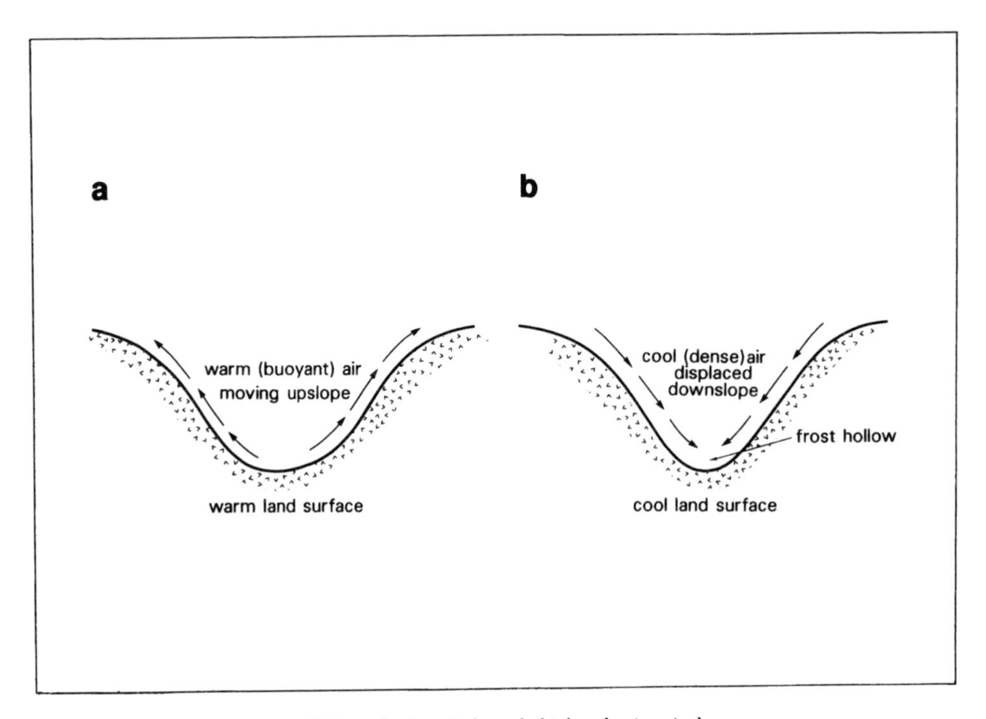

Figure 1.6 Simplified representation of (a) anabatic wind, and (b) katabatic wind systems

and quickly cooling to saturation point. Further elevation brings about cooling at the less marked *saturated adiabatic lapse rate* (Figure 1.7). Descent down the lee slopes, however, takes place exclusively at the more pronounced *dry adiabatic lapse rate*. On arrival back at low level, especially if it has had to negotiate the high ground of Scotland, the air may be 3 or more degrees C warmer than in its original state. This effect is most noticeable in stable airstreams in which vertical mixing is limited. Westerly airstreams having to negotiate the Scottish Highlands are those most likely to reveal this behaviour, and a good example is discussed in Chapter 10 (Figures 10.3a and b).

ATMOSPHERIC INTERACTIONS WITH LAND SURFACES

With the exception of those areas where the sea surface demonstrates a marked horizontal temperature gradient, such as are found on the northern margins of the North Atlantic Drift where it comes into contact with the cooler waters of the East and West Greenland currents, the ocean surfaces are undisturbed by rapid changes in temperature across distance or through time. Air masses become more or less stable depending on their source regions and their subsequent movements, and most will gain water vapour where they pass over the seas and oceans. Land surfaces are quite different. They lack the thermal and physical uniformity of the oceans and present abrupt changes in relief and other surface characteristics. Changes in the character of the lower levels of air masses and weather systems will take place over shorter distances when they negotiate the passage of such heterogeneous surfaces. More than any other element it is this variation that brings into play many of the factors that account for regional-scale climatic variation.

Variable relief is another important ingredient not found over the oceans that can be important

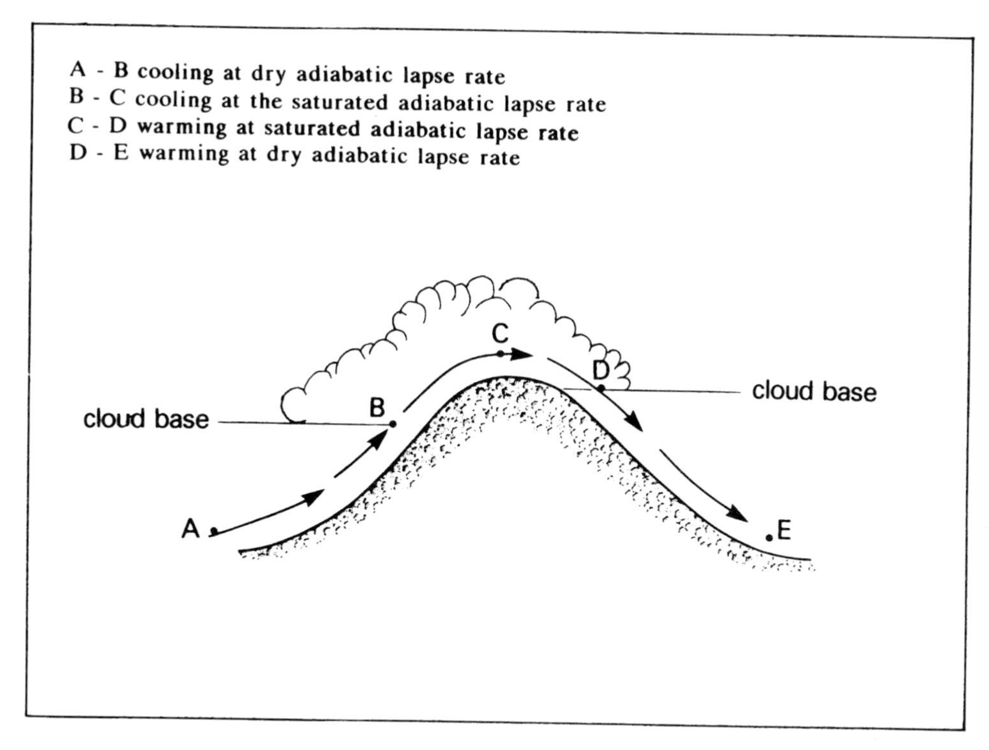

Figure 1.7 Schematic representation of the föhn effect

in determining the climate at global, synoptic and local levels (Barry 1981). Though the hills of Britain and Ireland are modest by many standards, they are nevertheless of sufficient height to bring about changes in the humid air masses which have to negotiate them. Wind fields suffer distortion when encountering irregular terrain, and while the relative smoothness of oceanic surfaces does little to subdue the vigour of the low-level airstreams, land surfaces exercise a much more marked effect and may reduce wind speeds by as much as 50 per cent. Conversely, channelling along valleys aligned parallel to the prevailing winds may induce locally stronger winds, but only on high and exposed ground such as is found in the Lake District or Scottish Highlands might winds speeds approach or even exceed their *geostrophic* levels. Other effects are less readily appreciated. The structural modification of warm and cold fronts is one of these and leads to loss of distinctiveness in the air mass boundary zone.

Relief has a further role, for in encouraging air to rise it brings induces *orographic* enhancement of precipitation. Adiabatic changes in the temperature of rising air initiate condensation and rain-making processes, and moisture gained over the oceans is now given up in the clouds over high ground. So consistent is the increase in rainfall with altitude that it has been summarised (Bleasdale and Chan 1972) by the regression equation

$$R = 714 + 2.42H$$

where R is mean annual precipitation in millimetres and H is altitude in metres above sea level. This model applies generally to the British Isles and indicates an average sea level annual precipitation of 714 mm rising by 2.42 mm for every metre of altitude gained. There are nevertheless marked regional variations about this median position, and the rainfall of western districts rises more rapidly with altitude than that in the east. Atkinson and Smithson (1976) have shown that increases in precipitation with altitude are gained only partly by longer durations and that increased precipitation intensities are an important feature of high ground. The fall in ambient temperatures ensures that the frequency of snowfall and its subsequent duration also increases with altitude. An additional element in areas of high relief is the convergence of air into exposed valleys, where once again precipitation might be enhanced (Browning *et al.* 1985). For this reason it is the west-facing valleys of the Lake District and of western Scotland that register some of the highest mean annual precipitations in the British Isles. Weston and Roy (1994) have made a detailed study showing how orographic enhancement of rainfall over Scotland differs between different wind directions.

The other mechanisms by which air can be encouraged to rise, bringing into play the same adiabatic/precipitation processes, are ascent along frontal zones and convective processes within unstable air masses. The latter are characterised by short-duration, often high-intensity events that usually display a regional and seasonal distribution reflecting the geography of summer land surface temperatures, which are highest in southern and low-lying regions. Instability showers may also result from polar air masses' having passed over relatively warm seas, for example during early winter in western Scotland.

Frontal rainfall is usually of a lower intensity but, especially when enhanced by orographic effects, can be prolonged. These combined processes have produced some notable 24-hour totals in highland areas (Douglas and Glasspoole 1947). Having lost some of its moisture the atmosphere has little scope for its replenishment as long as it remains over the land surface. The volumes of water vapour transpired by plants cannot match that lost by precipitation, and air masses generally become drier as they cross the British Isles, leaving windward slopes relatively wet and leeward slopes with less precipitation. By this means high ground casts a 'rain shadow' over land to its leeward. The distribution of windward and leeward slopes differs depending upon the prevailing wind direction but the dominance of westerly winds ensures that western areas are, at similar altitudes, more exposed and far wetter than eastern districts. The roles are reversed when winds from the east prevail.

Areas of subdued topography are nonetheless capable of inducing changes in the atmosphere. Land surfaces have a low specific heat that varies according to the character and depth of soil, extent of vegetation or extent of built-up area, but a value between 0.1 and 0.2 would describe most conditions. Thus while land surface temperatures can vary diurnally by as much as 20 degrees C it is uncommon to find sea surface temperatures changing by much more than 0.5 degree over the same interval. Because the temperature of the land surface changes more quickly and over a greater range than that of the oceans, the air in contact with it will behave in a similar fashion. Oke (1992) states that net terrestrial radiative cooling rates under ideal conditions approach 3 degrees C/hour. These are the very factors that underlie the local wind circulations described above.

ATMOSPHERE AND ENVIRONMENT AROUND THE BRITISH ISLES

Changes in the temperature and the humidity of the atmosphere take place in response to the character of the surfaces over which the air passes. At the large scale this gives rise to the well-known categories of air masses: maritime (humid), continental (dry), polar (cold) or tropical (warm). These air masses develop in the major anticyclonic zones from which they derive their distinctive temperature and humidity regimes. The two most frequently encountered over the British Isles are the tropical maritime and polar maritime air masses. The former reaches us from the Azores quasi-permanent anticyclone, the latter from the polar anticyclone. Both are humid, having been formed over either oceanic or ice surfaces, but whereas tropical air masses inherit the warmth of their source region, the polar masses possess the coolness of theirs. To these two air mass classes should be added their continental, and hence dry, counterparts from the seasonally active anticyclones over North Africa and Eurasia, the former giving rise to tropical air and the latter, active only in winter, to polar air. There are also polar maritime 'returning' and Arctic air masses. Both are of polar origin but differ in that the former reaches the British Isles only after having reached a lower latitude while the latter approaches directly from the north with consequent implications for their temperature and moisture regimes.

By dispersing outwards from their source regions, air masses come into contact with land and sea surfaces possessing quite different qualities (Figure 1.8). By the processes of heat and moisture exchange the lower levels of the air masses will respond to their new surroundings. These changes develop in the time required to travel hundreds of kilometres. Hence polar continental air from Eurasia might present a drier and cooler appearance in southern England, where it will have made a short sea crossing, compared to that which it offers in north-east England or eastern Scotland following a much longer passage over the North Sea. In a similar fashion, polar or Arctic air will become *unstable* as it passes southwards over the increasingly warm seas around the British Isles. In extreme cases *polar lows* may develop in Arctic air masses with the possibility of heavy snow showers. Tropical maritime air, conversely, will become more *stable* as it moves in the opposite direction. Belasco (1952) has found that between the Azores and the coast of south-west England the surface air temperature falls by an average of 5.0 degrees C in summer and 4.4 degrees C in winter. At approximately 900 m altitude the changes have fallen to 1.2 and 1.7 degrees C respectively, revealing the limited vertical extent of these modifications. Polar maritime returning air masses, having passed to more southerly latitudes before reaching the British Isles, will also acquire some tropical characteristics and tend towards stability in their lowest levels as they move back northwards, but the instability inherent in their polar origin

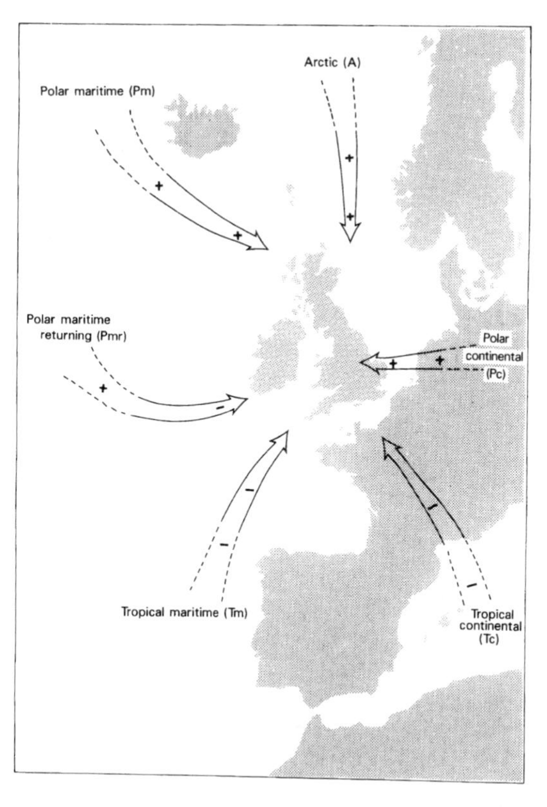

Figure 1.8 Sources and movement of air masses over the British Isles (positive signs indicate low-level warming of the air masses as they approach Britain, negative signs indicate cooling)

quickly becomes apparent as they rise over high ground with the forced ascent of the lower warm and humid layers.

In addition to contributing to the stability or otherwise of air masses and to determining the incidence of local sea breezes, the seas have a further important role to play. The high specific heat of sea water renders it slow to heat up but equally reluctant to cool down – a tempering characteristic which it imparts to the regime of coastal districts. This tempering role also has a regional aspect because the distribution of sea temperatures around the coasts is by no means uniform (Figures 1.9a and b). The warming effect is greater on western coasts, the east coast of Britain being subject to cooler

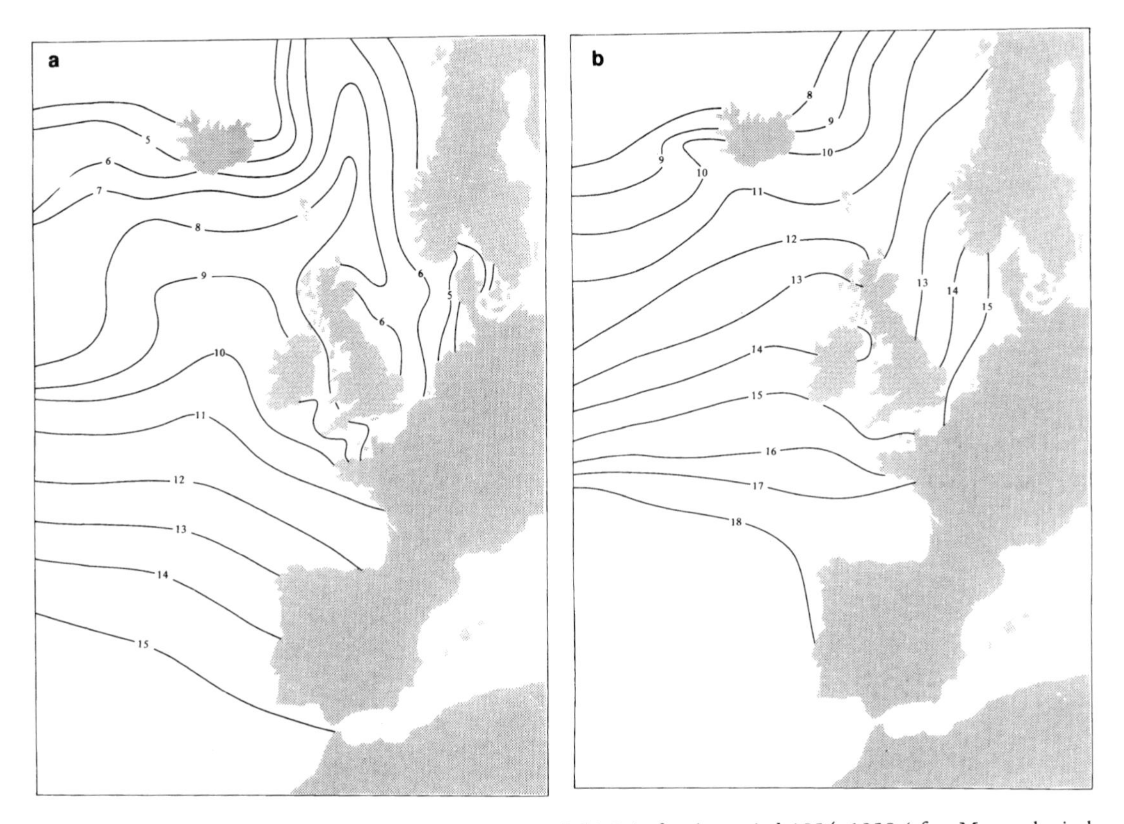

Figure 1.9 Mean sea surface temperature in (a) January and (b) July for the period 1854–1958 (after Meteorological Office 1975)

Table 1.3 Annual percentage frequencies of air masses and weather systems at Kew, Scilly and Stornoway

Location	*A*	*Pm*	*Pmr*	*Pc*	*Tm*	*Tc*	*F*	*H*
Kew	6.5	24.7	10.0	1.4	9.5	4.7	11.3	24.3
Scilly	4.2	27.5	10.0	0.9	13.5	2.7	11.8	22.1
Stornoway	11.3	31.5	16.0	0.7	8.7	1.3	11.8	13.8

Source: After Belasco (1952)
Key to air mass/weather systems: A = Arctic, Pm = polar maritime, Pmr = polar maritime returning, Pc = polar continental, Tm = tropical maritime, Tc = tropical continental, F = frontal, H = high pressure

conditions. Jones (1981) has shown that during any month the sea surface temperatures along the coast of north-east England will be between 1.5 and 2 degrees C cooler than off the south-west peninsula. The same is true, though to a lesser degree, of the Irish Sea.

Despite the relatively small area embraced by the British Isles the geographic range is nevertheless sufficient to reveal regional contrasts in the incidence of different air masses included in Table 1.3 and their associated weather systems. It follows that different temperature regimes may prevail between those regions experiencing such contrasts. Table

Table 1.4 The twenty-seven airflow types of the Lamb classification

Direction	*Anticyclonic curvature*	*Straight*	*Cyclonic curvature*
North-westerly	ANW	NW	CNW
Northerly	AN	N	CN
North-easterly	ANE	NE	CNE
Easterly	AE	E	CE
South-easterly	ASE	SE	CSE
South	AS	S	CS
South-westerly	ASW	SW	CSW
Westerly	AW	W	CW
Non-directional	A	Unclassified	C

1.3 shows air mass frequencies for three very different locations. The greater incidence of polar and tropical air does much to explain, for example, the warmer conditions at Kew and Scilly as compared with Stornoway. The higher incidence of anticyclonic weather at Kew is another factor in this distinction. Tropical continental air scarcely reaches the far northern latitudes, while Arctic air is nearly three times more abundant over Stornoway than over Scilly. On the other hand, frontal weather is equally abundant in all regions. It is against this background of evolving air masses with different regional frequencies that we must now turn to consider their more detailed geographical expression.

THE REGIONAL INFLUENCE OF THE ATMOSPHERIC CIRCULATION AROUND THE BRITISH ISLES

The previous sections show how environmental factors help to shape the pattern of regional climates in the British Isles. The actual weather and climate experienced at any location result from the interaction of these factors with the atmospheric circulation – in other words, by the interaction between environmental and meteorological factors. The part of climatology that relates local climates to the wider pattern of the atmospheric circulation is termed synoptic climatology. The purpose of this section is to show how the study of the synoptic climatology of the British Isles has built upon the concept of air masses to provide a comprehensive means of describing regional climatic variability.

We have already seen that the British Isles lie in the path of the mid-latitude westerly circulation. Belasco (1952) expressed this in terms of a prevailing polar maritime air mass, with the frequency of high pressure increasing south-eastwards (see Table 1.3). This showed that regional variations in the climate may result not just from local environmental contrasts but from a varying influence of the atmospheric circulation.

To explore further this varying influence of air masses, attention then turned to the concept of airflow types, a means of expressing the weather situation by reference to the prevailing wind direction (not necessarily just at the surface) or the presence of a pressure system. By far the most widely used airflow classification for the British Isles is that of H.H. Lamb (1972). This scheme succinctly expresses the airflow type of each day from 1861 to the present in a single code (see Table 1.4). Lamb's work constitutes a relatively homogeneous data set of climatic information extending over a sufficient period to be of value in the analysis of climatic variation and change (see Chapter 12).

The airflow type is determined subjectively from the character of both the surface and the mid-troposphere (500 mb pressure level) pressure distribution. As such, it conveys a more general indication of airflow behaviour than the detail of surface wind direction and is more akin to the direction of the steering forces acting on pressure systems. Lamb's study area encompasses the whole of the British Isles, extending from 50 to 60° N and from 10° W to 2° E. By capturing the daily configuration of the mid-latitude circulation, the Lamb classification 'provides the longest available perspective of one of the processes of climatic variation: the degree of prevalence of the westerlies vis a vis the frequency of blocking' (Lamb 1972, p. 3).

The driving force for the westerly circulation is derived from the temperature gradient found across the mid-latitudes. This usually strengthens in winter owing to polar cooling, resulting in an invigorated westerly circulation. By late spring, this temperature contrast has reached the annual minimum and, consequently, blocking patterns reach a peak in frequency. This seasonal cycle is depicted in the Lamb classification in terms of a higher daily frequency of the W (westerly) type in winter and a more varied distribution of types in late spring and early summer (Figure 1.10).

The Lamb classification has its origins in a classification of regional airflows in which daily airflow types were allocated to each of five regions within the British Isles (Levick 1949, 1975). Although Levick's work has not received the attention of Lamb's, the rationale for this regional distinction is pertinent here. Extending Belasco's regional contrasts, Levick argued that different parts of the British Isles sometimes experience contrasting airflow types. If Scotland and Ireland are being affected by a south-westerly airflow, England may lie under the influence of an anticyclone. Conversely, the weather in south-east England under the influence of a slow-moving depression appears all the more dismal when seen in conjunction with the fine weather in western Scotland and Ireland that often occurs at the same time. Such contrasts are an important influence upon regional climatic variations, as investigated more fully in the final chapter.

To acknowledge occasions when more than one airflow type or pressure pattern exists over different parts of the country, daily regional airflow types have been developed. Betts (1989) has produced a classification of airflow types for Northern Ireland to assist with the interpretation of rainfall events. A regional classification of airflow types has also been developed by Mayes (1991) which is based on an assessment of the surface airflow patterns over Scotland, Ireland, the South West (of England and Wales) and South East England. The latter classification runs from 1950 to the present, and is regularly brought up to date. It uses a notation similar to Lamb's catalogue but redefined according to the regional scale of the analysis.

The regional scale has also been adopted by El-Kadi and Smithson (1996) in the creation of an automated classification of fifteen surface air pressure types for the British Isles. Several of these pressure patterns incorporate regional contrasts; for example, type 11 is a combination of tropical maritime air over Scotland and Ireland (a SW airflow type in the

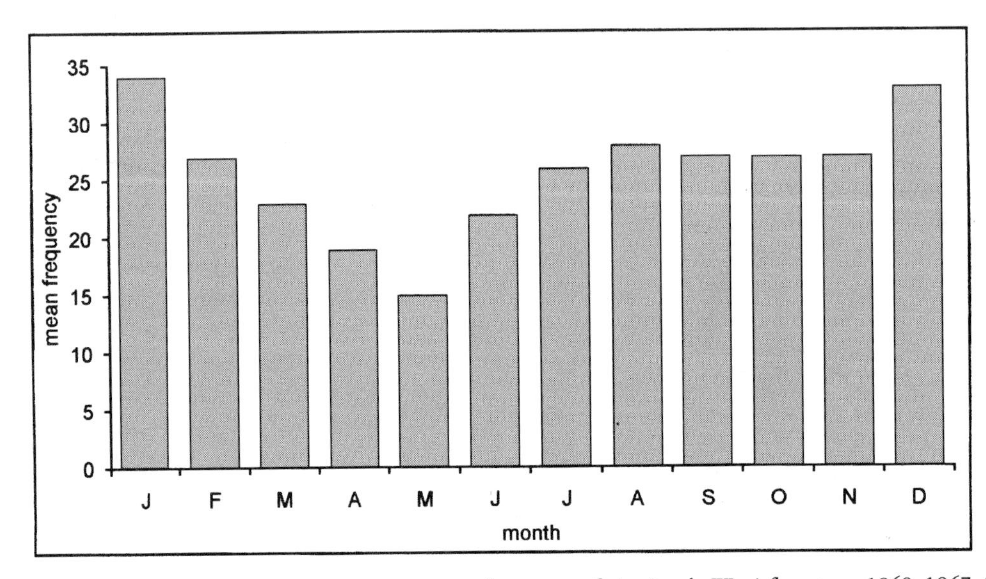

Figure 1.10 Monthly variation in the mean percentage frequency of the Lamb W airflow type 1869–1967 (after Lamb 1972)

Mayes classification) with tropical continental air over southern England (more usually a SE or E airflow type). In a significant departure from Lamb's classification, the authors devise three cyclonic and three anticyclonic pressure patterns, acknowledging thereby the variety of weather types and pressure distributions associated with these situations.

CHARACTERISTICS OF AIRFLOW TYPES AROUND THE BRITISH ISLES

An understanding of the regional diversity of climate depends upon a knowledge not just of the mid-latitude circulation but also of the way in which it interacts with environmental factors and topography. The airflow type determines which air masses reach any part of the British Isles at a given time. The route taken by each air mass influences the degree to which air is modified *en route* (for example, if an air mass of continental origin passes over the sea). When it reaches the British Isles, topography plays a critical part in determining the geographical distribution of weather according to the degree of exposure and shelter imparted to different areas.

In any typical month or year, the British Isles experience a succession of different airflow types. The fact that one may dominate a particular month helps to shape the character of the weather, but climatic data often mask regional weather contrasts when the weather has been created by a mixture of airflow types. For this reason, the illustration of regional weather types that follows is provided by means of both monthly and daily case studies.

Westerly and South-westerly Weather Type

The westerly and south-westerly weather type is the commonest airflow type, and is present on over 100 days in an average year. Indeed, Bilham (1938) referred to this as the 'normal' weather type. It results from a pressure distribution that conforms with the average pattern in which low pressure is situated to the north of the British Isles and high pressure to the south (pressure centres typified respectively by the Icelandic low and Azores high on maps of average air pressure).

Areas of low pressure are themselves usually driven by the westerly winds several kilometres above the surface, these being a product of a high-index weather situation in which the controlling upper winds maintain a west-to-east track around each hemisphere. Wind directions at the surface are subject to variation according to the location of the depression centres; to the north of the depression track surface winds will have an easterly component, and winds are usually variable in the vicinity of depression centres. Thus when 'westerly' weather covers much of Britain, Scotland may share neither the same airflow type nor wind direction.

The character of a month dominated by a westerly type depends upon the season. In winter, the warmth of the North Atlantic imparts a mild, maritime character to the weather. Conversely, in spring and summer the maritime influence tends to suppress daytime temperatures close to western coasts. Highest temperatures are observed further east, and areas downwind of upland areas such as the Pennines or Grampians may have temperatures locally elevated by a föhn-type wind at any time of year. Frequent rainfall and extensive cloud are most conspicuous close to western coasts. In January 1983, for example, Cape Wrath (in the extreme north-west of Scotland) recorded a monthly sunshine total of just 0.5 hour while eastern coastal districts of Scotland and England enjoyed their sunniest January since 1959 with about one-and-a-half times average sunshine. Rainfall under these conditions is often notable for its frequency. Though not achieving the short-term intensities associated with thunderstorms, heavy orographic rain is far more likely to persist, resulting in large accumulations of rainfall amounts over successive hours or even days. The general feature of western areas having the highest rainfall and the lowest sunshine is clearly shown by December 1974 (Box 1.1).

In some of the most unsettled westerly spells, high pressure fails to build to its usual degree behind the

Box 1.1

DECEMBER 1974 – A 'WESTERLY' MONTH

- An exceptionally vigorous westerly airstream covered Northern Europe. Mean monthly air pressure across the British Isles ranged from 996 mb in Shetland (over 9 mb below average) to nearly 1,024 mb in Jersey (9 mb above average).
- Westerly winds were recorded on over 20 days in the month; winds from between NE and SE were almost entirely absent. High pressure was sufficiently close to southern areas to weaken the influence of fronts there.
- Persistent moist westerly winds focused orographic enhancement of rainfall on west-facing hills, especially over northern areas, closest to the main depression tracks. Parts of the Western Highlands of Scotland recorded more than 800 mm of rain whilst parts of East Anglia had under 25 mm.
- There was a particularly large eastward increase in sunshine totals with many eastern areas having more than 50 hours of sunshine in contrast to under 15 over western upland areas – an exceptionally low total.
- This was the mildest December since 1934 over England, Wales and Northern Ireland and temperatures reached 15°C on the 28th. As with rainfall and sunshine, eastern districts were most favoured; anomalies ranged from around +1 degree C on western coasts to +4 degrees C in the Midlands. In many places the month was milder than October 1974 (see Box 1.5).
- December 1974 was a month of large contrasts in weather across small distances. Stonehaven, on the east coast of Scotland, in the rain shadow of the Highlands, recorded 78 hours of sunshine (176 per cent of average), the highest December total since 1931, while the west of Perthshire had barely 10 hours.

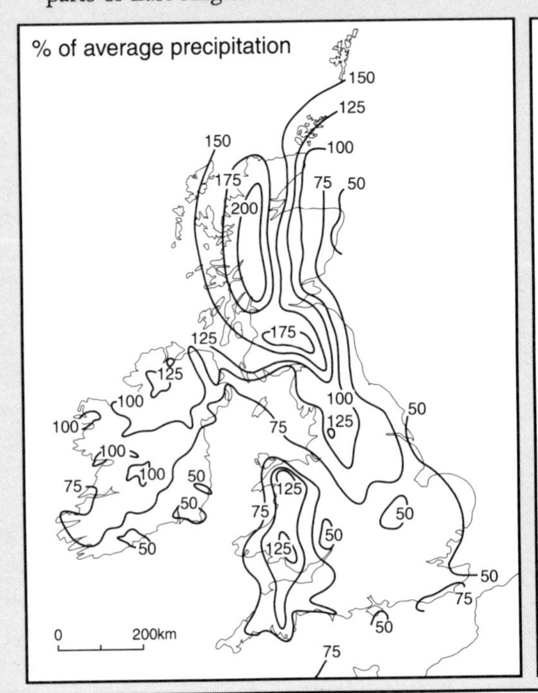

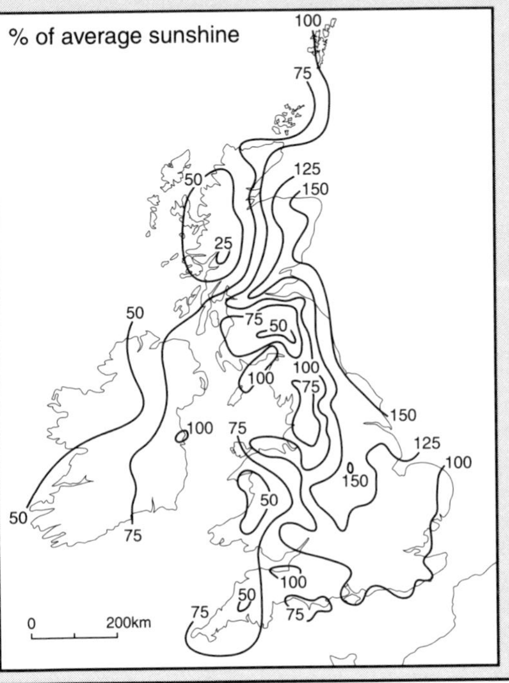

cold front and winds remain in the west or south-west. In the absence of a north-westerly wind to maintain its progress south-eastwards, the cold front may decelerate, delaying the clearance of any rain falling in the warm sector or from the front itself. Any such hesitation of fronts allows the warm conveyor belt to remain stationary, enhancing uplift of saturated warm air. This may give rise to very high rainfall totals, especially on high ground where these conditions are augmented by heavy orographic rain. This was well illustrated on 17 January 1974, when persistent heavy orographic rain was observed in western Scotland in an extensive warm sector. Loch Sloy (close to Loch Lomond) recorded 238.4 mm of rain, the highest January daily rainfall total in the United Kingdom. A fall of 169.7 mm in a similar synoptic situation was observed near Paisley on 10 December 1994 which led to widespread flooding in western Scotland (Black and Bennett 1995). Analyses of the geography of daily rainfall over Scotland by Smithson (1969) and Weston and Roy (1994) have shown how sharply rain shadow conditions become established downwind of the summits of the Western Highlands (Figure 1.11).

The highest daily fall observed in November was 211.1 mm at Lluest Wen Reservoir at the head of the Rhondda Fach valley in South Wales on 11 November 1929. This was associated with the eastward movement of warm and cold fronts. The high total was caused by the formation of secondary depressions on the cold front. By delaying the clearance of the front, the heavy rain and drizzle of the warm sector continued for many hours, augmented by the general uplift associated with the evolving secondary depression centre. Although orographic rainfall does not usually attain the exceptional hourly intensities of convectional rainfall in thunderstorms, it must be remembered that there may be

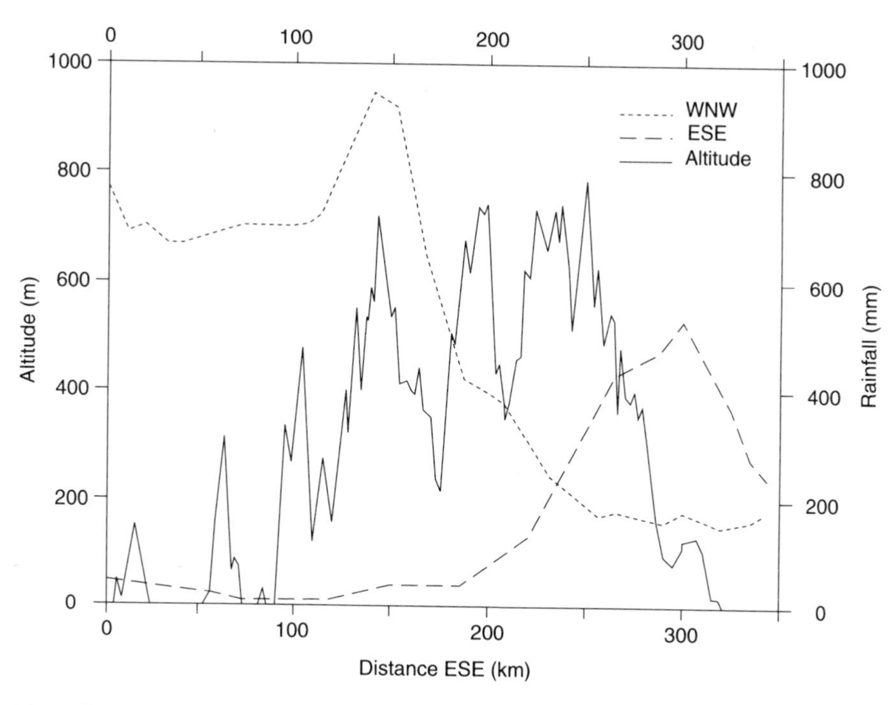

Figure 1.11 The influence of orography (solid line) on average precipitation over Scotland shown by the variation in precipitation across a transect over the Scottish Highlands between the Outer Hebrides and Arbroath under west-north-west and east-south-east air flows (after Weston and Roy 1994). See also this volume Figure 10.6 pp. 238–9

Box 1.2

HEAVY RAINFALL, A FÖHN WIND AND A WARM SECTOR

The combined effects of orographic enhancement of rain, rain shadow formation and warm föhn winds were illustrated by the weather of 13 November 1994. England and Wales lay in a warm sector. A small secondary depression formed just west of Ireland, thus setting up appropriate conditions for delaying the clearance of the cold front. Rainfall totals in north Wales ranged from 113 mm at Yspytty Ifan (in east Snowdonia) to under 10 mm in sheltered north-east Clwyd. The highest temperature of the day of 16.9°C was observed at Harwarden (in the same area of Clwyd), and the temperature reached 16°C at Point of Ayre, on the Isle of Man and downwind of Snaefell, and at Minehead, which lay downwind of Exmoor. A further example is discussed in Chapter 10 and describes the situation in early January 1992 when, with a well-developed warm-sector westerly air flow, temperature differences of as much as 3 degrees Celsius existed between Benbecula and Aberdeen. These are all typical locations for föhn type winds in stable south-westerly airstreams occurring in warm sectors of frontal depressions. A change of wind direction simply redirects the warm descending air to whichever areas lies downwind of high ground.

Box 1.3

THE INFLUENCE OF THE SEA IN SPRINGTIME

In autumn and winter, the relative mildness of the sea means that coasts exposed to westerly airstreams share the mildness of other parts of the country. However, in spring, the limited warming of the sea surface means that onshore winds tend to be relatively cool. This is especially noticeable in stable airstreams where cooling from the underlying surface may induce the formation of sea fog (haar) which may plague coastal districts with mist or persistent stratus cloud. While these conditions are often associated with eastern coasts because of the coolness of the North Sea (see Chapters 7 and 9), western coasts actually experience onshore winds more frequently. Over the first three days of April 1995 the British Isles were covered by an anticyclonic westerly flow of moist tropical maritime air. Most of South Wales and south-west England had less than 1 hour of sunshine over this period with coastal mist, fog and drizzle. By contrast, owing to the inherent warmth and subsequent land track of the air, east Kent had nearly 30 hours of sunshine, and maximum temperatures downwind of high ground were widely in the region of 20°C rather than the 10°C observed on western coasts.

many such frontal rainfall episodes in any given month. For example, in November 1929, the Rhondda had a further fall of 100 mm on the 24th/25th.

North-westerly Weather Type

A north-westerly type usually covers the British Isles when high pressure is situated to the west or south-west with low pressure over the northern North Sea or Scandinavia. It is most well-developed when the upper westerly winds are deflected into a trough just to the east of Britain with a corresponding ridge over the North Atlantic. This pattern was persistent during the summer of 1987 and, as a consequence, rainfall was often heavy in northern and eastern areas of England and Scotland while south-western districts closer to the upper ridge had a reasonably fine summer (Box 1.4).

The character of weather is markedly different from that of the westerly type since it is typically (but not always) associated with polar maritime air. This has a cool origin at all times of year and thus generally warms from below as it passes over the eastern North Atlantic. This warming increases the instability of the air, promoting the development of cumuliform cloud and vigorous convective uplift that often leads to showers the distribution of which is highly sensitive to the location of the warmest surfaces. In summer, when land surface temperatures exceed those of the adjacent seas, shower activity builds up inland and sometimes progressively increases towards the usually warmer south-east of England. Conversely, in autumn and winter, when the sea is warmer than the land, showers tend to form over the sea and then run on to windward coasts. Analysis of radar imagery has indicated the importance of horizontal convergence induced by topographic features such as coastline shape in enhancing showers into more active zones, such as over the North Channel (Browning *et al.* 1985) and the Bristol Channel. Exposed coasts are often notably cool during north-westerly airstreams; this characteristic of the Cardigan Bay coast is described in Chapter 6. By contrast, places sheltered from this direction can have temperatures well up to average and any briskness of the air is mitigated by the often lengthy spells of bright sunshine and excellent visibility; the Cardiff and Torbay areas are good examples.

Northerly Weather Type

High pressure to the west of the British Isles and low pressure over the North Sea or Scandinavia usually leads to a northerly airflow. Associated weather is usually crisp, bright and fresh with excellent visibility in relatively pure Arctic or polar air. It is associated with low-index weather situations, with large Rossby waves forcing the upper airflow into an upper ridge to the west of the British Isles and a trough to the east that may extend into the Continent. Pressure centres are often quite slow-moving. Northerly airstreams display similar characteristics to north-westerlies with a tendency to instability, showers and strong lapse rates. The drop in temperature with height is especially noticeable and important in spring in upland areas, when rain showers often turn to sleet or snow. For much of the year, showers may affect both west and east coasts simultaneously while localities sheltered by high ground to the north may enjoy bright sunshine; areas such as the southern coasts of Ireland, Wales and England together with central Scotland may be especially favoured. October 1974 provides an excellent example (Box 1.5). This weather type also reveals itself in a distinct rainfall peak around Moray identified by Weston and Roy (1994); see also Figure 10.6.

North-easterly Weather Type

North-easterly airstreams are usually indicative of a very blocked (low-index) circulation type in which depressions may be slow-moving close to the Low Countries and high pressure may extend over much of the north-east Atlantic, displacing the usual focus of low-pressure activity. The air reaching Britain and Ireland is usually unstable in winter owing to the relative warmth of inshore waters, and coasts exposed to this direction may experience heavy showers. In summer, the air may be less unstable, and windward coasts are more likely to experience stratus cloud, mist and fog (see next section). When low pressure or attendant frontal systems are situated close to south-east England, the weather of eastern districts as far north as the Scottish border can be an inclement combination of persistent rain, low temperatures and overcast skies while western coasts are often fine and clear (see Box 1.6). Places

Box 1.4

JULY 1987 – A 'NORTH-WESTERLY' MONTH

- High pressure was centred to the south-west of the British Isles and a WNW airstream covered all parts. This was associated with a weak upper ridge situated to the west of the area and a trough to the east.
- Much of the rainfall in southern Britain was regulated by instability rather than frontal or orographic processes. Consequently, there were several heavy daily rainstorms interspersed with spells of fine weather. Typically in a month of unstable north-westerlies in summer, many of the heaviest falls occurred in south-east England, with air becoming warmed from below after a land track. South-western districts were often fine, lying closest to high pressure.
- Rainfall in southern Britain ranged from around 25 mm in the South West to over 100 mm over much of the South East, a pattern that was repeated in similar circulation conditions in August 1987.
- North-west-facing coasts were often cloudy with a preponderance of onshore winds; the far north-west of Scotland had under 60 hours of sunshine.
- This was a month of dramatic weather in Essex. Severe thunderstorms caused major problems between the 18th and 20th and again on the 29th, when the Epping area had a severe storm that gave 71 mm of rainfall in total (including 54 mm in 21 minutes).

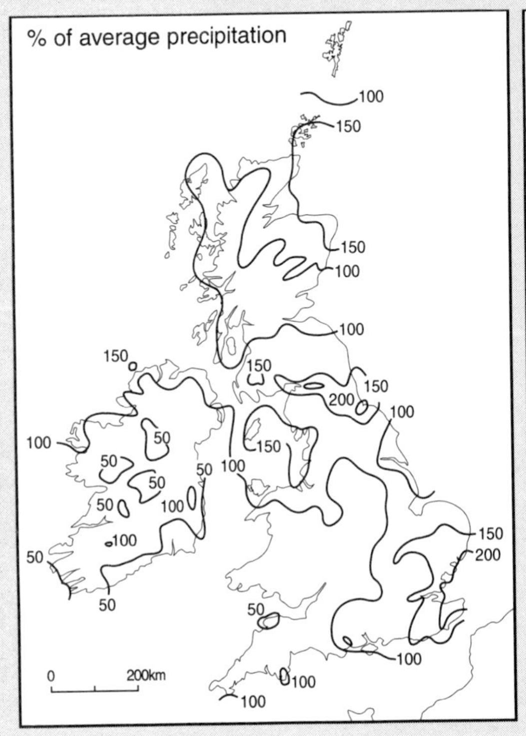

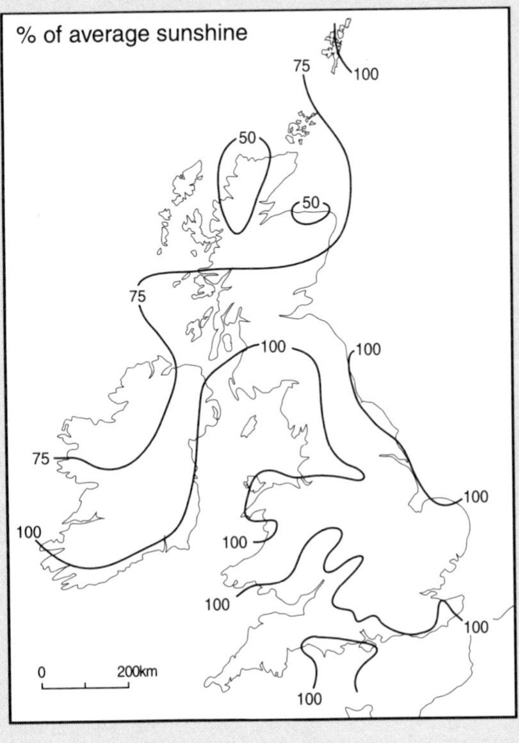

Box 1.5

OCTOBER 1974 – A 'NORTHERLY' MONTH

- Mean monthly air pressure ranged from around 1,012 mb in Shetland and the east of East Anglia to over 1,022 mb (+8 mb) in the west of Ireland. Northerly winds were observed on up to nineteen days in Scotland; further south, north-westerlies predominated.
- A cold and windy month with frequent showers in characteristically unstable air. Sleet and snow fell at times over northern hills and snowdrifts of more than a metre in depth were noted on the 28th.
- The rainfall distribution was controlled by proximity to the low pressure to the east, combined with exposure to northerly winds running off the North Sea along eastern coasts of Scotland and England, where most places had more than 100 mm. The driest areas were those with high ground to the north, illustrating the shift of rain shadow areas according to the prevailing wind.
- This was the coldest October of the century in Northern Ireland. Temperatures in coastal parts of south-east England were around 4 degrees C below average. The highest temperature recorded at Reading was as low as 12.9°C, less than the average maximum day temperature for October.
- The south coast of the Moray Firth caught the brunt of the northerlies. Nairn had the dullest October since 1906, and a little further east over twice the average rainfall was recorded. By contrast, the more sheltered south-west of Scotland had a notably fine, dry and sunny month.

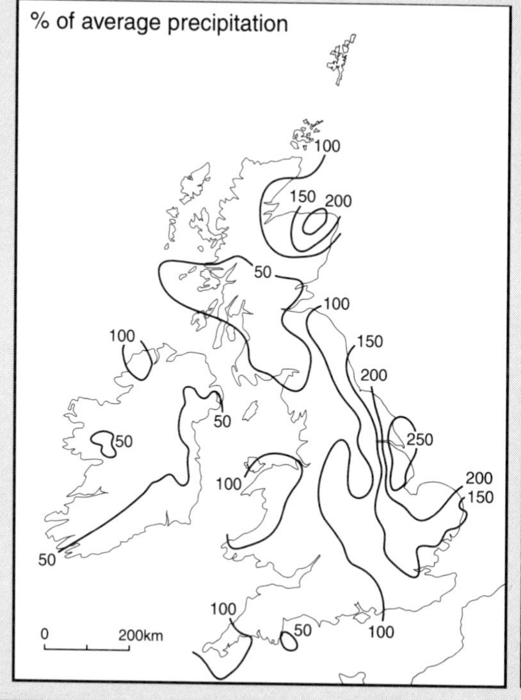

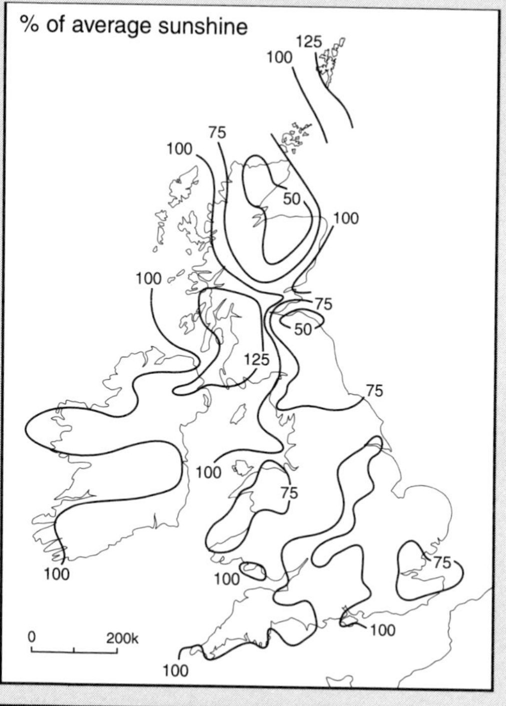

Box 1.6

MAY 1984 – A 'NORTH-EASTERLY' MONTH

- An exceptionally blocked month in which low pressure was often slow-moving over the Continent, and the Azores high was displaced rather to the north of its average position. A cyclonic north-easterly airstream covered much of the country with air pressure over the south-east of England being 1,011 mb (−11 mb).
- North and north-east winds dominated the weather – they were observed on twenty-two days in London, for example. At no time did fronts or depressions approach the British Isles from the west, a feature of extremely blocked weather patterns.
- Rainfall exceeded 50 mm south-east of a line from Yorkshire to South Wales. By contrast, Eskdalemuir in the Southern Uplands of Scotland and Nairn were the driest places in Great Britain with only 4 mm, whilst the wettest area was on the hills of east Hampshire where Liphook recorded 162 mm.
- Cloudy, wet weather and north-east winds kept temperatures below average over England and Wales, apart from on western coasts. Parts of western Scotland and Ireland were locally warmer than average, reflecting the fact that offshore winds are generally warmer than onshore winds at this time of year when sea surface temperatures are relatively low.
- Glasgow and the Western Highlands had a fine, sunny month. The mean temperature in Glasgow of almost 12°C compares with a value of only 9.1°C at Margate (anomalies from average being +0.6 degrees C and −2.7 degrees C respectively), whilst the average maximum temperature at Margate was only 11.4°C. The highest temperature of the month was at Onich, near Fort William, where the temperature reached 25.5°C on the 23rd. The Glasgow area had less than 20 mm of rain, in contrast to totals of between 60 and 70 mm in east Kent.

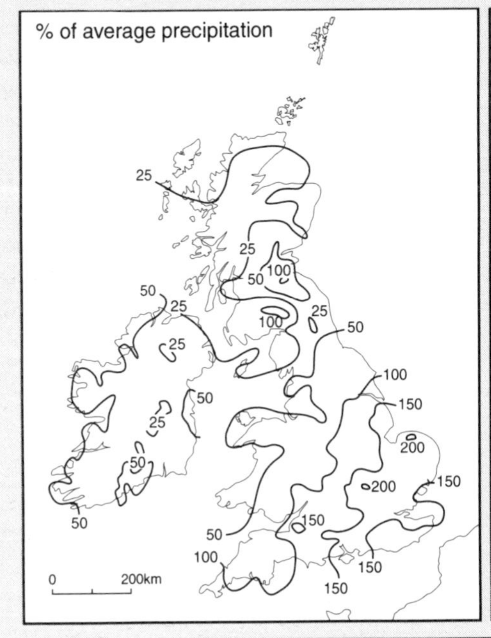

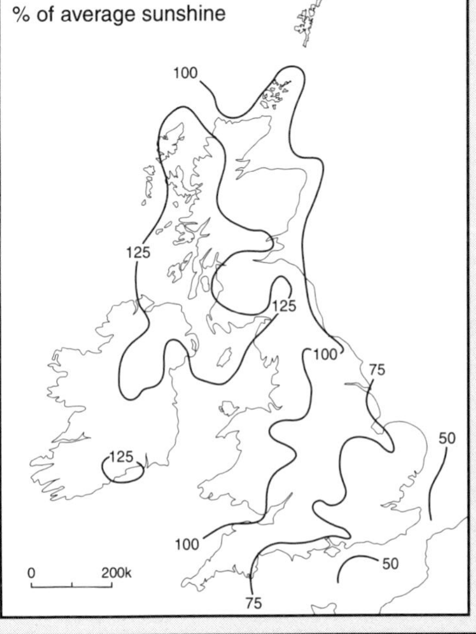

such as Manchester, Swansea and Plymouth may be particularly favoured in these situations.

Easterly Weather Type

Easterly winds at the surface are typically associated with a very blocked synoptic situation in which high pressure may be dominant over much of northern Europe. This feature may also extend upwards, disturbing the flow of the upper westerlies, which can become deflected both to the north and south of the high-pressure area as the polar jet stream splits into separate branches. When this occurs, the northern branch may initiate a depression track over the northern European seaboard while the southern branch may spread disturbed weather through the Mediterranean (as in the summer of 1976 and the winter of 1995/96) or further north, for example from Biscay into central Europe (as in February 1947).

The resulting weather is highly dependent on the time of year and the length of the sea track. In winter, this is often the coldest weather type, bringing polar continental air from Eastern Europe. The severity of the cold over northern Britain may be moderated by the lengthened track over the North Sea. In the extreme south of Britain and Ireland, the discomfort of the low temperatures can be increased by the strength of the wind and the possibility of wintry showers or longer periods of rain or snow when frontal systems edge north-eastwards across the Celtic Sea and English Channel (see Chapter 6 for a discussion of blizzards associated with this synoptic situation).

In spring and summer the temperature contrasts between lower atmosphere and sea are reversed and warm easterly airstreams off the European mainland are cooled by contact with the North Sea. This cooling increases the stability of the air, and exchanges the chance of showers for a fair probability of stratus or coastal fog patches. Coastal temperatures are then depressed to an extent determined by the sea surface temperature (see Chapters 7 and 9). Temperature contrasts are often most conspicuous when the easterly winds are on the margins of high pressure and inland areas experience long sunny periods and much higher temperatures. These points are discussed in Box 1.7 and in Chapter 7.

Southerly Weather Type

High pressure over central and northern Europe and low pressure to the west of the British Isles usually result in a southerly airstream over all or part of the British Isles. However, unlike the case of other directions, it is rather unusual for all districts to share the same wind direction. In a rather blocked weather situation in which an upper trough is found to the west and a ridge to the east, depressions and fronts that are moving east across the Atlantic may decelerate and then be steered north-east. Tropical maritime air may thus reach western areas while eastern Britain remains in a more continental south or south-easterly flow. This can result in a sharp west–east contrast in the weather. Slow-moving fronts may give persistent rain in Ireland and western Scotland, accentuated by the high moisture-carrying capacity of warm south or south-westerly air and also by orographic enhancement. Eastern England may have relatively little rain.

Many of the months dominated by a southerly bias to the circulation are actually characterised by a south to south-west airflow. A good example is provided by the exceptionally mild November of 1994 (Box 1.8). In April 1966 the continental influence was both extensive and persistent. The shelter afforded by the Scottish Highlands and lengthy land track allowed Skye and the Shetland Isles to become the sunniest part of the United Kingdom with more than twice the sunshine of much of southern England.

Temperature distributions and absolute levels depend on both the season and the stability of the air. While western districts have the benefit of milder, maritime air in autumn and winter, this is enhanced to the lee of high ground by the föhn effect, a process that operates most effectively when the airstream is stable, as southerlies often are. The localities warmed by this wind vary according to wind direction (Lockwood 1962), and the

Box 1.7

FEBRUARY 1986 – AN 'EASTERLY' MONTH

- This is the clearest example of a month dominated by easterly airstreams since the Second World War. High pressure was centred to the north of the British Isles with positive pressure anomalies exceeding 16 mb between Scotland and Iceland. The pressure at Jersey was 14 mb lower than that at Shetland (where the average air pressure was as high as 1,027 mb).
- Widely separated coastal districts of western Britain and Ireland recorded no measurable precipitation during the month. However, most areas were affected by light snow showers but these often produced a mere 'trace' of precipitation on many days in the west. Totals only reached 50 mm close to hills within about 80 km of the North Sea.
- This was the second coldest February of the twentieth century (only 1947 was colder). Frost was continuous over south-east England between the 8th and the 12th. Temperature anomalies ranged from −1 degrees C in the Western Isles of Scotland to −6 degrees C in Cambridgeshire. Mean daily maxima were below freezing in a few southern areas.
- Close to the high-pressure centre, Lerwick in the Shetland Isles had its sunniest February since records began in 1923. Coasts exposed to the easterly winds were much cloudier, especially further away from the high pressure. Sherkin Island off the coast of County Cork in Ireland had less than a third of the sunshine recorded at Belmullet (Sligo), and the lowest sunshine total reported by the Meteorological Office was at the Lizard in Cornwall, where the average duration was less than 1 hour per day.
- This was a severe month in the Cairngorms, which bore the brunt of the low temperatures and periodic heavy snowfalls. On the 6th, 49 cm of level snow was recorded at Braemar where snow lay on the ground throughout the month. On the 27th a temperature of −21.2°C was observed at Grantown-on-Spey.

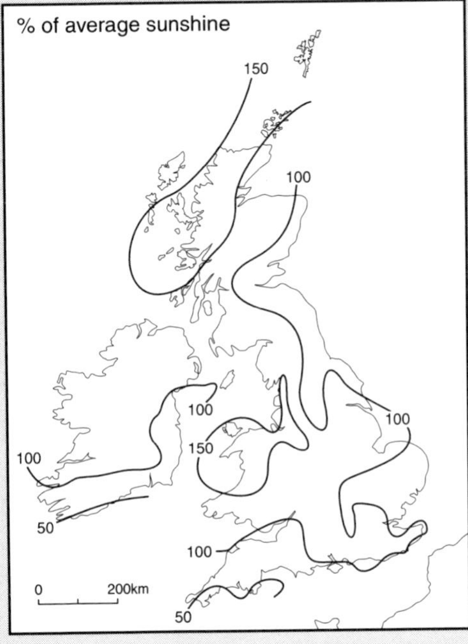

Box 1.8

NOVEMBER 1994 – A 'SOUTHERLY' MONTH

- This was a month of quiet, rather blocked weather patterns. Low pressure was situated generally to the west and north-west with high pressure lying over continental Europe. S and SW airflows were thus dominant over the British Isles. As a result, this was the mildest November on record. Air pressure was below average in all areas except south-east England, which was closest to the high-pressure centre.
- Rainfall amounts were rather variable owing to the slow movement of several frontal systems. Areas with open sea to the south were generally wettest since this is a factor in providing both instability and moisture in an airstream in autumn. Less than 20 mm was observed close to the East Anglian coast.
- This was the mildest November on the Central England series that starts in 1659. Monthly minima were between 3.5 and 4 degrees C above normal as a result of the mainly cloudy nights. In some southern coastal localities, the temperature reached 10°C on every day of the month; the mean maximum at Minehead was as high as 14.1°C, illustrating a föhn-effect-type warming to the lee of Exmoor.
- November was generally dull, as befits a month dominated by moist south and south-west winds. At Valentia, County Kerry, the prevalence of these onshore winds gave the station its dullest November in over a hundred years but at Malin Head, County Donegal it was the sunniest since 1976, illustrating the extent to which cloud may break up after passage over land.

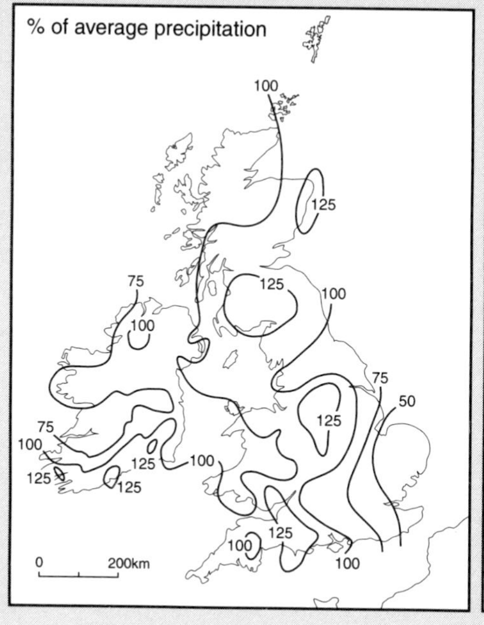

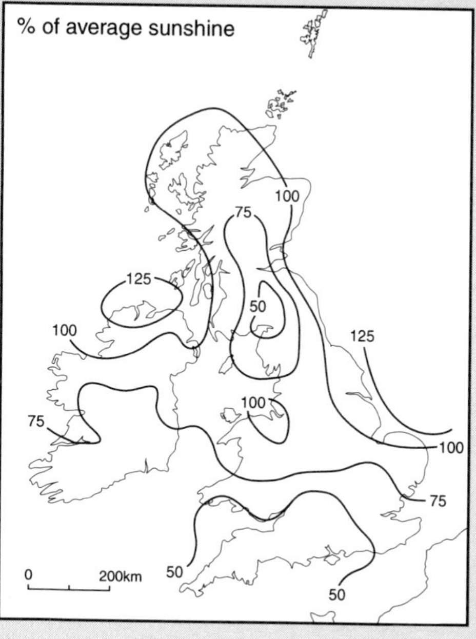

principal areas to benefit in this way include the North Wales coast, the Moray Firth, the Dublin area, the north Somerset coast around Minehead, the Yorkshire coast around Teesside and the Carlisle area. Lyall (1971) has reviewed the factors promoting warmth in the early part of the year. The föhn effect was clearly involved in two remarkable bursts of early warmth: the observation of 23°C at Cape Wrath on 12 March 1957, which, as Chandler and Gregory (1976) noted, is a value equivalent to the average annual maximum temperature, and 18°C at Aber on the North Wales coast on 10 January 1971, the latter being the highest January temperature on record in the United Kingdom.

Southerly airstreams are arguably most dramatic during warm spells in summer when they allow warm, potentially unstable air to reach the British Isles from Iberia or France – the 'Spanish plume' phenomenon (Young 1995). Air that has been subject to the intense surface heating of the Mediterranean summer is capable of triggering severe thunderstorms when it moves into the vicinity of colder air around the British Isles brought by an upper trough from the west. A more subtle phenomenon associated with southerly airstreams is the deposition of Saharan dust (usefully reviewed by Burt, 1995).

Anticyclonic Weather Type

Anticyclonic weather is generally associated with fine weather in which cloud development is inhibited by the gradual subsidence of air. This warms the air, enabling evaporation to disperse the cloud. Warmed, subsiding air may accumulate in a layer above the surface such that the temperature increases with height – a subsidence inversion. The presence of a warm layer at some distance above the ground is thus able to discourage instability even when air in contact with the ground is strongly warmed by sunshine. This accounts for the lack of cumulus cloud formation on some fine days.

The character of anticyclonic weather may be determined by the location of a high-pressure centre in relation to the British Isles. The British Isles are rarely covered uniformly by an area of high pressure. Belasco (1948) found that an anticyclone was most commonly situated to the south-east of the British Isles in January, to the north-east in May and to the south-west in July. These changes are indicative of the greater frequency of blocked weather situations in late spring when anticyclones are able to migrate to higher latitudes owing to the north–south extension of Rossby waves in low-index situations. The effects on local weather can be best illustrated by a series of examples. The drought of 1975–76 was caused by unusually persistent high pressure over and to the south-east of England and to which Scotland and Ireland were largely peripheral. These latter areas thus experienced periods of west-south-westerly winds and received near-average rainfall over the eighteen-month duration of the drought (see also Chapter 12). With high pressure centred to the south-east, much of north-western Britain and Ireland was quite wet in March, May and June 1976. In the latter month, an anticyclonic westerly flow helped to induce a contrast between monthly rainfall totals of 200 mm in the Western Highlands and less than 10 mm in Deeside, a very effective rain shadow event. Much of eastern England also had under 10 mm of rain, and a few localities were rainless in a memorably anticyclonic month. By contrast, Stornoway and Eskdalemuir were very dull, with only 67 per cent of average sunshine. High pressure became more generally dominant through July, and August was a truly anticyclonic month over the whole country as the high pushed further north.

If high pressure is centred to the north-west, fine weather is likely to prevail over much of Ireland and Scotland. South-east England may only be on the margins of its influence, with the added disadvantage of onshore winds. North-easterly winds are less likely to carry copious moisture and the subdued orography of the region is less likely to provoke significant orographic enhancement of any rain. However, a wind off the North Sea is an effective means of increasing cloud cover, inhibiting the diurnal range in temperature. This was shown clearly in the last week of June 1995, when a blocking anticyclone became slow-moving close to north-west Britain, allowing cool and cloudy north-easterly winds to reach the east coast of England. The resulting regional contrasts in temperature and sunshine are shown in Figure 1.12. Much of central England recorded between thirty and forty hours of sunshine as early-morning stratus spread in from the North Sea. Days, though generally starting cloudy, became brighter by mid- to late morning as the Sun's heat dispersed the thin, low-level cloud cover.

Figure 1.12 shows that although the western fringes had close to the maximum duration of sunshine possible for the week, early-morning visitations of low cloud spread westwards until they reached orographic barriers, often extending as far as the eastern sides of the Welsh mountains and Exmoor. The cloud cover persisted over many eastern coasts and as a consequence it remained very cool. The effect of onshore winds and varying cloud amounts were vividly illustrated by the contrast in maximum temperature on 25 June, when the temperature at Tynemouth reached only 11.1°C, but at Tulloch Bridge, north-east of Ben Nevis, a temperature of 28.6°C was logged at an automatic weather station.

The character of anticyclonic weather is strongly influenced by the season. Summer sunshine is usually sufficiently strong to warm the surface above dew point temperature, promoting evaporation of any low cloud within several hours of sunrise. In winter, however, stratus may last for several days, producing monotonous 'anticyclonic gloom'. When the inversion level intersects high ground, upland areas may lie in the zone of relatively warm, descending air above the inversion. These areas may then experience a remarkable contrast to the gloomy weather of lower-lying areas as brilliant sunshine may be combined with high temperatures. This was illustrated on 31 January 1992, when Manchester Airport had a

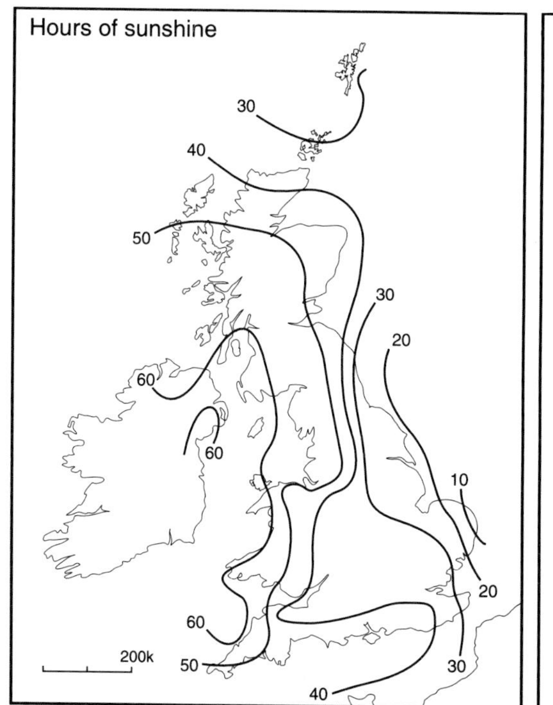

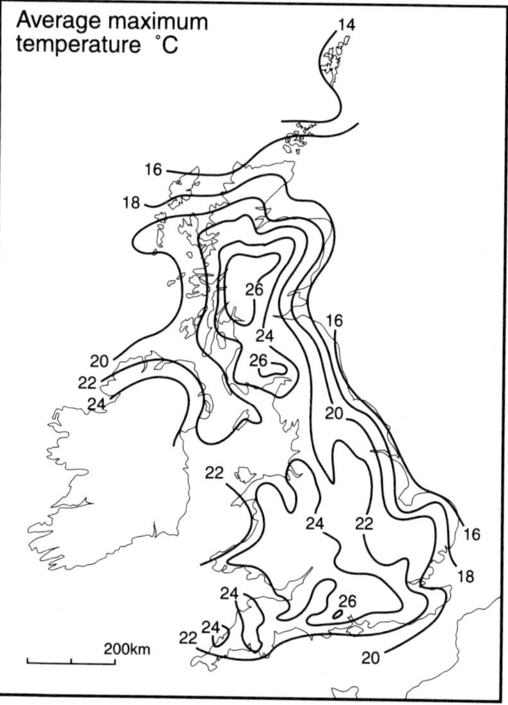

Figure 1.12 Distribution of (a) total sunshine, and (b) average maximum temperature for 22–29 June 1995

day of freezing fog with a maximum temperature of 0°C while Buxton (220 m higher and just 25 km distant) recorded six hours of sunshine with a temperature of 10°C (Eden 1992). The air above an inversion can also be notably dry, and relative humidities of less than 20 per cent were frequently observed at Ben Nevis Observatory (M.G. Roy, personal communication). On 19 February 1895 the observatory at the summit recorded a temperature 9.8 degrees C above that at Fort William (where the average lapse rate should give a temperature more than 8 degrees C higher than at the summit).

A further difference in the seasonal character of anticyclonic weather results from the development of sea breezes during the summer half-year. An anticyclone often provides the necessary pre-conditions of sunshine to stimulate the heating of inland areas and light winds to enable the movement of the breeze itself. As cooler air spreads inland, the local distribution of cloud is modified: uplift at the sea breeze front is marked by enlarged cumulus cloud. By contrast, places nearer the coast may experience a steady dispersal of cumulus as the incursion of cooler air increases stability. From coastal vantage points, it is possible to observe the simultaneous growth of cumulus inland around the sea breeze front and the disappearance of cumulus from the coast. Over peninsulas, where opposing sea breeze fronts may meet, uplift may be sufficient to induce a 'peninsula effect' of cloud growth and consequent showers. A sea breeze circulation will not be possible when there is a significant temperature inversion, especially when the warmed layer above the ground is deep in a well-established warm spell. This means that coastal locations may experience high temperatures, especially where an offshore breeze is observed.

Cyclonic Weather Type

The presence of an area of low pressure can result in two quite different weather types depending on the character of the atmospheric circulation. In a high-index situation, the circulation of air is energetic: mid-latitude depressions are carried from west to east giving unsettled, rapidly changing weather. The temperature contrasts that provide energy for the westerly circulation also enhance uplift of air around depressions, promoting their development. The deepest depressions thus result from the largest latitudinal temperature contrasts. Figure 1.13 shows the weather situation on 25 January 1990, which led to the 'Burns' Day Storm' on an occasion when the north–south temperature difference was clearly evident.

The unsettled nature of a cyclonic month is illustrated well by July 1988, demonstrating that the British Isles can lie in the path of vigorous depressions even in summer. This month completely changed the character of summer 1988, following a remarkably warm, dry June (Wheeler 1989). Figure 1.14 shows the shift in depression tracks between the two months.

July 1988 was thoroughly unsettled, combining a vigorous westerly airstream and an unusually southerly depression track. Frontal rain fell frequently and dry days were infrequent, especially in the west. The unseasonable character of the weather was emphasised further on the 25th, when parts of Scotland recorded their windiest July day in over fifty years. Rainfall exceeded three times the average on the Scotland–England border and there was a remarkable gradient of decreasing rainfall across southern Britain due to the high frequency of westerly winds (to the south of depression centres). Rainfall exceeded 200 mm on the coasts around the Bristol Channel but failed to reach 50 mm around the Thames estuary, where rainfall was a little below average. Westerly winds can be decidedly cool in summer on western coasts, and in parts of Cornwall the temperature failed to reach 18°C in July 1988. Even relatively sheltered East Anglia failed to exceed 22°C.

If some parts of the British Isles remain to the north of the depression track they may experience markedly different weather, remaining in polar air. Sometimes the local temperature gradient across the track of the depression between the two air masses may be abrupt. On 28 March 1995 a depression tracked east-south-eastwards across central Britain, when maximum temperatures ranged from 13°C in

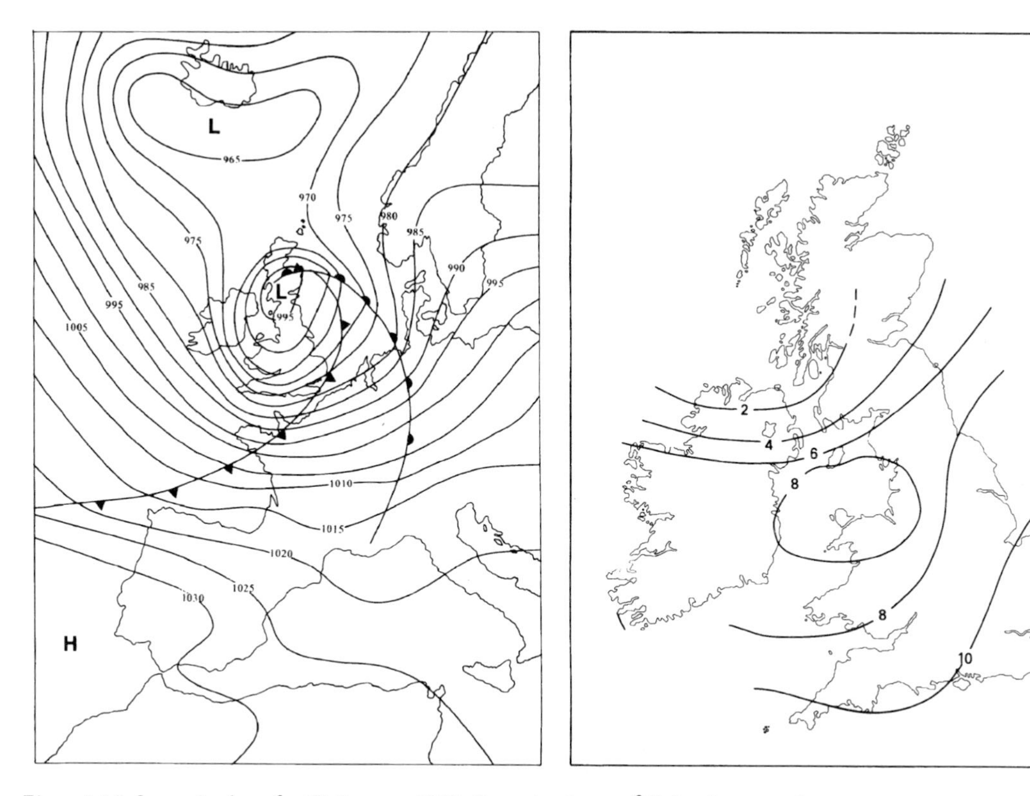

Figure 1.13 Synoptic chart for 25 January 1990. Inset: isotherms (°C) for the same day

Herefordshire in the warm sector to 2°C in Lincolnshire. It follows from such a temperature distribution that the depression track was also the demarcation line between rainfall on its southern side and a mixture of sleet and snow on the northern side.

When low pressure occurs in a blocked situation the resulting weather may be quite unpredictable. If a depression still has active fronts, rainfall may be highly persistent by virtue of the lighter wind speeds usually found on such occasions. Many of the largest 48-hour rainfalls have been due to persistent frontal rain from semi-stationary frontal systems (Jackson 1977). Serious flooding can result if rainfall intensities are enhanced by convectional activity. Notable examples include the Lynmouth flood disaster of 15 August 1952 and the south-east England floods of mid-September 1968 (see Chapter 2 for an account of the former).

Depressions may also enhance shower activity, especially when they coincide with upper troughs or 'cut-off' lows, areas of low tropospheric temperature. The distribution of showers is influenced by the time of year and the location of highest surface temperatures.

Regional Airflow and Weather Contrasts

Most of the regional weather contrasts noted so far result from the presence of a single airflow type that provides a variety of weather conditions in different parts of the country in response to different topographic or meteorological factors. Regional weather contrasts also occur when different areas are simultaneously affected by contrasting airflow types. Indeed, no single weather type ever has a completely uniform influence over the whole of the British Isles;

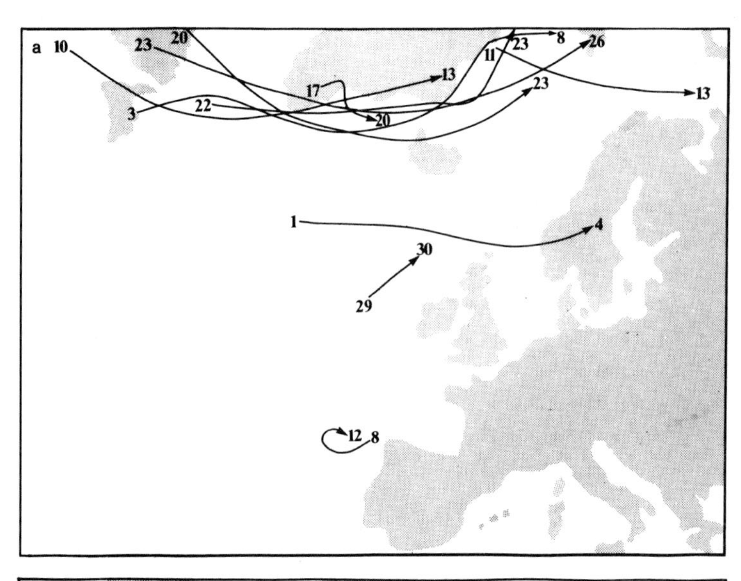

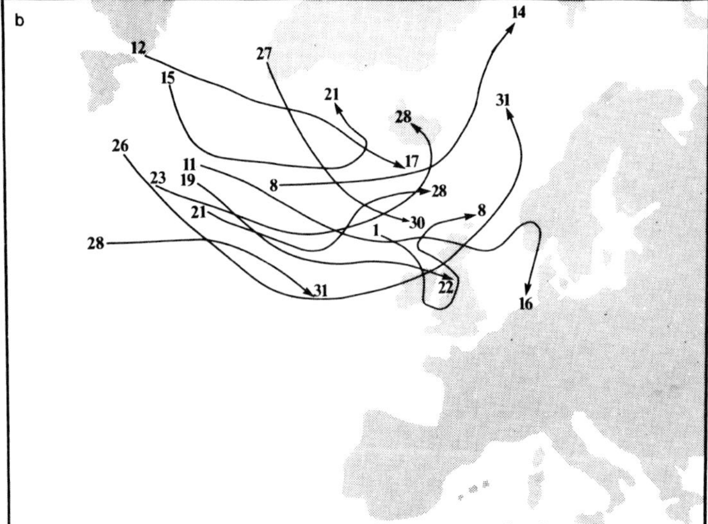

Figure 1.14 Map showing depression tracks with start and end dates in (a) June 1988 and (b) July 1988 (after Wheeler 1989)

a westerly type will usually be moister and more vigorous over northern districts closest to a low-pressure system. As one weather type is replaced by another, a transition period may be characterised by the occurrence of two irreconcilable types in different areas.

Three examples illustrate the origins of such contrasts. First, two different airflow types may persist over different parts of the British Isles. For much of the early 1970s, early in 1976 and again between late 1988 and 1991, depression tracks shifted to the north, allowing an anticyclonic influence to become established over southern parts of England in conjunction with a continuing westerly type over northern districts. The dichotomy of weather between a wet North West and a dry South East is illustrated by March 1990 (Box 1.9).

The second case illustrates the contrasts in

Box 1.9

REGIONAL AIRFLOW CONTRAST: A CASE STUDY FOR MARCH 1990

- A month of exceptionally vigorous and persistent westerlies in which depression tracks were displaced to the north of normal. Northern areas had a strong westerly airstream, whereas there was a distinct anticyclonic bias to the (weaker) westerly flow across southern parts of the British Isles. As in January and February 1990, North Atlantic depressions were unusually deep and a remarkably active westerly airstream was maintained. The north–south pressure difference over the British Isles averaged 22.5 mb.
- Frontal systems and secondary depressions gave frequent orographic falls of rain of great intensity over the mountains of Scotland. Western Scotland was exceptionally wet for the third consecutive month with over 1,000 mm of rain being observed locally. Some places recorded the wettest calendar month in over a hundred years of records. The strength of the rain shadow in eastern Scotland is demonstrated by the total of only 10 mm at Montrose. Under the influence of the high pressure, parts of Kent recorded as little as 2 mm.
- It was the warmest March in Central England since 1957 with anomalies exceeding 3 degrees C away from north-western coastal areas. Sunshine showed an enhanced gradient of increases to the south-east, which accords closely with the distribution of dryness. Herne Bay in Kent recorded nearly 200 hours of sunshine.

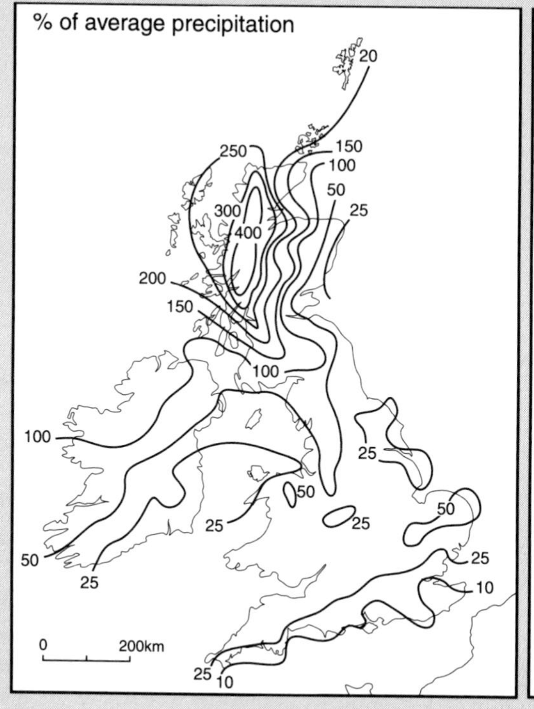

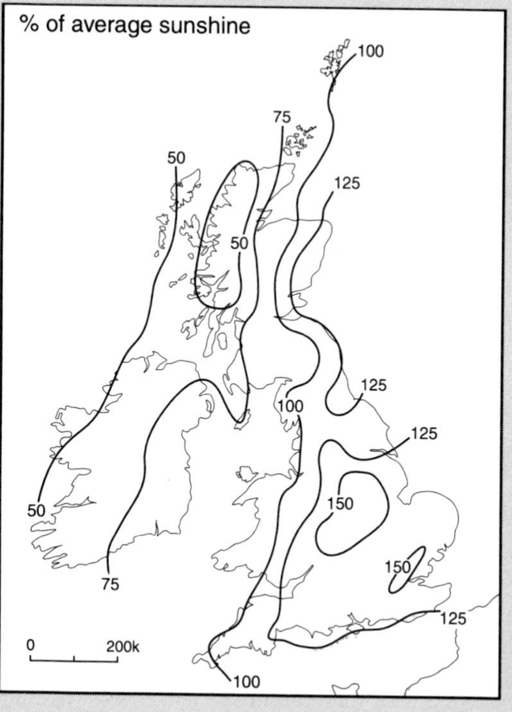

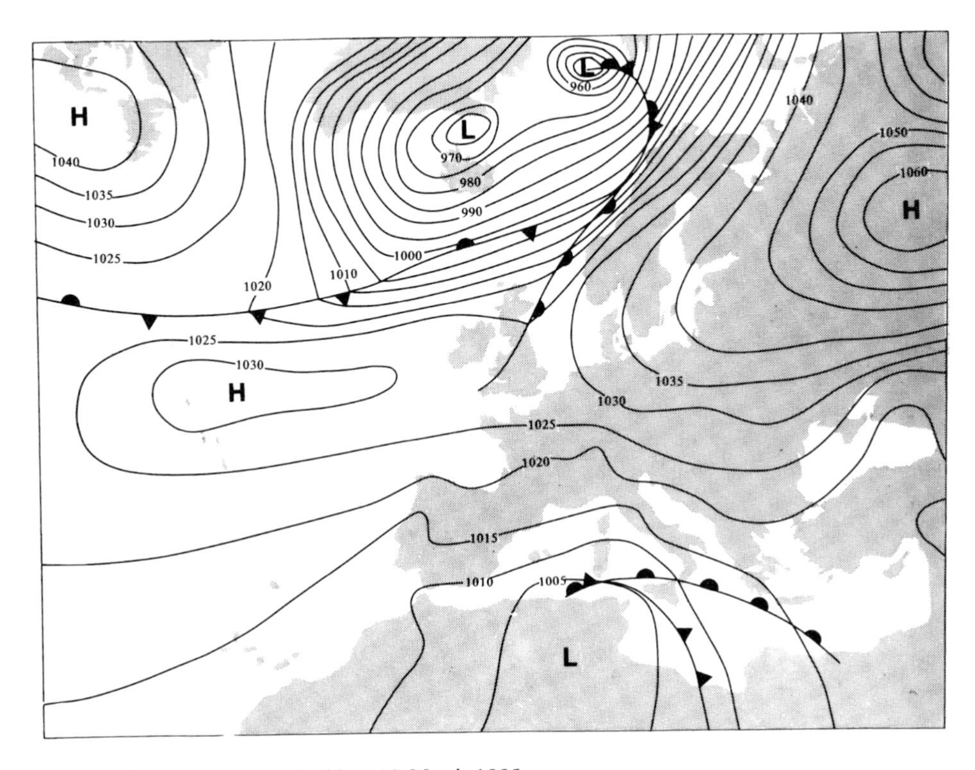

Figure 1.15 Synoptic chart for 1200 GMT on 13 March 1995

temperature that may arise from contrasting regional airflow types – in this case the juxtaposition of continental and maritime air masses. On 13 March 1995 an area of intense high pressure was situated over Russia and a deep depression was moving north-eastwards from Iceland (Figure 1.15). Between these systems lay a vigorous south-westerly flow introducing mild, tropical maritime air to Scotland and northern Scandinavia. East Anglia and south-east England remained mostly overcast and misty, and at Weybourne on the north Norfolk coast the temperature rose to no more than 4.3°C. In a large swathe of England from the South-West across the Midlands to Lincolnshire the day was sunny with light winds, and the temperature rose to 14.7°C at Finningley (near Doncaster, South Yorkshire). This temperature was almost equalled in the tropical maritime air over eastern Scotland, assisted by a slight föhn warming in the south-westerly airflow. A regional contrast of this type is not an isolated event; indeed, El-Kadi and Smithson (1996) have noted this particular combination of airflow types as one of fifteen characteristic pressure distribution patterns over the British Isles.

The third example of a regional weather contrast relates to meso-scale depressions (typified by the secondary depression). As already discussed, these features encourage rainfall to become heavier over a small area since they signify areas of enhanced uplift. Large local rainfall totals may result from the combined effects of this intensity and the persistence of the rain that is encouraged by the hesitation of the frontal movement. The scale of these features is such that if the front waves over north-western areas, the south-east of the British Isles may escape the rain and sometimes remain quite fine. Conversely, the front may cross Scotland and Ireland first and then hesitate over England and Wales. Figure 1.16 shows the pressure pattern on 18 September 1985 when a meso-scale depression crossed

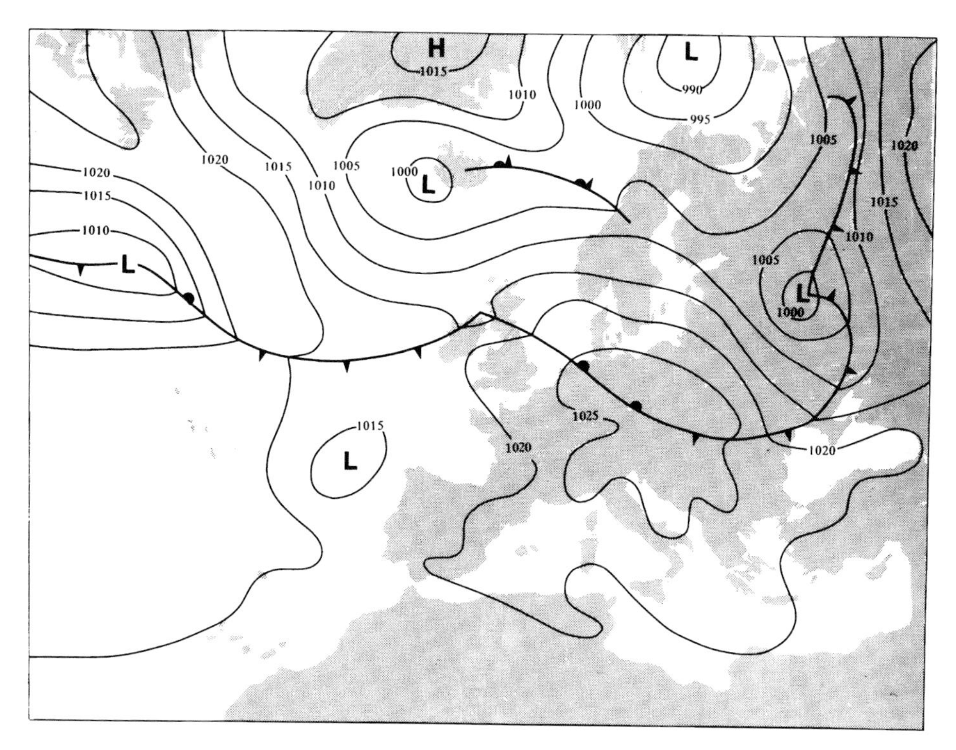

Figure 1.16 Synoptic chart for 1200 GMT on 18 September 1985

Scotland. Very heavy rain fell on the western side of Glasgow with Paisley recording 74 mm. Meanwhile, south-eastern parts of the British Isles remained warm and fine throughout the day in an anticyclonic south-westerly airstream.

CONCLUSION

Climate is a multi-dimensional variable: variations in a horizontal plane in a north–south direction relate to latitude and those in a west–east direction relate to such factors as oceanicity and moisture supply. In the vertical plane, lapse rates define the drop in temperature with height and influence the form in which precipitation may fall. The changing array of synoptic weather types outlined in this section dictates the variation in regional weather conditions, each type having a distinctive interrelationship with environmental factors. It is now time to examine the regional consequences of this behaviour by focusing on the weather and climate of each region of the British Isles.

REFERENCES

Atkinson, B.W. and Smithson, P.A. (1976) 'Precipitation', in T.J. Chandler and S. Gregory (eds) *The Climate of the British Isles*, London: Longman.

Barry, R.G. (1981) *Mountain Weather and Climate*, London: Methuen.

Barry, R.G. and Chorley, R.C. (1992) *Atmosphere, Weather and Climate*, 6th edn, London: Routledge.

Belasco, J.E. (1948) 'The incidence of anticyclonic days and spells over the British Isles', *Weather*, 3: 233–42.

—— (1952) 'Characteristics of air masses over the British Isles', *Meteorological Office Geophys. Memoir no. 87*, London: HMSO.

Betts, N.L. (1989) 'A synoptic climatology of precipitation in Northern Ireland', unpublished PhD thesis, Department of Geography, Queen's University of Belfast.

Bilham, E.G. (1938) *The Climate of the British Isles*, London: Macmillan.

Black, A.R. and Bennett, A.M. (1995) 'Regional flooding in

Strathclyde, December 1994', *Hydrological Data UK: 1994 Yearbook*, Wallingford: Institute of Hydrology.
Bleasdale, A. and Chan, Y.K. (1972) 'Orographic influences on the distribution of precipitation', *Distribution of Rainfall in Mountain Areas*; Geilo Symposium, Norway, Geneva: World Meteorological Organization.
Bodyko, M.I. (1971) *Climate and Life*, Leningrad (St Petersburg): Hydrological Publications.
Browning, K.A., Eccleston, A.J. and Monk, G.A. (1985) 'The use of satellite and radar imagery to identify persistent shower bands downwind of the North Channel', *Meteorol. Mag.*, 114: 325–331.
Burt, S. (1995) 'Fall of dust rain in Berkshire, 24 September 1994', *Weather*, 50: 30.
Chandler, T.J. (1965) *The Climate of London*, London: Hutchinson.
Chandler, T.J. and Gregory, S. (1976) 'Introduction', in T.J. Chandler and S. Gregory (eds) *The Climate of the British Isles*, London: Longman.
Douglas, C.K.M. and Glasspoole, J. (1947) 'Meteorological conditions in heavy orographic rainfall in the British Isles', *Q. J. R. Meteorol. Soc.*, 73: 11–38.
Eden, P. (1992) 'Weather Log' supplement to *Weather* for January 1992, Reading: Royal Meteorological Society.
El-Kadi, A.K. and Smithson, P. (1996) 'An automated classification of pressure patterns over the British Isles', *Trans. Inst. Br. Geogrs*, NS 21: 141–156.
Geiger, R. (1959) *The Climate near the Ground*, Cambridge, MA: Harvard University Press.
Hulme, M. and Barrow, E. (eds) (1997) *The Climate of the British Isles: Present, Past and Future*, London: Routledge.
Jackson, M.C. (1977) 'Mesoscale and synoptic-scale motions as revealed by hourly rainfall maps of an outstanding rainfall event: 14–16 September 1968', *Weather*, 32: 2–16.
Jones, S.R. (1981) 'Ten years of measurement of coastal sea temperatures', *Weather*, 36: 48–55.
Lamb, H.H. (1964) *The English Climate*, London: English Universities Press.
—— (1972) *Climate: Present, Past and Future*, vol. 1: *Fundamentals and Climate Now*, London: Methuen.
Levick, R.B.M. (1949) 'Fifty years of English weather', *Weather*, 4: 206–211.
—— (1975) 'British Isles – weather types', *Weather*, 30: 342–346.
Lockwood, J.G. (1962) 'Occurrence of foehn winds in the British Isles', *Meteorol. Mag.*, 91: 57–65.
Lyall, I.T. (1971) 'An exceptionally warm day in January', *Weather*, 26: 541–545.
Mayes, J. (1991) 'Regional airflow patterns in the British Isles', *Int. J. Climatol.*, 11: 473–491.
Meteorological Office (1975) *Weather in Home Waters*, vol. 2: *The Waters around the British Isles and the Baltic*, Part 1, London: HMSO.
Oke, T.R. (1992) *Boundary Layer Climates*, London: Methuen.
Orlanski, I.A. (1975) 'Rational subdivision of scales for atmospheric processes', *Bull. Am. Meteorol. Soc.*, 56: 527–530.
Sawyer, J.S. (1956) 'Rainfall of depressions which pass eastward over or near the British Isles', *Meteorological Office Professional Notes no. 118*, London: HMSO.
Simpson, J.E. (1994) *Sea Breeze and Local Winds*, Cambridge: Cambridge University Press.
Smithson, P.A. (1969) 'Regional variations in the synoptic origin of rainfall across Scotland', *Scot. Geogr. Mag.*, 85: 182–195.
Strahler, A.N. and Strahler, A.H. (1973) *Environmental Geoscience*, New York: Wiley.
Taylor, J.A. (1976) 'Upland climates, in T.J. Chandler and S. Gregory (eds) *The Climate of the British Isles*, London: Longman.
Weston, K.J. and Roy, M.G. (1994) 'The directional-dependence of the enhancement of rainfall over complex topography', *Meteorological Applications*, 1: 267–275.
Wheeler, D. (1989) 'Rainfall in NE England during the summer of 1988', *J. Meteorology (UK)*, 14: 131–138.
Young, M.V. (1995) 'Severe thunderstorms over south-east England on 24 June 1994: a forecasting perspective', *Weather*, 50: 250–256.

PART 2

THE CHARACTER OF REGIONAL CLIMATES ACROSS THE BRITISH ISLES

2

SOUTH-WEST ENGLAND AND THE CHANNEL ISLANDS

Allen Perry

INTRODUCTION

The tapering peninsula of south-west England, with sea on three sides, enjoys the highest annual mean daily temperatures in the British Isles. Not surprisingly, then, the sobriquet 'English Riviera' is often applied to the area in general and the sheltered shores of Torbay in particular. Cornwall is the most southerly county of Britain and the most westerly of England, and as a consequence it is subjected to maritime influences, but is also the most exposed and windy in England. While the mean winter temperatures on the southern coasts of Devon and Cornwall compare very favourably with stations on the northern shore of the Mediterranean there is much less sunshine and it rains on a greater number of days on average. The maritime nature of the climate, however, ensures an equability that is perhaps best expressed in terms of the annual temperature range or difference between the mean temperatures of the warmest and coldest months, which in west Cornwall is about 9 degrees C, a similar figure to that found along the western coast of Ireland and in the Outer Hebrides but one which contrasts with the 14 degrees C that characterises Midlands sites, where continentality is much greater. While a general climatic gradient exists across the region from south-west to north-east it is of far smaller significance than the more local climatic contrasts that are found between the exposed, mostly treeless uplands of Dartmoor and Bodmin Moor and the sheltered areas in the surrounding vales. A series of separated upland areas, which include the highest ground to be found in southern England (621 m on Dartmoor on the high ground around Yes Tor), run down the length of the peninsula, and their elevation, coupled with their proximity to the western seaboard, is conducive to producing a more rigorous climate than at lower altitudes: wetter, cooler and more windswept. Thus, while average annual rainfall over most of lowland Devon is not much above 1,000 mm, it is more than double this figure over the higher parts of Dartmoor. While the deeply cut river valleys, like that of the Tamar and coastal 'rias' of south Devon and south Cornwall, offer shelter, grass growth is almost continuous in the mild, moist winters, though the uplands are bleak and exposed. The scenic contrasts in Devon and their relation to weather and climate have intrigued several climatologists (e.g. Bonacina 1951; Manley 1952), and some amateur weather observers from an earlier era (Shapter 1862). Daphne du Maurier's (1967) keen observation on the climate of this most south-westerly extremity of the British Isles would pass unchallenged by most climatologists when she states, concerning the change of weather from place to place, 'this difference in temperature, this vagary of weather, varies from mile to mile with a kind of lunatic perversity'.

A marked contrast exists also across the peninsula between the northern and southern coasts. It was Brooks (1954) who drew a distinction between the 'bracing' or 'tonic' nature of the climate in places like Bude and Newquay and the very relaxing –

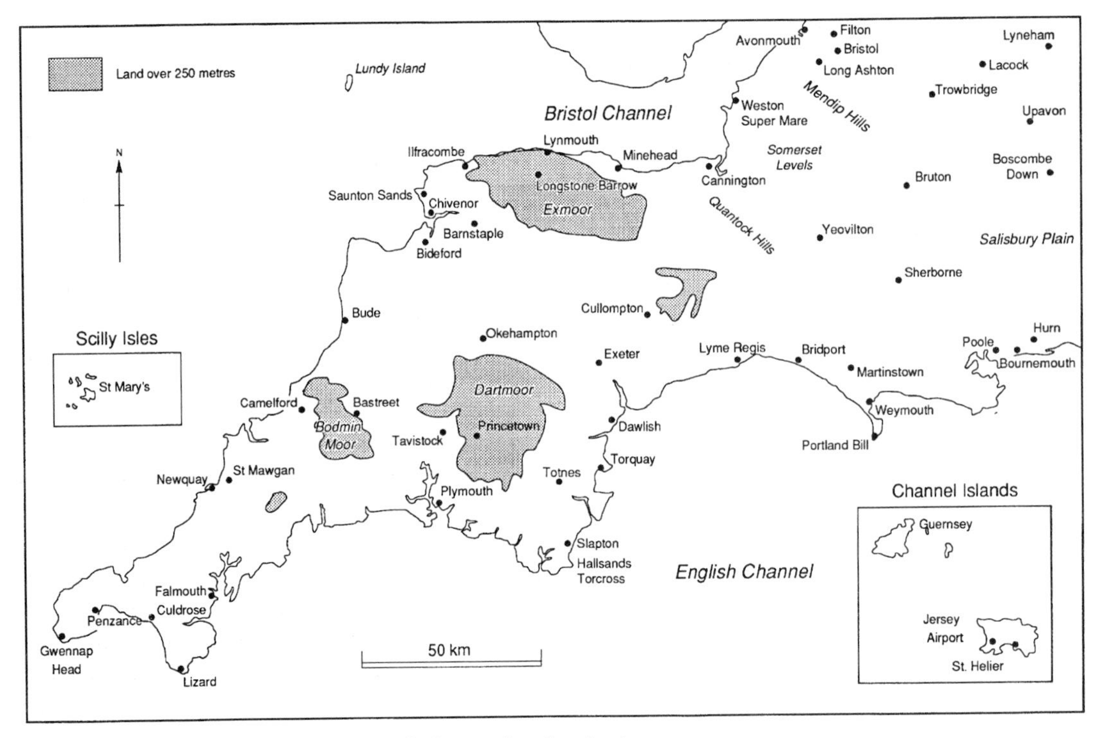

Figure 2.1 Location of weather stations and places referred to in the text

some might say 'somnolent' – climate of the south coasts of both Devon and Cornwall. Whether, as Davies (1954) believes, the short, mild winters and long growing seasons have led to an absence of hurry and rush by the local population, exemplified by Drake's finding time to complete his game of bowls before beating the Spaniards, is clearly an arguable point. Further east, Hannell (1955), writing on the climate of Bristol, notes that visitors frequently comment on its enervating character.

East of the narrow waist of the peninsula between Lyme Regis and the Bristol Channel, where just 56 km separates the Bristol and English Channels, the topography is very varied. The Somerset Levels is the largest area of low-lying ground, but this is bounded to the north and west by the Quantock and Mendip Hills and to the south and east by the chalk downlands of Dorset and Salisbury Plain. This area is much more sheltered than is the case further west, and secluded nooks like the Cheddar region are especially favoured compared with the exposed Mendip plateaux. Hence the area is a well recognised 'early' district, especially for strawberries. The mild, humid maritime tropical air that frequently cloaks the hills and coasts in Devon and Cornwall with fog and low cloud is often dried as it moves eastward, with the cloud thinning and breaking to give sunshine, especially in the afternoons in summer. Local subsidence to the lee of hills in warm-sector situations means that favoured areas like the west Somerset coast around Minehead often enjoy more sunshine and higher temperatures than elsewhere when the winds are south to south-west. On the other hand, in a cold winter easterly airstream, frosts are harder and more prolonged, and overall the growing season is shorter in the more eastern counties.

The two offshore island groups – the Isles of Scilly and the Channel Islands – have climates that are highly distinctive, both from those of the mainland and from each other. The Isles of Scilly lie 40 km from the Cornish coast, and enjoy one of the most oceanic climates to be found in the British Isles. Even in mid-winter, sea surface temperatures around the islands average close to 10°C, and as a consequence frost and snow are very rare, though not unknown when the normal maritime influence is temporarily ousted by cold easterly winds from the Continent.

The Channel Islands lie in the Gulf of Saint-Malo, just a few kilometres from the French coast, and have the sunniest climate in the British Isles (Blench 1967). Their nearness to the European mainland can, on occasions, lead to extremes of temperature imported on continental airstreams, although generally the climate is distinctly maritime. Overall, Channel Islands summers tend to be slightly sunnier, warmer and drier than those of the Isles of Scilly while winters in the Channel Islands are not quite so mild and frosts are more frequent than in Scilly. Nevertheless, as Hooper (1837) noted as long ago as the mid-nineteenth century, 'the two qualities of mildness and humidity gives the climate its most obvious peculiarities'.

The network of meteorological observing stations in south-west England (Figure 2.1) shows a considerable bias towards coastal sites. Many of them are run as Health Resort stations, with Torquay and Newquay operational since 1891 and Bournemouth (1902) and Penzance and Weymouth (1908) having the longest records. Princetown has records since 1912 at an elevation of 414 m on Dartmoor and is the most important upland station in the area. At Plymouth observations were taken on the Hoe from 1874 to 1979 (Wood 1994). The largest gaps in the network are in central Devon and in the north Dorset–south Somerset area, and there is also a lack of observations from the chalk downland 'hill climates', the highest station being at Upavon in Wiltshire at 179 m (Figure 2.1) in the surprisingly remote desolation of Thomas Hardy's Great (Salisbury) Plain. Records have been kept on the Isles of Scilly since 1871 (though observations there are now automated) and in Jersey since 1901, while in Guernsey daily records began in 1843.

TEMPERATURE

The dominating influence of the sea plays an important role in the temperature regime of the area. Mean annual temperatures reach 11.5 to 12°C in the Channel Islands and the Isles of Scilly and are only slightly lower along the coastal strip of Devon and Cornwall. Further east, in Somerset and Avon, 10°C is a more representative figure. In these eastern areas of the peninsula January is on average the coldest month, but in the west February is a little colder, since sea temperatures reach their lowest levels in late February or early March. From November to February the Isles of Scilly and the Penwith peninsula are the warmest parts of the region, and the early cut-flower industry of the area capitalises on this warmth. Using a 6°C base temperature, Broad and Hough (1993) have shown that the average length of the growing season exceeds 300 days per year in west Cornwall, and in sheltered locations is effectively year-round (Hogg 1967). In mid-winter mean minima range from around 4.5°C at coastal stations to only a little above freezing (0.7°C) on the grim heights of Dartmoor around Princetown. Ratsey (1975) notes that minima at Slapton seem to be the most moderate in south Devon. Blench (1967) commented that the mildness and equability of Jersey's climate are the island's abiding characteristic and a consequence of its oceanicity. Mean minimum and maximum temperatures for the 1961–90 period at a number of stations are shown in Table 2.1.

On the mainland, winter conditions can be more extreme than on the islands. A good example occurred on 12 January 1987, which was probably the coldest day of the century in terms of maximum temperature. At Okehampton the maximum did not exceed −8.5°C, the lowest maximum recorded in the UK on that day, and at all stations in the South-West, including those of Scilly and the Channel Islands, temperatures remained below freezing. Rather longer ago Bonacina (1961) drew attention to the minimum of −15°C at Exeter on 23 January 1958, only 16 km inland from the English Channel. Extremely cold nights with a deep snow cover and

Table 2.1 Mean monthly maximum and minimum temperatures (°C) for the period 1961–90

Location	*Alt. (m ASL)*	*Jan.*	*Feb.*	*Mar.*	*Apr.*	*May*	*June*	*July*	*Aug.*	*Sept.*	*Oct.*	*Nov.*	*Dec.*	*Year*
St Mawgan	104	8.2	7.9	9.5	11.6	14.1	16.8	18.6	18.7	17.1	14.4	10.9	9.2	13.1
		3.2	2.8	3.9	5.3	7.9	10.7	12.7	12.8	11.2	9.2	6.0	4.5	7.5
Torquay	8	8.8	8.5	10.3	12.4	15.4	18.5	20.6	20.2	18.2	15.2	11.6	9.8	14.1
		3.4	3.3	4.1	5.6	8.4	11.2	13.1	13.0	11.4	9.1	5.7	4.4	7.7
St Helier	54	8.2	7.9	10.2	12.7	16.0	18.8	21.0	21.1	19.1	15.8	11.6	9.3	14.3
		4.1	3.9	5.0	6.5	9.4	12.0	14.0	14.4	13.2	10.9	7.4	5.3	8.8
Bude	15	8.6	8.4	9.9	11.9	14.6	17.2	19.0	19.2	17.6	15.1	11.4	9.6	13.5
		3.3	2.9	3.9	5.0	7.5	10.4	12.3	12.3	10.7	8.7	5.7	4.3	7.3
Lyneham	145	5.9	6.3	8.9	11.9	15.3	18.5	20.6	20.1	17.5	13.8	9.2	7.0	12.9
		0.0	0.5	1.8	3.7	6.6	9.6	12.5	11.4	9.6	7.0	3.5	1.7	5.7
Hurn	10	7.6	7.8	10.1	12.7	16.1	19.3	21.4	21.1	18.6	15.1	10.8	8.6	14.1
		1.1	0.8	2.1	3.4	6.5	9.5	11.3	11.1	9.1	6.9	3.2	1.7	5.6

Note: ASL = above sea level

intense radiational cooling often result in much lower temperatures in sheltered valleys than in the uplands. The night of 14 January 1982 is a good example, with a minimum of −19.5°C at Lacock while at Upavon the temperature fell only to −8.5°C and at Boscombe Down to −9.3°C.

More often it is winter mildness which is the predominant theme of the season. Maximum temperatures over 15°C have occurred at some time in all three winter months, and the value of 19.4°C at Barnstaple, Devon, in February 1891 is the highest on record for the month anywhere in Britain. Just as remarkable was the 20.2°C on 2 March 1977 at Exeter Airport, the earliest date that 20°C has been recorded in the British Isles. In some mild winter months daily maximum temperatures may be close to 10°C on western coasts if winds blow persistently from between south-west and west. In January 1989, for example, Falmouth recorded temperatures above 10°C on twenty days during the month. The south-west winds that bring winter mildness are frequently accompanied by misty rain and drizzle and generally muggy conditions, so characteristic of stable, tropical maritime air masses in this area.

The fluctuation of winter mean temperature is much greater than in other seasons of the year, with Plymouth, for example, having a standard deviation in January that is more than twice as great as the July figure. January mean temperatures at Plymouth have ranged from a high of 8.5 in 1990 to a low of −0.2°C in 1963. A consequence of these temperature fluctuations is an eightfold difference in air frost frequency between very mild and very severe winters (Tout 1987). The topographic dimension of the Devon climate is summarised in Table 2.2.

By March, as the land begins to warm up, south-west England ceases to be the warmest part of Britain by day, and maximum temperatures are on average higher in the more eastern counties of the region than near the western coasts. Coastal stations in the south-west are cooler than those inland from mid-March onwards as measured by daily mean temperatures. Night-time minima continue to be highest in west Cornwall and the Channel Islands, a pattern that continues through the summer. Although we think of spring as the season when temperatures progressively rise, in practice few springs conform to this idealised concept, and war-

Table 2.2 Lowland and upland climates in Devon for the period 1921–50

Location	*Altitude (m ASL)*	*Mean annual precipitation (mm)*	*Mean January temperature (°C)*	*Mean July temperature (°C)*
Plymouth Hoe	37	1,010	6.0	15.9
Princetown	414	2,226	3.0	13.7

mer weather typically arrives in fits and starts punctuated by often drastic setbacks. In April 1981, for example, air frost occurred in several inland parts of Devon and Cornwall late in the month, while temperatures had reached 20 to 21°C around Bristol on the 10th. The year after, frost damage occurred to apple and pear blossom in Somerset after air temperatures fell to −2°C on 5 May, and to −5°C at Bastreet, Cornwall, but by the end of May 1982 maximum temperatures reached 25.6°C at Yeovilton. As elsewhere in Britain, these wild swings of temperature are a result of the meandering jet stream in the upper atmosphere and consequent alternations of cold northerly and warm southerly winds at the surface.

Spring is the season when the prevailing southwest winds are least frequent, but for the southwest this can sometimes be an advantage, since with seas at their coldest (about 9°C in the South-West Approaches), mild air in April from the south-west is quickly cooled to its dew point and widespread sea fog can plague the south Devon and south Cornwall coasts as well as the Channel Islands and lead to maximum temperatures of 10 to 12°C even well into April, whereas just inland there may be pleasant warm sunshine.

Whilst July is the warmest month of the year at most stations in the eastern part of the peninsula, in the more maritime west August is slightly warmer, reflecting the delayed warming of the surrounding seas. Mean maximum temperatures in summer range from about 19°C along the north coast of Cornwall and in north-west Devon to just over 21°C inland around Bristol and in lower-lying parts of Wiltshire and Dorset. Warm summer days with a maximum temperature above 25°C occurred on average three times per summer in the period 1961–90 at Plymouth and five times at Bristol and Poole. Extreme maximum temperatures by county are shown in Table 2.3. Temperatures in excess of 32°C have been recorded as far west as eastern Cornwall and on the north Cornwall coast (Shaw 1977), while in the eastern parts of the peninsula, away from coasts, the all-time high is close to 35°C, as it is in the Channel Islands. In west Cornwall it appears that 30°C might be expected about once in fifty years, while in the Isles of Scilly the highest recorded temperature is 27.8°C, although this has now been recorded twice during the twentieth century.

Perhaps the most remarkable extreme value is the 35.4°C at Saunton Sands on the north Devon coast in the extremely hot spell of August 1990. The very warm continental tropical airstream from the southeast brought this great heat from a direction that can produce some remarkably high temperatures at this site, as a result of the sheltering effect of Exmoor. Shelter can also be important in less extreme weather situations. With north-west winds there is often a considerable temperature difference between the cool Bristol Channel coast of Devon with its onshore winds, and Torbay, lying in the lee of Dartmoor, which often enjoys long hours of sunshine and temperatures up to 5 degrees C higher. Table 2.1 shows that on average Hurn in summer is about 2 degrees C warmer than Bude.

Revesz (1969) notes that the fall of temperature after the summer peak is gentle at first in September, becoming much steeper through October and November. In many years the lingering warmth prolongs the holiday season, as in 1989, when Jersey recorded 27°C as late as 21 September. Recently the exceptionally mild October of 1995 gave many days,

Table 2.3 Absolute maximum and minimum temperatures (degrees C) by county in south-west England

County	*Absolute maximum*	*Location*	*Date*	*Absolute minimum*	*Location*	*Date*
Avon	35.0	Clifton	12 July 1923	−16.1	Bristol Weather Centre	15 Jan. 1982
Somerset	34.9	Yeovilton	3 Aug. 1990	−16.1	Yeovilton	14 Jan. 1982
Dorset	35.0	Dorchester	3 Aug. 1990	−13.4	Hurn	23 Jan. 1963
Wiltshire	35.6 35.6	Trowbridge Marlborough	2 July 1976 9 Aug. 1911	−19.5	Lacock	14 Jan. 1982
Devon	35.4	Saunton Sands	3 Aug. 1990	−16.7	Cullompton	21 Jan. 1940
Cornwall	33.9	Ellbridge	3 Aug. 1990	−10.9	Culdrose	13 Jan. 1987
Isle of Scilly	27.8	St. Mary's	19 Aug. 1932	−7.0	St. Mary's	12 Jan. 1987
Channel Islands	35.6	Jersey	19 Aug. 1932	−11.7	Jersey	12 Jan. 1987

Sources: Meaden (1983); Webb and Meaden (1983, 1993)

even in the last week of the month, with temperatures in the range of 16 to 19°C, especially along the north Devon coast. Although November is a month in which a variety of inclement drab weather is likely, in 1969 the temperature reached 20.6°C at Totnes, Devon, on the 3rd. The warmest November for more than three hundred years in 1994 with persistent mild and moist south-west winds led to several balmy nights with minima of 12 to 14°C, and at Minehead the mean minimum was 9.8°C. Not all autumns are so genial, offering extensions of summer. For example, in October 1992 temperatures remained below 16°C by day and early frosts occurred in vulnerable places, like Hurn Airport, Bournemouth, where the temperature fell below −5°C. This station is situated on sandy soils which have rapid drainage and low heat conductivity, and is prone to early and late frosts.

Frosts in the South-west of England

Frost frequency is highly variable across the region, and despite the region's reputation for the temperate nature of its climate can be surprisingly high in some locations. The Isles of Scilly will experience subzero air temperatures only very occasionally (Figure 2.2), usually with outbreaks of easterly polar continental air, although even under those extreme conditions the long sea passage has the effect of reducing the probability of frost. The mainland enjoys less such protection and even Plymouth can expect over twenty air frosts a year, much of Devon as many as forty, and Yeovilton, in the fastness of central Somerset, has an annual average of fifty-three air frosts – a figure which compares with many north British locations. Small-scale differences also exist between the adjacent coasts; for example, minimum temperatures at Penzance are on average about 2 degrees C higher than those for Bude. Manley (1944) has noted that Bude lies at the mouth of a valley which appears to be a favourable site for cold air drainage from an extensive inland plateau area.

The first air frost can be expected about the middle of October inland in Wiltshire and Somerset but not until early in November over much of central Devon. The continuing warmth of the sea (still about 12 to 13°C in mid-November) usually prevents coastal autumnal frosts. Average monthly numbers of air frosts for the 1961–90 period are shown in Figure 2.2, which reveals clearly the distinction between the frost frequencies at inland

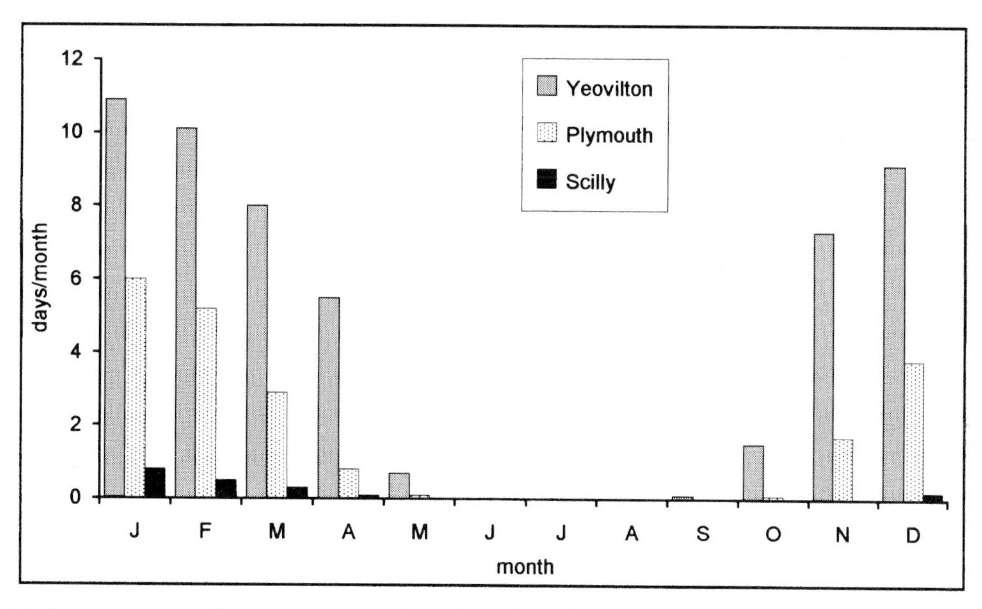

Figure 2.2 Average number of air frosts per month for the period 1961–90

locations such as Yeovilton and extreme maritime situations such as Scilly (with an annual average of only 1.8 air frosts).

Most air frosts are caused by radiative cooling, but in the most maritime locations, such as the Isles of Scilly, temperatures below freezing are likely to occur only when very cold (polar continental) air is advected from the east, usually on the north side of depressions working through the English Channel. Tout (1985) has shown that for the period 1956–81 the median number of days with air frost was only one. The absolute minimum at St Mary's of −7.0°C occurred on 12 January 1987 and was accompanied by a 40 to 50 knot easterly gale (Webb 1987) giving a wind-chill equivalent temperature of −30°C. Despite these rare extremes the Isles of Scilly probably hold the record for freedom from frost in Britain, although the intermittent records from Lundy in the Bristol Channel in the late 1980s suggest a similar or greater immunity. On the other hand the Channel Islands' greater exposure to polar continental outbreaks reveals itself in Jersey's average air frost frequency of only slightly less than ten per year.

PRECIPITATION

Stamp (1941) has noted that 'an appreciable part of Devon may be classed as one of the wet areas of the British Isles'. Certainly the local expression 'Devon, Devon, six days of rain and one in heaven' echoes this theme, and most of the region experiences a distinctly damp climate in which rainfall is well distributed through the year. Figure 2.3 shows the decisive imprint of topography on rainfall distribution, with average annual totals ranging from over 2,000 mm over large parts of Dartmoor to below 650 mm on the Somerset levels and around Weymouth. Close to sea level, precipitation varies little through the region with about 800 mm at the western tip of the Land's End peninsula and on the low-lying Isles of Scilly, and very similar figures are recorded on the Somerset coast. In Jersey, average annual precipitation ranges from about 800 mm in the south-west of the island to over 900mm on the higher ground near the north coast (Butler *et al.* 1985). Guernsey Airport's annual precipitation is 833 mm, ranging from 37 mm in July to just over 100 mm in November and December. More

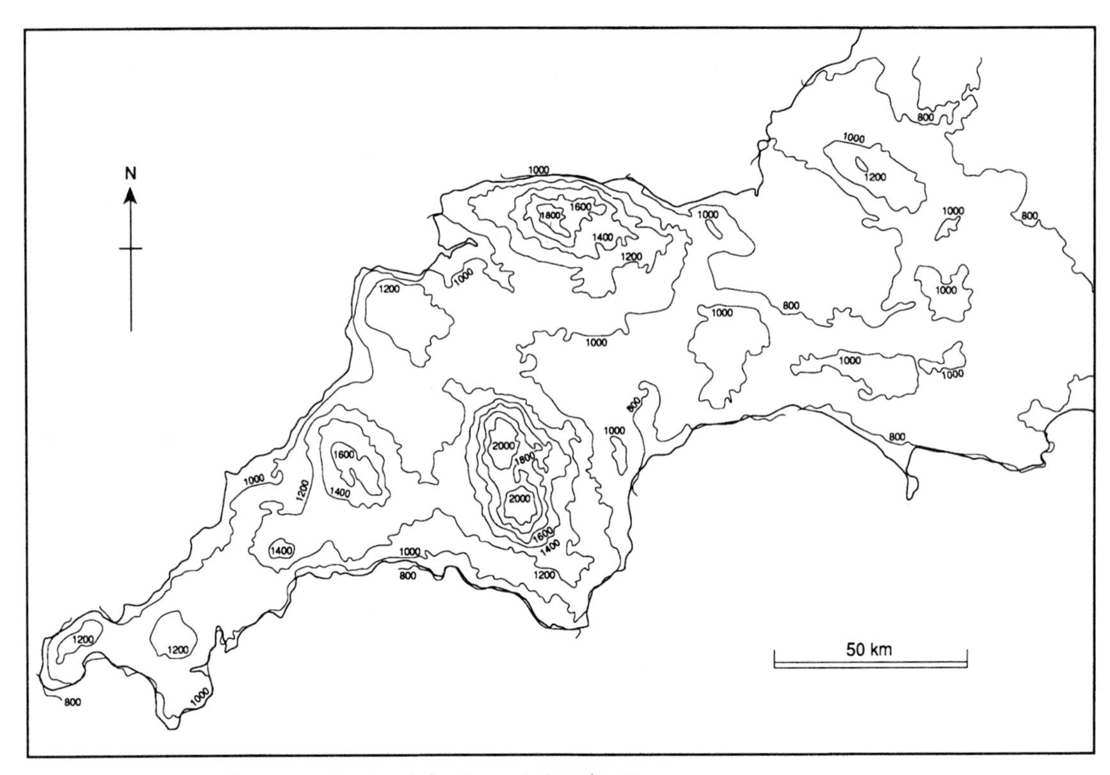

Figure 2.3 Mean annual precipitation (mm) for the period 1961–90

than 2,500 mm will be recorded on Dartmoor in the wettest years and Princetown's annual average is well over 2,000 mm (Table 2.4). Exmoor will occasionally record similar totals (Archer 1956). One effect of the major upland areas of Dartmoor (and Exmoor) is to create a dry zone on the northern and eastern leeward side, extending from Bideford Bay south-eastward to the east Devon coast. Over most of the peninsula rain falls on about 200 days per annum, and Dancey (1981) has shown that in January 60 per cent of all days will be wet in Cornwall and about 40 per cent in Dorset. At Chivenor rain falls on average for about 680 hours a year, but only for about 550 hours at Exeter (Atkinson and Smithson 1976). On the other hand, Ilfracombe on the north Devon coast and backed by the heights of Exmoor, which rise precipitiously from the Bristol Channel, suffers because of its exposed location with over 1,000 mm annually. The occasional propensity for this area to experience extremely heavy falls of rain is exemplified in the Lynmouth flood of 1953 (reviewed in Box 2.1 of this chapter). Precipitation averages for stations in the region are shown in Table 2.4

All the region has less than half the annual average rainfall in the summer half-year (April–September), with the smallest percentages in the Channel Islands (38 to 39 per cent), rising to about 49 per cent in the Bristol area. The wettest individual month is, in many places, October, but in others lies between November and January. The late autumn–early winter period, when sea temperatures are still relatively high, favours the formation of vigorous depressions accompanied by moisture-laden air masses moving in from the west and yielding copious amounts of rain, especially over the hills.

Rainfall amounts are extremely variable, as Table 2.5 suggests, on both the annual and the monthly

Table 2.4 Mean monthly and annual precipitation totals (mm) for the period 1961–90

Location	*Alt. (m ASL)*	*Jan.*	*Feb.*	*Mar.*	*Apr.*	*May*	*June*	*July*	*Aug.*	*Sept.*	*Oct.*	*Nov.*	*Dec.*	*Year*
Marlborough	129	78	54	68	52	57	57	50	65	66	69	73	85	774
Poole	5	89	64	66	47	52	51	38	53	64	80	85	92	781
Bude	15	95	67	68	51	52	53	58	68	78	90	102	98	880
Sidmouth	10	87	71	67	48	54	48	49	57	58	73	74	87	773
Torquay	15	114	91	78	53	58	52	46	64	66	82	89	109	902
Princetown	741	246	176	181	119	121	121	124	152	162	204	215	257	2,077
Ilfracombe	8	116	79	82	64	68	65	65	84	96	124	125	120	1,088
St Mawgan	103	116	91	80	52	58	62	61	75	83	80	114	116	988
St Mary's	51	101	83	74	50	58	49	51	62	68	87	95	100	878
Jersey	98	97	73	72	53	54	44	38	42	65	86	109	103	836

Table 2.5 Rank order of the three driest and wettest seasons (mm) at Plymouth for the period 1920–95.

	Spring	*Summer*	*Autumn*	*Winter*	*Year*
Driest					
	81 (1994)	43 (1995)	77 (1978)	120 (1976)	557 (1921)
	86 (1938)	53 (1976)	107 (1941)	125 (1934)	687 (1933)
	103 (1956)	69 (1983)	148 (1934)	127 (1932)	695 (1953)
Wettest					
	325 (1947)	339 (1968)	518 (1960)	506 (1966)	1,334 (1960)
	311 (1981)	330 (1958)	436 (1976)	451 (1978)	1,272 (1924)
	299 (1964)	309 (1971)	430 (1969)	444 (1960)	1,252 (1974)

Note: The year of occurrence is given in parentheses

scales. Wood (1989) has shown that since 1921 rainfall at Plymouth has shown annual peaks centred on 1957 and 1967 and an increase in summer variability in recent years with a decrease in winter variability. Outstanding anomalous seasons in south-west England include the extremely wet autumn of 1960, which led to repeated flooding, especially in the Exe valley, and the two extremely dry summers of 1976 and 1995, the earlier of which led to extreme drought conditions (Booker and Mildren 1977). Individual rainless months have occurred in places: for example, Bideford and Cullompton, Devon, had no rain at all in February 1965; neither did a small area in east Cornwall in April 1976 and Dawlish, Devon, in June 1976. In Jersey, June 1976 gave less than 1 mm of rain. Precipitation yields over south-west England from different synoptic circulation types have been investigated by Sweeney and O'Hare (1992) using two stations, Bude and Bournemouth. Mean daily precipitation totals were highest at Bude with cyclonic southerly and south-westerly types, but cyclonic easterly, south-easterly and north-easterly types produced the largest daily totals on average at Bournemouth.

Heavy falls on any one day are usually the result

Box 2.1

THE LYNMOUTH FLOOD – 15 AUGUST 1952

The most notorious flash flood in Britain in living memory was at Lynmouth, Devon, on 15 August 1952. Thirty-four people were killed and 400 made homeless by the flooding that resulted when 200,000 tonnes of boulders swept down the East and West Lyn rivers into the town. Upward of 230 mm fell over a period of a few hours in the afternoon in a series of violent storms over Exmoor (Bleasdale and Douglas 1952; Dukes and Eden, 1997). It has been estimated by Kidson (1953) that two-thirds of the total rainfall fell in five hours. The very heavy rainfall was restricted to a relatively small area on the northern and north-eastern slopes of Exmoor, suggesting the importance of orographic uplift of moist air.

The small depression responsible for the great rainstorm moved north-eastwards from the Bay of Biscay to a position near Exeter at the time of the storm. The main area of thundery rain reached Plymouth shortly before 0700 hours on the 15th and moved slowly northwards, lasting all day over the rest of Devon and Cornwall. Over Exmoor the heaviest rain occurred in the afternoon and evening, and the flooding at Lynmouth reached its height and destroyed most property after dark. Among the contributing factors to the disaster was the unusually wet and saturated soils in the area following previous heavy rainfalls (Marshall 1952; over 100 mm had already fallen in that fateful August) and the topography of the northern edge of Exmoor, which is drained by short, steep valleys. It is all but impossible to calculate the precise levels of rainfall but at Longstone Barrow on Exmoor the day's yield was 228 mm, and that was for a site that may have been removed from the centre of precipitation activity (Holford 1976)! This acute catastrophic flood is comparable with the worst floods ever recorded in these islands and deservedly remains renowned as a weather disaster.

Since the Lynmouth floods, research into the probable maximum precipitation or the largest depth of precipitation that can be expected for a given duration at a given place has revealed that a value of around 300 mm over 24 hours increasing to perhaps 500 mm over Exmoor and Dartmoor is possible (Clark 1995). From such work it is apparent that the Lynmouth storm was by no means the most severe possible in the area.

Table 2.6 The largest daily falls of rain (mm) on record in south-west England

Location	*Total fall*	*Date*
Martinstown, Dorset	279.4	18 July 1955
Bruton, Somerset	242.8	28 June 1917
Cannington, Somerset	238.8	18 Aug. 1924
Longstone Barrow, Devon	229.5	15 Aug. 1952
Camelford, Cornwall	203.2	8 June 1957
Chew Stoke, Avon	173.0	10 July 1968
Isles of Scilly	65.2	10 July 1968
Channel Islands	96.8	10 July 1968

of thunderstorms in the summer and early autumn months. Of interest are the number of large individual daily falls of rain that have occurred in the Somerset–Dorset region in particular and the south-west peninsula in general (Table 2.6), including the largest fall ever measured in 24 hours in the British Isles (Paxman 1957). Bleasdale (1963) has noted the remarkable cluster of extremely heavy daily rainfalls in the area and suggested that three factors – a long sea fetch to the west, several upland areas and a location within reach of thundery outbreaks originating over the Continent – occur in a combination not found elsewhere in the United

Figure 2.4 Severe flash flooding at Cheddar on 10 July 1968
Photo by courtesy of Barry Horton

Kingdom. To produce such large falls usually requires a sluggish flow of warm, humid air from the south or south-west and probably a measure of orographic uplift, but the 18 July 1955 fall occurred in a very light northerly flow. Blocked cyclonic weather situations with slow-moving depressions and fronts are most likely to produce the requisite conditions (Mayes, forthcoming). Among the worst floods of the past forty years were those of autumn 1960 in east Devon, caused by repetitive heavy rainfalls, and the 10 July 1968 event, which badly affected Bristol and north Somerset (Salter 1969). Figure 2.4 shows the flooding at Cheddar on this day.

The average number of days with thunder in the year ranges from less than six locally on the coasts to more than ten inland (Figure 2.5). In summer, when thunder is most frequent at all locations, storms generated over western France and the Bay of Biscay can drift northwards at the conclusion of a heatwave, sometimes accompanied by large hailstones and often arriving in the late evening or at night. Such storms can occur as early as May, as the flash floods in the Bridport area of Dorset in 1979 showed, although (Figure 2.5) June and July are the most thunder-prone months. The notably severe thunderstorms in the early hours of August 1938 which gave 145 mm of rain at Torquay in just over five hours were an excellent example of these severe storms. One of the most severe thunderstorms, probably accompanied by a tornado, struck Widecombe-in-the-Moor on the eastern edge of Dartmoor on 31 October 1638 when four people were killed and about sixty others injured (Bonacina 1946). It is worth noting the exceptional case of the Cannington storm in 1924, which occurred in a cold air mass with nocturnal convergence and instability over the Bristol Channel.

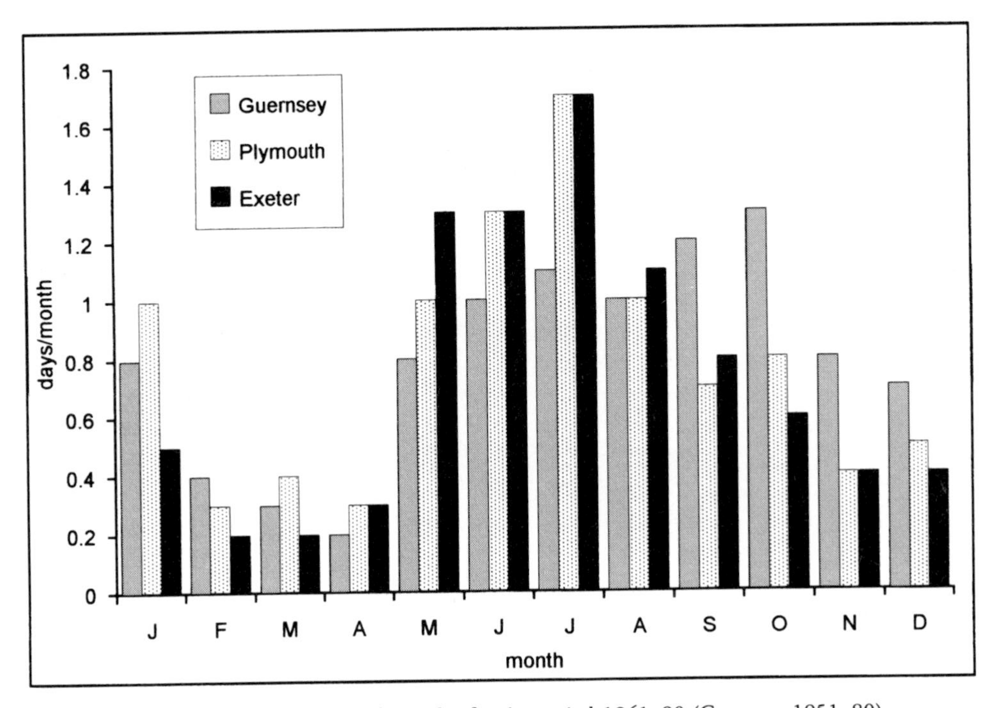

Figure 2.5 Mean monthly frequency of days of thunder for the period 1961–90 (Guernsey 1951–80)

Thunder with hail in winter is usually caused by cold (polar maritime) air moving around a large North Atlantic depression with frequent blustery showers resulting from surface heating of the air over the eastern Atlantic. Such showers usually pass by quickly, but a spectacular and unusual example of such a weather type occurred on 13 December 1978 (Owens 1980) when the Kingsbridge area of south Devon received large hail which caused considerable damage. Stirling (1982) has suggested that the high hail frequencies in the Channel Islands are the result of the already unstable north-westerly airstreams crossing the warm waters of the English Channel. Damaging hail in summer is more frequent in the eastern part of the peninsula. For example, in Wiltshire and Dorset, where warmed land surfaces are more extensive, one of the most severe hailstorms in the past thirty years has been described by Hardman (1968). A trail of damage from Trowbridge north-eastwards to the northern borders of Wiltshire occurred as storms developed after hot, sunny weather on 13 July 1967.

The peninsular nature of south-west England ensures that the thermal relationship between sea and land is extremely important in influencing precipitation distribution, especially in unstable airstreams. In winter, when the sea is warmer than the land, convection can be potent over the warm sea, and frequently with onshore unstable airstreams showers affect coastal locations while inland they die out, especially at night (Sims 1960). In a winter northerly airstream, the western end of the north Cornwall coast attracts frequent showers that are generated over St George's Channel. In summer the warmth of the land compared with the sea encourages daytime convection and this effect, combined with convergence associated with sea breezes on both coasts of the peninsula, can produce groups of showers on a summer's day well inland, sometimes with degenerate showers reaching the coast in the evening (Douglas 1960).

Spring is clearly the season when thunder is least evident (Figure 2.5). At this time of year the delayed cooling of the sea typical of all maritime settings is most effective in damping down instability, while the land surface has yet to recover fully from the effects of winter.

Snowfall

While snow is normally a fleeting visitor to most of south-west England and an ephemeral element of the landscape, this is not always the case, and over the past century a number of notable blizzards have affected the area. At low levels snow rarely lies on the ground before December or after March, although the establishment of a temporary cover in Jersey on 6 November 1980 in a cold, easterly airstream is worth noting. Jersey is, of course, the southernmost of our islands, so it may seem odd that it should have experienced one of the most severe early cold snaps. The reason is that only in the Channel Islands can a bitter easterly wind arrive after crossing only a very short stretch of sea, and a similar event occurred in November 1985 when 12 cm of snow fell on the 20th. Along western and southern coasts many winters pass with little or no snow cover, but over the higher parts of the moors snow can be expected to lie for ten to fifteen days, but with a great deal of year-to-year variability. Over the 1961–90 period snow lay on average for two to three days per year at Plymouth, but on thirteen days at Princetown. In 1963 snow cover persisted on Dartmoor for sixty days, whereas at Plymouth no lying snow has occurred in about one-third of winters in the past fifty years (though in 1962/63 there were fifteen days with a snow cover). It is in severe winters like those of 1947 and 1963, when a series of depressions move up the English Channel, that south-west England receives unusual amounts of snow, and Burt (1978) has noted that 'despite being the mildest region of the British Isles, the south-western peninsula has experienced some of the most severe blizzards ever to affect the country'. For such conditions to occur, an active front separating warm moist air from bitterly cold easterlies, perhaps arriving from as far afield as eastern Europe or Siberia, should lie just off the South-West Approaches with active depressions forming on it and moving eastward into central France. In the past twenty years the greatest snowfalls occurred in such situations in 1978, 1982 and 1987. In February 1978 large areas of inland Devon lay under 50 cm of snow on the morning of the 19th with more than 90 cm on both Exmoor and Dartmoor (see Box 2.2).

A repetition occurred on 8/9 January 1982, with conditions especially severe in the Bristol area, where drifts were locally 1 m deep. In Wiltshire, main roads were impassable for up to a week. In January 1987 the Isles of Scilly and west Cornwall experienced their most severe weather this century with snow 39 cm deep at Penzance and 30 cm at Falmouth on 12th, with much disruption to communications and the isolation of many rural communities for several days.

WIND

The presence of deformed, stunted and distorted trees in the landscape, particularly on exposed coasts and over the moors, is a continual reminder of a windy climate, where shelter, especially from the west, is highly valued and of paramount importance. Flower fields on the Isles of Scilly and farm buildings elsewhere are tucked into folds in the ground while high hedges, and in the west stone walls, provide shelter. The position of the peninsula, jutting out into the Atlantic, ensures that in the west the most exposed conditions in England occur with the highest mean wind speed and gale frequency. A wind zonation map based on Miller (1987) is shown in Figure 2.7 and suggests that western Cornwall is about as windy as large parts of the western mainland of Scotland. Further east the degree of exposure declines, until in eastern Dorset and Avon conditions are similar to those prevailing across the Welsh borders. At Plymouth about thirteen gales occur in an average year and at Scilly about twenty but in a windy year there can be three times as many gales.

Box 2.2

TWO LATE NINETEENTH-CENTURY SNOWSTORMS: JANUARY 1881 AND MARCH 1891

These two late-nineteenth-century storms probably remain unequalled in terms of their severity and the depth of snow that they brought, particularly to Dartmoor, to the present day. Bonacina (1928) described them as 'standing in a class absolutely by themselves in the whole fifty years 1876–1925' and they have probably not been exceeded since, even in 1947 or 1963. Manley (1952, p. 205) notes that 'Dartmoor snowfalls are occasionally extremely heavy and are the more noticeable in the relatively mild south-west'. The severity of the events has continued to fascinate up to the present day (Lewis 1990). On 18 January 1881, after days of intense cold, a blizzard struck which killed large numbers of people. In Exeter snow depth was 35 cm, in Kingsbridge, Devon, 45 cm, and at Sherborne, Dorset, 66 cm, while on Dartmoor 120 cm was recorded. On Exmoor Hurley (1972) has described the calamitous visitation of this storm on a population equipped with little more than shovels to dig itself out of the enormous snowdrifts. A depression had moved from the Bay of Biscay to the Isle of Wight and then turned south into France.

The great snowstorm of 9–13 March 1891 was at least as bad as that of 1881 and was caused by a deepening depression that moved from west of Biscay towards Brest and subsequently to the Low Countries. Severe east to north-east gales affected the whole of southern England with continuous heavy snow. The snow was dry and finely powdered, suggesting that temperatures were below freezing. A second depression, following a track similar to that of the first, brought renewed snow and gales. The disruption in Devon and Cornwall was colossal, with ten trains snowed up at one time, one of which was effectively lost for 36 hours near to Princetown. Tawy Cleave, a deep ravine on the edge of Dartmoor north-east of Tavistock, was filled up with snow as a result of mass drifting off the moor; while in the English Channel well over fifty ships were wrecked.

These great West Country blizzards stand as a reminder of just how inclement winter weather can be in this the mildest part of the United Kingdom under certain synoptic situations. The similarity of the weather maps on the occasion of the 1891 and of the more recent 1978 snowstorms is striking confirmation of the significance of the prevailing synoptic state (Figures 2.6a and b).

In the Channel Isles Guernsey averaged ten days a year with gales in the 1951–80 period while Alderney, more exposed in the mouth of the English Channel, averaged twenty-two days. Not surprisingly, both the strongest mean wind speeds and the maximum gusts have been recorded in Cornwall, the Isles of Scilly and the Channel Islands, and these are listed in Table 2.7. Over much of Cornwall and south-west Devon the gale of 13 December 1979 produced wind speeds which exceeded previous maxima in records extending back to 1921. In this region, the notorious storm of 16 October 1987 severely affected only the area east of Portland Bill and the Channel Islands. In Jersey the highest gust was 85 knots and at Portland Bill, 78 knots. Another severe gale on 16–17 December 1989 caused nearly £2 million worth damage, mainly in coastal towns and villages as high tides and exceptionally rough seas accompanied the strong winds (Jarvis 1992).

Coastal sites along the English Channel and areas in the east of the region experienced the most extreme winds during the storm of 25 January 1990 with records broken at Plymouth (gust of 84 knots), St Mawgan (89 knots) and Boscombe Down (79 knots). The latter station, situated on Salisbury

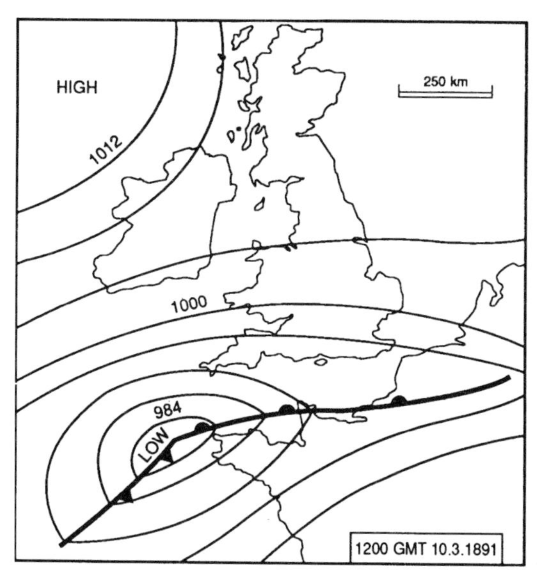

Figure 2.6a Synoptic chart for 10 March 1891 at 1200 GMT

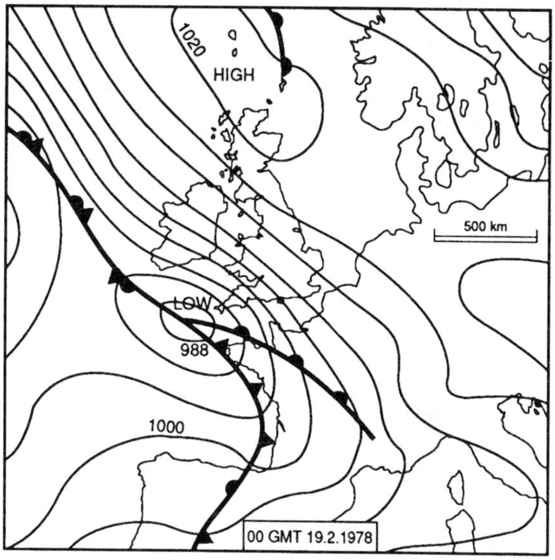

Figure 2.6b Synoptic chart for 19 February 1978 at 0000 GMT

Table 2.7 Mean wind speeds and gust speed extremes (knots) at four locations

Location	*Alt. (m ASL)*	*Maximum mean hourly speed*	*Maximum gust speed*	*Date*
Jersey	98	68	94	9 Oct. 1964
St Mary's	70	66	99	15 Jan. 1979
Gwennap Head	66	60	103	15 Jan. 1979
Avonmouth	28	45	73	11 Jan. 1962

Plain, is extremely exposed and recorded a gust of 70 knots in October 1945. Analysis of the return period of the 1990 gale by Hammond (1990) suggests that in a narrow corridor across east Dorset the return period was of the order of 150 years and elsewhere in the South-West was about fifty years. In several areas over 30 per cent of local authority housing was damaged (Buller 1994) and many other structures were affected, while on the roads seventy-five accidents were attributed to the severe winds, twenty-nine of them in Dorset, mainly as a result of vehicles overturning. Table 2.7 shows that the most severe gales across the region occur in autumn and winter, but no season is totally immune. Even in summer storm-force winds can occur occasionally, the most severe storm on record being probably that of 29 July 1956 when the Lizard had a gust of 87 knots and St Mawgan 76 knots.

While the strongest winds are often from a west or south-west direction, there have been a number of occasions when east and north-east gales have caused large amounts of damage on east-facing parts of the English Channel coast. In January 1917 such a gale wholly destroyed the village of Hallsands in south Devon, and on 30 December 1978 the nearby village of Torcross, also on Start Bay, suffered near-destruction (Emberson 1988). Further east,

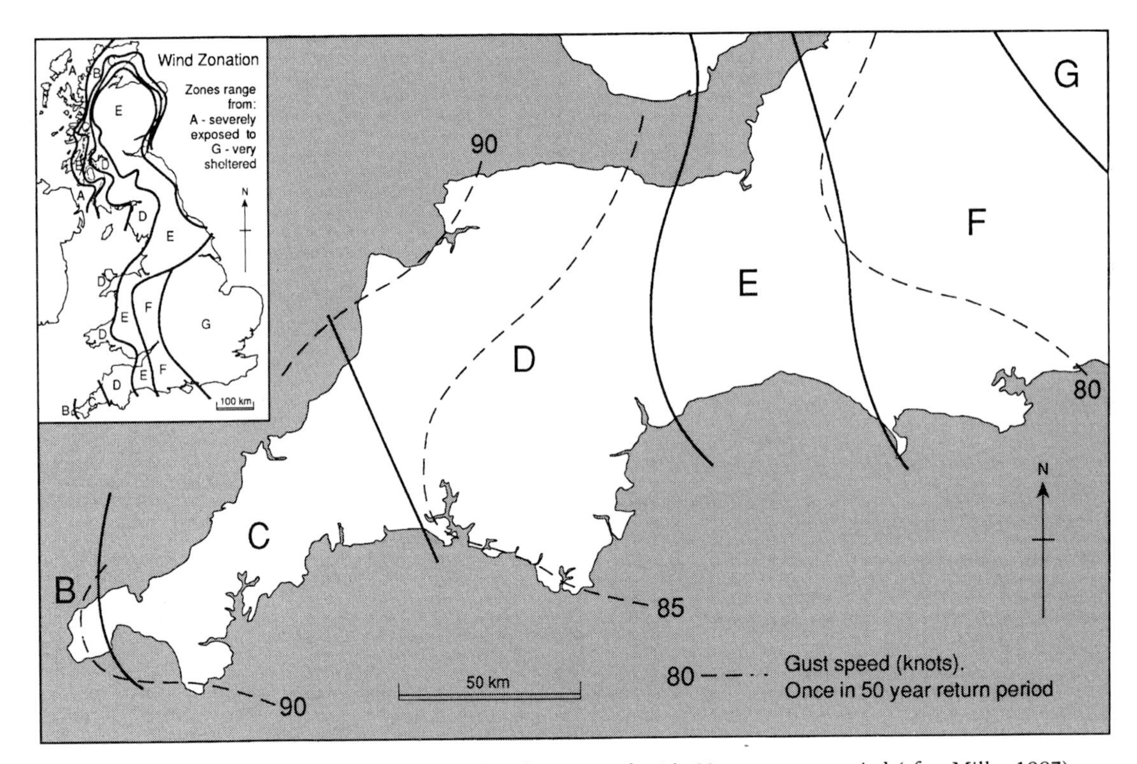

Figure 2.7 Wind zonation in south-west England and gust speed with 50-year return period (after Miller 1987)

the low-lying part of the Isle of Portland was badly damaged in mid-February 1979, as it was in 1942.

SUNSHINE

St Helier, Jersey, is the sunniest town in the British Isles with an average 1,915 hours of sunshine per year. In a sunny year, like 1959, more than 2,250 hours have been noted. Between 1961 and 1990 St Helier was the British Isles' sunniest location on no fewer than eleven occasions, and it was in second or third place on a further ten occasions. The worst year, by a notable margin, in that period was 1981, when only 1,587 hours of sunshine were recorded (the plaudit of Britain's sunniest place went that year to Leuchars in Fife!). Otherwise, annual totals were always better than 1,690 hours. Blench (1967) has observed that Jersey's August mean sunshine is 55 per cent of the possible, and it is the only site in the British Isles where 50 per cent is exceeded. On the mainland, highest totals in the region are on the Dorset coast with 1,771 hours at Weymouth, a figure that is almost matched in the extreme south-west of Cornwall and the Isles of Scilly. At Plymouth the annual average is 1,675 hours. Over the upland areas and further inland, totals fall, and in Wiltshire and Somerset 1,480 to 1,550 hours per year is the normal. Days which benefit most the southern and western coasts are those with a south-westerly airstream in summer when the sea is clear of cloud but cumulus forms inland as a result of the convection process. The coast may then be slightly cooler than inland but it enjoys extra hours of sunshine.

The sunniest months on average are June and July, and the least sunny are December and January (Table 2.8). This seasonal rhythm reflects not only the changing hours of daylight but also the corresponding fluctuations in the Azores anticyclone,

Table 2.8 Mean monthly and annual sunshine totals (hours) for the period 1961–90

Location	*Alt. (m ASL)*	*Jan.*	*Feb.*	*Mar.*	*Apr.*	*May*	*June*	*July*	*Aug.*	*Sept.*	*Oct.*	*Nov.*	*Dec.*	*Year*
St Helier	54	62	90	141	194	239	248	261	235	176	132	77	60	1,915
Guernsey Airport	104	53	79	131	183	234	240	251	226	168	117	72	51	1,805
St Mawgan	103	58	76	126	189	215	207	209	194	157	108	76	54	1,669
Lyneham	145	53	70	113	157	192	201	205	184	143	101	71	51	1,541
Hurn	10	59	77	123	171	210	214	219	201	154	109	76	57	1,670
Penzance	19	60	80	135	191	222	208	219	207	165	113	76	52	1,728

which is at its most powerful in summer but weakens each winter. South-west England, more than most other regions, is open to the influence of the Azores system, which does much to provide the high summer sunshine totals. In June 1925 Falmouth recorded 381.7 hours of sunshine and in the same month Long Ashton had 334.8 hours. In August 1976 Ilfracombe had 333 hours, a near-record for the month. In the dullest winter months totals of less than 20 hours have been noted. About one day in three can be expected to have no sunshine during December and January but in mid-summer only one or two days each month are on average sunless. This is, of course, due in part to the differences in the length of day. Average daily sunshine in January is a little over 2 hours, rising to over 8 hours in June in the Channel Islands. In winter most weather stations have recorded spells of eight to ten sunless days but in summer a period of two or three consecutive days without sunshine is unusual.

VISIBILITY AND FOG

Two main types of fog affect the area. The first is radiation fog, which forms mainly inland in the winter half-year at night or in the early morning, though it does occasionally persist all day, more especially in Somerset and Wiltshire. With a light surface easterly flow, fog can get trapped on the east side of Exmoor. Kidd (1995) has described an occasion of widespread radiation fog on 23 December 1994 when high ground, such as Salisbury Plain and the Mendip Hills together with the smaller hills around Lulsgate (Bristol Airport), protruded above the fog, which stubbornly affected the Somerset Levels, suggesting the fog reached between 100 and 150 m in altitude on this occasion. Indeed, such events can be interpreted as visual evidence for the persistence of winter temperature inversions in Somerset, leading to a monotonous sequence of overcast days. Low-lying ground on both banks of the Bristol Channel can be an effective reception area for subsiding cold dense air under a winter temperature inversion, with consequent heightened fog risk. The second type of fog, sea fog or advection fog, is formed by the passage of warm, moisture-laden air over a relatively cold sea and is most frequent from late winter to early summer.

Figure 2.8 gives information on the frequency of fog across the region. Over the 1961–90 period, Plymouth averaged 5.6 days per year with fog and Scilly 14.0. Jersey, in contrast, has an average closer to thirty-three. In such maritime settings sea fogs dominate the statistics. Inland locations such as Lyneham are principally affected by radiation fog, however, while at Exeter both types occur. Figure 2.8 also shows with notable clarity the contrasting monthly fog regimes of coastal locations. Thus at Scilly late spring and early summer sea fogs dominate the picture and result from the cooling of tropical maritime air masses over the still cool sea

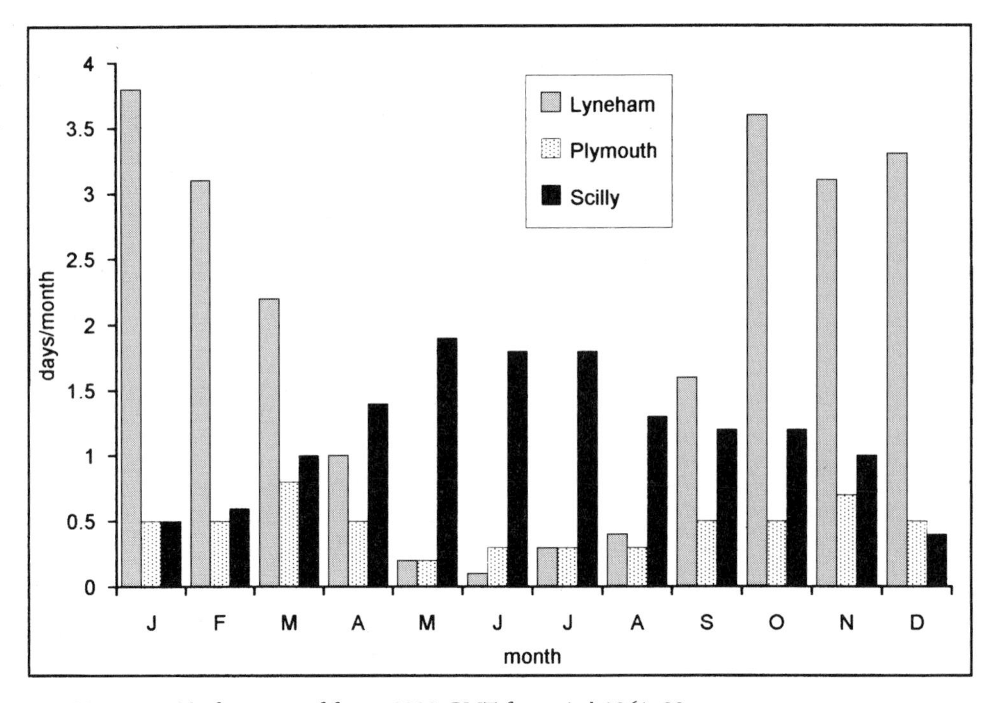

Figure 2.8 Mean monthly frequency of fog at 0900 GMT for period 1961–90

surfaces. At Lyneham, the typical inland location, nocturnal cooling creates frequent radiation fogs at any time between October and March. A note of caution should, however, be sounded, and Saunders (1957) showed that at Exeter thick fog could form quickly but normally dispersed two to three hours after sunrise, as it does in many such areas, giving the 0900 GMT observations a partially unrepresentative influence.

Douglas (1960) has shown that further east in the Axe Valley, radiation fogs outflow seawards into the English Channel after sunrise. Anticyclonic spells in early summer often produce patchy sea fog which drifts on to the beaches, but it is especially in warm sectors that such fog is most persistent and also cloaks the upland areas. Data on the percentage frequency of good visibility (greater than 21 km) show that in summer about half of all days have such conditions, but in winter only 14 per cent of days at Boscombe Down and 6.5 per cent at Plymouth enjoy such good visibility. Visibility of more than 40 km occurs on average on thirteen mornings at Plymouth and twenty-three at Boscombe Down (Stirling 1982).

CONCLUSION

Whilst it is the natural endowment of a favourable climate with the greatest potential for plant growth in the United Kingdom that is the most obvious attribute of south-west England, the intra-regional differences of climate are also striking. The open, windswept moors in Devon and Cornwall and further east the hills and plateaux of the Mendips and the Quantocks seem on many days to belong to a different climatic province from the sheltered vales and river valleys. The geographical propinquity of these varied local climates is particularly close in the narrow peninsula of Cornwall. Whilst adjectives like soft, muggy and equable might seem appropriate to describe the frequent days, especially in winter, of mild, grey weather, a further marked character of the climate is its changeability. The location of the South-West peninsula makes it particularly conducive to importing air from a wide range of sources with different temperature and moisture characteristics. Height above sea level, inland or coastal situation, aspect, slope and exposure all combine to produce a particularly complex and distinctive range of topoclimates which, in practice, means that climate varies considerably and appreciably from place to place and over small areas as well as over quite short time intervals.

REFERENCES

Archer, B. (1956) 'The wettest areas of S.W. England', *Weather*, 11: 35–39.

Atkinson, B.W. and Smithson, P.A. (1976) 'Precipitation', in T.J. Chandler and S. Gregory (eds) *The Climate of the British Isles*, London: Longman.

Bleasdale, B.A. (1963) 'The distribution of exceptionally heavy daily falls of rain in the U.K. 1863–1968', *J. Inst. Water Engineering*, 17: 45–55.

Bleasdale, B.A. and Douglas, C.K.M. (1952) 'Storm over Exmoor on August 15 1952', *Meteorol. Mag.*, 81: 353–366.

Blench, B.J. (1967) 'An outline of the climate of Jersey', *Weather*, 22: 134–139.

Bonacina, L.C.W. (1928) 'Snowfall in the British Isles during the half century 1876–1925', in *British Rainfall*, London: HMSO.

—— (1946) 'The Widecombe calamity of 1638', *Weather*, 1: 123–126.

—— (1951) 'The scenery of Devonshire in relation to the weather and climate', *Weather*, 6: 131–135.

—— (1961) 'Comparative climatological data for Exeter, Ross-on-Wye and Cambridge in the decade 1950–59', *Weather*, 16: 187–193.

Booker, F. and Mildren, J. (1977) *The Drought in the South West*, Department of Economic History, University of Exeter.

Broad, H.J. and Hough, M.N. (1993) 'The growing and grazing season in the United Kingdom', *Grass and Forage Science*, 48: 26–37.

Brooks, C.E.P. (1954) *The English Climate*, London: English Universities Press.

Buller, P.S.J. (1994) *The Gale of January and February 1990*, Watford: Building Research Establishment Report.

Burt, S.D. (1978) 'The blizzards of February 1978 in south-west Britain', *J. Meteorology U.K.*, 3: 261–278.

Butler, A.P., Grundy, J.D. and May, B.R. (1985) 'An analysis of extreme rainfalls observed in Jersey', *Meteorol. Mag.*, 114: 383–395.

Clark, C. (1995) 'New estimates of probable maximum precipitation in S.W. England', *Meteorol Applications*, 2: 307–312.

Dancey, D.W.G. (1981) 'Wet working days in the United Kingdom', *Meteorol. Mag.*, 110: 12–27.

Davies, A. (1954) 'The personality of the south-west', *Geography*, 39: 243–249.

Douglas, C.K.M. (1960) 'Some features of local weather in S.E. Devon', *Weather*, 15: 14–18.

du Maurier, D. (1967) *Vanishing Cornwall: The Spirit and History of Cornwall*, London: Victor Gollancz.

Dukes, M.D.G. and Eden, P. (1997) 'Climate records and extremes', in M, Hulme and E. Barrow (eds) *The Climate of the British Isles: Present, Past, Future*, London: Routledge.

Emberson, L. (1988) *Storm Siege*, Dartmouth: Tozer & Co.

Hammond, J.M. (1990) 'Storm in a teacup or winds of change?', *Weather*, 45: 443–449.

Hannell, F.G. (1955) 'Climate', in C.M. McInnes and W.F. Whittend (eds) *Bristol and Its Adjoining Counties*, Bristol: British Association for the Advancement of Science.

Hardman, M.E. (1968) 'The Wiltshire hailstorm 13 July 1967', *Weather*, 23: 404–414.

Hogg, W.H. (1967) 'Meteorological factors in early crop production', *Weather*, 22: 84–94.

Holford, I. (1976) *British Weather Disasters*, Newton Abbot: David & Charles.

Hooper, G.H. (1837) *Observations on the Climate and Topography of Jersey*, London.

Hurley, J. (1972) *Snow and storm on Exmoor*, Dulverton, Somerset: Exmoor Press.

Jarvis, J. (1992) 'An analysis of the storm of 16–17 December 1989 in Devon and Cornwall', *J. Meteorology (UK)*, 17: 120–125.

Kidd, C. (1995) 'Images of widespread radiation fog over southern England on 23 December 1994', *Weather* 50: 370–374.

Kidson, C. (1953) 'The Exmoor storm and the Lynmouth floods', *Geography*, 38: 1–9.

Lewis, R.P.W. (1990) 'The winter of 1890–91', *Weather*, 45: 438–443.

Manley, G. (1944) 'Topographical features and the climate of Britain: a review of some outstanding effects', *Geogrl J.*, 103: 241–258.

—— (1952) *Climate and the British Scene*, London: Collins.

Marshall, W.A.C. (1952) 'The Lynmouth flooding', *Weather*, 8: 338–342.

Mayes, J. (forthcoming) 'Orographic influences on local weather: a Bristol Channel case study, *Geography*.

Meaden, G.T. (1983) 'Britain's highest temperatures: the county records', *J. Meteorology (UK)*, 8: 201–203.

Miller, K.F. (1987) 'The assessment of wind exposure for forestry in upland Britain', *Forestry*, 60: 179–192.

Owens, R.G. (1980) 'Severe winter hailstorm in S. Devon', *Weather*, 35: 188–199.

Paxman, D.J. (1957) 'The exceptional rainfall of 18 July 1955', *Weather*, 12: 246–251.

Ratsey, S. (1975) 'The climate at Slapton Ley', *Field Studies*, 4: 191–206.

Revesz, T. (1969) 'Climate', in *Exeter and Its Region*, Exeter: British Association for the Advancement of Science.

Salter, P.R.S. (1969) 'A further note on the heavy rainfall of 10 July 1968', *Meteorol. Mag.*, 98: 92–94.

Saunders, W.E. (1957) 'Variations of visibility in fog at Exeter airport and the time of fog dispersal', *Meteorol. Mag.*, 86: 362–368.

Shapter, T. (1862) *The Climate of the South of Devon*, 2nd edn, London: John Churchill.

Shaw, M.S. (1977) 'The exceptional heatwave of 23 June to 8 July 1976', *Met. Mag.* 106: 329–346.

Sims, F.P. (1960) 'The annual and diurnal variation of shower frequency at St. Eval and St. Mawgan', *Meteorol. Mag.*, 89, 293–297.

Stamp, L.D. (1941) 'Devonshire', in *Report of the Land Utilization Survey: South West England*, vol. 9, London: Geographical Publications.

Stirling, R. (1982) *The Weather of Britain*, London: Faber & Faber.

Sweeney, J.C. and O'Hare, G. (1992) 'Geographical variations in precipitation yields and circulation types in Britain and Ireland', *Trans. Inst. Br. Geogrs*, NS 17: 448–463.

Tout, D. (1985) 'A century of climate and its effect on the flora of the Isles of Scilly', *Weather*, 40: 374–379.

—— (1987) 'The variability of days of air frost in Great Britain, 1957–1983', *Weather*, 42: 268–273.

Webb, J.D.C. (1987) 'Britain's highest daily rainfalls, the county and monthly records', *J. Meteorology (UK)*, 12: 263–266.

Webb, J.D.C. and Meaden, G.T. (1983) 'Britain's lowest temperatures: the county records', *J. Meteorology (UK)*, 8: 269–272.

—— (1993) 'Britain's highest temperatures by county and by month', *Weather*, 48: 282–291.

Wood, N.L.H. (1989) 'Rainfall variability at London and Plymouth', *Weather*, 44: 202–209.

—— (1994) 'The Plymouth Hoe Observatory 1874–1979', in B. Giles and J. Kenworthy (eds) *Observatories and Climatological Research*, Department of Geography, University of Durham Occasional Publication no. 29, 117–129.

3

SOUTH-EAST ENGLAND

Julian Mayes

INTRODUCTION

South-east England is usually regarded as comprising the counties to the south of the Thames, together with most of the Thames catchment itself and Greater London. For the purposes of this chapter, the region also includes much of Hertfordshire, Buckinghamshire, Oxfordshire and south Essex (Figure 3.1). This is an area of varied topography but modest relief: the highest point is Leith Hill, Surrey, at 294 m. While the physical features might at first seem rather muted, consideration of the region's climate must also acknowledge the degree to which the area has been subject to human settlement and

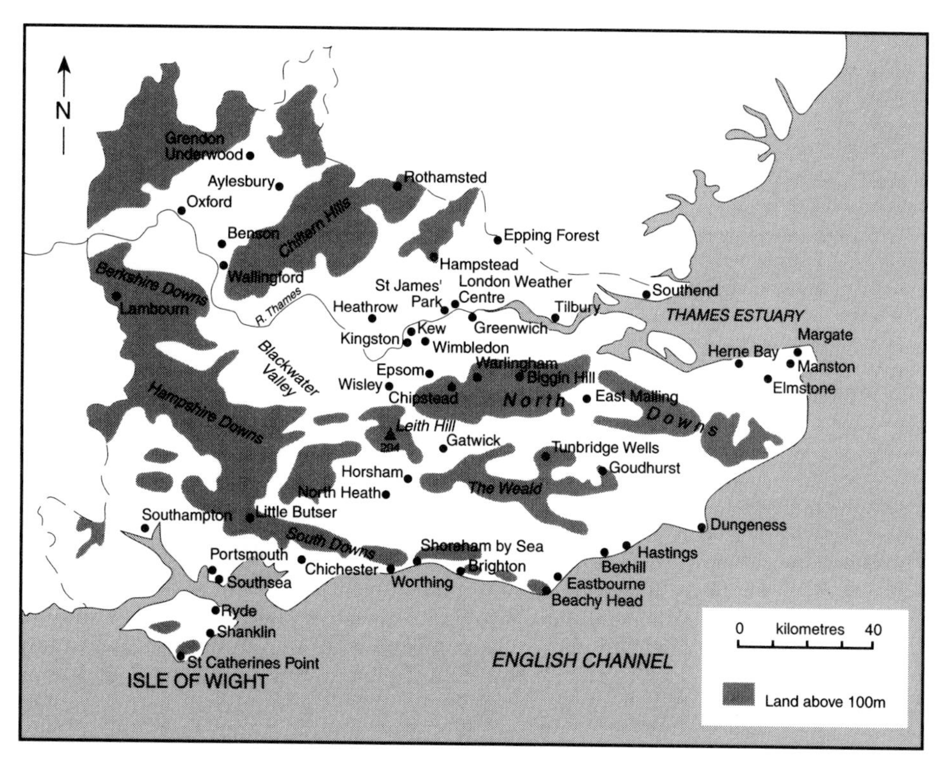

Figure 3.1 Location of weather stations and places referred to in the text

land use change. The relatively high population density influences the demands placed upon natural resources (such as water) and the response of the land surface to extreme climatic events, such as flooding. The spread of urban and suburban landscapes through much of the region in the twentieth century possibly challenges our notions of discrete rural and urban climates.

Despite the low-lying character of the region, the distinction between the Chalk downlands and the Thames valley is reflected in subtle climatic variations. The region has the dual benefit of relative distance from the mean position of Icelandic depressions and a certain amount of shelter from the prevailing west-south-west winds. It also occupies the favoured end of the north-north-west to south-south-east climatic gradient identified by Hatch (1973). The sheltering factor increases north-eastwards towards the Thames estuary, accentuated by the fact that the more exposed south-west of the region contains much of the chalk hill country of the South-East (Harrison 1977). The significant length of coastline introduces further diversity since it comprises an unusual variety of directional aspects and exposures. For example, the north coast of Kent is more sheltered from the prevailing west-south-west airstreams than the other coasts, and perhaps has more climatic affinity with London and the Thames valley.

Continuity of Past Observations

It might be expected that south-east England would offer a profusion of climate stations (Figure 3.1). In practice, the value of past and present records is affected by the continuity of observing sites. This issue was brought into focus in 1980 with the closure of Kew Observatory (Mayes 1994), a site for which a composite monthly rainfall series is available back to 1697 (Wales-Smith 1980). Other key stations, such as Greenwich and Kingsway/London Weather Centre, have been affected to varying degrees by site changes. These events have reduced the availability of homogeneous records in the London area. Records from the long-term stations such as St James's Park and Hampstead Observatory (where observations started in 1903 and 1910, respectively) have thus acquired extra value. The former is a now all-too-rare example of a ground-level site close to the centre of London's heat-island, whilst the latter is an excellent contrast, lying as it does at London's highest point of 137 m.

On London's western margins Heathrow provides observations back to 1946, though even this important site has been subject to small changes in location (Webster 1984). Beyond London, the outstanding record is that of the Radcliffe Observatory at Oxford (Smith 1974), where regular observations of rainfall and temperature are available from 1815. Perhaps unsurprisingly, a small urban warming effect has been detected in the records (Smith 1994). Maybe climatologists should be grateful that the growth of north Oxford has been constrained by the nearby flood plain of the Thames, thereby inhibiting heat-island formation.

The region still possesses long-running records in nearly unchanged rural environments, such as Rothamsted (where records began in 1872), Wisley and East Malling. Finally, this section would not be complete without acknowledgement of the long-established 'Health Resort' stations, many of which now have records exceeding a hundred years in duration. The termination of the record at Worthing in 1993 is a major loss. It is to be hoped that other municipally run stations with even longer records (such as Eastbourne, from 1867) have a more secure future.

TEMPERATURE

South-east England is well known for having the highest summer day temperatures in the British Isles. As Tout (1976) has shown, values of Conrad's continentality index peak just to the west of London. By permitting a greater diurnal range of temperature, this relative continentality means that daytime temperatures compare favourably with those of the rest of Britain, even the South-West. This also holds true for much of the year, not just

Table 3.1 Mean monthly maximum and minimum temperatures (°C) for the period 1961–90

Location	*Alt. (m ASL)*	*Jan.*	*Feb.*	*Mar.*	*Apr.*	*May*	*June*	*July*	*Aug.*	*Sept.*	*Oct.*	*Nov.*	*Dec.*	*Year*
Rothamsted	128	5.8	6.1	9.0	11.8	15.6	18.8	20.9	20.8	18.0	14.0	9.0	6.7	13.0
		0.3	0.0	1.6	3.4	6.2	9.0	10.9	10.9	9.2	6.7	2.9	1.3	5.2
St James's Park	9	7.4	7.7	10.3	13.2	17.1	20.4	22.3	21.9	19.2	15.4	10.5	8.3	14.5
		2.5	2.6	3.8	5.7	8.8	11.7	13.8	13.5	11.5	9.1	5.2	3.4	7.6
East Malling	33	6.8	7.1	9.8	12.4	16.3	19.5	21.6	21.5	18.9	15.1	10.2	7.8	13.9
		1.2	1.2	2.4	4.2	6.9	9.8	11.9	11.6	9.5	7.0	3.6	2.1	5.9
Southampton	3	7.7	8.0	10.4	13.3	16.8	19.9	21.9	21.6	19.3	15.6	11.1	8.7	14.5
		2.2	2.4	3.3	5.1	8.1	11.0	13.0	12.9	10.9	8.5	4.7	3.1	7.1
Hastings	45	6.7	6.7	8.8	11.4	15.0	17.8	19.8	20.0	18.0	14.8	10.3	7.8	13.1
		2.5	2.3	3.4	5.2	8.4	11.2	13.3	13.6	12.0	9.6	5.5	3.5	7.5

Note: ASL = above sea level

the summer. The region's setting at the south-eastern extremity of the British Isles not only implies a greater proximity to the influence of the European landmass, but also places the area relatively close to the mean position of subtropical anticyclones and relatively distant from mean depression tracks. Tropical air masses can persist longer here than further north in Britain; tropical continental air often reaches this region first, and tropical maritime or continental air often lingers here before the arrival of a south-eastward-moving cold front, such as in early August 1990.

Table 3.1 shows mean monthly maximum and minimum temperatures. The warmth of St James's Park is clearly shown, showing that London's heat-island counteracts the inland winter coolness quite early in the year. Delayed seasonal warming is evident at Hastings, where the maxima are lower than those at Rothamsted from March to August, despite altitudinal and latitudinal differences. The coolness of Kent in spring is usually attributable to the greater frequency of north-east winds cooled by passage over the North Sea. These situations are likely to give bright sunshine and higher temperatures in a sheltered location such as Southampton. The latter city manages to rival London's temperatures right through the year; it is sufficiently close to the English Channel to enjoy relatively mild winters but lacks the open exposure to the sea that would inhibit the rise of daytime temperatures in spring and summer. By contrast, eastern coasts often enjoy the warmest temperatures in early autumn as a combined result of maritime heat retention and the increasing frequency of westerly winds.

The region holds the United Kingdom maximum temperature record for the months of April to July, together with that for October (Table 3.2). Burt (1992) has pointed out that the record July temperature at Epsom in 1911 was almost certainly recorded in a Glaisher stand and that the maximum temperature that day was probably close to 35°C. The highest temperature observed under standard conditions was 36.5°C at Heathrow, on 3 August 1990.

The varying 'synoptic origin' of high day temperatures is depicted in Figure 3.2. The regional airflow type on the warmest day of each year from 1961 to 1990 was noted. The anticyclonic type occurs frequently at each site, probably as a result of subsidence of air within an anticyclone, suppressing sea breezes and thus promoting general warmth, even on coasts. This was particularly

Table 3.2 Locations in south-east England holding the United Kingdom monthly maximum temperature records, 1875–1994

Month	*Temperature °C*	*Location*	*Date*
April	29.4	Camden Square, London	16 April 1949
May	32.8	Camden Square, London	22 May 1922
June	35.6	Mayflower Park, Southampton	28 June 1976
July	36.0	Epsom, Surrey	22 July 1911
October	29.5	Waddon, south London	1 October 1985

Source: Adapted from Webb (undated)

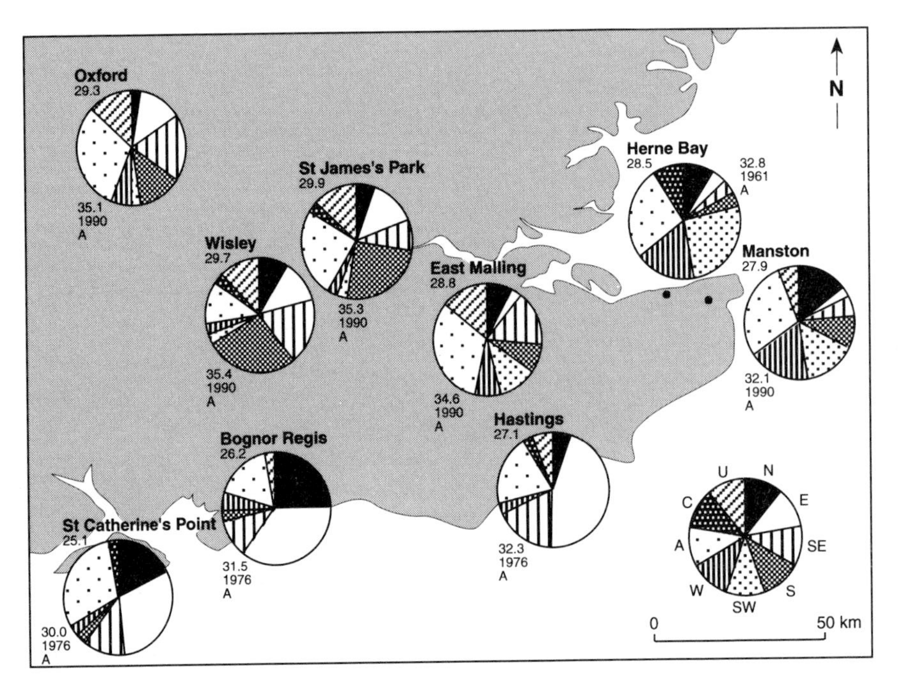

Figure 3.2 Synoptic origin of maximum temperatures at different locations in south-east England (the average annual extreme maximum temperature is shown below each station name; the highest temperature 1961–90, year of occurrence and prevailing airflow type are shown below each circle)

noticeable in 1976, when south coastal resorts recorded their highest temperatures of the period. Contrasts between stations illustrate the effect of local shelter. While SW and W types account for as little as one day at Hastings and St Catherine's Point (usually indicative of onshore winds), they provided 43 per cent of warm days at Herne Bay, where the S type was surprisingly insignificant. SE and E types were significant at all stations except Herne Bay, and Manston. The days with northerly airstreams were characterised by light and variable winds ahead of cold fronts approaching from the north-west.

Although the south-east did not attain the extremė highest temperatures of 1976 and 1990, summer warmth is often more consistent and reli-

Table 3.3 Hottest months at Heathrow since 1950, in order of mean maximum temperature (° C), with details of the other highest UK temperatures at other sites

Month	*Year*	*Heathrow mean daily maximum*	*Heathrow monthly mean*	*Other UK sites, absolute maximum*	*Other UK sites, highest monthly mean*
July	1983	27.6	21.8	27.1 (SJP)	21.9 (SJP)
July	1976	26.6	20.7	26.8 (NHth)	21.1 (SJP)
July	1995	26.3	20.8	—	—
July	1994	26.2	20.7	—	—
August	1990	26.0	20.3	26.1 (Kew)	21.1 (LWC)
August	1975	25.9	20.5	25.5 (SJP)	20.5 (SJP)
July	1989	25.8	20.3	26.1 (NHth)	20.4 (SJP)
July	1976	25.5	19.6	25.6 (SJP)	19.8 (SJP)

Key: SJP, St James's Park; NHth, North Heath; Kew, Kew Royal Botanic Gardens; LWC, London Weather Centre

able here than elsewhere in the British Isles. Figure 3.2 shows that inland stations can expect the temperature to exceed 30°C one year in two, and the temperature in London has exceeded 25°C in each year since 1860. However, in the hottest months, mean maxima reach 25°C with mean temperatures close to 20°C. Table 3.3 shows that the highest monthly means are usually found in London, and that Heathrow is not significantly cooler than central London. The exceptional heat of July 1983 is clearly shown, and this was also the occasion of the highest mean minimum (of 16.6°C at St James's Park). The warmest night on record at a ground-level site in the British Isles is probably that of 3 August 1990, when the minimum at Brighton was 23.9°C (Burt 1992).

Causes and Incidence of Frost

When a cold air mass affects Britain in winter, the region is often one of the coldest parts of the country, especially by day. Even coastal areas feel the effects of the proximity of the Continent, which is usually markedly colder than south-east England during spells of polar continental weather types. Figure 3.3 shows that Hastings, for example, with an annual mean of 24.6 air frosts, is just as vulnerable as St James's Park (mean annual 24.8 air frosts). Such inland locations are generally thought of as being more prone to frosts than coastal sites but on this occasion the consequences of the inland location are probably offset by the urban heating discussed in Box 3.1. The area does not have extensive areas into which cold air can drain (unlike large frost hollows such as those of north Shropshire). The best example is probably the plain below the scarp face of the Chilterns in Oxfordshire and Buckinghamshire where Grendon Underwood recorded the region's lowest temperature of −21.6°C in January 1982. This site is one of the region's frostiest, with an annual mean of 75.9 days, and only June, July and August are frost-free. Nearby Benson and Wallingford are also frosty, and record yet lower temperatures following the relocation of the thermometer screen at Benson early in 1995 into a more frost-sensitive location. Sandy hollows, where soils are predominantly dry, such as the Blackwater valley and dry valleys on the Chalk downland, are also frost-prone.

In the mildest years, coastal stations record little, if any, air frost. Southsea escaped air frost in 1974, and several locations had only one in 1991. While Ryde was frost-free in 1984 (along with Eastbourne), Hosking (1978) noted that in the severe months of January and February 1963, the town recorded air frost on fifty-five out of fifty-nine nights. The variability of frost from severe to mild winters is thus more marked in the mildest, coastal

Box 3.1

LONDON'S HEAT-ISLAND

London's heat-island was identified by temperature measurements by Luke Howard at the beginning of the nineteenth century. Since then, it has been the subject of extensive research, most notably by Chandler (1965). In general, heat-island intensity is related to city size, with urban morphology and population being other important factors. It is therefore reasonable to assume that it is stronger and more persistent around London than any other British city. Mean temperatures are on average about 1.5 degrees C higher in central London than in the surrounding countryside. St James's Park is now the only conventionally sited (ground-level) station in inner London, whilst published 'central London' temperatures are roof-top observations at successive sites of the London Weather Centre (since 1994 from a roof-top automatic weather station). This raises the question of which type of site is more representative of the urban environment. Mobile surveys by Chandler (1965) indicated that the peak of the heat-island usually lies north-east of central London in Hackney and Islington, a reflection of the dense urban development in these districts, possibly augmented by a north-eastward displacement of the heat-island by south-westerly winds.

The urban heating effect is caused by changes in the energy and radiation balance as vegetation is replaced by buildings and the emission of anthropogenic heat. The feature is most conspicuous overnight, partly because of the retention of heat in the urban fabric; Chandler (1965) quoted an average daytime heating effect of +0.9 degrees C compared with a nocturnal temperature difference of +1.9 degrees C. A further influence is the general weather situation; light winds, a stable atmosphere and lack of cloud maximise temperature contrasts. The heat-island is thus most conspicuous in cold spells in winter and in summer heatwaves. Indeed, perhaps the most inconvenient facet is the discomfort of warm nights in summer. The warmest night on record was that of 3/4 August 1990 with a minimum temperature of 24.0°C at London Weather Centre. However, at St James's Park the temperature fell to 19.7°C, and the warmest night here remains that of 29/30 July 1948 with a minimum of 74°F (23.3°C). Lee (1992) has shown that the heat-island is becoming stronger at night but weaker during the day (though this may in part be a consequence of a small site change at his 'rural' site – Wisley). Lee also found that the urban–rural temperature difference nearly doubles (in relation to average) in the warmest summers, a finding that may have implications for future nocturnal heat-stress. On a more positive note, Chandler (1965) estimated that the urban heating led to a 20 per cent saving of fuel costs in central London.

parts of the region. Of the three decades since 1961, the 1960s was the frostiest and the 1970s the least frosty; at St Catherine's Point the decadal average dropped from twenty-five nights a year to less than twelve. Frost frequency was also below the 30-year average in the 1980s.

The area has its share of smaller frost hollows, usually found on the dip slopes of either the Chilterns or the North Downs. Ironically, both slope towards London, allowing cold air to accumulate on the margins of the capital. In the famed (or notorious) Rickmansworth frost hollow, scrutinised by Hawke (1944) in the 1930s, air frost could be observed in any month of the year. This site also boasted the largest daily temperature range observed in the British Isles from a minimum of 1.1°C to a maximum of 29.4°C within nine hours on 29 August 1936.

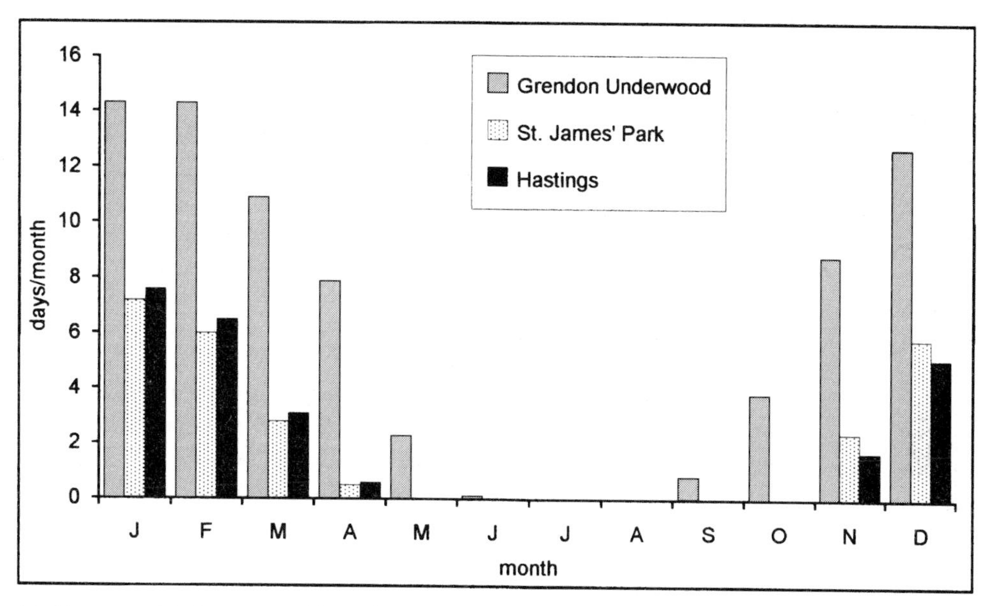

Figure 3.3 Mean monthly frequency of frosts for the period 1961–90

The dry valleys of the North Downs around London's southernmost suburbs have received attention (Harrison and Currie 1979), and more recently a notable frost-hollow has been discovered at Chipstead, in a partially urbanised but sheltered dry valley, where the frequency of air frost rivals that of the Scottish Highlands (Currie, 1994; Currie, personal communication). Surprisingly, Chandler (1965) did not report on any frost hollows within the London area. However, where the largest open spaces coincide with valleys, the potential for cold air drainage will be realised. This has recently been identified where cold air drains northwards over Wimbledon Common and Richmond Park, resulting in sharp temperature contrasts with the densely settled adjacent district of Putney.

There have been some noteworthy cases of low daytime maximum temperatures at coastal resorts in the coldest winters, even when overnight minima are unremarkable. On 12 January 1987, for example, the temperature remained below −9°C all day at both Southend and Warlingham; the presence of a 38-cm snow cover at Southend was clearly significant in minimising surface heating by reflecting away most incoming solar radiation.

PRECIPITATION

Not only is south-east England relatively distant from the average route of Icelandic depressions, but much of it is sheltered from prevailing west-south-west winds. On average, this shelter reaches its full potential around the Thames estuary (often the driest part of the whole country), as loss of moisture is aided by a moderate rain shadow effect accompanying the descent from the North Downs. Average annual rainfall totals thus range from a little over 1,000 mm over the South Downs to half this amount around the Thames estuary. Much of the chalk hill country has 700–900 mm a year, whilst the Thames valley averages about 600–650 mm. The South Downs are wetter than the North Downs as a result of greater proximity to the moisture source of the English Channel. There is little evidence for an eastward decrease in rainfall along the south coast, where precipitation is more a function of exposure and relief. For example, the driest parts of the south coast are found to the lee of the Isle of Wight, where Southsea averages 685 mm, and the low-lying Dungeness Point has a similar total.

The significance of exposure to the south-west and relief were confirmed by Stone (1983), whose studies revealed that the mean daily rainfall receipt from cyclonic SW airstreams increased from 2 mm around the Thames estuary to over 7 mm around the New Forest. Cyclonic W and S airstreams also gave twice as much rain over the Weald of East Sussex as around the Thames estuary. By contrast, in cyclonic northerly and north-easterly situations the driest areas were south Hampshire and around Dungeness.

The greater the prevalence of south-west winds, the more conspicuous the wetness of the southern coastal counties. A blocked situation, with abundant northerlies or easterlies, gives a more uniform distribution to rainfall across the region. It follows that the seasonal variation in the frequency of westerly winds helps to determine the seasonal variation in rainfall. Frontal rainfall, for example, is more frequent, heavier and more likely to approach from the south-west in autumn and winter. This seasonal intensification of the westerly circulation leads to a subtle backing of the surface wind towards the south-west, increasing the exposure of the south coast. As Table 3.4 shows, this leads to a peak in monthly rainfall in autumn and winter for southern coastal locations. Summer rainfall, more dependent on the contribution from convectional sources, however, shows a slightly different pattern, and tends to decrease southwards despite an increase in annual totals.

A feature of the 1961–90 averages is the large area of southern England that has a minimum in July. Further north, in the Thames valley, the driest month is still February, by a margin of over 10 mm, whilst Goudhurst, on the high ground of the Weald of Kent, is as dry as London in July, despite being 32 per cent wetter over the year as a whole. Bexhill is 58 per cent wetter than London in November but nearly 10 per cent drier in July. Summer rainfall (June to August) contributes over 25 per cent of the annual fall in London and Oxfordshire northwards but as little as 18 per cent on the Isle of Wight, the latter being the lowest proportion in the British Isles. Conversely, the 61 per cent contribution of the winter half-year (October to March) on the Isle of Wight is the largest such value, and exposed coastal sites such as Ryde (Table 3.4) can record annual totals more commonly associated with higher ground in the region. The proximity of high pressure over or to the south of the area in many summers in the 1970s and 1980s has had a marked effect in decreasing summer rainfall (Mayes 1991). Consequently, the melancholy scene of parched landscapes noted by Southern (1976) has reappeared in many subsequent summers, several of which have accumulated significant water deficits (Woodley 1991).

Notable Rainfall Events

The relative importance of different types of rainfall is markedly different from that in north-west

Table 3.4 Average monthly and annual precipitation (mm) for the period 1961–90.

Location	*Alt. (m ASL)*	*Jan.*	*Feb.*	*Mar.*	*Apr.*	*May*	*June*	*July*	*Aug.*	*Sept.*	*Oct.*	*Nov.*	*Dec.*	*Year*
St James's Park	8	53	36	48	47	51	50	48	54	53	57	57	57	611
East Malling	33	62	41	49	46	47	50	45	48	60	60	67	65	640
Goudhurst	85	86	57	65	55	54	55	48	58	70	87	88	86	809
Bexhill	4	77	50	56	47	45	48	44	49	64	82	90	77	729
Southsea	2	74	50	57	45	47	42	37	52	60	71	75	75	685
Ryde (IoW)	4	86	57	62	48	50	45	40	55	66	82	86	84	760

Britain. The dominance of frontal rainfall is tempered by distance from the average route of Atlantic depressions, whilst the frequency of convectional rainfall is increased by local heat sources. Conditions favouring the heaviest frontal rainfall are a southward shift of depression tracks, onshore winds and relatively high sea surface temperatures. This was illustrated in October 1939, a month in which a series of depressions tracked along the English Channel, resulting in a bias towards easterly winds. Consequently, the area was exceptionally wet, with over 250 mm of rainfall in east Kent, and more than 150 mm in most other areas. Such a track appears to have become less frequent over recent decades with a consequent reduction in the number of heavy daily rainfalls in the region after the 1920s or 1930s (Tyssen-Gee 1981; Howells 1985).

Two exceptionally heavy rainfall events occurred in autumn 1980, with daily rainfall totals reaching a maximum in north Worthing of 112 mm and 133 mm (Potts 1982). Both falls exceeded the previous record 24-hour fall for the area and were associated with slow-moving occluding depressions. The intensity of such storms in autumn can be attributed to the interplay of high sea surface temperatures, uplift over the South Downs and slow movement of weather systems. The 1980/81 winter season actually produced four such storms in that part of West Sussex, and autumn 1982 produced sustained heavy rainfall over a wider area. This led to rill and gully erosion and sheetwash of exposed arable soils on the South Downs (Browne and Robinson 1984), bringing into question the appropriateness of modern farming practices on sloping terrain.

The cumulative effects of persistent wet weather in the winter of 1993/94 had serious flooding consequences that focused public attention on the wisdom of human flood plain development. After a particularly wet week at the end of December 1993, the river Lavant flooded Chichester in early January 1994; the cost of the flood damage to West Sussex County Council was £1.9 million (Holmes 1995; Taylor 1995). It was the wettest December in Sussex since 1959; 229.1 mm of rain fell at Plumpton, where the September 1993 to January 1994 total was 807.8 mm.

The outstanding example of convective activity accentuating frontal rainfall was the three-day storm of 14–16 September 1968 (Bleasdale 1970). This comprised a series of convectional storms that developed *in situ* whilst anchored to a stationary occluded front that provided larger-scale convergence. The combined effects of these processes resulted in a rainfall distribution that had a disregard for altitude; the highest totals, of more than 200 mm, occurred around Tilbury in Essex, where such a fall represents nearly 40 per cent of the average annual rainfall. This peak may have been caused by the warmth of the adjacent Thames estuary and local airflow convergence (Jackson 1977).

The seasonal distribution of convectional rainfall shows a well-developed peak in the summer half-year (Figure 3.4). Frequencies of thunderstorms are rather greater in the north of the region and lower in coastal localities, and there is also a marked diurnal peak in the afternoon (Davis 1969). Thunder is heard more often towards the coast in the latter half of the year in association with thundery showers in unstable polar maritime air masses.

The importance of thunderstorm rainfall is demonstrated by the proportion of Britain's heaviest short-period falls that have occurred within the region. Jackson (1979) found that five of the twenty-five heaviest 5-hour falls from the start of the century were in south-east England and as many as eight of the top twenty-five 3-hour falls. The second highest fall in the shorter period was the Hampstead storm of 14 August 1975, when 169 mm fell in 2.5 hours (Keers and Wescott 1976), three times the previous highest daily fall at Hampstead (Tyssen-Gee 1976). This was an excellent example of a single-cell (non-travelling) storm developing on a convergence line. The location of the primary storm was crucially related to topography, additional stimuli being London's heat-island (Atkinson, 1968) and the uplift provided by London's highest point, which is at Hampstead itself.

Despite the dense network of rain gauges around

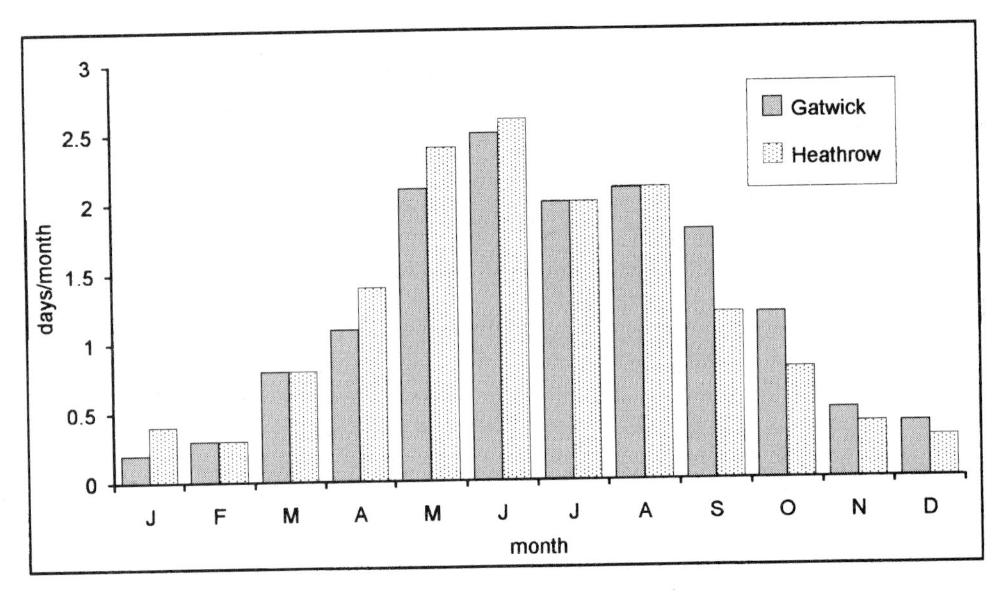

Figure 3.4 Mean monthly frequency of days with thunder for the period 1961–90

London, it remains possible that many other storms may have had cores giving totals well over 100 mm that have gone unrecorded. Slow-moving thundery showers on 22 September 1993 gave a gradient of 46–48 mm on the Wimbledon ridge area to 3 mm two miles to the north. A dramatic storm formed over the Thames valley on 24 May 1989. The satellite image reproduced in Figure 3.5 clearly shows the localised storm cell over the area at this time. Rainfall amounts were highly variable, and appear to have been influenced by warm south-facing slopes initiating small cells of very intense rain (Davis 1991), possibly augmented by urban heating effects. There is even evidence for a possible micro-burst near Farnborough, a downward eruption of diverging air associated with a thunderstorm downdraught (Waters and Collier 1995). The escarpments of the Berkshire and Oxfordshire Downs may have accentuated rainfall totals in the thunderstorms on 26 May 1993 (Pike 1994).

There are many cases of thunderstorms breaking out quite locally over or close to north London (Haggett 1980). An interesting cluster of severe, often localised, storms has also been noted around Epping Forest, in south-west Essex. Prichard (1987) has listed conditions that appear to favour the development of storms in this area, which include a slack surface trough and the presence of stagnant hot air of southerly origin (possibly suggesting storm initiation over London in some cases). Hot southerly airstreams can also advect travelling storms across the area after formation over France. Such events are often associated with the phenomenon now known as the 'Spanish plume', the name given to areas of warm, moist air on the eastern side of upper troughs. Recent case studies include the storms of 20/21 July 1992 (McCallum and Waters 1993) and 24 June 1994 (Young 1995; Prichard 1994). However, the summers of 1989 and 1990 showed that warmth does not necessarily induce widespread thunderstorms if the summers are anticyclonic.

In late spring and early summer, the region can have particularly unstable air when polar maritime air reaches the region and when surface temperatures remain relatively high. This instability not only makes the area prone to convectional rainfall but can also lead to damaging outbreaks of hailstorms, especially when a cold front introduces colder air from the south-west or west. Webb *et al.* (1984) have found that London and Surrey have the highest

Figure 3.5 NOAA infra-red image at 1244 GMT for 24 May 1989 showing storm cells over southern England
Photo by courtesy of University of Dundee

risk of severely damaging hail in the country. Two outstanding storms both occurred in June: at Tunbridge Wells in 1956 in a polar maritime airstream (Ogley *et al.* 1993) and along the south coast in 1983 under an upper cut-off low (Wells 1983). The heaviest known hailstone in Britain, with a weight of 191g, fell at Horsham on 5 September 1958.

A more dramatic manifestation of instability is the formation of tornadoes. The most damaging tornado for which eye-witness accounts are available occurred on 8 December 1954 when a tornado tracked across west London from Brentford to Acton. Numerous vehicles and houses were damaged, and twelve people were injured (Brazell 1968). A tornado that crossed the Royal Horticultural Society's Garden at Wisley on 21 July 1965 at least partly uprooted 179 fruit trees, and another fifty-six were badly damaged (Gilbert and Walker 1966).

Snowfall

Long-term averages suggest that in a normal year the number of days with lying snow (at 0900 GMT) varies from less than five at some coastal sites to over fifteen in the Chilterns, where Rothamsted's annual mean is 13.8 days (Figure 3.6). The lack of lying snow close to the English Channel is noteworthy and varies between only 2.9 days at St Catherine's Point on the Isle of Wight to 7.6 days at Hastings. Kew is generally representative of London, with an annual mean of 7.6 days of snow. However, the usefulness of such data is restricted by the variability of snow from year to year and between different localities. Between 1939/40 and 1990/91, no snow cover was reported in ten winters at Kew and Hastings (including the three consecutive winters 1987/8 to 1989/90) but in only three at Goudhurst. By contrast, in the winter of 1962/63, snow lay continuously from 26 December to 8 March over most of the region, even at Southampton (Barry 1964). The lack of snow in the late 1980s contrasts with the relatively lengthy snow cover in the three winters from 1984/85 to 1986/87. Two of the five snowiest winters since 1940 occurred in this spell in the mid-1980s. The data suggest that snowfall has been erratic rather than scarce in the 1980s.

Clarke (1969) analysed the synoptic origin of snowfall in south-east England and found that 40 per cent was associated with easterly winds, 27 per cent fell ahead of warm fronts, and the remainder was brought on north-easterly airstreams in showery form. Since the early 1970s, the region has escaped many of the major frontal snowfall events. This is in marked contrast to the late nineteenth century when depressions tracking close to the English Channel caused widespread blizzards. The worst of the century was probably that of 18/19 January 1881.

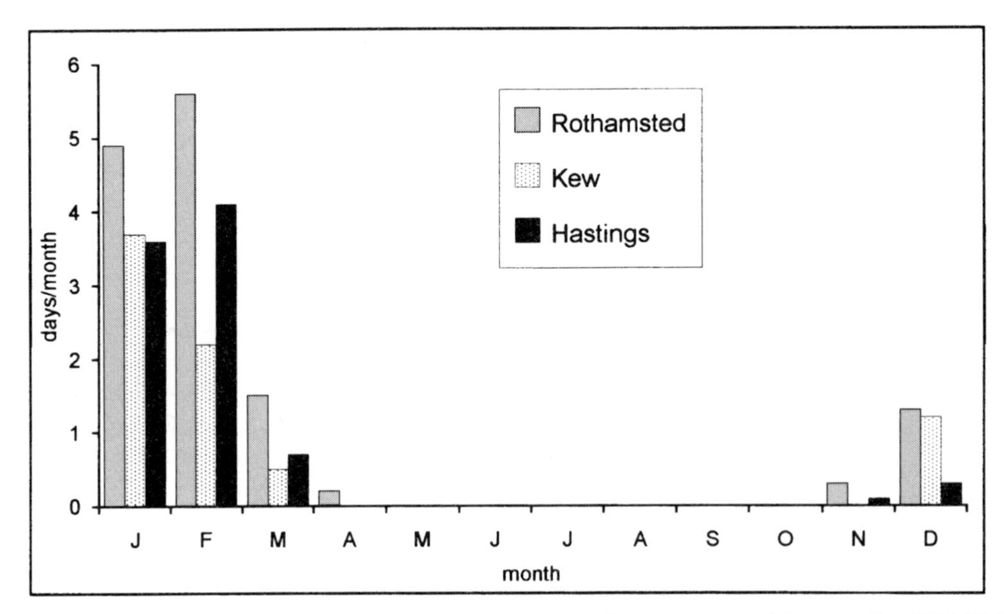

Figure 3.6 Mean monthly frequency of snow lying and covering more than 50 per cent of the ground at 0900 GMT for the period 1961–90

Drifts of snow were reportedly as great as 4.6 m in central London. A similar depression track in December 1927 gave drifts of 5.5 m at Biggin Hill on the North Downs, and local villages were cut off for a week. The conclusion to the snow period was equally disruptive and fourteen people drowned in the flooding that followed, caused by the combination of snow melt and high tides.

The blizzards of February 1978 and January 1982 in the south-west (see Chapters 2 and 6) failed, however, to bring significant snow across the region as a whole. Snow showers can produce significant falls, especially in Kent, which is exposed to unstable north-easterly winds blowing onshore. Between 11 and 14 January 1987 frequent heavy showers gave level depths of more than 35 cm on both sides of the Thames estuary and drifts to 6 m near Sittingbourne. The greatest level depth recorded at that time was 52 cm at East Malling, probably the greatest depth at this low-level station (33 m) for forty years. This event illustrated how effectively snow showers can accumulate, possibly as a result of local convergence and instability over relatively warm estuarine waters. Comparison of the snowfall records of Goudhurst (85 m) with those of Little Butser in Hampshire (183 m) for 1981–90 indicates that the lower-lying Kent station recorded an annual average of 12.6 days with snow whilst Little Butser averaged 12.1 days. Exposure to onshore winds carrying snow showers is thus seen as a critical factor in the distribution of snow cover in the region, which suggests that Kent possibly deserves its wintry reputation.

WIND

Inland parts of south-east England have the lowest average wind speeds and frequencies of gales of any part of the United Kingdom (Shellard 1976). The climate could thus be regarded as being the least dynamic of any in the country. Closer study of data from anemograph stations indicates another facet of the south-west to north-east division within the region. Parts of the south coast average over 10–15 days with a gale per year, whilst most inland areas average well under one per year. The extent of the contrast arises from the fact that on the windiest days of the year, speeds may reach gale force at the most exposed coastal stations whilst inland stations will experience just gusts of wind to gale force (a gale being defined as a period of at least ten minutes with a wind speed of 34 knots or more). Data for contrasting sites are summarised in Table 3.5. While Heathrow had only three gales within this period, the highest gusts were often comparable to those at coastal locations.

Shoreham-by-Sea has a frequency of gales and a mean wind speed comparable to those of several coastal sites in northern Britain; it recorded the strongest gust in the October 1987 storm. Most years in the 1980s were actually calmer than the

Table 3.5 Summary of anemograph records at Heathrow, Manston and Shoreham-by-Sea

	Heathrow	*Manston*	*Shoreham-by-Sea*
Period of record	1968–90	1968–90	1981–90
Mean annual wind speed	8.2/6.5	10.8/10.0	11.8
maximum gust, 1968–90/1981–90	53/55	61/63	71
Average number of hours with gales, 1968–90/1981–90	0.2/0.3	8.4/6.3	111.1

Note: All data in knots

Figure 3.7 Storm damage at Templeton Hall of Residence, Roehampton Institute London, 25 January 1990. One of a pair of cedar trees felled by the westerly gale at a site having a greater exposure to this direction; it was relatively sheltered from the southerly gale of 16 October 1987
Photo by courtesy of Julian Mayes

1970s; although 1987 has acquired a certain notoriety for strong winds in the region, it was actually the calmest year of the study period at Heathrow and Manston.

The frequency of north to north-east winds peaks sharply in April as the vigour of the mid-latitude westerlies abates. While this feature is found elsewhere in the British Isles in spring, it has greater significance for the east of the region owing to its exposure to winds from this direction, an effect amplified by the coolness of the North Sea at this time of year. Unlike in north-east England, where easterly winds have a long sea track, the influence of the North Sea for Kent is maximised by northerly and north-easterly winds. By virtue of their greater instability, these winds are more likely to produce showers, but early-morning stratus is sometimes a feature of the more stable airstreams. These northerly and north-easterly winds are usually associated with a blocking anticyclone at a relatively high latitude, typically centred just to the north-west of the British Isles. The resulting lack of warmth and sunshine can reverse the usual north-west to south-east climatic gradient across the country, as happened in the summers of 1968 and 1977 (Hughes 1979).

The land–sea temperature difference in spring means that coastal temperatures are very sensitive to changes of wind direction. The long land track of west or south-west winds means that these can be relatively warm by the time they reach the east coast. Spells of springtime westerly weather can thus produce high temperatures in those same areas that bear the brunt of the chilling north-easterlies. For example, Elmstone, near Canterbury, reached 20.7°C on 28 March 1989.

The occasional savagery of north and north-west winds was starkly demonstrated by the North Sea surge of 31 January 1953, when a depression crossing the North Sea started to deepen. The resulting tidal surge led to flooding of up to 2.7 m in the coastal towns of north Kent; it is fortunate that the high tide in London did not coincide with the storm surge as it did on the east coast (Eden 1995; Chapter 4, this volume). The severest loss of life was on the northern side of the Thames estuary at Canvey Island, where fifty-eight people were killed.

Sea Breezes

The dominant sea breeze in the region is that from the south coast, which in its most well-developed form can penetrate inland to affect much of the region (Simpson 1964). Analysis of seasonal wind direction frequencies at Portsmouth showed that southerly flows between 1100 and 1700 hours were about three times as frequent in summer as in winter (Harrison 1976). Conversely, the frequency of northerly winds at 0700 hours in summer was more than double that of winter, indicative of the nocturnal land breeze, possibly reinforced by katabatic drainage from the South Downs.

Evidence for the westward advance of sea breezes over the Thames estuary is scarce, though it is often

Box 3.2

THE LEGACY OF THE STORMS OF 1987 AND 1990

There can be few single weather events that have provoked such a response from both the meteorological community and the public as the 'Great Storm' of 16 October 1987. Remarkable for its concentrated impact over a small part of the United Kingdom, it was said to be the severest storm in the south-east since that of 1703.

A key feature of the storm was the rapid deepening of a depression on 15 October as it passed north-eastwards over Biscay (Burt and Mansfield 1988). The energy of the system was related to the warmth of both the sea surface and of the air in the warm sector: 18°C was observed at St

Catherine's Point at 0000 GMT on 16 October and Heathrow had a record rise in temperature of 7 degrees C over the next hour (Advisory Services Branch 1988). Highest gusts were observed just after the clearance of the cold front; the highest in the United Kingdom was 100 knots at Shoreham-by-Sea with an hourly mean of 74 knots. The resulting swathe of destruction included the loss of 24 per cent of the woodland of East Sussex (Quine 1988) and caused the most widespread loss of electrical power since the Second World War.

Despite the extensive damage at Kew Gardens in 1987, the maximum gust of 59 knots was exceeded as soon as 1990 when 71 knots was logged in the 'Burns Day' storm of 25 January. After the loss of so many vulnerable trees in the first storm, the 1990 storm did not produce as much devastation of the natural environment as the 1987 storm, and hence failed to achieve the same notoriety in public perception. However, an important difference was that it was at its worst during daylight hours; as a result, road traffic was severely disrupted, leading to near-'gridlock' conditions in London.

While the clearance of felled timber from the 1987 storm continued well into the 1990s, the legacy of broadened vistas provided a compensation for the damage endured. There have also been associated ecological opportunities in modified habitats; for example, it was reported in 1994 that the population of nightjars had increased by over 50 per cent since the mid-1980s.

Figure 3.8 Damage in Meadvale, Surrey, in the aftermath of the Great Storm of 16 October 1987
Photo by courtesy of Ian Currie

noted that east London is a little less warm than west London when sea breezes are present. As Manley (1952) noted, a weak sea breeze is sometimes noticeable in the late afternoon in London, providing welcome, if subtle, relief from the heat of hot summer days.

SUNSHINE

South-east England is, on average, the sunniest part of mainland Britain. Average annual sunshine values are, however, greatest in the Channel Islands, followed by the Isle of Wight (Simmons 1978). The sunniest place in the British Isles over the 1961–90 period has been Jersey in 11 years, Guernsey in six years, various stations on the Isle of Wight in five years. The 2,242 hours at Bognor Regis in 1990 appears to have been the first value exceeding 2,200 hours to have been recorded on the mainland. This forms part of a remarkable series of years in which that resort has been the sunniest place on the mainland each year from 1983 to 1991. Other sunshine peaks this century have included 1911, when Eastbourne recorded 2,158 hours (including the sunniest month on record in July when 384 hours were observed), and 1949, when the Isle of Wight recorded its sunniest year with 2,263 hours at Shanklin.

London sunshine shows a remarkable increase through the twentieth century owing to reduced atmospheric pollution (see Box 3.3, below). Notable studies of individual records include Jenkins (1969) for central London, Tyssen-Gee (1982) for Hampstead and Hatch (1981) for Kew Observatory. Bilham (1938) reported losses of bright sunshine in central London of 80 per cent in relation to Kew for winters from 1881 to 1885. This was followed by London's dullest month on record – December 1890 – when totals included zero at Westminster and only 0.1 hour at Bunhill Row in the City (Brazell 1968). As recently as 1921–50, central London showed a loss of over 50 per cent for winter. Indeed, the main improvement in winter sunshine coincided with the introduction of smokeless zones in the 1950s in inner London (Chandler 1965). The Kingsway/London Weather Centre record shows an increase in winter sunshine of 41.5 per cent between 1931–60 and 1958–67, whilst Rothamsted actually had a marginal decrease of 9 per cent (Jenkins 1969).

The 100-year record of sunshine at Kew Observatory analysed by Hatch (1981) has been updated by the observations at Kew (Royal Botanic Gardens). The new site is officially described as 'shaded' for the months of September to March, and this would tend to underplay the effects of increasing sunshine. In addition, owing to its suburban location, Kew did not suffer the loss of sunshine to the extent of central London. Despite these factors, the increasing trend of sunshine has continued. Monthly sunshine records were equalled or exceeded in no fewer than five months of the year in the first nine years of the Kew (Royal Botanic Gardens) record (Table 3.6).

The 1961–90 average annual value for the composite Kew record is as high as 1,564 hours, and this includes values for 1981 that were anomalously low in relation to surrounding stations. This compares with 1,496 hours for 1881–1980 at the observatory. Table 3.7 shows the extent to which the increase in sunshine has been concentrated in the autumn and winter. For November, the 1961–90 average of 69.3 hours is 20 per cent greater than that of 1941–70. Up to 1965 the highest November sunshine total on record was only 79 hours, a value that had subse-

Table 3.6 Months for which the 1881–1980 sunshine records (hours) of Kew Observatory were exceeded at Kew (Royal Botanic Gardens) for the period 1981–90

Month	*Total sunshine*	*Year*	*Previous record*	*Year*
January	92	1984	82	1952
February	115	1988	110	1970
April	245	1984	239	1909
May	320	1990	315	1909
August	279	1989	279	1947
November	104	1989	103	1971

Table 3.7 Mean monthly and annual sunshine totals (hours) for the period 1961–90

Location	*Jan.*	*Feb.*	*Mar.*	*Apr.*	*May*	*June*	*July*	*Aug.*	*Sept.*	*Oct.*	*Nov.*	*Dec.*	*Year*
Hastings	60	81	124	168	217	226	220	215	160	123	81	56	1,731
Heathrow	52	63	111	147	193	200	194	187	145	107	68	46	1,513
Kew	52	70	113	152	202	204	202	198	148	110	69	48	1,568
% change at Kew 1880–1980 to 1961–90	+9	+15	+3	+1	0	−1	+2	+4	+2	+11	+23	+19	+4.6

Table 3.8 Average annual sunshine totals (hours) in south-east England

Period	*Heathrow*	*Kew*	*Kingsway/London Weather Centre*	*Rothamsted*	*East Malling*
1931–60	—	1,516	1,362	1,472	1,556
1951–80	1,498	1,560	1,541	1,443	1,563
1961–90	1,513	1,568	1,538	1,461	1,567
1982–90	1,582	1,611	1,600	1,552	1,629

quently been exceeded on ten occasions by 1995. The improvement in central London has been even more conspicuous, from an average of 38 hours in 1921–50 to 70 hours in 1961–90, an increase of 84 per cent (Eden 1995).

Eight of the Februaries between 1981 and 1990 exceeded the 1881–1980 average of 61 hours. Such a concentration of 'sunny' months within a fairly short period is likely to be caused partly by natural variations of weather types. Similarly, the absence of increased sunshine in June may be due to such synoptic changes. A decrease in the clarity of the summer atmosphere, perhaps because of pollution, especially from vehicle exhausts, may also be influential here (Bonacina 1960; Lee 1985; Gomez and Smith 1987).

Annual totals at Kew for both 1989 and 1990 exceeded 1,900 hours, beating the previous sunniest year of 1959, which had only 1,852 hours. The long summer (May–September) of 1989 had 1,290 hours compared with just 1,212 hours in 1959. The total for the three-month summer was 838 hours, beating the previous record of 824 in 1976 and substantially above the 1881–1980 average of 589 hours. Both 1989 and 1990 were clearly exceptional years for sunshine, and nowhere was this more applicable than in south-east England – both in absolute and relative terms (Brugge 1991).

London's sunshine has now 'caught up' with that of the rest of the region; sunshine now simply increases steadily towards the southern and south-eastern coasts (Table 3.8). From 1931–60 to 1961–90, sunshine increased by 13 per cent in central London but remained unchanged in the rural south-east.

An excess of sunshine in central London occurs in cold foggy months when the warmth of the heat-island encourages the evaporation and dispersal of fog. For example, in November 1988 the London Weather Centre recorded 13 per cent more sunshine than Kew. On 15 November, fog cleared central London in the early morning to give 5.7 hours of sunshine. By contrast, the suburbs and environs remained foggy for most of the day, with visibility below 100 m till midday at Roehampton. Central London was also the sunniest non-coastal location in the south-east in the cold January of 1982 and the foggy December of 1991.

Box 3.3

LONDON'S CHANGING AIR QUALITY

The damaging effects of atmospheric pollution in London were described by the seventeenth-century diarist John Evelyn. In *Fumifugium* (1661) he noted the effects of decreased visibility, damage to the built environment and impaired plant growth (Agnew 1986). The principal cause of pollution was the widespread adoption of coal as a domestic and industrial fuel. In an early application of the principle that 'the polluter pays', Brimblecombe (1987) reported that Archbishop Laud (1573–1645) fined brewers at Westminster for burning coal, though unfortunately this was used as evidence against him in his trial for treason in 1640! Sulphur dioxide (SO_2) and smoke have thus long been present in London's air and the effects were sometimes noted far afield; in the eighteenth century, the naturalist Gilbert White recorded that he could smell the smoke at Selborne in Hampshire.

By the nineteenth century, increasing population density and industrial activity in the wake of the Industrial Revolution resulted in a continuing deterioration in air quality. The improvement during the twentieth century has been a consequence of progressive changes in fuel use, as well as the Clean Air Acts of 1956 and 1968 and the spread of smokeless zones away from inner London after the mid-1950s. Since then the focus of London's air pollution has shifted to the contribution of motor transport. Photochemical smog is a regular feature of London's summer atmosphere, involving the generation of ozone in the presence of nitrogen oxides and hydrocarbons from vehicle exhausts and the catalysing influence of sunshine. Lee (1993) estimated that a 1 degree C rise in summer temperature is associated with a 14 per cent rise in surface ozone concentrations. Such climate sensitivity suggests that even if traffic volumes could be stabilised in the future, ozone and the hazard of photochemical smog could increase according to any greater incidence of warm summers. The yellow-brown haze of a photochemical smog has become a familiar feature of the London sky in recent summers, and such incidents have been linked with reported increases in the numbers of asthma cases, though links remain unproven. However, Prichard (1994) reported how an oppressive and thundery evening (24 June 1994) coincided with a localised 'epidemic' of asthma cases involving over 1,200 people in north-east London.

An important variable in determining air pollution is the mobility of the atmosphere. Mild, 'westerly' winters increase the dispersal of pollutants and also reduce the need for fuel use itself. A spell of quiet, anticyclonic weather in December 1991 gave a vivid illustration of the potential hazards. From 12 to 15 December concentrations of NO_2 peaked at 423 ppb, twice the maximum threshold recommended by the World Health Organization. These were the highest values observed in London since comparable observations began in the early 1970s (Bower *et al.* 1994). Unlike summer ozone episodes, the pollution was concentrated in inner London, close to areas of high traffic volumes. However, smoke and sulphur dioxide levels were an order of magnitude lower than in the smogs of the 1950s and 1960s. Although the event was thought to have been responsible for about 150 extra deaths, this compares with about 8,000 in the smog of December 1952.

FOG

The growth and industrial development of Victorian London created ideal conditions for dense winter fogs, which appear to have been at their worst around the 1890s (Brimblecombe 1987). Air pollution impaired visibility even when 'natural' fogs were not present. This was illustrated by a fog survey instituted by London County Council in 1901, the results of which showed that between 20 December 1901 and 17 January 1902, St Paul's Cathedral was not once visible from the Houses of Parliament, a distance of only 2.4 km. (Thornes 1978). By the 1920s, sixty-two occasions of persistent fog lasting throughout the day were noted by Bilham (1938) for London. The last serious 'smog' event was that of 5–8 December 1952 (Box 3.3). Over a two-week period the average death rate doubled as over 8,000 people died, and the incidence of pneumonia and bronchitis increased dramatically.

It is difficult to compile a data-set of fog frequency covering the whole of the last century because of changes in observation sites. However, the frequency of dense and thick fog in central London (Kingsway) dropped by 52 per cent and 24 per cent respectively between 1947 and 1954 and between 1955 and 1962, straddling the passage of the Clean Air Act in 1956 and the first smokeless zones. At Heathrow, the reductions were only 12 and 5 per cent respectively, indicating that the focus of the improvement was over the inner urban area (Brazell 1968). Evidence for the amelioration of fog in central London is also provided by Manley (1952), who noted that fog was often denser around the margins of the area's parks than in Oxford Street – a contrast attributable to local variations in building density, moisture supply and temperature. By the late twentieth century it has become not unusual to find fog clearing away from a much wider area of inner London, reversing the previous urban effect on visibility. This is an interesting consequence of a less smoky urban heat-island aiding the evaporation and dispersal of fog.

The lack of a homogeneous series of observations for central London is unfortunate, and the average values for the 1960s are partially estimated. However, whilst it can be seen that the highest fog frequencies in the 1950s were within London itself, the pattern is now inverted, following a dramatic reduction in fog frequency within the built-up area.

Hastings is, of course, affected by coastal (advection) fog in addition to radiation fog. The contrasting seasonal incidence of advection fog along the south coast means that St Catherine's Point (Figure 3.9) on the southern tip of the Isle of Wight has a distinctive peak in fog frequency in the first half of the year, when sea temperatures are at their lowest, whereas further inland the peak is in winter owing to both low stratus (especially at Long Sutton) and radiation fog (more particularly at Shinfield). In both these locations fogs are overall more abundant than on the coast. The Long Sutton and Shinfield annual means are 26.2 and 22.7 fogs at 0900 GMT. The St Catherine's Point annual mean is only 9.3 although there is a greater chance of summer fogs to spoil the holiday setting.

CONCLUSION

The climate of south-east England is relatively sheltered from the dynamic influences of mid-latitude depressions, lying as it does at the 'favoured' end of the north-north-west to south-south-east climatic gradient. A suggested north–south subdivision of the region expresses a contrast between the moister, milder and more maritime area to the south of the North Downs and the drier, more continental area of the Thames valley to the north. The latter area has a climatic affinity with East Anglia, modified in the London area by greater warmth and lighter winds.

Despite the apparent quiescence of the climate, the region has experienced numerous severe weather events over recent decades, including the rainstorms of September 1968, the droughts of 1975–76 and 1988–92 and the Great Storm of October 1987. Given the problems caused by warm weather in recent summers, it is perhaps appropriate to classify

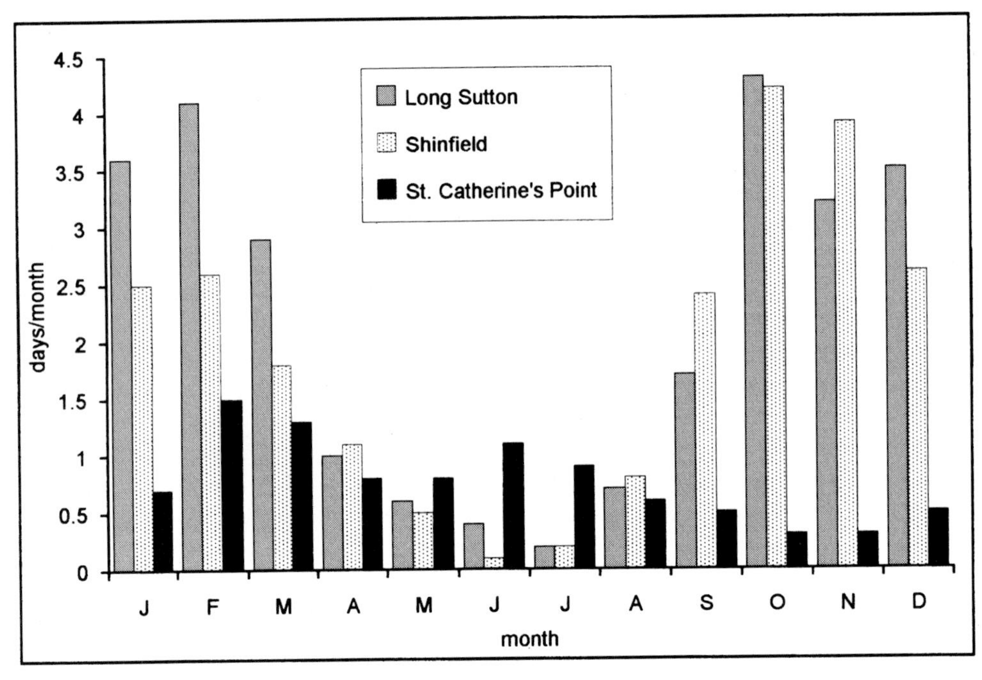

Figure 3.9 Mean monthly frequency of fog at 0900 GMT for the period 1961–90

summer heat and drought as regional climatic hazards – especially when aggravated by the urban heat-island and photochemical pollution. This provides an interesting distortion to the notion of a north-north-west to south-south-east climatic 'favourability' gradient over the British Isles.

REFERENCES

Advisory Services Branch (1988) 'A detailed description of the wind and weather during the passage of the storm of 15/16 October 1987 across southern England', *Meteorol. Mag.*, 117: 104–109.

Agnew, C. (1986) 'Air pollution control', in H. Clout and P. Wood (eds) *London: Problems of Change*, Harlow: Longman.

Atkinson, B.W. (1968) 'A preliminary examination of the possible effect of London's urban area on the distribution of thunder rainfall 1951–60', *Trans., Inst. Br. Geogrs*, 44: 97–118.

Barry, R.G. (1964) 'Weather and climate', in F.J. Monkhouse (ed.) *A Survey of Southampton and Its Region*, Southampton: British Association.

Bilham, E.G. (1938) *The Climate of the British Isles*, London: Macmillan.

Bleasdale, A. (1970) 'The rainfall of 14th and 15th September 1968 in comparison with previous exceptional falls of rain in the United Kingdom', *J. Instn Water Engrs*, 24: 181–189.

Bonacina, L.C.W. (1960) 'Summer evening visibility over London in 1959; and ten year retrospect', *Weather*, 15: 127–130.

Bower, J.S., Broughton, G.F.J., Stedman, J.R. and Williams, M.L. (1994) 'A winter NO_2 smog episode in the U.K.', *Atmospheric Environment*, 28: 461–475.

Brazell, J.H. (1968) *London Weather*, London: HMSO.

Brimblecombe, P. (1987) *The Big Smoke: a History of Air Pollution in London since Medieval Times*, London: Methuen.

Browne, T.J. and Robinson, D.A. (1984) 'Exceptional rainfall around Lewes and the South Downs, autumn 1982', *Weather*, 39: 132–136.

Brugge, R. (1991) 'Two remarkably sunny years in southern Britain – 1989 and 1990', *Weather*, 46: 384–390.

Burt, S.D. (1992) 'The exceptional hot spell of early August 1990 in the United Kingdom', *Int. J. Climatol.*, 12: 547–567.

Burt, S.D. and Mansfield, D.A. (1988) 'The great storm of 15–16 October 1987', *Weather*, 43: 90–114.

Chandler, T.J. (1965) *London Weather*, London: Hutchinson.

Clarke, P.C. (1969) 'Snowfalls over south-east England, 1954–69', *Weather*, 24: 438–447.

Currie, I. (1994) 'A remarkable Surrey valley: a frosty reputation confirmed', *Weather*, 49: 77–78.

Davis, N.E. (1969) 'Diurnal variation of thunder at Heathrow Airport, London', *Weather*, 24: 166–172.

Davis, R.A. (1991) 'The storms of 24 May 1989: the rainfall in the Thames Valley area', *Meteorol. Mag.*, 120: 11–15.
Eden, G.P. (1995) *Weatherwise*, London: Macmillan.
Gilbert, E.G. and Walker, J.M. (1966) 'Tornado at the Royal Horticultural Society's garden, Wisley', *Weather*, 21: 211–214.
Gomez, B. and Smith, C.G. (1987) 'Visibility at Oxford, 1926–1985', *Weather*, 42: 98–106.
Haggett, C.M. (1980) 'Severe storms in the London area – 16–17 August 1977', *Weather*, 35: 2–11.
Harrison, S.J. (1976) 'The land and sea breezes of south Hampshire: a climatological viewpoint', *J. Meteorology (UK)*, 1: 383–387.
—— (1977) 'Hill climates in southern England', *J. Meteorology (UK)*, 2: 326–329.
Harrison, S.J. and Currie, I. (1979) 'A severe frost-hollow in the North Downs, Surrey', *J. Meteorology (UK)*, 4: 265–270.
Hatch, D.J. (1973) 'British climate in generalised maps', *Weather*, 28: 509–516.
—— (1981) 'Sunshine at Kew Observatory, 1881–1980, *J. Meteorology (UK)*, 6: 101–113.
Hawke, E.L. (1944) 'Thermal characteristics of a Hertfordshire frost-hollow', *Q. J. R. Meteorol. Soc.*, 70: 23–48.
Holmes, C.G. (1995) 'The West Sussex floods of December 1993 and January 1994', *Weather*, 50: 2–6.
Hosking, K.J. (1978) 'Warm days, air frosts and heavy 48-hour rainfalls at Ryde, Isle of Wight, 1918–1977', *J. Meteorology (UK)*, 3: 205–10.
Howells, K.A. (1985) ' Changes in the magnitude–frequency of heavy rainfalls in the British Isles', unpublished PhD thesis, University of Wales, Swansea.
Hughes, G.H. (1979) 'The summers of 1968 and 1977: some similarities and contrasts', *Weather*, 34: 319–325.
Jackson, M.C. (1977) 'Mesoscale and synoptic-scale motions as revealed by hourly rainfall maps of an outstanding rainfall event: 14–16 September 1968', *Weather*, 32: 2–16.
—— (1979) 'The largest fall of rain possible in a few hours in Great Britain', *Weather*, 34: 168–175.
Jenkins, I. (1969) 'Increase in averages of sunshine in Greater London', *Weather*, 24: 52–54.
Keers, J.F. and Wescott, P. (1976) 'The Hampstead storm – 14 August 1975', *Weather* 31: 2–10.
Lee, D.O. (1985) 'Trends in summer visibility in London and southern England, 1962–79', *Atmospheric Environment*, 17: 151–159.
—— (1992) 'Urban warming?: an analysis of recent trends in London's heat island', *Weather*, 47: 50–56.
—— (1993) 'Climatic change and air quality in London', *Geography*, 78: 77–79.
McCallum, E. and Waters, A.J. (1993) 'Severe thunderstorms over south-east England, 20/21 July 1992: satellite and radar perspectives of a mesoscale convection system', *Weather*, 48: 198–208.
Manley, G. (1952) *Climate and the British Scene*, London: Collins.
Mayes, J.C. (1991) 'Recent trends in summer rainfall in the United Kingdom', *Weather*, 46: 190–196.
—— (1994) 'Kew Observatory 1769 to 1980: climatological implications of an observatory closure', in J. Kenworthy and B. Giles (eds) *Observatories and Climatological Research*, Department of Geography, University of Durham.
Ogley, B., Currie, I, and Davison, M. (1993) *The Kent Weather Book*, Westerham: Froglets.
Pike, W. (1994) 'The remarkable early morning thunderstorms and flash-flooding in central southern England on 26 May 1993', *J. Meteorology (UK)*, 19: 43–64.
Potts, A.S. (1982) 'A preliminary study of some recent heavy rainfalls in the Worthing area of Sussex', *Weather*, 37: 220–227.
Prichard, R. (1987) 'Severe thunderstorms in south-west Essex: 29 July and 22 August 1987', *J. Meteorology (UK)*, 12: 340–343.
—— (1994) 'TORRO thunderstorm summary: June 1994', *J. Meteorology (UK)*, 19: 360–362.
Quine, C.P. (1988) 'Damage to trees and woodlands in the storm of 15–16 October 1987, *Weather*, 43: 114–118.
Shellard, H.C. (1976) 'Wind', in T.J. Chandler and S. Gregory (eds) *The Climate of the British Isles*, London: Longman.
Simmons, D.J. (1978) 'The weather of Vectis', *J. Meteorology (UK)*, 3: 15–16.
Simpson, J.E. (1964) 'Sea-breeze fronts in Hampshire', *Weather*, 19: 208–220.
Smith, C.G. (1974) 'Monthly, seasonal and annual fluctuations of rainfall at Oxford since 1815', *Weather*, 29: 2–16.
—— (1994) 'The Radcliffe Observatory, Oxford', in J. Kenworthy and B. Giles (eds) *Observatories and Climatological Research*, Department of Geography, University of Durham.
Southern, G.A. (1976) 'The bronze summer of 1976', *J. Meteorology (UK)*, 2: 11–13.
Stone, J. (1983) 'Circulation type and spatial distribution of precipitation over central, eastern and southern England', *Weather*, 38: 173–177, 200–205.
Taylor, S.M. (1995) 'The Chichester flood, January 1994', *Hydrological Data UK, 1994 Yearbook*, Wallingford, Institute of Hydrology/British Geological Survey.
Thornes, J.E. (1978) 'London's changing meteorology', in H. Clout (ed.) *Changing London*, London: University Tutorial Press.
Tyssen-Gee, R.A. (1976) 'Recording Hampstead's weather', *J. Meteorology (UK)*, 1: 210–214.
—— (1981) 'Hampstead's 70 years of rainfall recording', *J. Meteorology (UK)*, 6: 3–8.
—— (1982) 'Hampstead's sunshine from 1910 to 1979', *J. Meteorology (UK)*, 7: 14–17.
Tout, D.G. (1976) 'Temperature', in T.J. Chandler and S. Gregory (eds) *The Climate of the British Isles*, London: Longman.
Wales-Smith, B.G. (1980) *Revised Monthly and Annual Totals of Rainfall Representative of Kew, Surrey for 1697–1870 and an updated analysis for 1697–1976*, Hydrological memorandum No. 43, Met O 8, Bracknell: Meteorological Office.
Waters, A.J. and Collier, C.G. (1995) 'The Farnborough Storm: evidence of a microburst', *Meteorological Applications*, 2: 221–230.
Webb, J. (undated) 'Temperature extremes in Britain: maxima', *Tornado and Storm Research Organisation Information Leaflet*, Oxford: TORRO.
Webb, J., Rowe, M.W. and Elsom, D.M. (1994) 'The frequency and spatial features of severely-damaging British hailstorms, and an outstanding 20th-century case study monitored by the T.C.O. on 22 September 1935', *J. Meteorology (UK)*, 19: 335–345.

Webster, F.B. (1984) 'The climatological station at London's Heathrow Airport', *Weather*, 39: 311–315.
Wells, N.C. (1983) 'The south coast hailstorms of 5 June 1983,' *Weather*, 38: 369–374.
Woodley, M.P. (1991) 'Were the dry spells of 1988–90 worse than those in 1975–76?', *Meteorol. Mag.*, 120: 164–169.
Young, M.V. (1995) 'Severe thunderstorms over south-east England on 24 June 1994: a forecasting perspective', *Weather*, 50: 250–256.

4

EASTERN ENGLAND

Julian Mayes and Geoffrey Sutton

INTRODUCTION

Eastern England as defined in this volume extends from the Humber estuary to the coast of Essex, a distance of 260 km, and as such has one of the largest latitudinal extents of any of the regions in this study. An obvious justification for this can be found in the climatic unity that might be expected in a region of unusually homogeneous orography – in simpler terms, it is the flattest part of the British Isles. The main aim of this chapter is to ascertain the degree to which this generalisation linking orography and climate can be verified. The region is very similar to the Eastern England district as defined by the Meteorological Office, and was also the largest of the areas served by single regional climatological memoranda (Meteorological Office 1989).

The northern edge of the region is formed by the Humber estuary, the western (inland) boundary at first by the lower reaches of the river Trent and subsequently by county boundaries, such that the whole of Cambridgeshire and Bedfordshire are included. The latter county provides a physical contrast between the low-lying plain of the Great Ouse in the north to the Chiltern Hills in the south, where the highest point in the region is found at Dunstable Downs (243 m). The southern parts of Hertfordshire and Essex are placed in south-east England rather than in this region as a result of their location in the Thames basin. This follows the pattern of drainage by placing most of the catchment of the river Lea in the South-East, as well as the north coast of the Thames estuary. This corresponds also to their links with London in terms of human and economic activity; in this sense, they are more compatible with other parts of south-east England in terms of both the physical and the human environments. The eastern boundary of the region, the North Sea, plays a fundamental role in the climate as a result not only of its distinct thermal characteristics (see also Chapter 7) but also because of the varied orientation of this vulnerable coastline.

The region thus defined (Figure 4.1) contains a variety of landscape; in addition to the Chilterns, the East Anglian 'Heights' and the Lincolnshire Wolds both provide extensive areas above 100 m, while Norfolk provides smaller areas of hills from the Norfolk Edge in the west to the Cromer ridge in the north. The whole of the Fens is included, comprising the largest area of land lying below sea level in the British Isles.

The Legacy of Meteorological Observations

It might seem unlikely for a region of possible climatic uniformity and low rainfall to provide a rich legacy of observations. However, archive material ranges from an exceptionally early personal diary of weather made by the Reverend Father Merle in Lincolnshire between 1337 and 1344 (Lamb 1995) to the famous Marsham phenological record of the impact of weather on plant growth and animal behaviour, made between 1736 and 1747. Instrumental records in the twentieth century have been

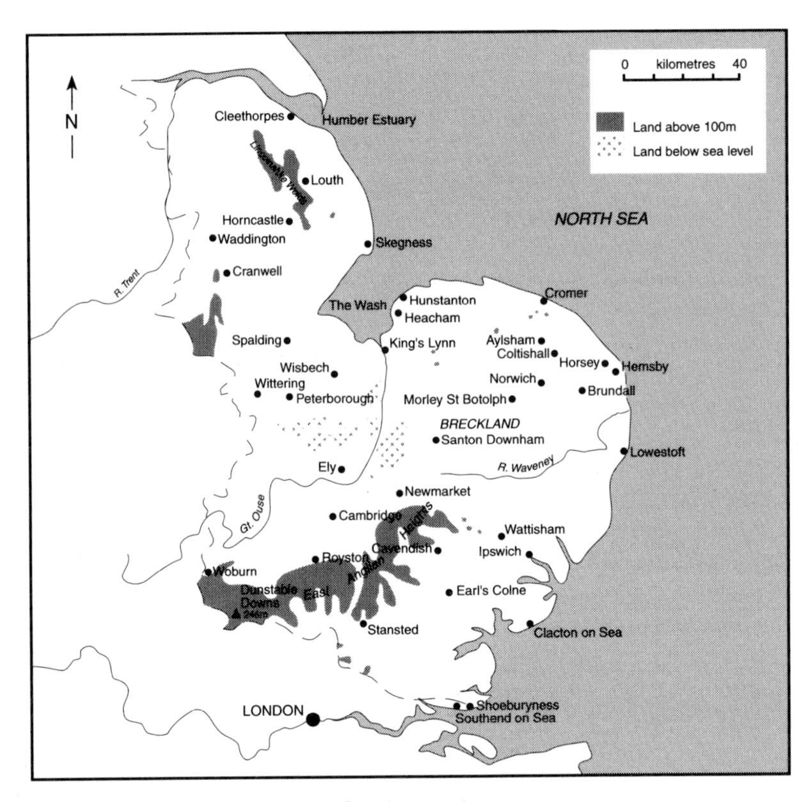

Figure 4.1 Location of weather stations and places referred to in the text

supplemented by the establishment of meteorological observing stations at airfields such as Waddington, Wattisham and Stansted. Further evidence for a keen interest in climate comes from the number of amateur observers who contribute observations to bodies such as the Climatological Observers Link, the Rainfall Organisation of Suffolk and Essex and the Norfolk Rainfall Organisation (Norgate 1970).

An obvious and likely explanation for this interest lies in the importance of agriculture in a region that, despite rapid urban expansion, still retains its rural ambience. Apart from being the chief cereal-growing area of the British Isles, eastern England has sugar beet and barley as important crops, and significant local specialisms exist in market gardening in Bedfordshire together with bulbs, fruit and lavender in the vicinity of Spalding, Wisbech and Heacham respectively. The weather and the progress of the seasons undoubtedly plays a more subtle role as an influence on everyday life in an agricultural region. Evidence for this comes from the range of terms to describe the weather in the Lincolnshire dialect (Robinson 1968), and it could even be suggested that East Anglian people find a substitute for mountain scenery in the majesty of cloud-scapes in their extensive skies, as some of the paintings of John Constable confirm. This is nicely expressed in a weather proverb that originates from the Peterborough area:

> When mountains and cliffs in the sky appear, some sudden and violent showers are near.
>
> (quoted in Glenn 1987)

It is therefore not surprising to find that a comprehensive study of the region's climate has been written (Glenn 1987) as well as a more historical account by Professor Hubert Lamb (1987), founding Director of the Climatic Research Unit at the University of East Anglia.

TEMPERATURE

The region is famous for having a climate that is described as continental by British Isles standards. Here more than elsewhere this influences the whole character of the climate, not just temperature. However, such effects are controlled to some extent by the proximity of the North Sea. Bonacina (1949) has pointed out that the continental character of easterly airstreams is mitigated by the moderating effect of this water body, leading to a heightened maritime influence towards the coast, in contrast to much of Cambridgeshire and inland Norfolk and Suffolk. Conversely, more generally maritime airstreams from the west or south-west might be 'continentalised' by the lengthy land passage over the Midlands and western Britain, especially as the longest land track of the prevailing west-south-west wind lies from Cornwall to Norfolk. The influence of the sea is clearly marginalised when synoptic conditions create such south-westerly airstreams.

Geography also determines the duration of continental and maritime air masses. If a frontal system is attempting to replace continental air with maritime from the west, any deceleration of the system over the Midlands (as often happens on the edge of a blocking anticyclone located over Scandinavia) will prolong the exposure of East Anglia to summer warmth and winter cold. Alternatively, a ridge of high pressure over south-east Britain may allow continental air to cover the region while Ireland and Scotland experience maritime south-westerlies.

Mean Monthly Maximum Temperatures

Table 4.1 shows the variation of monthly mean maximum and minimum temperatures for 1961–90 at contrasting locations. The stations at Lowestoft and Clacton are both situated within 1 km of the sea and exhibit a coastal thermal regime with some clarity, the latter resort having a marked tendency to achieve low day maxima since the observing site is at the sea front. The station at Cambridge Botanic Gardens combines a relatively continental climate with a tendency for high daytime temperatures which are comparable with those of London in summer. Cranwell lies 43 km west of the Wash at a rather exposed site in the Lincolnshire countryside. The fact that the mean annual maximum temperature is the same here as at Clacton (180 km to the south-east) is testimony to the cooling effect of the sea at the latter site, and to the dominance of the

Table 4.1 Mean monthly maximum and minimum temperatures for the period 1961–90

Location	*Alt. (m ASL)*	*Jan.*	*Feb.*	*Mar.*	*Apr.*	*May*	*June*	*July*	*Aug.*	*Sept.*	*Oct.*	*Nov.*	*Dec.*	*Year*
Cranwell	62	5.8	6.1	9.0	11.6	15.4	18.6	20.5	20.3	17.8	14.0	8.9	6.6	12.9
		0.3	0.2	1.5	3.4	6.2	9.3	11.2	11.2	9.3	6.6	3.0	1.1	5.3
Lowestoft	25	5.9	6.2	8.3	10.6	14.2	17.6	19.8	20.0	18.2	14.4	9.7	7.2	12.7
		1.1	1.2	2.4	4.4	7.6	10.4	12.5	12.7	10.9	8.5	4.4	2.4	6.5
Santon	24	6.0	6.5	9.5	12.6	16.4	19.7	21.3	21.2	18.9	14.8	9.6	7.0	13.6
Downham		−0.6	−0.7	0.4	2.0	4.9	7.9	9.7	9.4	7.4	5.1	2.2	0.1	4.0
Cambridge	12	6.5	6.8	9.7	12.7	16.4	19.6	21.5	21.5	18.8	14.9	9.7	7.3	13.8
(Bot. Gdns.)		0.8	0.8	2.0	3.9	6.7	9.6	11.7	11.5	9.8	7.1	3.5	1.5	5.7
Clacton	16	6.1	6.3	8.6	11.0	14.8	18.2	20.3	20.3	18.1	14.5	9.7	7.2	12.9
		1.7	1.9	1.3	5.0	8.3	11.2	13.3	13.5	11.7	9.3	5.0	2.0	7.0

Note: ASL = above sea level

coast rather than latitude as an effective control on temperature. Table 4.1 shows that Clacton is warmer than Cranwell in autumn and winter, but the cooling effect of the sea is apparent as early in the year as March, despite the difference in latitude and altitude. This effect is shown even more starkly when comparison is made with Santon Downham, in the quite continental environment of the Breckland of south-west Norfolk, in which sandy soils predominate. The site of the climatological station here merits special attention, and the daily rise in temperature even in winter is sufficient to counteract the remarkably low nocturnal temperatures of this location. The range of annual mean maximum and minimum temperature varies from 5.9 degrees C at Clacton to 9.6 degrees C at Santon Downham.

Mean Monthly Minimum Temperatures

Lamb (1964) noted that whereas the incidence of warm summer days is relatively high in inland areas, warm humid nights are relatively infrequent. South-east England has about four times as many nights with a minimum temperature exceeding 15°C as eastern England. This is a reflection of the fact that relatively calm, clear nocturnal conditions are common during anticyclonic spells in summer. Under such conditions, the light, glacial soils of much of the region are able to cool at a faster rate than the damper, heavier clays of south-east England. There is little variability in the mean minimum temperatures of successive summers. For example at Cranwell, the highest mean minimum of any month for the period 1961–90 was as low as 13.0°C (August 1975); the 30-year average minimum for July was 11.2°C with a standard deviation of only 0.8 degrees Celsius. The main contrast between coastal and inland sites lies in the frequency of very low minima. For example, at Clacton, where the mean minimum exceeds 13°C in both July and August, it is rare for any night at this time of year to have a minimum temperature of less than 10°C.

Extreme Monthly Temperatures

A daily maximum temperature of 25°C is often interpreted as representing a warm summer day and Lamb (1964, 1987) has shown how the average annual frequency of this temperature varies from around four days at coastal locations to around twelve days at Cambridge. The most outstanding occurrence of this temperature was on 29 March

Box 4.1

THE COOLING EFFECT OF AN ONSHORE BREEZE IN SUMMER – A CASE STUDY

A dramatic illustration of the cooling effect of the North Sea in summer was seen on 30 June 1995. A slack area of high pressure covered the region during the day. A weak but hot southerly airflow slowly changed to south-westerly, helping to lift temperatures to between 29 and 31°C away from the coast. These high temperatures, combined with the slack pressure gradient, led to the formation of a 'heat-low' over East Anglia. This in turn induced a weak north to north-easterly onshore breeze that caused a dramatic drop in temperature as the cool maritime air arrived late in the day. The temperature at Wittering in Cambridgeshire dropped by 11 degrees C within one hour and unofficial observing sites elsewhere in East Anglia recorded hourly falls of more than 15 degrees (*Climatological Observers Link Bulletin*, June 1995). The change of airflow also brought about a sharp jump in air pressure, gusty winds and a cover of stratus cloud. An example of a similar cooling influence on the coast of north-east England is described in Box 7.3 (see also Figure 1.12 for a summary of some aspects of the weather at this time).

1968 when it was logged at Santon Downham and Cromer, equalling the highest temperature recorded in the United Kingdom in March (at Wakefield, three years before to the day!). This was an exceptional but not isolated example of how the sandy heathland soils of the Breckland can warm rapidly in response to sunshine and how shelter from the south and south-west can have a warming effect upon ('continentalised') offshore winds at Cromer. An offshore wind on fine, warm days will become progressively drier as it approaches the coast with a lengthening land track and it may prevent a sea breeze from commencing if it is sufficiently strong.

During a prolonged heatwave the depth of atmosphere warmed by insolation may prevent a sea breeze circulation from being initiated. In these rare cases, most of East Anglia and Lincolnshire may reach 30°C, as happened on 3 August 1990. In addition to being a day of general heat which embraced coastal as well as inland areas, Cambridge was a close rival to Cheltenham (which observed the highest ever recorded temperature in the British Isles of 37.1°C) with a maximum of 36.5°C (Burt 1992).

The apparently subtle changes in coastline orientation can have marked effects on temperature by altering the distribution of offshore and onshore winds. This can be particularly effective in southerly or easterly periods in spring and early summer when the potential for high temperatures is curtailed near the coast by the coldness of the North Sea (Box 4.1). The cooling effects may be widely distributed in an easterly spell, a southerly sometimes providing a cool onshore breeze at Skegness (after some land warming) with conditions cooler still at Clacton and Lowestoft, while places experiencing a southerly wind as an offshore breeze will be much warmer, especially where this is accompanied by a degree of descent from hills, such as at Cromer and Cleethorpes (Figure 3.2 in the previous chapter).

Figure 4.2 shows that the monthly maximum temperature at Clacton is usually lower than that for Cambridge even in winter, while in May it averages 4.2 degrees C cooler and over 3.5 degrees C in April and July. Indeed, of the region's climatological stations,

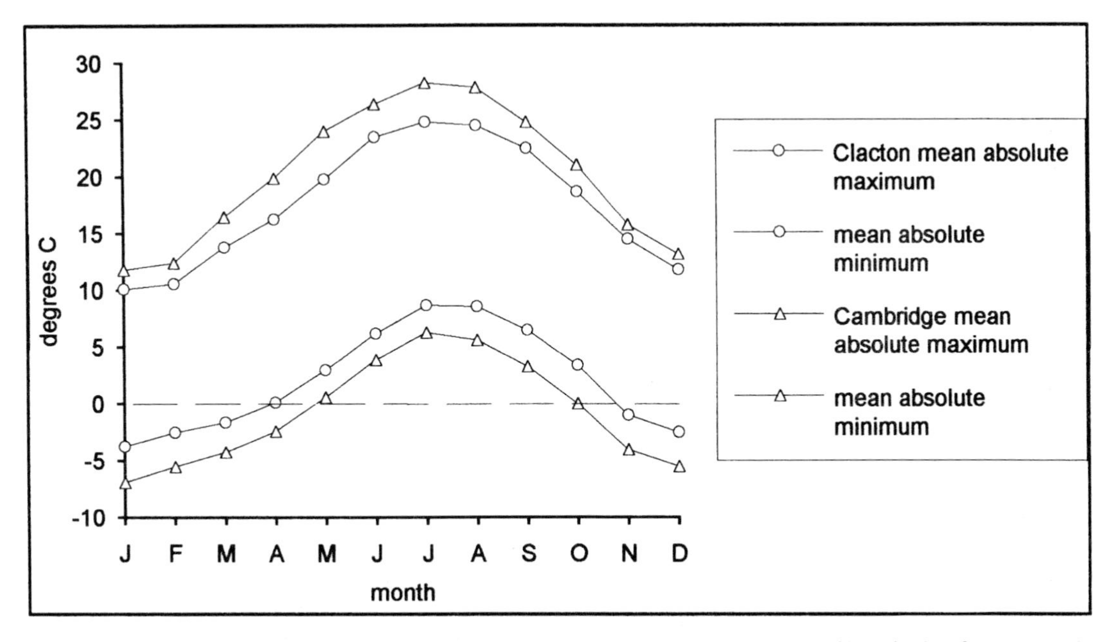

Figure 4.2 Mean absolute monthly maximum and minimum temperatures at Clacton and Cambridge for the period 1961–90

Clacton has the lowest extreme monthly maximum in about one month in every four (including the period February to July 1990 inclusive). The favourable location of Cambridge for achieving high extreme maxima is common to much of Cambridgeshire; the nearby town of March recorded 29°C on 1 October 1985, just short of the highest October temperature in the British Isles (also recorded on this day; see Chapter 3). The highest temperature of the year in the United Kingdom is observed in this region in about one year in four (Webb 1984). For 1961–90, the average extreme annual maximum was 29.6°C at Cambridge, 29.1°C at Santon Downham and 25.9°C at Clacton. The two former values are comparable with those of the London area and the Home Counties (see Table 3.3). Temperature variability measured by the standard deviation ranges from 2.7 degrees C at Cambridge to 1.8 at Clacton, highlighting the greater variability of maximum temperatures at sites having a more continental thermal regime. The monthly mean minima for Cambridge and Clacton are also consistently different but Cambridge records the lower temperatures throughout the year in agreement with its more 'continental' setting.

In terms of wide range of temperature extremes this region has much in common with the Midlands (the temperature ranges of which are reviewed in Chapter 5) but here temperatures are also characterised by variability within individual months. Figure 4.3 shows the monthly distribution of the means of the absolute monthly maxima and minima at Santon Downham together with the individual extremes for each month. The average difference between the highest and lowest temperature of each month reaches 29.8 degrees C in July and 29.0 degrees C in August, highlighting the tendency for the surface temperature of the Breckland to fluctuate rapidly on the locally dry and light soils, especially during the summer months. The extreme temperature range in individual months has quite often exceeded 30 degrees C. On two occasions this has been noted within a single week; a maximum temperature of 33.3°C on 1 July 1961 was followed by a minimum of −1.1°C on the 6th. Similarly, in August 1964, 32.2 was logged on the 26th, followed by −1.7°C on the 31st, this being one of many months when the lowest temperature in England was logged at this station. In considering these individual extremes, it is relevant to note that they are derived

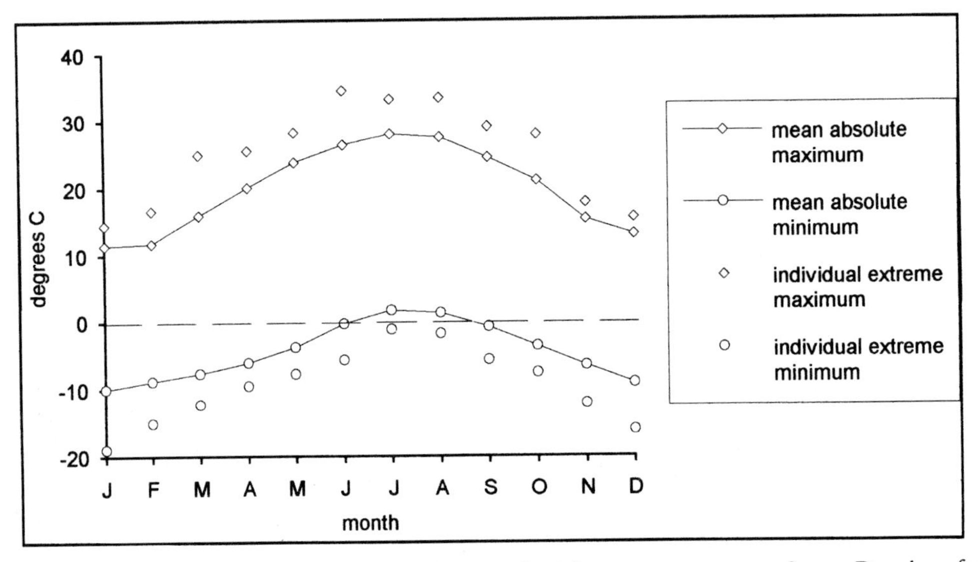

Figure 4.3 Mean absolute and extreme monthly maximum and minimum temperatures at Santon Downham for the period 1961–90

from only a 28-year record within the study period, observations having terminated in 1988. As a result, the warmth of the summers of 1989 and 1990 is excluded. The average absolute annual minimum is as low as −12.0°C at Santon Downham, −9.1°C at Cambridge but only −5.0°C at Clacton while the average annual absolute temperature range varies from 41.2 degrees C at Santon Downham to 31.0 degrees C at Clacton.

The Incidence of Frost

With relatively subdued orographic features, the principal control on the distribution of frost is proximity to the coast. Of course, even gentle slopes can influence the tendency of air to descend to hollows when nocturnal cooling takes place, generating distinctive patterns of frost hollows. Local soil type and land cover play a more important role here than in some other regions, as exemplified by the Breckland, and similar heathland areas.

Figure 4.4 shows the frequency of air frost at three locations: Clacton and Santon Downham (representing contrasting settings) and Woburn in Bedfordshire, which is representative of much of the rather 'frost-prone' inland part of the region. Thus Clacton's annual frequency of air frosts is only 25.0, in contrast to Santon Downham's 89.1 (the Woburn mean annual frequency is 57.6). The mean annual total of ground frosts is naturally greater with 58.8 at Clacton. Meanwhile 165 nights of the year (45 per cent) at Santon Downham will suffer a ground frost. The corresponding Woburn mean is 113.9.

The warming effect of the sea delays the start of the frost season at coastal sites, and in October at Clacton, for example, even ground frost is unusual, occurring in only eight Octobers out of thirty while at Santon Downham nearly half of October nights record a frost on the grass. In addition to the unusual phenomenon of air frost in the summer months, the data for the latter station also reveal a frequency of December ground frost that exceeds that of February and nearly approaches that of January. A feature of all three sites is the limited reduction in

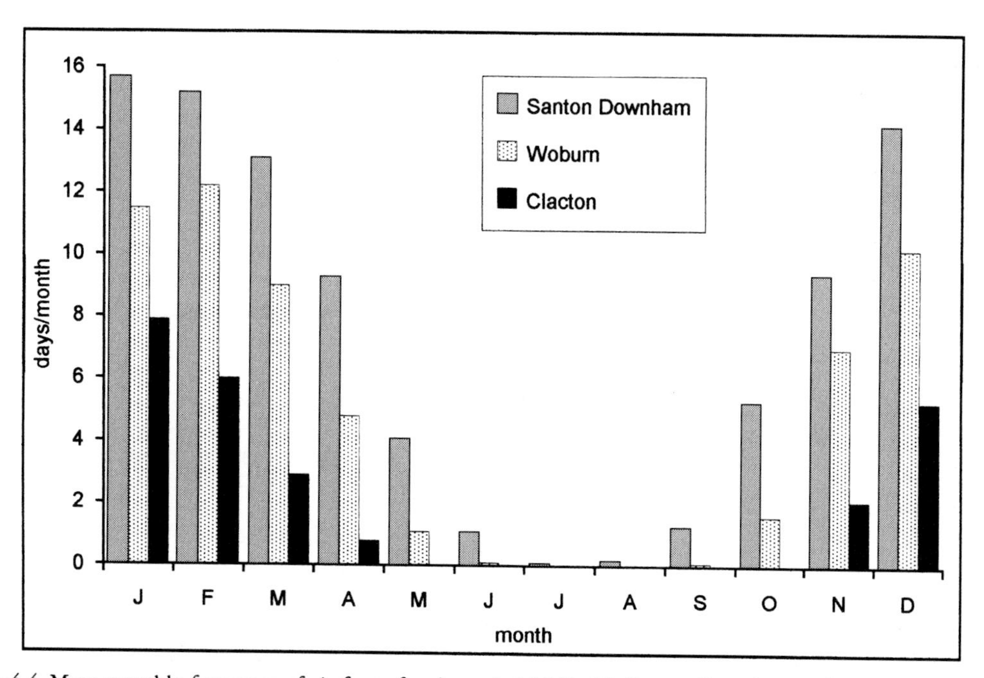

Figure 4.4 Mean monthly frequency of air frosts for the period 1961–90 (Santon Downham 1961–88)

ground frost incidence in March and April. This is a consequence of low subsurface temperatures at this time of year due to the delayed seasonal thermal regime of soil. The frequency of lighter winds after the end of winter may also be sufficient to encourage ground frost by encouraging thermal stratification of air close to the surface.

PRECIPITATION

Eastern England has long been recognised as the driest region of the British Isles. The actual location of the station with the lowest average annual rainfall has varied between the Essex coast and parts of Cambridgeshire according to which average period is used. Of more practical importance is the fact that many parts of the region have an annual average rainfall of less than 600 mm and nowhere does it exceed 750 mm. In practice, annual rainfall totals fail to reach 500 mm in at least some parts of the region in most years. This large region of modest rainfall is a notable outlier to the general pattern of low rainfall in Europe; the other area of sub-600 mm averages in northern Europe is well to the east in Germany and Poland (where it is due to drier, colder winters), whereas much of Southern Europe has a higher annual total.

While this relative scarcity of annual rainfall has consequences for water resources and agriculture, the seasonal distribution of precipitation and the relationship to evapotranspiration are of greater relevance. The region has a slightly 'continentalised' rainfall regime in which rainfall over the summer months is usually boosted by heavy convectional falls in short periods.

Precipitation is evenly distributed through the seasons (Table 4.2), with coastal districts having a little more than 50 per cent of the annual fall in the winter half-year and inland areas a little less; percentages range from 47 at Cranwell to over 53 on the Chiltern Hills in the south of the region. Percentages in summer (June, July and August) range from 28 at Cranwell to 25 per cent on the coast of Suffolk and Essex. This shows the extent to which convectional rainfall responds to the heating of inland areas in summer and the influence of latitude on summer rainfall, whereby a lower proportion of the annual precipitation occurs towards the south as a result of proximity to extensions of the Azores anticyclone.

With an increased frequency of drier summers in the 1970s and 1980s, the former peak in monthly rainfall in July and August is less conspicuous a feature of the 1961–90 averages than hitherto. Indeed, annual rainfall has declined over much of the region since the mid-twentieth century. Average summer (June to August) precipitation at Belstead, near Ipswich, declined from 160 mm over the period 1902–69 to only 129 mm in the decade 1970–79. July precipitation declined from 57 to 42 mm over the same periods. The dryness of the 1970s is emphasised further by the frequency of drought, which was greater than in any previous decade of the century (Glenn 1987). Even before the recent decline in summer rainfall, Terry and Jackson (1962) demonstrated the benefits of irrigation for crops such as sugar beet, carrots and potatoes. Irri-

Table 4.2 Mean monthly and annual precipitation totals (mm) for the period 1961–90

Location	*Alt. (m ASL)*	*Jan.*	*Feb.*	*Mar.*	*Apr.*	*May*	*June*	*July*	*Aug.*	*Sept.*	*Oct.*	*Nov.*	*Dec.*	*Year*
Skegness	5	50	37	47	44	46	54	47	57	46	50	60	58	596
Cranwell	62	48	37	46	47	51	54	49	61	46	45	53	50	587
Gorleston	4	47	35	40	44	44	45	51	50	52	54	62	53	577
Lowestoft	25	50	38	40	45	41	42	49	52	53	56	64	54	584
Cambridge	12	43	32	42	42	48	48	48	53	47	48	50	50	551
Clacton	16	49	31	43	40	40	45	43	43	48	48	55	50	535

gation was deemed to be beneficial in at least eight years out of ten across most of the region.

Synoptic Origin of Precipitation

As south-westerly and westerly winds pass between south-west England, Wales and eastern England, they can lose much of their rain-bearing potential, although other factors such as embedded instability may trigger heavier falls on occasions. The rainfall receipt from these directions is not therefore always matched by the usually high frequency of rainfall. By contrast, the region generally has a greater exposure to north-westerly and northerly airflows. Indeed, for individual depression systems, Norfolk sometimes receives more precipitation from showers following the clearance of a frontal system (in a NW or N airflow) than from the fronts themselves. Just as Sweeney (Chapter 11, this volume) has noted a strong dependence on W airflow types for much of the precipitation in County Limerick, so much of East Anglia is sensitive to the frequency of NW and N airflow types (Stone 1983). The driest conditions are often experienced when a westerly airstream develops an anticyclonic characteristic, as happened often between 1988 and 1991. This drought was notable for the continuation of dry weather through the summer of 1991 in Lincolnshire and Norfolk when south-westerly airstreams had brought a return to wetter weather in areas further south and west.

Cyclonic situations are likely to promote the development of widespread heavy rain when depressions track over the region. A track across southern England will have a similar effect, owing to the likely presence of easterly winds and persistence of rainfall from slow-moving fronts. The saying '*When the rain is from the east, four-and-twenty hours at least*' is particularly apt for this area, especially when slow-moving fronts stagnate over, or just to the south of, the region. A noteworthy case was between 23 and 25 April 1981, when over 125 mm of rain fell in parts of Norfolk. Persistent and heavy rain may also accompany a front that hesitates or returns westwards after clearing the region, providing two 'doses' of rain from the same system.

A noteworthy example of exceptional precipitation over a longer period is provided by the months of September and October 1993. Persistent blocking with quite frequent onshore south-east winds associated with slow-moving depressions provided some exceptionally large daily precipitation totals. Over the thirty-one-day period from 12 September to 12 October over 250 mm was observed near the Suffolk coast. This equates to nearly six months' average rainfall. While October rainfall reached three times the average in a few localities, parts of upland Wales and north-west England recorded under one-third of the local average rainfall.

The Duration and Frequency of Precipitation

In months of heavy short-period precipitation, duration and frequency statistics can give a more realistic impression of the 'wetness' of the month. The lowest published average for annual duration is 407 hours at Ipswich (based on 1951–60 data, quoted by Atkinson and Smithson 1976). Of this total, only 22 per cent (83 hours) fell in the three summer months; at Cranwell the summer duration was 116 hours, but even this represented only 22 per cent of the duration for the whole year. The rainfall regime therefore conspires to give an impression of summers being roughly as wet as the other seasons whereas in reality summers are characterised by short-period rainfall which is sometimes of high intensity.

Rainfall frequency is conveniently denoted by the number of 'rain days', defined as days with at least 0.2 mm of rain. The geographical distribution of annual totals of rain days was mapped by Glasspoole (1926) and revealed a remarkable peak over much of Norfolk of more than 200 days a year. This is more than for some western areas of Britain. This anomaly can be attributed to the susceptibility to showers in north-westerly and northerly airstreams. It can thus be said that rain falls on a large number of days but often for only short periods and in small amounts. Figure 4.5 shows that there has been some reduction

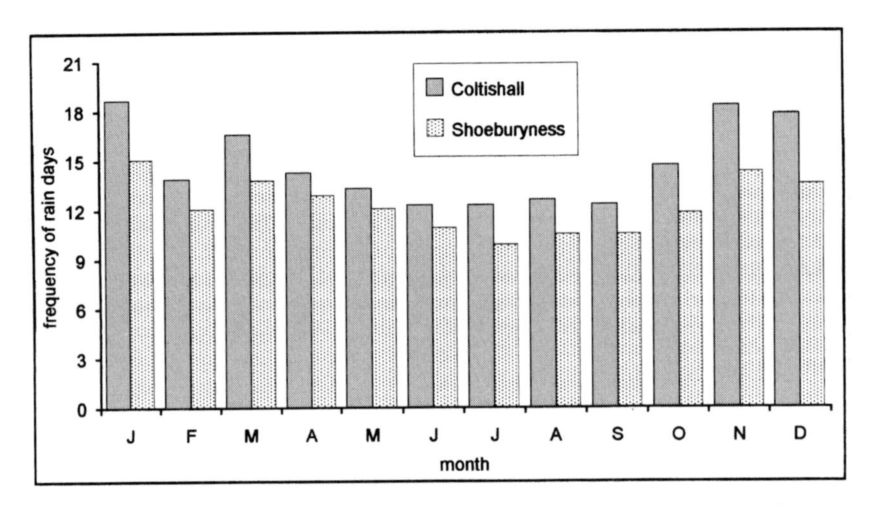

Figure 4.5 Mean monthly frequency of rain days at Shoeburyness (1961–90) and Coltishall (1964–90)

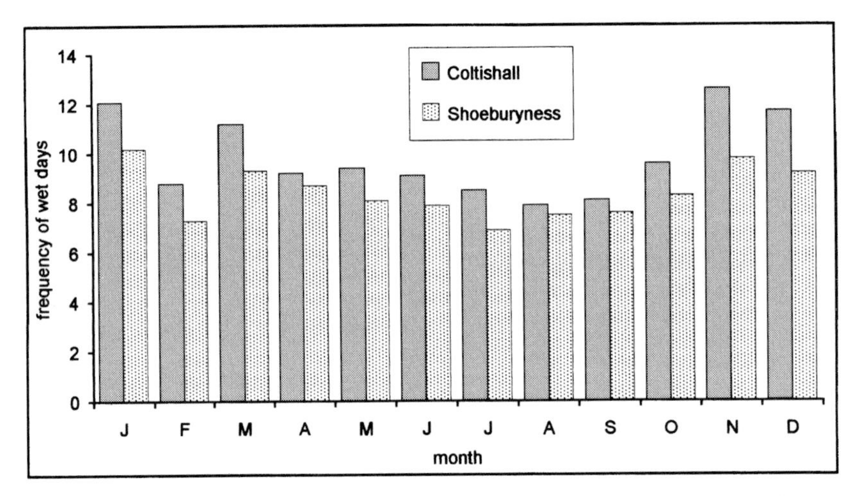

Figure 4.6 Mean monthly frequency of wet-days at Shoeburyness (1961–90) and Coltishall (1964–90)

from this frequency in the second half of the twentieth century (Coltishall being on the edge of the area of high frequency identified by Glasspoole), and evidence is marked for a southward reduction in the frequency of rain days and wet-days. The relative dryness of July is shown by the fact that as many as 10 'wet-days' (at least 1.0 mm of rainfall) were registered on only six occasions during the thirty years at Shoeburyness. The difference between the incidence of rain days and that of wet-days, the latter being far less frequent, emphasises the fact that rainfall in this region is often light, accumulating to only small daily totals that make a negligible contribution to summer moisture availability for agriculture and horticulture. The lowest values were found on the Essex coast, where Shoeburyness and adjacent stations averaged less than 150 rain days and only 101 wet-days per year. Coltishall, in comparison, has respective annual means of 177 and 147. The monthly variation in wet-days for these two sites is summarised in Figure 4.6, where the summer minima are clearly depicted.

Thunderstorms

With a combination of modest relief features and a tendency for high daytime temperatures over much of the year, it is not surprising to find that convectional, rather than orographic or frontal, rainfall accounts for much of the high-intensity precipitation in eastern England. Much of the region, especially away from the cooler coastal districts, averages around or slightly more than fifteen days with thunder a year (Meteorological Office 1952; Figure 4.7); Waddington, for example, averages 14.1 thunder days a year and Coltishall 15.2 days. The peak in the summer half-year is characteristic of those areas where thunder is usually associated with summer storms rather than winter thundery showers. It is not unusual for the more thundery years to have twenty to twenty-five thunder days, of which several may be major thunderstorms that make a significant contribution to the total rainfall of the year.

The distinction between days with thunder heard and the frequency of major thunderstorms was analysed by Crossley and Lofthouse (1964), who found a tendency for the severest storms to be more common in the south of the region, closest to the storm-generating areas of warmth over France and the London area. Even though thunder may be heard on at least as many days in the north-west of the region (towards the relatively thundery lower Trent valley), they suggested that severe storms occur less frequently here. This could be accounted for by once severe storms having decayed by the time they reach this far north. A tendency for nocturnal storms to regenerate after around 0300 hours has also been noted, possibly as a result of airflow motion above the temporary night-time low-level temperature inversions.

Despite the suggestion that Lincolnshire experiences severe storms rather less often than places further south, Jackson (1974) has shown that the largest two-hour rainfall total to have occurred in the region was of 126 mm at Cranwell on 11 July

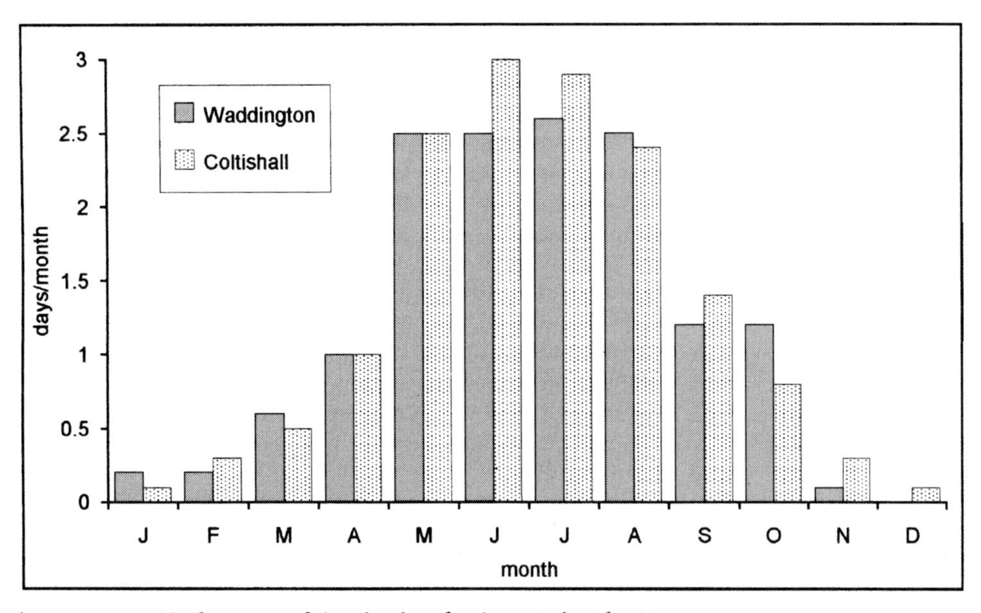

Figure 4.7 Mean monthly frequency of thunder days for the period 1961–90

1932. A fall of 117 mm was observed nearby at Horncastle on 7 October 1960, but this was part of a longer-lasting rainfall event that yielded 183 mm in 5 hours. Eleven of the fifty largest two-hour storms surveyed by Jackson occurred in the region. Statistics such as these, emphasising the peak period of a storm, often mask the fact that such events are often quite isolated in both space and time. The hydrological effectiveness of a sudden, localised inundation (much of which will run off the land) may be limited.

Some of the thunderstorms that occur within the region are the result of activity initiated over France or the Low Countries and are 'imported' in the prevailing airstreams. Others may be triggered within the region as local (often single-cell) storms that may show little movement, leading to the possibility of very localised inundation. Local storms may also be triggered by convergence of air originating from the inland movement of sea breeze fronts (see later section), a contributory factor in the Louth storm in 1920 (see Box 4.2).

There have on occasions been more widespread intense daily rainfalls when convectional activity has augmented frontal rainfall. On 25/26 August 1912, a rainfall of 186 mm was observed at Norwich over a 29-hour period. At Brundall, to the south-east of the city, the total was as high as 205 mm. This was an exceptional event in many respects, for it was associated with a slow-moving depression rather than a localised convectional rainfall. As it approached Norfolk from the Thames estuary it deepened and the rainfall rate became enhanced by convectional activity.

Hailstorms

The most damaging outbreaks of large hailstones are often associated with warm, unstable south or south-easterly winds bringing warm air over the surface ahead of an approaching colder south-westerly or westerly airstream aloft, producing thereby a vertical wind shear. Outbreaks of severe hailstorms have been logged by the Tornado and Storm Research Organisation (TORRO), and a catalogue of storms over the past hundred years shows that the risk of damaging hail is greatest in the southern parts of East Anglia, the south-east Midlands and the London area (Webb *et al.* 1994). Another feature of the severest hailstorms is that they often occur in well-defined linear tracks, sometimes extending for over 100 km. Webb (1993) listed the location of major hailstorm tracks on thirty-two days having severe outbreaks since the late nineteenth century. Severe hailstorms were noted in eastern England on fourteen of these occasions (though many of these days had multiple storm tracks). Oddly, there were no occurrences listed between 1959 and 1985. On 12 August 1938, on the other hand, nineteen separate storms were identified. The worst hailstorms inflict severe damage to roofs and glass surfaces, and greenhouses are especially vulnerable to damage. In addition, damage to crops is sufficiently common to necessitate hail insurance cover by farmers in the region.

A severe thunderstorm on 9 August 1843 caused a trail of hailstone damage along a track from Cambridge north-eastwards to Aylsham, Norfolk (a typical orientation for severe hailstorms, steered by the incoming colder airstream aloft). Damage to the cost of £25,000 was reported to have occurred at Cambridge, while in surrounding districts crops were destroyed over a wide area. A more bizarre consequence was the death of 100 sparrows at Thetford, and bird deaths have featured in accounts of other storms in the region. Peabody (1993) reported that the damage to buildings was so severe in the small Norfolk town of Aylsham that a voluntary rate of three pence in the pound was levied by the council. Similarly, Prichard (1993) noted that a subscription fund was started to help those with damaged property after a hailstorm crossed Essex in June 1897.

Snowfall

With some notable exceptions, snowfall in eastern England is remarkable more for its frequency than its severity. The frequency of days with snow or sleet falling is greater than for some places of similar latitude and altitude in western Britain or Ireland

Box 4.2

THE LOUTH FLOOD OF 29 MAY 1920

On 29 May 1920 a low-pressure area was situated over Wales with a trough extending north-eastwards towards Lincolnshire. A thunderstorm developed around 1400 hours over the Lincolnshire Wolds, a location that suggests some triggering of convectional activity by the combined effects of an advancing sea breeze and uplift stimulated by slight orographic effects of the Wolds themselves. Within one hour of commencement, the rainfall intensity was sufficient to damage bridges over watercourses upstream of the town of Louth in the catchment of the river Lud. Rainfall totals from the storm were estimated to have reached about 150 mm within three hours at the centre of the storm, just west of Louth. Gullies cut into the chalk were reported to have been 1.85 m in depth. However, the drainage network in this area (including stream confluences) was such that the flooding was amplified just where the Lud flows through the town. Flood water within the town reached depths of 4.57 m across a 15.2-m-wide area, in effect cutting the town in two. Roads quickly filled with a tangled mass of debris (often uprooted trees) and several houses collapsed. Twenty-three people died, and a relief fund was established to assist those made homeless.

Figure 4.8 After the flood: Wellington Street, Louth as it appeared following the flood of 29 May 1920
Photo by courtesy of the Louth Antiquarian and Literary Society
Source: Robinson (1995)

and not far short of figures for more northerly locations. Waddington (Lincolnshire) and Wattisham (Suffolk) have respectively 33.5 and 30.0 days of snow falling while Buxton in the Peak District is scarcely more subject to snowfall with an annual mean of 35.9 days (see Chapter 8). A value of 33.3 days a year with snow or sleet falling is quoted for Coltishall (Meteorological Office 1989), a figure which is notably high for a site at an altitude of just 17 m. The explanation probably lies in the fact that north-westerly and northerly airstreams reach this part of north-east Norfolk as onshore winds, with the passage over the North Sea providing an effective source of instability and showers, especially in winter, when the sea surface is relatively warm. The susceptibility of the region to significant snow depends upon the synoptic situation and it can sometimes avoid the excesses of other regions. In January 1982, for example, snow-bearing fronts stagnated over south-western Britain bringing exceptionally deep and persistent snowfall to those areas (see Chapters 2 and 6); the snow failed, however, to reach most of eastern England. The region is nevertheless exposed to the north and is prone to snow showers or larger areas of snow from this direction, and an early snowfall on 22 November 1965 has been reviewed by Pedgley (1968).

It is a snow-laden easterly airstream that will usually give the heaviest falls. The North Sea is an effective means of charging easterly airstreams with snow by providing additional instability to the air during the winter. This direction is also commonly associated with depressions passing to the south of the region with, in winter, their precipitation potential being expressed in snowfall. When the sea surface temperature exceeds that of the air, the lowest layers of air become further warmed, buoyant and therefore unstable. Heavy snow showers can thus be created over the sea which then drift over coastal districts before dying out inland, unless local convergence effects take place. Figure 4.9 shows for two sites the dominance of January and February in the annual snow regime. At this time of year sea and land temperatures approach their lowest and, from

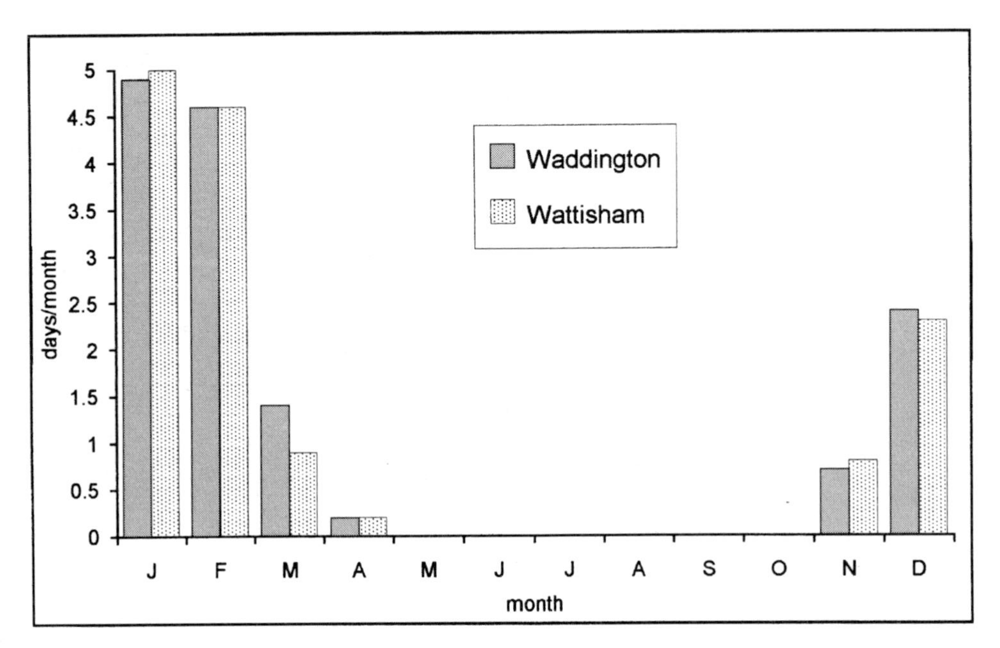

Figure 4.9 Mean monthly frequency of days of snow lying and covering more than 50 per cent of the ground 0900 GMT for the period 1961–90

the synoptic point of view, outbreaks of easterly, often polar continental, air are most probable.

It might be expected that since the region has lighter rainfall than these western districts, so the risk of deep snowfall is likewise reduced. This is confirmed by the snow depth measurements for Waddington, which indicate that most of the occurrences of deep snow (more than 16 cm) since the 1940s were actually in the cold winter of 1947. Persistence of low temperatures in winter sometimes compensates for the lack of deep snow by preserving a snow cover for lengthy periods. Snow covered at least half the ground on sixty mornings in both 1946/47 and 1962/63, roughly equivalent to the average annual frequency at Braemar (see Chapter 10). Even in the south of the region in the Ipswich area, snow lay for sixty-three days in 1962/63 (Glenn 1987). The mildest winters (for example, 1960/61) have little or no snow cover, partly because the region is often thermally favoured in mild south-westerly airstreams.

After the persistent snow cover of 1962/63, the next snowy winter was that of 1978/79. The heaviest snow in the eastern counties occurred in February, and Glenn (1987) has described the effects of this moderate snowfall in dislocating road traffic. While winter road weather conditions may appear to be eased by the infrequency of steep gradients, a different feature of the landscape worsened the effects upon travellers: the susceptibility of roads to blowing snow in the absence of hedges to provide shelter.

The next widespread major snowfall took place during a spell of intense easterly winds in January 1987. A very full analysis of this event was prepared by Bacon (1987). Between 10 and 16 January, cold easterly winds brought a succession of heavy snow showers on to the East Anglian coast. The area of unstable showery weather is evident in Figure 4.10,

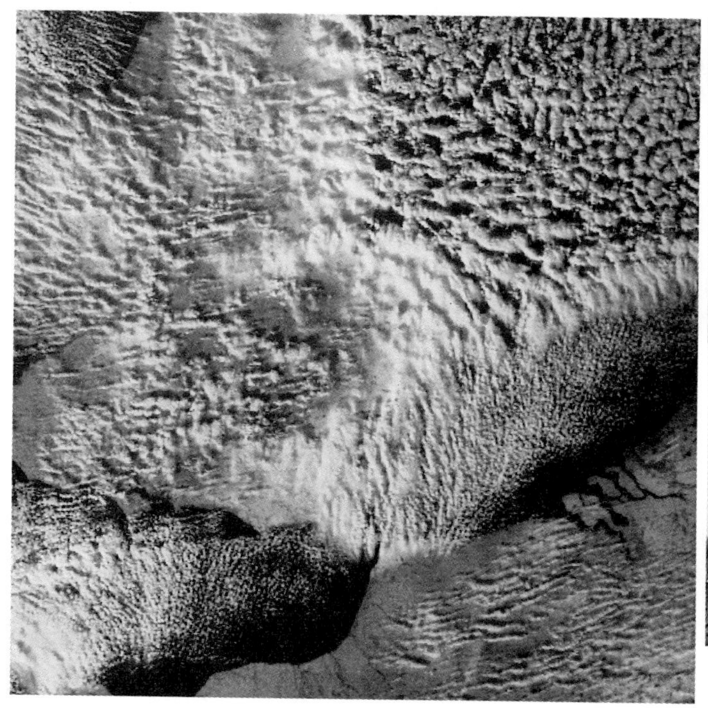

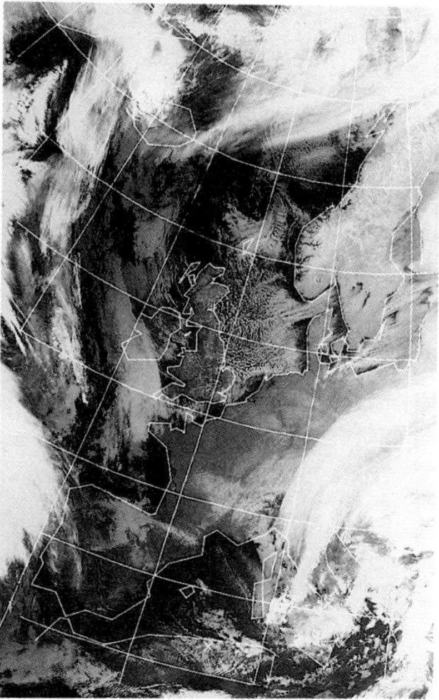

Figure 4.10 NOAA infra-red image for 11 January 1987 taken at 1415 GMT showing instability cumulus cloud over the North Sea
Photos by courtesy of Mr J. Bacon and University of Dundee

where the development of cumulus clouds within the easterly airstream can be easily identified over the North Sea. Continuation of the strong winds then allowed the lying snow to drift, worsening the effects for the communities affected and intensifying the wind-chill effect. On 12 January the temperature at Royston failed to exceed −7.3°C and the maximum at Belstead, near Ipswich, of −6.1°C was the lowest day maximum since 1902. The heavy snow showers were again caused by the air flowing over the North Sea becoming unstable owing to warming from below. While temperatures in the source region in Scandinavia were between −30 and −20°C, the surface of the North Sea was as warm as 7–8°C. Although this had the effect of ameliorating the severity of the cold close to the coast, it allowed deep cumulus cloud to develop. The showers were especially heavy a short distance inland, where a convergence zone aided the ascent of air. At the northern end of this zone, depths of undrifted snow in inland parts of Norfolk reached 45 cm. The satellite picture for 11 January (Figure 4.10) also shows some evidence for a convergence line enhancing cloud formation over the North Sea. This cloud was directed towards the area of Norfolk that received the deepest snowfall over the following 24 hours. The absence of significant development of cumulus cloud further south is notable and a consequence of the shorter sea track between the Netherlands and the Thames estuary, and of the origin of this (originally dry, continental) air over the northern European landmass. The rate at which the snow showers degenerated over the colder land surface is indicated by the fact that snow depths were less than 5 cm across much of west Suffolk and Cambridgeshire.

WIND

A distinction can be made between the climatology of gales and strong winds in eastern England and their observed consequences. In strict climatological terms, this is not one of the windiest parts of the British Isles, belonging as it does to the quieter south-eastern districts furthest from mean depression tracks. The frequency of gales increases towards the north and east with increasing proximity to the coast, and the Norfolk coast in particular is exposed to a range of wind directions, from north-west, through north to south-east. The region is nonetheless unusually sensitive to strong winds, owing to a range of non-climatic factors. These include the storm surge characteristics of the North Sea and the vulnerability of a low-lying (and largely retreating) coastline to flooding and erosion. Across the region as a whole, it might be argued that the flat landscape generates a greater impression of exposure to strong winds as well as having a reduced frictional effect on the wind itself.

Table 4.3 shows that the annual frequency of gales and strong winds is much lower at Stansted Airport in the south of the region than for Hemsby on the Norfolk coast and Cranwell inland in Lincolnshire, despite the fact that Stansted is one of the highest meteorological stations in the region. The fact that the wind vanes at the other sites are at an 'effective' height above their surroundings of more than the standard 10 m probably has some effect on the data, emphasising the northward and coastward increase in wind speed. Comparison with longer period averages at Cranwell and Stansted suggests

Table 4.3 Average annual frequency of gales and strong winds (knots)

Location	*Height of vane*		*1979–91*				*1965–91*
	(m ASL)	*(ground)*	*Gale days*	*Gale hours*	*Hours > 22 knots*	*Mean wind speed*	*Mean wind speed*
Cranwell	76	12	2.2	7.2	355	10.0	10.0
Hemsby	26	13	2.0	7.8	341	10.3	–
Stansted	107	10	0.2	0.5	69	8.2	8.3

that the more recent period covered by the Hemsby data is representative of longer-term wind speeds.

Data published by the Meteorological Office (1989) for Waddington indicate that south-west winds are more likely to exceed speeds of 4, 11 and 22 knots at this site, suggesting that for many inland areas at least, these can be regarded as both the prevailing and the dominant wind type, at least as far as speed is concerned.

The region is particularly sensitive to strong winds when a depression crosses the North Sea from north-west to south-east. Such a situation is likely to lead to a strong north to north-west wind on the western flank of the depression (i.e. over eastern England), a direction to which the region is notably exposed. Furthermore, significant reductions in air pressure over the North Sea as a whole can induce a storm surge, the sea acting like an inverted barometer (Bowden 1953). On 1 January 1922 Southend recorded a surge to 3.3 m above the expected tide level, though with great fortune this coincided with low water. Another aspect of the storm surge is the upward gradient of sea water towards the southern end of the North Sea due to frictional drag in the direction of the wind. This may lead to water piling up along the low-lying coasts of the Netherlands, northern Germany and (to a certain extent) East Anglia. There are several accounts of serious inundations of the coast during the nineteenth century (Jensen 1953). For example, in 1897 the Horsey Gap was breached and sand dunes were extensively flooded. The most famous flood in recorded history is probably that of 31 January–1 February 1953. A depression crossed the central North Sea reaching a central pressure as low as 968 mb at 1200 GMT on the 31st. Strong north-west winds then coincided with a high tide late in the day forcing the water level at Southend to rise 2.5 m above the expected height. The rise was probably greater still around the Wash, where King's Lynn was flooded to a greater depth than previously recorded. Cliff retreat following this single storm exceeded 10 m in places and across England as a whole the death toll reached 350 (though the largest number of lives lost was just outside the region at Canvey Island on the Thames estuary).

If a depression takes a more southerly track across the British Isles (rather than turning south-eastwards over the North Sea as in 1953), the region is likely to suffer from gale damage on land. Shaw *et al.* (1976) have reviewed the geographical distribution of gale damage associated with a depression that crossed the British Isles on 3 January 1976. As it deepened to a central pressure of 962 mb over the North Sea it brought a vigorous trough eastwards across England following the cold front. The strongest winds in the British Isles were recorded in the region (together with the south Midlands), and Wittering in Cambridgeshire recorded the highest gust of 91 knots. Tides were higher along the Lincolnshire coast than in the 1953 flood. On 11/12 January 1978 a similar storm resulted in more widespread flooding along the coast as an Arctic flow followed a depression tracking over the Netherlands (Lamb 1991). The sea bank and wall at Wells in north Norfolk were damaged, but owing to the extensive improvements of coastal defences since 1953 only one fatality was recorded (at Wisbech). By contrast it was south-east of the region that bore the brunt of the October 1987 storm (see Chapter 3) when gusts approached 100 knots close to the coast.

Sea Breezes

Sea breezes have a profound effect upon temperature and can also act as triggers for showers or even thunderstorms. Although they are most common over the coastal districts of East Anglia and Lincolnshire, they can occasionally penetrate inland to cover the whole region, albeit weakly and briefly. Only about half of the sea breezes noted at Coltishall actually reach Norwich. Simpson (1978) noted the arrival of weak sea breezes at Cambridge on seven days in the fine summer of 1976, five of which later progressed as far west as Bedfordshire. This is a distance of more than 100 km inland from the coast of Essex and south Suffolk, where the airflow was initiated.

Owing to the curvature of the coastline, the pattern of sea breeze movement is quite complicated and it is possible for sea breezes of differing direc-

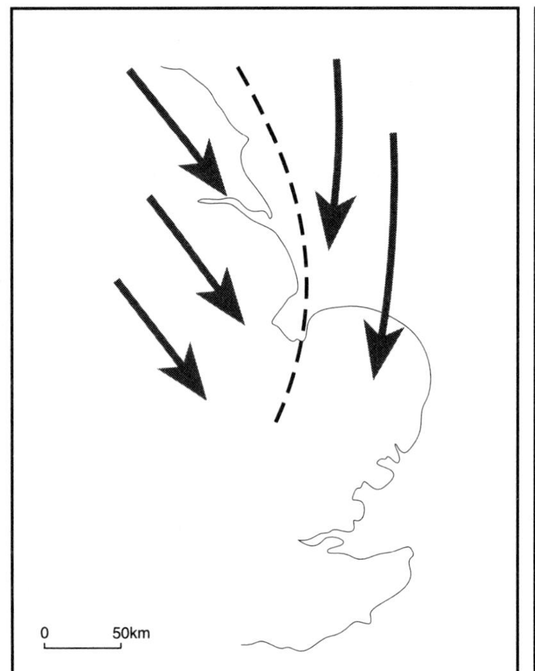

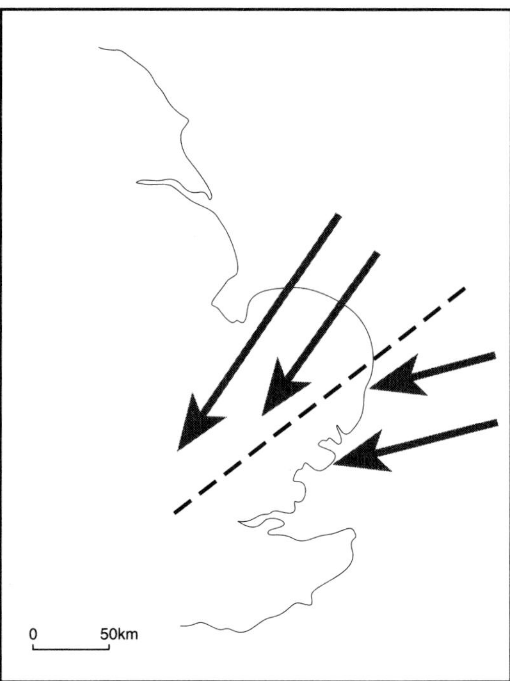

Figure 4.11 Distribution of convergence zones on days of northerly and easterly airstreams

tions to occur simultaneously. A southerly breeze may assist the inland progress of a sea breeze on the Essex coast (a process aided by the estuaries found there) but is likely to retard the formation of any onshore breeze along much of the coast of north Norfolk, and temperature differences of up to 10 degrees C are possible in spring when the cooling power of the sea breeze is especially strong. An east-south-east sea breeze may occur on the Suffolk and Essex coasts in opposition to a south-west pressure gradient wind. In such situations a convergence line may form around the meeting point of the two airstreams (Figure 4.11 and Box 4.3). This enhances the uplift at the landwards limit of the sea breeze (a point known as the sea breeze front) and increases the likelihood of creating uplift to generate large cumulus clouds and possibly showers. Indeed, Findlater (1964) has shown how this process can be a precursor to thunderstorms later in the day, and Glenn (1987) cites the case of a thunderstorm at Earl's Colne in Essex forming in this way, giving a localised rainfall of 33 mm.

Tornadoes

The most dramatic manifestation of the wind regime in eastern England is the occasional formation of tornadoes. The area of greatest frequency in the British Isles extends from the Humber estuary to Kent, thus embracing the whole of eastern England (Elsom 1985). Many of the reported tornadoes occur in humid, warm-sector air just ahead of an approaching cold front in generally cyclonic weather situations. The undercutting of the warm air by the cold leads to deep convection that can on occasion initiate tornado formation. A particularly long-lived and damaging tornado crossed Buckinghamshire on 21 May 1950 (Bonacina 1950). It passed through the centre of the small town of Wendover on the Chiltern Hills and severe roof damage was reported both there and further north at Linslade, where

Box 4.3

CONVERGENCE ZONES AND LOCAL WEATHER

In a northerly airflow, frictional and stability effects over Lincolnshire will lead to a backing of the surface wind towards north-westerly (see Figure 4.11). This effect is much less over the adjacent North Sea and therefore the northerly direction is maintained across the northern coasts of East Anglia. A convergence zone is thus created between these areas extending from the Lincolnshire coast across western parts of Norfolk and Suffolk as well as Cambridgeshire. Convergence leads to localised uplift and forced ascent, resulting in enhanced cloud development and perhaps precipitation (typically a line of showers). Showers formed in this way can persist through the night when they would otherwise be expected to die down. This is particularly important in winter, when road surface temperatures may drop close to freezing point.

A convergence zone may also be set up a short distance inland from the North Sea when a general easterly airstream backs to north-easterly over the land. Eastern districts of Norfolk, Suffolk and Essex may then experience heavy showers. This situation was responsible for the extensive snow in these areas in January 1987, December 1995 and January 1996.

about fifty houses were unroofed (Rowe 1985). A double-decker bus was overturned near Ely, Cambridgeshire. This tornado (or a successor) reached the north Norfolk coast later in the day (Simmonds 1950). A depression was situated off south-west Ireland and a humid south-easterly warm sector flow covered the region.

Elsom (1993) reported widespread tornado damage in association with the gust front of a thunderstorm. A 100-m-wide damage track occurred in the vicinity of the south Norfolk village of Long Stratton. Elsom noted that a man having a drink in the village pub witnessed the winds carry his car past the window first in one direction and then the other way as the wind direction suddenly changed! Again, this occurred in a very cyclonic situation with cold air undercutting the warm sector air.

Tornadoes sometimes form as an active cold front sweeps across East Anglia; indeed, many of these tornadoes seem to affect eastern England. The most famous outbreak is that of 23 November 1981 when 105 known tornadoes were reported during the day from Anglesey to Essex. They reached their greatest density in eastern England and as many as ten were reported from the small county of Bedfordshire alone (Rowe 1985). Eastward-moving cold fronts were also responsible for the Newmarket tornado of January 1978 (Buller 1978; Meaden 1978), when some modern houses lost most of their roof tiles, and the Winslow and Ely tornadoes of January 1981 (Buller 1982). More unusually, Briscoe and Hunt (1977) reported a multiple tornado outbreak on 1/2 December 1975 that took place overnight. Pike (1992) provides a detailed survey of damage caused by two tornadoes on 12 November 1991 that formed close to a fast-moving 'triple-point' of a depression (the point of frontal occlusion).

SUNSHINE

Average annual sunshine across the region varies from under 1,500 hours per year (little more than 4 hours per day) along the western margins of the region to around 1,650 hours (about 4.5 hours per day) on the coasts of east Norfolk, Suffolk and Essex (Table 4.4). In the absence of significant orographic influences upon cloud amount, this eastward gradient shows that coastal influences counterbalance latitudinal variation, which has little effect.

There are two reasons for higher sunshine in the

Table 4.4 Mean monthly and annual sunshine totals (hours) for the period 1961–90

Location	*Alt. (m ASL)*	*Jan.*	*Feb.*	*Mar.*	*Apr.*	*May*	*June*	*July*	*Aug.*	*Sept.*	*Oct.*	*Nov.*	*Dec.*	*Year*
Waddington	68	57	68	110	141	199	198	186	178	140	105	68	51	1,501
Hemsby*	14	54	74	119	163	210	218	210	202	157	112	69	50	1,638
Cambridge	26	54	68	109	146	193	195	185	177	144	109	67	47	1,494
Lowestoft	25	46	69	116	155	205	204	193	190	148	106	62	40	1,534
Clacton	16	58	76	117	156	208	212	200	199	154	118	75	54	1,627

* Data for Hemsby derived from observations between 1979 and 1991, reduced to 1961–90 equivalents according to the relationship between the two periods at other sites

coastal regions. First, heating of the land during daylight hours encourages the development of cumulus cloud. The formation of sea breezes has a stabilising effect on the air, discouraging cumulus growth as the breeze advances inland. Thus clear skies over the sea have a tendency to spread inland with the sea breeze. While these effects are clearly more likely in spring and summer, winter sunshine may be affected by the formation of stratus over inland areas (in response to cooling) while coastal areas may be clearer.

It is also possible to interpret variations in sunshine according to the synoptic situation, as for precipitation. North, north-east and east winds may spread either shower cumulus or stratus across coastal districts, both of which may clear with distance inland, except for summer showers, which may intensify with travel over the warmer surfaces inland. These directions may lead to a reversal of the average gradient of sunshine between the coast and inland areas. By contrast, a period of persistent south-west winds may lead to higher sunshine totals towards the Lincolnshire and Norfolk coasts. This characteristic is a clear example of the progressive drying of air as it passes over the land surface, and places such as Cromer, Cleethorpes and Hunstanton may be sunnier than the resorts of the south coast of England, which may suffer from periodic advection fog in months dominated by south-westerly winds in tropical maritime air masses.

Lamb (1987) has drawn attention to the fact that inland East Anglia (as exemplified by west Suffolk) is one of the sunniest inland parts of the British Isles. Indeed, average annual sunshine values for 1969–91 include 1,554 hours at Morley St Botolph (the Norfolk agricultural station in the south of the county), an average of 1,602 hours at Wattisham (in central Suffolk) and 1,572 further inland at Cavendish (for 1979–91). While this is probably a consequence of the absence of high ground over such an extensive area, the occasional dryness of the atmosphere in summer can discourage the formation of extensive stratocumulus or cumulus, and broken fair-weather cumulus alone has little effect on sunshine duration.

FOG

The foggiest parts of eastern England are inland areas where rapid nocturnal cooling encourages the formation of radiation fog on calm, clear nights. Despite the low-lying nature of the region, fog may also be observed at ground level on occasions of very low stratus cloud associated with cold winter conditions and heavy rainfall. Higher sites such as Stansted Airport thus compensate for a reduced frequency of radiation fogs as compared with adjacent valleys by a higher frequency of hill fog. Figure 4.12 shows how fog day frequencies (based on 0900 GMT observations) and total hours of fog vary through the season at Stansted. The greater potential for winter nocturnal cooling emerges clearly in respect of both variables, and only occasionally do summer fogs disturb the activities of the airport.

Whilst coastal areas average less than twenty-five hours of thick fog a year, this increases to over 150

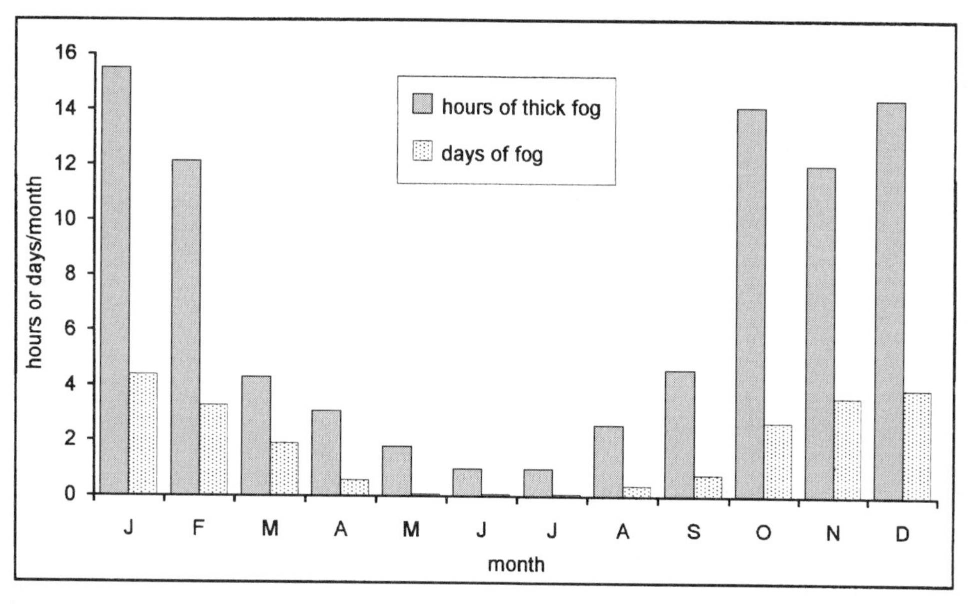

Figure 4.12 Mean hours of thick fog (visibility less than 200 m) and mean monthly frequency of days of fog (visibility less than 1,000 m) at 0900 GMT for Stansted for the period 1961–90

hours inland towards the north of the region, especially where stratus forms over higher ground. However, stratus is also an occasional but distinctive feature of the coastal climate when sea fog (or haar) is advected on to the land from the cool surface of the North Sea. As in other parts of the country, this is most likely in spring when the sea surface temperatures are less than the dew point of the air above, leading to condensation and fog formation (Sparks 1962). Freeman (1962) showed that the inland movement of stratus across the Fens is facilitated by the Wash, in effect giving the stratus a 'head-start', especially with onshore north-easterly winds. However, it must not be forgotten that the densest fogs are those associated with radiation cooling, especially over low-lying inland districts such as south-west Cambridgeshire. The propensity for fog in this area is of considerable importance in view of the M11 motorway and the risk of road accidents. An additional factor, especially in areas such as the Fenland, is the role of surface water bodies such as dykes, marshes, ponds and drainage ditches in raising the local frequency of fog.

CONCLUSION

The flatness of the region leads not to a simple pattern of climatic uniformity but instead to a climate that draws its fundamental influences from distance from the coast rather than orographic effects. Although there is evidence for relative continentality affecting many aspects of climate (including precipitation as well as temperature, for example), this does not mean that the influence of the sea is muted. The thermal properties of the North Sea oppose those of a relatively dry land surface, resulting in marked local temperature contrasts in many different weather situations. Overall, the region probably owes its climatic distinctiveness both to continentality and to its related facet of comparative dryness. The importance of this to local agriculture – particularly in view of the mid-twentieth-century decline in rainfall, especially in summer – provides further evidence of the regional differentiation of climate across the British Isles.

REFERENCES

Atkinson, B.W. and Smithson, P.A. (1976) 'Precipitation', in T.J. Chandler and S. Gregory (eds) *The Climate of the British Isles*, London: Longman.

Bacon, J. (1987) *The Heavy Snowfall in East Anglia January 1987*, Norwich: Anglia Television.

Bonacina, L.C.W. (1949) 'British weather and continentality', *Weather*, 4: 189–191.

—— (1950) 'The Buckinghamshire tornado', *Weather*, 5: 254–255.

Bowden, K.F. (1953) 'Storm surges in the North Sea', *Weather*, 8: 82–84.

Briscoe, C.E. and Hunt, M. (1977) 'Severe outbreak of nocturnal thunderstorms in East Anglia', *J. Meteorology (UK)*, 15: 69–73.

Buller, P.S.J. (1978) 'Damage caused by the Newmarket tornado of 3 January 1978', *J. Meteorology (UK)*, 3: 229–231.

—— (1982) 'Damage caused by the Winslow tornado 13 January 1981', *J. Meteorology (UK)*, 7: 4–8.

Burt, S. (1992) 'The exceptional hot spell of early August 1990 in the United Kingdom', *Int. J. Climatol.*, 12: 547–567.

Crossley, A.F. and Lofthouse, N. (1964) 'The distribution of severe thunderstorms over Great Britain', *Weather*, 15: 172–177.

Elsom, D.M. (1985) 'Tornadoes in Britain: where, when and how often', *J. Meteorology (UK)*, 10: 203–211.

—— (1993) 'The thunderstorm gust front as a trigger for tornado formation: the Long Stratton tornado of 14 December 1989', *J. Meteorology (UK)*, 18: 3–12.

Findlater, J. (1964) 'The sea breeze and inland convection: an example of their interrelation', *Meteorol. Mag.*, 93: 82–89.

Freeman, M.H. (1962) 'North Sea stratus over the Fens', *Meteorol. Mag.*, 91: 357–360.

Glasspoole, J. (1926) 'The distribution over the British Isles in time and space of the annual number of days with rain', *British Rainfall*, 66: 260.

Glenn, A. (1987) *Weather Patterns of East Anglia*, Lavenham: Terence Dalton.

Jackson, M.C. (1974) 'Largest two-hour falls of rain in the British Isles', *Weather*, 29: 71–73.

Jensen, H.A.P. (1953) 'Tidal inundations past and present', *Weather*, 8: 85–89.

Lamb, H.H. (1964) *The English Climate*, London: English Universities Press.

—— (1987) 'Some aspects of climate and life in East Anglia down the ages', *Trans. Norfolk Norwich Nat. Soc.*, 27: 385–397.

—— (1991) *Historic Storms of the North Sea, British Isles and Northwest Europe*, Cambridge: Cambridge University Press.

—— (1995) *Climate History and the Modern World*, London: Routledge.

Meaden, G.T. (1978) 'Tornadoes in England on 3 January 1978', *J. Meteorology (UK)*, 3: 225–229.

Meteorological Office (1952) *Climatological Atlas of the British Isles*, London: HMSO.

—— (1989) *'East Anglia and Lincolnshire': The Climate of Great Britain*, Climatological Memorandum 133, Bracknell: Meteorological Office.

Norgate, T.B. (1970) 'The Norfolk rainfall organisation', *Weather*, 25: 121–122.

Oliver, J. (1966) 'Lowest minimum temperatures at Santon Downham, Norfolk', *Meteorol. Mag.*, 95: 13–17.

Peabody, R. (1993) 'Awful and destructive tempest: the Aylsham hailstorm of 9 August 1831', *J. Meteorology (UK)*, 18: 15–17.

Pedgley, D.E. (1968) 'A mesoscale snow system', *Weather*, 23: 469–476.

Pike, W.S. (1992) 'The two East Anglian tornadoes of 12 November 1991 and their relationship to a fast-moving triple point of a frontal system', *J. Meteorology (UK)*, 17: 37–50.

Pollard, E. and Miller, A. (1968) 'Wind erosion in the East Anglian fens', *Weather*, 23: 415–417.

Prichard, R.J. (1993) 'Great hailstorm in Essex, 24 June 1897', *J. Meteorology (UK)*, 18: 3–12.

Robinson, D.N. (1968) 'Describing the weather in Lincolnshire dialect', *Weather*, 23: 72–74.

—— (1995) *The Louth Flood*, Louth: Louth Naturalists', Antiquarian and Literary Society.

Rowe, M.W. (1985) 'Britain's greatest tornadoes and tornado outbreaks', *J. Meteorology (UK)*, 10: 212–220.

Shaw, M.S., Hopkins, J.S. and Caton, P.G.F. (1976) 'The gales of 2 January 1976', *Weather*, 31: 172–183.

Simmonds, J. (1950) 'The English tornadoes of 21 May 1950', *Weather*, 5: 255–257.

Simpson, J.E. (1978) 'The sea breeze at Cambridge', *Weather*, 33: 27–30.

Sparks, W.R. (1962) 'The spread of low stratus from the North Sea across East Anglia', *Meteorol. Mag.*, 91: 361–365.

Stone, J. (1983) 'Circulation type and spatial distribution of precipitation over central, eastern and southern England', *Weather*, 38: 173–177, 200–205.

Terry, G.C. and Jackson, C.I. (1962) 'Irrigation in Eastern England: a case study', *Weather*, 17: 153–157.

Webb, J.D.C. (1984) 'Highest authentic maximum temperatures in British counties', *J. Meteorology (UK)*, 9: 82–84.

—— (1993) 'Britain's severest hailstorms and "hailstorm outbreaks" 1893–1992', *J. Meteorology* (UK), 18: 313–327.

Webb, J.D.C., Rowe, M.W. and Elsom, D.M. (1994) 'The frequency and spatial features of severely-damaging British hailstorms, and an outstanding 20th-century case study monitored by the T.C.O. on 22 September 1935', *J. Meteorology (UK)*, 19: 335–345.

5

THE MIDLANDS

John Kings and Brian Giles

INTRODUCTION

The English Midlands can be viewed as that part of the centre of England which remains when other, perhaps more distinct, regions have been assigned. Uniquely among the regions of this volume, its boundaries are not even partly aligned with obvious and striking physical discontinuities such as a major watershed or a coastline. However, this fact should not detract from the importance of a region which has a clear recognition in the public consciousness.

There is, of course, more than one way of defining a region. For this chapter, the periphery of the Midlands is marked by a line running from the Severn estuary, across the Cotswolds to the upper Thames valley in the south, by the Jurassic escarpments of the Northamptonshire uplands in the east, to the Trent valley and then taking in the southernmost parts of Derbyshire and the Welsh Marches. However, for the purposes of this study, the region may be identified with the administrative counties which surround Warwickshire – Shropshire, Hereford and Worcester, Gloucestershire, Northamptonshire, Nottinghamshire, Derbyshire and Staffordshire. The Midlands has a vague unity brought about by the hilly region of the Birmingham plateau which separates its major river systems, the Severn, Trent and Avon. High ground emerges again as an important topographic element only towards, but not necessarily forming, the region's limits: the Welsh Marches to the west, the hills of the Staffordshire moorlands and lower parts of the Peak District to the north, the Cotswolds to the south and the Jurassic hills to the east. As such, the region is analogous to a shallow bowl with a central, if subdued, central dome. Finer detail is added by the free-standing prominences in the west of the Malvern Hills, the Clee Hills and the ridge of Wenlock Edge. It is a region of varied physiography and it is this which has important implications for the climate.

Observing Stations and Data

Within the Midlands there is a long history of weather observing and recording. The most notable and earliest contribution to an understanding of the region's climate was made by Thomas Barker (1722–1809), who was squire of Lyndon Hall in Rutland. His weather journal spans the period from 1733 to 1795 and includes not only narrative accounts but a notable accumulation of instrumental data (Kington, 1988). Today several stations figure proudly in the region's network: Birmingham, Ross-on-Wye (see Box 5.1), Raunds and Cheltenham all have records from the 1850s. The later part of the nineteenth century was a popular time for weather recording, and other locations became part of a developing network. There are therefore other stations with a century or more of weather data. These include Malvern, Coventry, Leicester and Northampton, though contineuous observations in all but the first of these towns have now ceased. In Birmingham, members of the Lunar Society designed and built a number of self-registering instruments (Robinson, 1957). They began routine weather observations in April 1793,

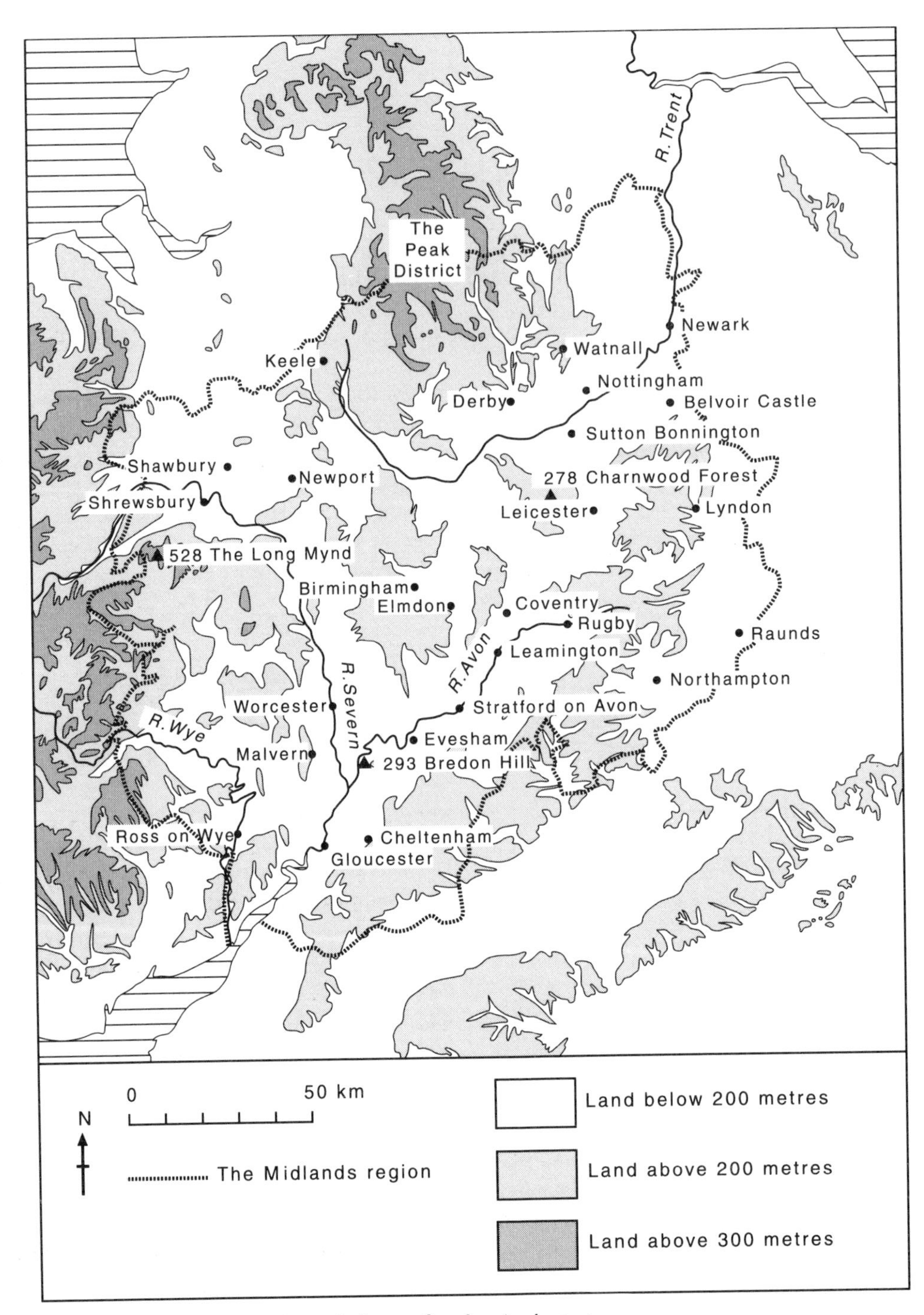

Figure 5.1 Location of weather stations and places referred to in the text

Box 5.1

CLIMATOLOGICAL OBSERVATIONS AT ROSS-ON-WYE: AN EXEMPLARY RECORD

The small Herefordshire town of Ross-on-Wye has an outstanding record of weather observation which is possibly without equal in terms of individual dedication. Observations of rainfall were begun in 1859 by a Mr Henry Southall, with temperatures being recorded from 1875. A separate station was set up in 1914 by Mr F. Parsons, who continued observing until 1974. The resulting 116 years of observations had been provided by just two observers – an apparently unprecedented feat anywhere in the world (Parsons 1975).

This dedication was a fitting tribute to the owner of the land on which the new station stood – Mrs Edith Purchas – who stated in her will that the land was to continue to be used for meteorological purposes. The station flourished under Mr Parsons's care; it was designated a Health Resort station and appeared in the *Daily Weather Report* from its opening in 1914. It then progressed to be an auxiliary station after 1920, and by 1938 observations were being sent to the Meteorological Office six times a day. The centenary celebrations were marked by an account by John Glasspoole (1959), at which Mr Parsons made the comment: 'I feel confident that in the near future some practical scheme can be devised whereby the future of the Observatory can be established in a firm basis for observations to be continued for yet another 100 years.'

Alas, this was not to be. As a result of a problem that is all too common, the station closed in 1975 just a few months after the retirement of Mr Parsons – a result of the impossibility of finding observing staff. Because of the continuing appreciation of the value of this station for climatology, the station was reopened in 1985 and continued into the 1990s as a Health Resort station giving daily weather observations to the Meteorological Office.

which makes the Birmingham record the longest in the UK (Giles and Kings 1996). A count of stations published in the *Monthly Weather Report* from the 1880s onwards gives about ten sites in that decade increasing to about sixty in 1960. Since then the number has remained fairly constant. In common with other regions, climatological stations are relatively plentiful in comparison with the networks of synoptic, auxiliary and anemograph stations reporting to the Meteorological Office (UKMO). Between 1960 and 1992 the synoptic network was reduced from ten to three. Other networks of observing sites include those of the Climatological Observers Link (COL) with around 50 stations and the Road Weather Information Sensors (RWIS). The latter is used mainly for winter road maintenance. Despite the proliferation of stations, few site histories and summaries are available in published form. The literature that is available focuses on Nottingham (Barnes 1966), Keele (Beaver and Shaw 1970), Leicester (Pye 1972), Birmingham (Kings 1985; Giles and Kings 1996), Rugby (Burt 1975, 1976) and Ross-on-Wye (Parsons 1975). By far the most comprehensive climatology is that published for Birmingham, where synoptic observations ceased in 1994 but, together with anemograph records, are continuous since 1885 (Giles and Kings 1996).

A Midlands Climate: Towards a Definition

So what constitutes the climate of the Midlands? There are a number of ways of defining climate and various methods of classifying the climate of a region. This may include some combination of temperature and rainfall, with a loading towards rainfall seasonality and water availability such as that devised by Gregory (1976). In this the Midlands is assigned to his class BD2, which translates into the following characteristics:

- a growing season of 7–8 months;
- a probability greater than 0.3 of annual rainfall being less than 750 mm;
- a rainfall maximum in the latter half of the year.

It could be suggested in addition that a distinguishing feature about the climate of the Midlands is its transitional character. It is probably a truism to note that it features a thermal transition between the north and the south of England. Any tendency towards east-to-west contrasts in precipitation is probably moderated by the fact that the western side of the region in particular lies in the rain shadow of the Welsh mountains. The overall isolation and distance from the sea also imparts to the region a more continental character than is found elsewhere except perhaps for the area around the mouth of the Thames. The general climate and weather of the region is, however, controlled by the same synoptic-scale features which dominate and persist to varying degrees over the rest of the British Isles. It is the interaction of these air masses with the extensive land surface which finally contributes to the climate of the Midlands.

TEMPERATURE

Temperatures within the region are governed by the prevailing air masses. Here, topography often plays an important role in modifying air mass characteristics both before and after they cross the regional boundary. Furthermore, in a region where the influence of the sea is minimal, temperatures will respond with more than usual readiness to changes in radiation input and losses. This aspect is most evident where large-scale frost hollows are found in the river valleys. The Severn, Wye and Avon valleys

Box 5.2

THE RECORD-BREAKING HEAT OF 3 AUGUST 1990 – WHY SO HOT?

The pattern of maximum temperatures on this memorable day has been comprehensively reviewed by Burt (1992), who showed that they exceeded 34°C across almost all of the region and exceeded 35°C in most of the lower Severn and Avon valleys. Highest readings were 37.1°C at Cheltenham and 37.0°C at Barbourne, Worcester. These are at least half a degree Celsius higher than the next highest maxima, which were recorded in Cambridge and the London area. Why was this area so warm?

Figure 5.2 shows the importance of anticyclonic conditions with the slow motion of tropical continental and subsiding air moving towards and over the British Isles. Both sites are subject to a modest degree of urban warming, which would accentuate the warmth of all low-lying, inland and sheltered areas in such heatwave conditions. A more speculative explanation is that there may have been a slight south-east flow of air off the Cotswolds which would have warmed adiabatically on its steep descent of the scarp face into the Severn valley, thus qualifying for description as a föhn wind. However, there was a col between principal anticyclonic centres over the country with only light winds. But such an occurrence was by no means improbable, and it is interesting to note that Cheltenham (Figure 5.3) also recorded the highest temperatures during the remarkable summer of 1976 with a maximum of 35.9°C on 3 July, suggesting that some local factor may occasionally favour high temperatures.

Table 5.1 Record absolute maximum and minimum temperatures (°C) for the region

County	*Maximum*	*Date*	*Location*	*Minimum*	*Date*	*Location*
Nottinghamshire	36.1	3 Aug. 1990	Clarborough	−21.1	7 Dec. 1879	Hodsock
Staffordshire	34.5	3 Aug. 1990	Penkridge	−20.0	14 Jan. 1982	Penkridge
Leicestershire	35.0	3 Aug. 1990	Belvoir Castle Caldecott	−22.2	8 Feb. 1892	Ketton
Shropshire	34.9	3 Aug. 1990	Shawbury	−26.1	10 Jan. 1982	Newport
Warwickshire	36.1	3 Aug. 1990	Wellesbourne	−21.0	14 Jan. 1982	Stratford-upon-Avon
Northamptonshire	36.7	9 Aug. 1911	Raunds	−23.9	14 Jan. 1982	Higham Ferrers
Herefordshire	35.0	3 Aug. 1990	Ross-on-Wye	−22.4	13 Dec. 1981	Preston Wynne
Gloucestershire	37.1	3 Aug. 1990	Cheltenham	−20.1	14 Jan. 1982	Cheltenham
Worcestershire	37.0	3 Aug. 1990	Barbourne	−19.2	14 Jan. 1982	Pershore

Sources: Meaden (1983), Webb and Meaden (1983 and 1993)

are the areas most likely to attract cold air drainage and they enhance thereby the general frostiness of the western half of the Midlands. On the other hand, the topography of the East Midlands is less conducive to frost resulting from cold air drainage. The potential for very low minima in the west of the region was fully realised by the extreme conditions of December 1981 and January 1982 when cold, slow-moving or calm Arctic continental air accumulated over the already snow-covered Midlands. Two remarkable temperature records for England were established in separate spells: −25°C at Shawbury on 13 December and −26.1°C at Newport (Shropshire) on 10 January. Roach and Brownscombe (1984) obtained a more comprehensive picture of the temperature distribution on these exceptional nights by analysis of satellite imagery. Very low temperatures were also observed on 14 January across a wider area of the south Midlands, with a particularly extensive pool of cold air in the upper Avon valley in Warwickshire, where several sites reported air minima below −20°C (Burt 1984).

Just as the inland location of the Midlands reduces the moderating effect of the sea in cold winter weather, it also promotes high temperatures in summer, especially when combined with subsiding tropical air within an anticyclone or a downslope föhn wind off the surrounding 'rim' of the region. Both of these factors were active 3 August 1990, when the heat exceeded 34°C across most of the region and the highest temperature ever recorded in the British Isles (37.1°C) was observed at Cheltenham (see Box 5.2). These extreme events nicely illustrate the large range of temperatures possible within this, Britain's only, land-locked region. Burt (1992) noted that the small Shropshire town of Newport has experienced the largest range in temperature of any station in the British Isles. More remarkably, this has occurred within a ten-year period from the −26.1°C in January 1982 to the 34.9°C on 3 August 1990. Table 5.1 summarises these and similar extremes county by county within the region.

Until 1990 the highest temperatures in the region were often observed in the lower-lying valleys of the East Midlands. For many years the highest maximum temperature recorded in the British Isles was 36.7°C at Raunds in the Nene valley, and temperatures still tend to be higher on this side of the Midlands, which is less exposed than the Severn valley to the cloud and rain of an unsettled south-westerly airstream. Table 5.2 shows that the mean monthly maximum temperatures in the summer half-year at Raunds compare quite favourably even with the more southerly stations at Malvern, Ross-on-Wye and Cheltenham. Hamilton (1987) has reviewed, for Birmingham, spells of exceptional warmth and coolness.

Table 5.2 also reveals a general south-to-north temperature gradient. Mean annual maximum temperatures reflect this, with the 14°C isotherm to the south of Birmingham and the 12°C isotherm

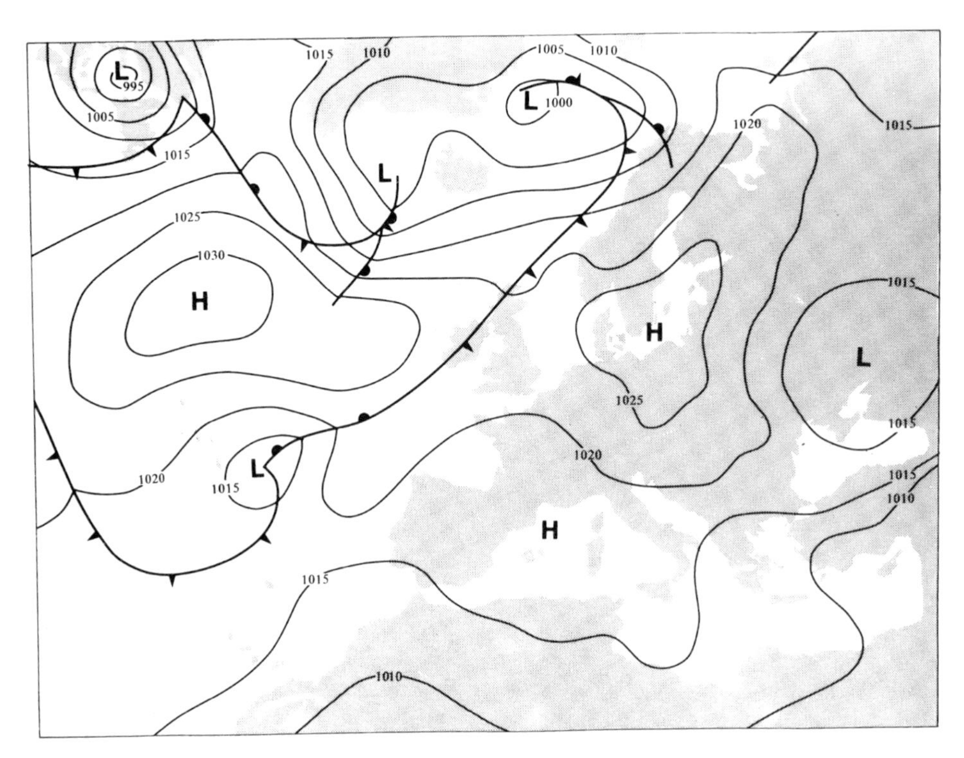

Figure 5.2 Synoptic chart for 1200 GMT on 3 August 1990

running across the north-east of the region. Mean minimum temperatures are dictated more forcefully by the local relief, and lowest mean readings occur more generally across the region with no corresponding overall trend. The warmth of Birmingham, most clearly seen in the minimum temperatures, is associated with the urban heat-island effect, although, as Unwin (1980) has demonstrated, this influence is highly dependent upon the prevailing synoptic situation at any given time.

On average the temperature can be expected to reach 26°C on at least one day by June in the warmer counties of Northamptonshire, Worcestershire, Gloucestershire and Warwickshire, but will not attain this level until July elsewhere in the region. The warmth persists through September, especially in southernmost districts where heavier, clay soils will retain the summer heat. In Staffordshire, Nottinghamshire and Derbyshire the more sandy soils will already have started to dissipate the residual heat of summer. It is in these northern and eastern areas that the autumn fall in temperature is most marked, rendering them liable to frost from October onwards. Bonacina (1964) has also related the occurrence of autumn warmth to geography, showing the local importance of geological and soil conditions.

Frost in the Midlands

The preceding paragraphs will have made it clear that frost is a widespread, occasionally marked, characteristic of the region. Air frosts are by no means restricted to the winter months, and the combination of a dry soil, radiational cooling and cold air drainage leaves many areas vulnerable as late in the season as May or, on the other hand, as early as September. This essentially leaves only a sixty-day

Figure 5.3 Where Britain's highest official temperature was recorded: a view of the Cheltenham weather station
Photo by courtesy of Julian Mayes

frost-free period around mid-July. Ground frosts are approximately twice as frequent as air frosts and there is no time of year when these have not been known to occur at many sites. Shawbury can expect as many as 123 ground frosts a year and 62 air frosts, while Elmdon records averages of 109 and 50 respectively. Keele records an average of 112 ground frosts and 59 air frosts but avoids both forms in summer as a result of its hilltop location, which excludes the possibility of the accumulation of cold air drainage from nearby higher ground.

PRECIPITATION

Most of the Midlands has an annual total within the range 600 to 850 mm with the wettest parts of the region having more than twice the average annual precipitation of the driest. The driest areas are in the lowest parts of the eastward-draining valleys where precipitation is less than 600 mm annually in parts of the lower Nene valley (as exemplified by Raunds in Table 5.3) and falls close to 550 mm downstream from Newark-on-Trent in east Nottinghamshire.

Table 5.2 Mean monthly maximum and minimum temperatures (°C) for the period 1961–90

Location	*Alt. (m ASL)*	*Jan.*	*Feb.*	*Mar.*	*Apr.*	*May*	*June*	*July*	*Aug.*	*Sept.*	*Oct.*	*Nov.*	*Dec.*	*Year*
Watnall	117	5.9	6.0	8.8	11.6	15.6	18.7	20.4	20.2	17.5	13.7	8.9	6.7	12.8
		0.6	0.5	1.8	3.6	6.4	9.4	11.3	11.2	9.3	6.8	3.1	1.5	5.5
Keele	179	5.5	5.6	8.0	10.7	14.6	17.1	19.3	19.1	16.3	12.6	8.5	6.7	12.0
		0.6	0.4	1.7	3.2	6.3	9.1	11.3	11.2	9.0	6.5	3.1	1.8	5.3
Birmingham	132	5.6	5.9	8.6	11.4	15.1	18.3	20.2	19.8	17.2	13.4	8.7	6.4	12.5
		1.8	1.5	2.7	4.2	7.1	10.1	11.9	11.8	10.0	7.6	4.1	2.7	6.3
Raunds	59	6.0	6.5	9.6	12.4	16.3	19.6	21.6	21.3	18.6	14.4	9.3	6.9	13.5
		0.4	0.4	1.6	3.3	6.1	9.1	11.1	11.0	9.1	6.5	2.9	1.3	5.2
Malvern	62	6.8	7.0	9.7	12.6	16.4	19.6	21.6	21.0	18.3	14.4	9.9	7.7	13.7
		1.7	1.5	2.9	4.7	7.4	10.5	12.5	12.3	10.2	7.6	4.1	2.5	6.5
Ross-on-Wye	67	6.9	7.0	9.7	12.5	16.1	19.2	21.2	20.8	18.1	14.4	9.9	7.8	13.6
		1.4	1.2	2.5	4.1	6.9	9.8	11.7	11.6	9.6	7.1	3.7	2.3	6.0
Cheltenham	65	6.7	6.9	9.6	12.5	16.4	19.7	21.7	21.0	18.3	14.6	9.8	7.7	13.7
		1.4	1.3	2.7	4.4	7.4	10.5	12.4	12.1	10.0	7.6	3.8	2.2	6.3

Note: ASL = above sea level

Table 5.3 Mean monthly and annual precipitation totals (mm) for the period 1961–90

Location	*Alt. (m ASL)*	*Jan.*	*Feb.*	*Mar.*	*Apr.*	*May*	*June*	*July*	*Aug.*	*Sept.*	*Oct.*	*Nov.*	*Dec.*	*Year*
Watnall	117	62	53	57	53	56	60	56	63	57	59	63	68	707
Raunds	59	51	40	47	50	49	53	48	54	46	49	53	56	596
Keele	179	71	51	56	55	63	69	65	78	69	71	77	71	796
Stratford-upon-Avon	49	52	40	48	44	49	54	51	65	50	49	53	61	616
Ross-on-Wye	67	70	51	53	44	54	51	42	58	60	59	60	69	671
Malvern	62	70	55	58	49	59	57	48	62	61	60	62	74	715
Cheltenham	65	62	50	57	51	61	64	52	66	62	56	62	74	717
Newport (Shropshire)	64	56	43	50	48	57	54	49	60	56	52	62	64	651

Nowhere in the British Isles has significantly less precipitation, and this emphasises the degree to which the eastern fringe of the Midlands has totals similar to those of the Fens.

The precipitation of the Midlands reflects the influence of altitude, with the higher totals being limited to small areas of high ground. Those parts of the Peak District included within this region, their extension into the Staffordshire moorlands, and the Welsh Borders have totals widely exceeding 1,000 mm. The wettest part of the Cotswolds above Cheltenham reaches about 950 mm, but even this is drier than the western side of the Forest of Dean, which lies on the south-west limits of the region and is one of the few areas of the region where the degree of wetness is influenced by proximity to the sea; this is a good illustration of how proximity to moist south-westerly airstreams can produce higher totals than for some more elevated districts. Away from these peripheral zones the highest averages are only just over 800 mm in small outliers mostly on or near the Birmingham plateau. Whilst this total is attained over the rising ground of Cannock Chase in Staffordshire, it is interesting to note that the more easterly Charnwood Forest area in Leicestershire (which is similar in terms of elevation) has a lower annual average of around 750 mm. Keele occupies an elevated site in north Staffordshire close to the region's boundary. In this location it is unusually exposed to the north-west and its precipitation is accordingly greater.

Apart from relatively small areas, therefore, most of the Midlands has a notably moderate rainfall regime with annual totals of less than 750 mm, although from the national point of view it should be noted that the 750 mm isohyet makes a notable westward deviation towards the Welsh border leaving much of southern and northern England at similar altitudes in a wetter zone. This regional dryness finds particular expression in the relatively low annual duration of precipitation in the west of the region (Atkinson and Smithson 1976). The reason for this is primarily the fact that the western Midland counties lie in the rain shadow of the Welsh mountains and therefore offer a parallel with the eastern parts of north-east England (Chapter 7) and Highland Scotland (Chapter 10). This effect is enhanced at a more local scale by the topography of the Wye and Severn valleys where values fail to exceed even 700 mm in the lower-lying stretches around Hereford and the mid-Shropshire lowlands to the east of Shrewsbury. The analysis of the synoptic origin of precipitation at Shawbury (Mathews 1972) highlights the relative dryness of warm-sector situations, so characteristic of rain shadow areas.

The Avon valley emphasises the region's tendency to dryness yet again with values less than 620 mm. For example, Stratford-upon-Avon's annual average precipitation is only 616 mm and some other small areas register less than 600 mm. This is an exceptionally low value for such a westward location and is only about three-quarters of the average for the Birmingham plateau to the north and the Cotswolds to the south. Further south-westward progress brings us into the fertile market gardening area around the Vale of Evesham. An interesting feature of this valley is the way in which average annual rainfall increases towards the south-west as the coast and the exposed lowlands of the lower Severn are approached. Below the Avon's confluence with the Severn at Tewkesbury the region embraces an area that is more exposed to the moisture of south-westerly, often tropical maritime, airstreams, but the Vale of Evesham itself lies in the lee not only of the Welsh mountains under westerly regimes but also, when southerly winds prevail, of the Cotswolds. Thus the south-west is by dint of geography the only direction in that general quarter from which airstreams can approach unobstructed. This particular source of moisture is hinted at in the following piece of Worcestershire weather lore, which refers to stratus gathering on one of the most striking topographic features in the local landscape:

> When Bredon Hill puts on his hat,
> Ye men of the vale, beware of that.
>
> (quoted in Inwards 1893)

Precipitation is generally well-distributed through the year, though western and central areas have a tendency for a maximum in the second half of the year, and December is often the wettest month. This is most clearly expressed in Hereford and Worcester and Gloucestershire. The tendency for the February to April period to be the driest in the year is less marked in this region than in many others and is so diminished in the cases of Ross-on-Wye and Malvern (Table 5.3) that July is the driest month at both sites. The 1961–90 averages identify quite a large area of the south-west (and the east) Midlands with less than 50 mm of rainfall in July, and the value of 42 mm at Ross is one of the lowest July totals in the country. December and January are clearly the wettest months in the hillier areas, where frontal rainfall exerts a strong influence, as well as more generally in the south-west of the region. The driest time of the year is usually spring especially in the south-west, and dry Februaries are usual in eastern and central sites such as Birmingham, Nottingham and Raunds.

Rain day and wet-day statistics give a general

indication of the frequency of rainfall during a month. There are, over the region as a whole, some 180 days on which measurable rain falls and of these about two-thirds are designated 'wet', with at least 1 mm of rain. Rainfall in the summer half-year is concentrated into a smaller number of days than that of the winter months. This confirms the tendency for summer rainfall to occur at higher intensities and over shorter periods than at other times owing to an increase in the contribution from convectional activity and a decrease in frontal rainfall. This contrast is a statistical expression of the response to the heating of the region's extensive land area in summer. One outstanding case of heavy and persistent frontal-cyclonic rainfall in summer occurred on the August Bank Holiday (26 August) of 1986 as the remnants of ex-hurricane Charley crossed England and Wales. Although the heaviest rain fell in Wales (see Chapter 6), the storm nevertheless preserved enough vigour to produce daily rainfall totals of more than 50 mm over several parts of the Midlands. It was made a more memorable event in the region because of the fact that it coincided with the first Grand Prix to be held in the City of Birmingham, an event that was eventually abandoned owing to surface water on the roads. Several other former tropical storms have produced similar events at about the same time of year. These storms occur frequently during August and September, at the end of the Atlantic hurricane season.

Large daily totals result most commonly, however, from the accumulation of extended periods of moderate rainfall (up to 4 mm per hour). For Birmingham, hourly totals rarely exceed 10 mm. This observation is consistent with the findings at Keele made by Mathews (1972), and the distribution of daily rainfall totals shows a marked reduction in frequency above 9 mm per day. In this connection, Kings (1982) has shown that a daily rainfall threshold of 15 mm identifies the 98th percentile of daily rainfalls, i.e. only 2 per cent of all daily totals exceed this value. Although occurring infrequently, in combination such extreme events can give up to 20 per cent of the annual total. Seasonally, these events reach a peak during the summer months of July and August, and again suggest the importance of convectional rainstorms. Their frequency exhibits an inevitable intra-annual variability that is dictated by the dominant synoptic character of the season, though overall their frequency has remained quasi-stationary. A high proportion of these events in the Midlands are associated with non-progressive weather types, including a substantial number of anticyclonic and easterly situations (the occasional importance of the latter is exemplified in the following section on the region's snowfall).

Snowfall

In climatological terms snow events are easily identified (Kelly 1986) but their areal distribution is a forecaster's nightmare, reducing both public and utilities to panic (see Box 5.3). They are often associated with well-defined synoptic situations. These include unstable Arctic or polar maritime airstreams, especially in conjunction with polar lows (Lyall, 1970, 1977) and areas of low pressure moving east to the south of the region (Fairmaner 1987). Such situations are capable of producing copious snowfall in eastern England (see Chapters 4 and 7), but the subdued topography of the east Midlands allows the occasional vigour of such systems to extend their grip as far west as the Welsh Marches.

The snow season in the region occurs during the months November to April, although snow has ocurred in October, May and June. In the west and north altitude plays an important role, whilst in the east the snow season is sometimes prolonged by incursions from the North Sea. The best snow events are a combination of shallow convergence under anticyclonic conditions and an easterly flow. Whiteout conditions then persist on the higher ground. The Watnall and Elmdon sites (Figure 5.5) are representative of the region's snow season.

Birmingham has the longest snow record in the UK (Jackson 1978). Here, the annual median number of days on which snow is observed to fall is near thirty and reaches its monthly maximum during January and February, though any month from

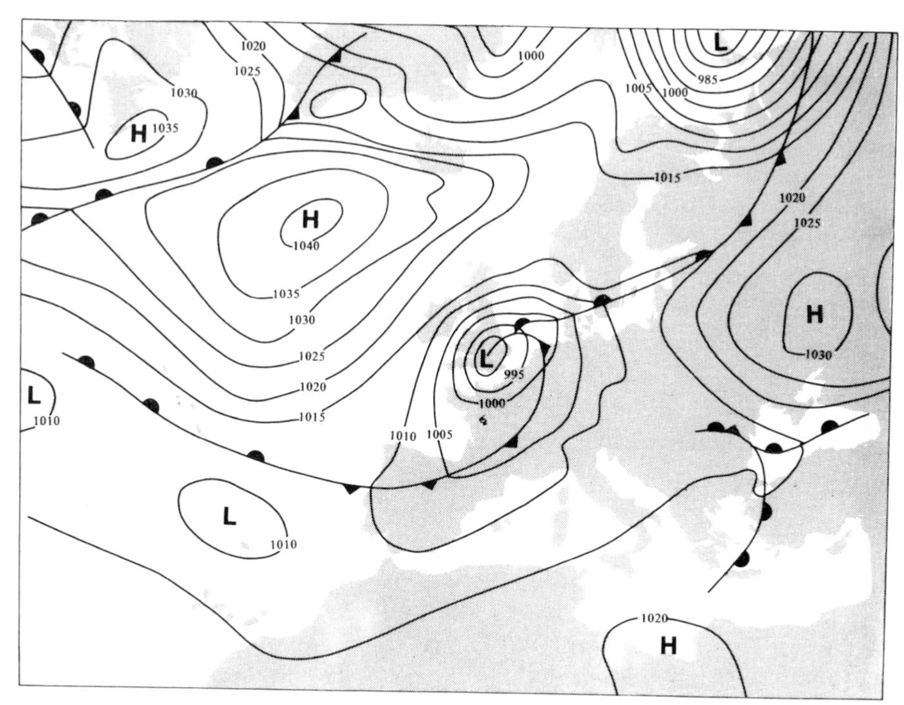

Figure 5.4 Synoptic chart for 1200 GMT on 8 December 1990

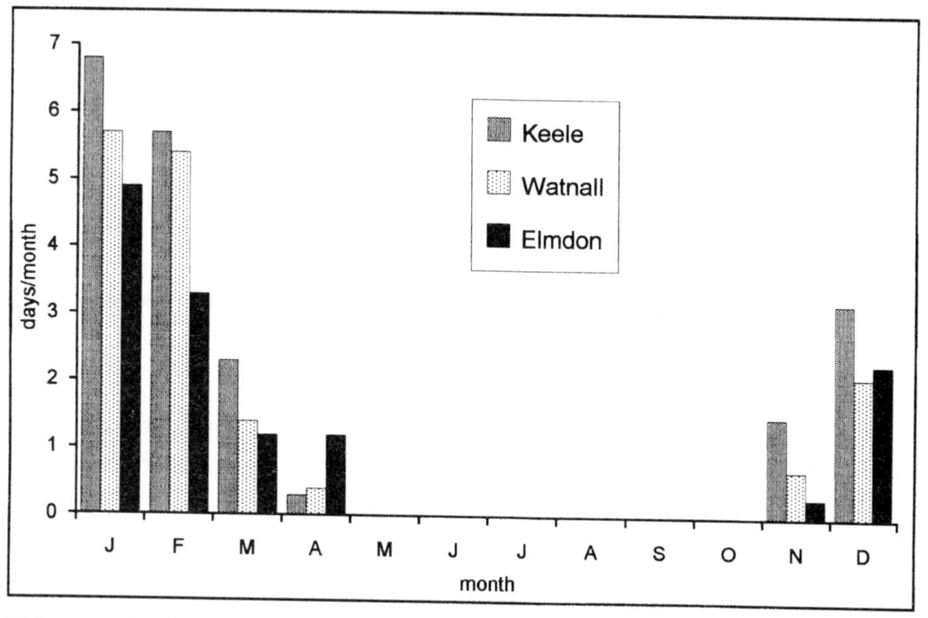

Figure 5.5 Mean monthly frequency of days with snow lying and covering more than 50 per cent of the ground at 0900 GMT for the period 1961–90

November to April can have up to twenty days. October and May falls are not unknown, and Beaver and Shaw (1970) noted a fall on 18 May 1968 at Keele. The events of 2 June 1975 at nearby Buxton are reviewed in Chapter 8. For low-level sites across the region October and May snowfalls occur about once every five to ten years.

The accuracy of snow day data often depends on the enthusiasm of the local observer, and individual falls can easily be missed. A more reliable and consistently observable element is snow cover at the 0900 GMT daily observation. In the Midlands, as in many low-lying parts of other regions, there is a relatively high frequency of days when snow is observed to fall but a relatively low frequency of days when it lies. Depths of between 1 and 10 cm are most common. Only on higher ground, such as the southernmost parts of the Pennines or the desolate ridge of the Long Mynd, are air temperatures low for a sufficient length of time for the snow to persist on the ground for any length of time. Because of its greater exposure to tropical maritime air masses, the south-west of the region experiences less snowfall than other areas. The Gloucester annual average number of days with snow lying is only 6.7, but northwards and eastwards this figure rises to 11.5 at Shawbury, 12.2 at Birmingham (Elmdon) and 15.7 at Watnall (the respective number of days of snow falling are 19.5, 30.7, 33.0 and 36.4). At most stations January is the month when snow is most often to be found lying, but February most often witnesses it falling. Snowfall is, however, a

Box 5.3

THE GENERAL PUBLIC IGNORE THE DANGERS OF MIDLANDS SNOW

The events of 7 and 8 December 1990 provide a good example of one of the synoptic conditions likely to lead to heavy snowfall in the Midlands region. They also demonstrate that the population at large can often underestimate the threat posed by some aspects of our seemingly benign climate.

A low-pressure system formed over southern England on 7 December, producing a fall of pressure of 40 mb in just 36 hours as the system then moved away to the east and south-east. The resulting easterly airflow produced heavy rain over north-east England but in the Midlands it was cold enough to produce snow, which began to fall on the night of 7 December. By the following evening much of the region had a level covering of 20 cm and in the Birmingham area snow accumulated to more than 40 cm. It was the heaviest fall of snow in the area for nearly ten years. The strong winds made matters worse by whipping the snow into drifts of 60 cm, most seriously over the motorways. Figure 5.4 depicts the situation at midday on 8 December. The rapidly deepening low dominates the picture with fresh north-east winds to the north of the occluding front providing the driving force behind the rain and snow. It represents a near-ideal situation in which heavy precipitation will occur in exposed eastern districts. In the case of winds from this particular quarter there was also little to obstruct the south-westwards progress of the rain, which turned to snow in the cold air of the Midlands region.

The local forecaster's view of events is described by Galvin (1991). Perhaps one of the most remarkable aspects of the situation was provided by the public: despite the frequent warnings issued by the Meteorological Office many people insisted on making largely unnecessary journeys, their snow-bound cars merely adding to the confusion on the motorway. Galvin (1991) reported that some trips were made merely to look at the snow! The extent to which the conditions disrupted the emergency services is also reviewed by Speakman (1994).

highly variable phenomenon from year to year; the Birmingham record shows, for example, seventy-five days of snow lying in the winter of 1962/63 but only two in 1960/61 (Jackson 1978), the corresponding figures for Keele were fifty-nine and four (Beaver and Shaw 1970). This important characteristic should be remembered when studying average figures.

Although winter depressions located to the south of the region are the model for snowfall situations (Box 5.3), frontal systems do not have the sole claim in this regard. Beaver and Shaw (1970) note the importance of polar lows and troughs embedded within northerly airstreams in producing snow in the northern part of the region. Snow will often fall during spring from deep cumulonimbus clouds as a result of increasing convective activity and steeper lapse rates in the atmosphere that encourage the required degree of instability to initiate these events. Often under these circumstances, snow will be accompanied by hail. Counts of hail days are vulnerable to the same criticisms as those of snow days; nevertheless, the annual averages vary between a little over four at Shawbury to nearly seven in such disparate locations as Gloucester, Birmingham and Nottingham, but everywhere reveals a peak of activity in springtime (Meteorological Office 1982). For reasons of exposure, Nottinghamshire and Staffordshire, which mark the open north-east and north-west corners of the region, are most vulnerable to hail-producing unstable northerly airflows.

Thunderstorms

Thunder is most likely to occur when surface temperatures are highest, and for this reason the thunder season is confined very much to the period from May to September, reaching its peak in July and August. Their average annual frequency across the region varies from fifteen days in the east to eight days elsewhere. Eastern districts tend to be more thundery than those in the west, and this contrast is seen in Figure 5.6, where the monthly data for Watnall (annual mean 14.2 thunder days), Keele and Elmdon are presented. Barnes (1966) has suggested that Belvoir Castle in Nottinghamshire is one of the

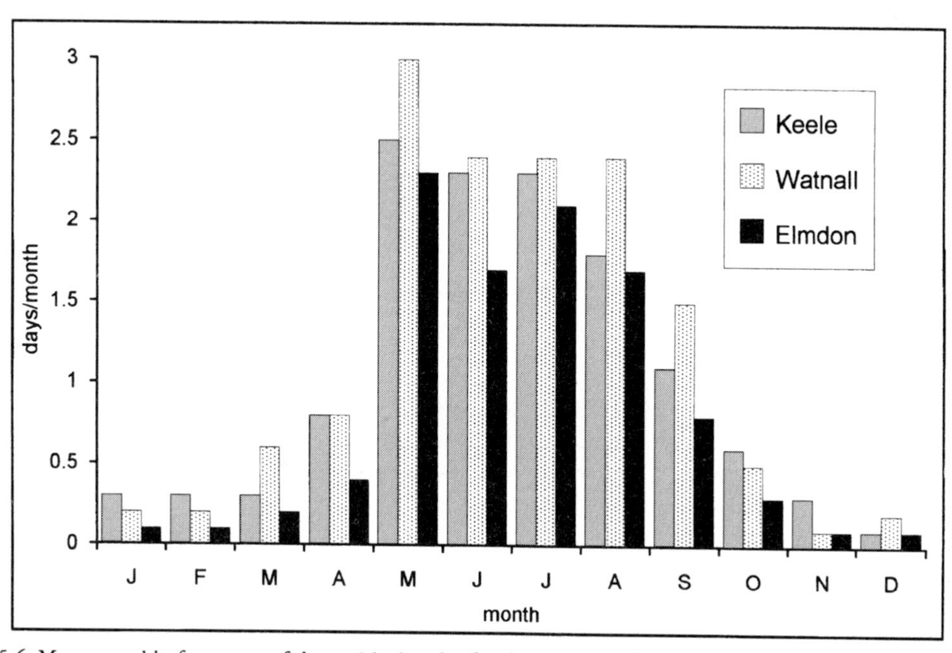

Figure 5.6 Mean monthly frequency of days with thunder for the period 1961–90

most thunder-prone areas of the British Isles. In support of this claim he cited the 1896–1930 average annual figure of 19.9 thunder days with a maximum of 4.0 days in July. The same author also reported on a severe thundery outbreak in July 1952 (Barnes 1960) when nine thunder days were recorded. He showed that there existed a tendency for such activity to be enhanced as storm cells passed over the urban heat islands of the major towns and cities in the north-east of the region.

Together with hail, thunderstorms can produce substantial damage to crops and property, and may also be accompanied by squalls and modest, though potentially damaging, tornadoes. The Midlands region has a considerable number of these perilous weather phenomena. They require a strong updraft of air to initiate and sustain them, and are hence often associated with thunderstorms and usually occur with warm, moist air advected from the south. Having formed, they often pass in a north-east direction on the right side (in terms of forward motion) of thunderstorm cells. A notable tornado struck Birmingham on 14 June 1931 during an outbreak of thundery weather and left a trail of damage twenty kilometres long in its wake. The city was again the subject of tornado damage on 4 February 1946. It was the turn of Coventry on 21 April 1968 in what was to become known as the Barnacle storm (Roach 1968) after the name of the nearby village which it narrowly avoided.

As noted above, the thunderstorms can be thermally driven by intense solar heating in an already conditionally unstable and moist tropical air mass. They can also be associated with vigorous cold and occluded fronts advecting polar maritime air against air of tropical origin. This latter situation helps to account for the continued, if limited, incidence of thunder during the winter months. Thunderstorms are not, however, stationary and the region has two favoured storm tracks. Thunderstorms being generated over Biscay and moving northwards favour at first the Severn valley, then track along the Trent valley before dissipating. Those which develop *in situ* over southern England often travel into the Trent basin or cross the low hills of north Staffordshire between the high ground of the southern Pennines and the north Welsh mountains to reach the Cheshire plain. Those associated with frontal activity are often confined to Staffordshire, Shropshire and Warwickshire.

Lacey (1982) assessed storm characteristics in terms of amount of rainfall and duration. Intensities within summer showers or thunderstorms can reach as high as 100 mm per hour, but usually last less than 60 minutes. Two notable examples in Birmingham have been examined: the Birmingham Thunderstorm in July 1982 with peak intensities near 250 mm per hour (Kings *et al.* 1983), and the two thunderstorms which affected the city in July and August 1994 (Galvin *et al.* 1995). In the latter case the different mechanisms responsible for the two events are contrasted. The earlier of the two was the result of a heat-low developing over the area caused by ground-level warming. As sea breezes were drawn into the low a convergence line formed, drawing air into cooler, unstable conditions above, thereby triggering the thunderstorm. The August storm was the result of the northwards movement of warm and moist low-level air ahead of a front advancing from the west (a situation often referred to as a 'Spanish plume'). In this instance it was the over-running cold air behind the front that triggered the storm development and not the heating effect of the ground surface. The former type of system will develop more readily over the wide expanses of the Midlands and eastern England than elsewhere; the latter type result from wider-scale, non-regional elements. Another example, from north-west England, is reviewed in Chapter 8.

WIND

Birmingham possesses the region's longest and most detailed anemograph record, dating from the early days in the instrument's design in the late nineteenth century. The inspection of this record lends modest support to the observation that the city is the 'Chicago of the Midlands', which is to say it is a windy, rather exposed, city. Certainly, given its inland, low-level siting, the average wind speeds bear comparison with those of other, coastal centres of population.

Measured wind speeds vary over short intervals of time but the highest values are usually recorded ahead of warm fronts and to the rear of a cold front together with rapidly deepening depressions (Hill 1982). Recognition of such variability notwithstanding, Birmingham's mean daily wind speeds average around 8 knots. Table 5.4 shows the percentage of hours each month with wind speeds in Beaufort force classes. From this it can be seen that calm conditions prevail for just under 4 per cent of the year, the percentage being slightly higher in summer than winter. Around 60 per cent of hours have speeds above 7 knots but below 21. Higher hourly speeds do occur and those in excess of 22 knots account for 4 per cent of the total. Such high hourly speeds are most frequent during the December to March period when depressions are most often at their most active. Hours exceeding 34 knots average under ten during the year with peaks in December and February.

Using a statistic such as daily mean wind speed, though useful, hides the true range of speeds during an hour or day. A typical anemogram shows a mixture of peaks and troughs, the result of friction and turbulent eddies within the boundary layer. Peak speeds, or gusts, are noted each hour and this affords a first step in an analysis. Gusts are the instantaneous speeds, well in excess of the mean, but short-lived (generally around five seconds in duration). Squalls, lasting for nearer ten minutes, are also significant events, but not generally tabulated for climatological purposes. It is often the gust which is ultimately responsible for damage, destruction of property and even occasional loss of life. The gustiness ratio is a useful indicator of the likely gusts which can be expected given a mean wind speed. Another indicator is the gust factor, which is a measure of the ratio of the maximum gust to the mean wind speed in a time period. This has a value ranging from 1.5 at coastal locations to just over 2.0 within urban areas, but subject to diurnal, seasonal and synoptic variation. For Birmingham, whilst in general the gust ratio is between 2.0 and 3.0, the observed daily gust factors vary from zero on calm days to near six on a day with squally showers, or near a thunderstorm. Gusts up to 43 knots are relatively frequent, though with a seasonal decline through the summer. Above 52 knots obstacles including houses, trees and cars are at risk. This high speed is very much a winter half-year event with January and February the most susceptible months.

Since wind is a vector, an assessment of direction is also required. Direction is far less variable than speed and most days have a steady flow from a single direction. The prevailing direction in the region is from the south-west, which accounts for just over 50

Table 5.4 Percentage distribution of mean hourly wind speed in Birmingham for the period 1980–95

Speed (knots) and Beaufort force	*Jan.*	*Feb.*	*Mar.*	*Apr.*	*May*	*June*	*July*	*Aug.*	*Sept.*	*Oct.*	*Nov.*	*Dec.*	*Year*
Calm (0)	2.0	2.6	2.3	3.5	3.9	4.6	3.1	4.0	5.0	5.1	3.9	3.7	3.6
1–3 (1)	9.0	9.5	8.1	11.8	16.5	15.7	8.8	13.5	14.0	15.4	14.9	10.8	12.8
4–6 (2)	12.5	12.9	13.8	16.9	18.3	18.1	9.5	13.1	17.8	15.5	17.1	12.0	15.3
7–10 (3)	24.5	25.7	27.9	34.5	30.3	31.1	22.3	34.1	30.2	26.2	27.8	26.3	29.1
11–16 (4)	32.9	27.6	28.4	27.5	25.6	26.1	25.1	27.9	26.1	25.8	26.5	28.1	27.9
17–21 (5)	11.3	11.1	12.1	4.1	4.1	3.5	2.4	6.1	5.0	7.9	6.5	11.5	7.2
22–27 (6)	6.4	7.5	5.8	1.6	1.1	0.7	0.6	1.2	1.6	3.5	3.1	6.1	3.3
28–33 (7)	1.0	2.4	1.4	0.2	0.1	0.2	0.0	0.1	0.2	0.6	0.2	1.4	0.6
Over 34 (>7)	0.4	0.6	0.2	0.0	0.0	0.0	0.0	0.0	0.0	0.0	0.0	0.1	0.1

per cent of all hours. North-east winds are common in April and May, but not in September and October. Continental winds blow (pollution as well) for about 12 per cent of the time in February and 10 per cent in October.

SUNSHINE

The stimulus to the measurement of sunshine in the Midlands in the late nineteenth century was the fashionability of inland spa towns such as Cheltenham, Leamington, Buxton, Tenbury and Malvern. By the end of the century interest turned to the problem of air pollution in the growing industrial towns and cities of the region and the effect this had on recorded levels of sunshine and human health (this issue is also reviewed for London and Manchester in Chapters 3 and 8 respectively). Long-term sunshine records are available for Birmingham, Nottingham and Coventry, and these indicate a substantial improvement in sunshine during the twentieth century, with the clearance of much of the former winter air pollution since the Clean Air Act of 1956. Average monthly sunshine totals for 1961–90 are shown in Table 5.5. Annual totals range from around 1,200 hours in the hills of Derbyshire to 1,400 hours in many of the main Midland cities and 1,500 hours on the southern margins of the region.

In summer, the tendency for convective cloud to develop over inland areas accounts for the absence of monthly totals exceeding 200 hours. Overall, the Midlands cannot claim to be one of the sunniest regions in the British Isles, and these data can be compared with those presented in Tables 7.6 and 9.7, for example where east-coast sites, although having lower summer temperatures, are less troubled by convective cloud over the cool North Sea. However, with the improvement in air quality, the region can achieve some quite respectable sunshine totals in favourable weather conditions. In the 1980s and early 1990s, there have been several years when totals across the region have exceeded 1,500 hours. This has occurred not just in the exceptionally sunny years such as 1989 and 1990 but also in periods when westerly airstreams have predominated. It is in this situation that the region benefits from lying in the rain shadow of the Welsh mountains.

A particular feature of winter anticyclonic weather is the effect that meteorology, topography and even soils have on sunshine. Low-lying and rural areas may often be shrouded in persistent fog, sometimes over several days, while urban locations such as Birmingham enjoy fine weather. This reversal of the well-established urban to rural contrast in sunshine reflects the increase in urban sunshine but provides also an insight into the way in which climates have changed at the local scale. Low-lying areas such as the Trent and Severn valleys provide a favourable setting for mist and fog because of their damp, flood plain soils and ever-present rivers. Shropshire, Herefordshire and Worcestershire may, on the other hand, encounter their gloomiest conditions when a veil of stratus lying over the Mid-

Table 5.5 Mean monthly and annual sunshine totals (hours) for the period 1961–90

Location	*Alt. (m ASL)*	*Jan.*	*Feb.*	*Mar.*	*Apr.*	*May*	*June*	*July*	*Aug.*	*Sept.*	*Oct.*	*Nov.*	*Dec.*	*Year*
Nottingham	117	46	55	93	119	170	176	168	162	121	87	58	41	1,296
Keele	179	44	58	87	129	179	152	170	161	117	88	59	38	1,281
Birmingham	132	52	62	102	130	177	182	182	170	134	97	66	47	1,401
Raunds	59	48	59	99	125	171	176	169	159	129	93	59	43	1,330
Malvern	62	50	61	107	139	178	185	191	173	134	91	62	45	1,415
Ross-on-Wye	67	52	63	106	139	175	181	182	167	131	92	65	47	1,401
Cheltenham	65	50	64	106	141	179	185	186	176	137	99	65	45	1,434

Table 5.6 Average percentage frequency of sunshine duration (classified in hours) for Birmingham for the period 1945–90

Month	*0.0*	*<1*	*1.0–1.9*	*2.0–2.9*	*3.0–3.9*	*4.0–4.9*	*5.0–5.9*	*6.0–6.9*	*7.0–7.9*	*8.0–8.9*	*9.0–9.9*	*10.0–10.9*	*11.0–11.9*	*>11.9*
January	46.0	13.9	10.3	8.9	7.1	6.3	4.5	2.9	<1					
February	35.0	16.6	10.2	8.4	7.1	8.2	5.0	4.8	3.4	1.3				
March	27.1	12.9	9.4	6.6	7.3	6.6	6.3	6.1	6.8	4.7	4.7	1.5	<1	
April	10.5	12.6	8.0	8.4	8.2	8.7	6.3	7.7	7.7	7.7	7.3	3.7	3.0	<1
May	9.0	10.0	7.6	7.3	5.6	6.8	7.3	7.3	7.6	7.3	6.1	4.5	4.8	8.9
June	7.3	11.3	7.3	6.5	7.0	8.0	3.8	5.8	6.3	8.3	5.2	5.2	5.3	12.7
July	6.0	10.6	8.4	7.7	9.5	5.6	8.2	6.2	8.7	7.0	4.7	6.8	3.9	6.7
August	7.6	11.4	9.2	8.9	8.4	6.1	9.0	7.7	8.5	5.8	4.8	5.5	3.2	3.9
September	10.8	14.3	10.8	10.2	8.0	7.7	9.2	8.8	6.3	5.3	4.5	2.7	1.3	
October	20.6	16.8	11.1	8.9	9.5	7.6	9.0	6.2	5.2	3.6	1.5			
November	42.8	17.3	8.7	7.8	7.3	4.7	5.5	3.8	2.0					
December	47.1	16.0	9.5	8.4	7.4	5.5	5.2	<1						

lands starts to blow westwards; it then has a tendency to 'pile up' against the Welsh mountains, progressively reducing the receipt of even diffuse radiation as it thickens.

There are a number of ways to explore the nature of daily sunshine. The frequency distribution of daily sunshine for Birmingham over the period since 1945 is shown in Table 5.6. By far the highest number of dull days (less than 1.0 hour) occur during the winter half-year and, as expected, the greatest frequency of sunny days (more than 6.0 hours) is found during the summer. About 5 per cent of days from May to September have 12 hours or more sunshine and the best days may exceed 15 hours. Winter is not, fortunately, a season of unremitting gloom, and under anticyclonic conditions days in December, January or February can receive proportionally more sunshine than many spring and summer days. In addition, consecutive days are likely to remain sunny, since the high pressure is often a persistent synoptic condition.

FOG

The inland location of the region means that advection fog is associated with low stratus cloud in winter rather than with the arrival of coastal mist and fog in spring as noted in many coastal regions. Figure 5.7 shows the monthly means for three sites well distributed across the region. All three reveal a marked concentration of fog into the winter months, indicating an additional contribution made by radiation fogs on cold winter nights when settled, anticyclonic conditions prevail. The topography of the region further aids this process, and the broad valleys of the Severn and Trent in particular afford, as noted above, favourable locations for mists and fog to gather over wide areas.

Other factors may also come into play. Watnall, in the Trent valley, has a relatively high annual frequency of fog days (25.6 fog days). Unsworth *et al.* (1979), when looking at another lower Trent valley site at Sutton Bonington, suggested that pollution from the towns, cities and power stations of the Trent valley may be an important factor. The authors contrasted Sutton Bonington with Newport (Shropshire), where the annual frequency is closer to fifteen fog days in a less polluted setting. This latter figure, they suggested, was a 'natural' background frequency for the region. The trend at Watnall and Elmdom (located on the fringes of Nottingham and Birmingham respectively) shows a notable if erratic decrease during the period 1961–90 towards this background level (Figure 5.8). This emphatic change in local climate has clearly contributed to

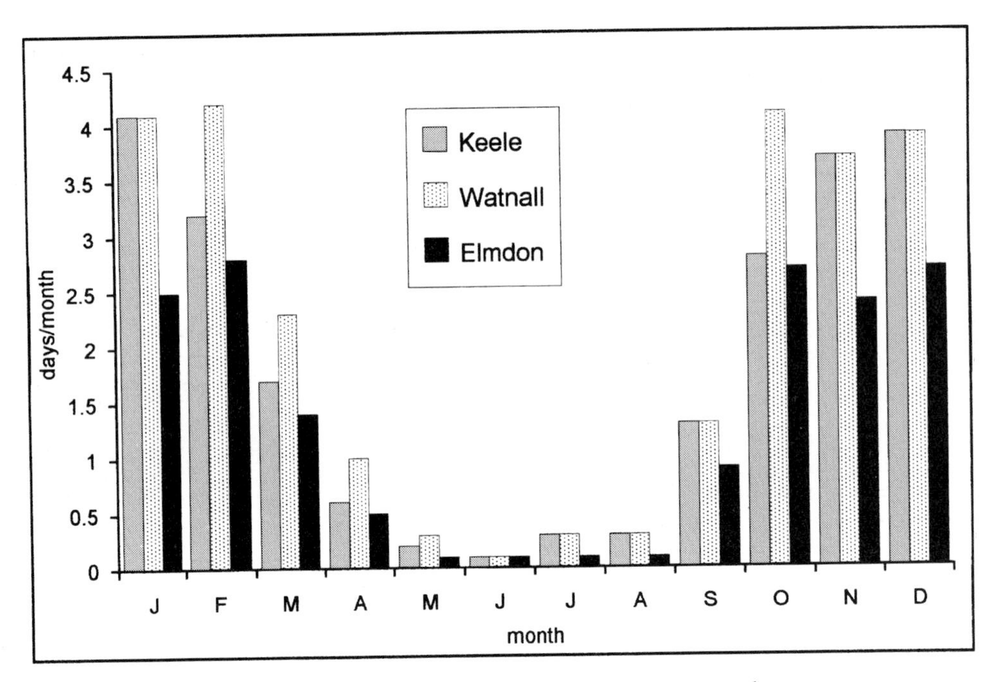

Figure 5.7 Mean monthly frequency of days with fog at 0900 GMT for the period 1961–90

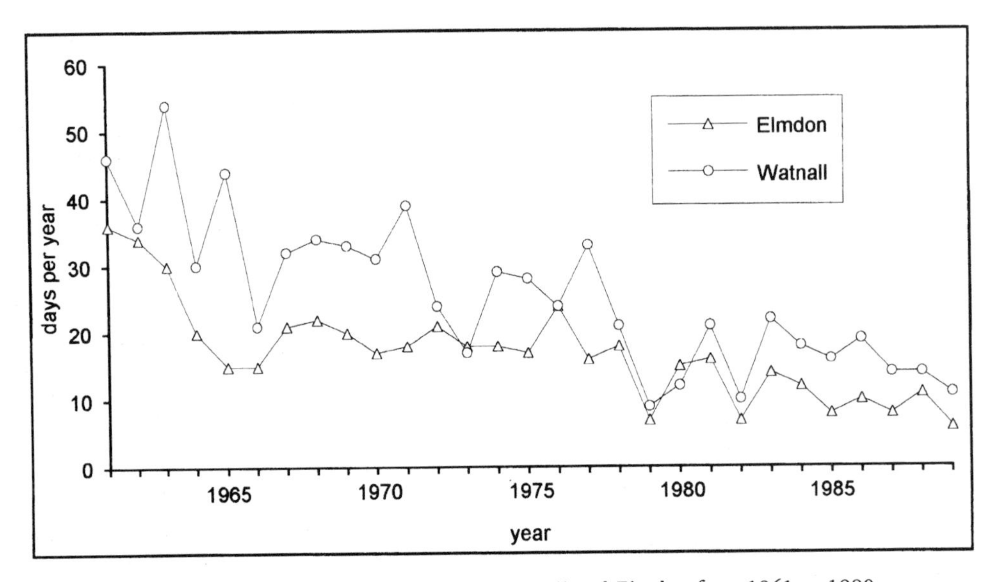

Figure 5.8 Decline in the annual frequency of fog days at Watnall and Elmdon from 1961 to 1990

the increase in sunshine around these urban areas of the Midlands (as also noted in Chapter 8 for the Manchester area). However, data for Birmingham (Edgbaston) show that fog frequency has remained at present levels throughout the twentieth century.

CONCLUSION

This region was introduced as one that possessed indistinct physical borders and it was suggested that this might lead to its having a climate that is essentially transitional between the north and south of England in thermal terms and between Wales and eastern England as regards rainfall. It might be argued conversely, perhaps perversely, that lying as it does in the geographic heart of England, the region represents the country's quintessential climate, less 'disrupted' by external elements than other, more peripheral, regions.

In the absence of coastal modification, variations in weather and climate within the region are governed to a large degree by orography. This is emphasised at the meso scale by the presence of extensive river valleys which act as attractors for the diverse elements of cold air, fog and low cloud and high day temperatures at various times of the year. At a larger scale, the rain shadow of the Welsh mountains produces a broad imprint on the regional distribution of rainfall and sunshine, moderating the effects of changeable, westerly-type weather for much of the region.

Until the middle of the twentieth century, the public perception of the industrial Midlands was one which emphasised the frequency of cloud, fog and smoke pollution. With the passing of the era of heavy industry in most Midlands cities the climate has acquired a brighter aspect but also increased vehicular pollution, made more dangerous by increasing sunshine. The problems of lack of precipitation and high summer temperatures are almost as much a feature of the eastern and southern Midlands as they are in the South-East. Indeed, it stands out as a region of significantly lower precipitation than the neighbouring regions to the north, west and south. The weather of the late twentieth century has shown that if the region stands out at all climatologically, then it does so in terms of temperature extremes in both winter and summer.

REFERENCES

Atkinson, B.W. and Smithson, P.A. (1976) 'Precipitation', in T.J. Chandler and S. Gregory (eds) *The Climate of the British Isles*, London: Longman.

Barnes, F.A. (1960) 'The intense thunder rains of 1st July 1952 in the northern Midlands', *East Midlands Geographer*, 14: 11–26.

—— (1966) 'Weather and climate', in K.C. Edwards (ed.) *Nottingham and Its Region*, Nottingham: British Association.

Beaver, S.H. and Shaw, E.M. (1970) *The Climate of Keele*, Keele University Library Occasional Publication no. 7, Keele: University of Keele.

Bonacina, L.C.W. (1964) 'Hot spells in September in relation to equinoctial sunshine', *Weather*, 19: 287–289.

Burt, S.D. (1975) 'The climate of Rugby: Part 1, Temperature', *J. Meteorology (UK)*, 1: 9–14.

—— (1976) 'The climate of Rugby: Part 2, Rainfall', *J. Meteorology (UK)*, 149–55.

—— (1984) 'Minimum temperatures, 14 January 1982', *Weather*, 39: 372–373.

—— (1992) 'The exceptional hot spell of early August 1990 in the United Kingdom', *Int. J. Climatol.*, 12: 547–567.

Fairmaner, W.D. (1987) 'An analysis of the snowfall of 7th March 1987 in the Midlands', unpublished MSc dissertation, School of Geography, University of Birmingham.

Galvin, J.F.P. (1991) 'The long night shift at Birmingham', *Weather*, 46: 231–233.

Galvin, J.F.P., Bennett, P.H. and Couchman, P.B. (1995) 'Two thunderstorms in summer 1994 at Birmingham', *Weather*, 50: 239–250.

Giles, B.D. and Kings, J. (1996) 'Birmingham weather through two centuries', in A.J. Gerrard and T.R. Slater (eds) *Managing a Conurbation: Birmingham and Its Region 1996*, Studley: Brewin Books.

Glasspoole, J. (1959) 'Centenary celebrations at Ross-on-Wye Observatory', *Weather*, 14: 328–333.

Gregory, S. (1976) 'Regional climates', in T.J. Chandler and S. Gregory (eds) *The Climate of the British Isles*, London: Longman.

Hamilton, M.G. (1987) 'Very warm and very cold spells at Edgbaston, Birmingham during 1920–1979', *Weather*, 42: 2–8.

Hill, J. (1982) 'Bombs around Britain: a study of rapidly deepening depressions', unpublished MSc dissertation, School of Geography, University of Birmingham.

Inwards, R. (1893) *Weather Lore*, London: Elliot Stock (reprinted in 1994 by Senate/Studio Editions, London).

Jackson, M.C. (1978) 'Sixty years of snow depths in Birmingham', *Weather*, 33: 32–34.

Kelly, D. (1986) 'Forecasting snow in the Birmingham area', unpublished MSc dissertation, School of Geography, University of Birmingham.

Kings, J. (1982) 'Extreme daily rainfall in Birmingham since 1940', *Climatological Memorandum no. 1*, Department of Geography, University of Birmingham.
Kings, J. (1985) 'The meteorological record at Birmingham (Edgbaston), England', *Weather*, 40: 388–395.
Kings, J., Beresford, A.K.C. and Wenman, P. (1983) 'The Birmingham thunderstorm, 14 July 1982', *Weather*, 38: 34–37.
Kington, J. (1988) *The Weather Journals of a Rutland Squire: Thomas Barker of Lyndon Hall*, Oakham: Rutland Record Society.
Lacey, P. (1982) 'Spatial and temporal variation in thunderstorm frequency and intensity around Birmingham', unpublished MSc dissertation, School of Geography, University of Birmingham.
Lyall, I.T. (1970) 'Recent trends in spring weather', *Weather*, 25: 163–166.
—— (1977) 'Spring snowfalls in southern Britain', *J. Meteorology (UK)*, 2: 97–102.
Mathews, R.P. (1972) 'Variation of precipitation intensity with synoptic type over the Midlands', *Weather*, 27: 63–72.
Meaden, G.T. (1983) 'Britain's highest temperatures: the county records', *J. Meteorology (UK)*, 8: 201–203.
Meteorological Office (1982) *The Climate of Great Britain: The Midlands*, Climatological Memorandum no. 132, Bracknell: Meteorological Office.
Parsons, F.J. (1975) 'Sixty years of weather recording: reminiscences from the Great War and Ross-on-Wye Observatory', *J. Meteorology (UK)*, 1: 51–54.
Pye, N. (1972) 'Weather and climate', in N. Pye (ed.), *Leicester and Its Region*, Leicester: Leicester University Press for the British Association.
Roach, W.T. (1968) 'The Barnacle tornado', *Weather*, 23, 418–423.
Roach, W.T. and Brownscombe, J.L. (1984) 'Possible causes of the extreme cold during winter 1981–82', *Weather*, 39: 362–372.
Robinson, E. (1957) 'The Lunar Society and the improvement of scientific instruments', *Annals of Science*, 12: 296–301 and 13: 1–8.
Speakman, D. (1994) 'Ambulances adrift: the impact of the snowstorms of the winter of 1990/91 upon the services of the City of Birmingham', *Weather*, 49: 96–101.
Unsworth, M.H., Shakespeare, N.W., Milner, A.E. and Ganendra, T.S. (1979) 'The frequency of fog in the Midlands of England', *Weather*, 34: 72–76.
Unwin, D.J. (1980) 'The synoptic climatology of Birmingham's urban heat island 1965–1974', *Weather*, 35: 43–50.
Webb, J.D.C. and Meaden, G.T. (1983) 'Britain's lowest temperatures: the county records', *J. Meteorology (UK)*, 8: 269–272.
—— (1993) 'Britain's highest temperatures by county and by month', *Weather*, 48: 282–291.

6

WALES

Graham Sumner

INTRODUCTION

Wales's position as an area of extensive upland exposed to prevailing rain-bearing winds from the Atlantic Ocean gives it a climate which, though often mild, may be wet and windy. Its weather and climate are, as elsewhere in Britain and Ireland, dependent on the intrinsically changeable character of the westerly regime, but the twin effects of its location on the ocean fringe and its topography give the country a reputation for being wet. When active frontal depressions prevail, pre-existing precipitation is enhanced and warm-sector precipitation is notably responsive to orographic uplift. Outbreaks of polar maritime air are often conditionally unstable, so that cloud development may be initiated, or pre-existing clouds given additional stimulus to produce frequent showery outbreaks 'anchored' to the higher ground, with clearer conditions up- and downwind. At such times, unless there is a marked ridge of high pressure, the difference locally between the forecast of 'rain' linked to frontal depressions and 'showers' associated with the intervening air masses may appear subtle in the extreme. Crowe (1940) remarked, with some apparent surprise, that Welsh Augusts were almost as wet as Welsh Octobers: perhaps August *is* a bad time for summer holidays! When, on the other hand, the more typical westerly flow is blocked by an anticyclone over the near Continent, fronts approaching from the west may only affect the west of Britain and Ireland, so that again Wales will often experience cloud and precipitation while areas further east remain largely dry.

Wales is, though, transitional between the full maritime climates of western Ireland and Scotland and the more continental regimes of the south and east of England. Most of Wales may be classified as 'oceanic' under Conrad's (1946) index of continentality, but only the far west of Pembrokeshire and the Lleyn peninsula (Figure 6.1) in Gwynedd approach the degree of 'hyperoceanicity' attained in western Ireland or north-west Scotland. Maritime climates are generally wet and cloudy. That of Wales is no exception, though there are important qualifications which must be made. While Wales's upland character provides the major trigger for cloud and precipitation, the shape of the country's coastline and that of the Irish Sea introduce an important secondary veneer to cloud and precipitation development. Writing in a political context, Halford Mackinder (1925) once called the northern part of the Irish Sea 'Britain's inland sea' (Figure 6.2). The sea's configuration and near-landlocked nature are important to the climatologies of the land areas which enclose it. Wales is sheltered by Ireland and south-west Scotland from Atlantic winds between west and north. Many parts in the south of Cardigan Bay are also sheltered from south-westerly winds by Pembrokeshire.

The Irish Sea possesses two openings on to the Atlantic Ocean. The first and smaller, the North Channel, is sometimes important in allowing a long sea fetch for airflows between north-west and north. Airflows from this direction are generally

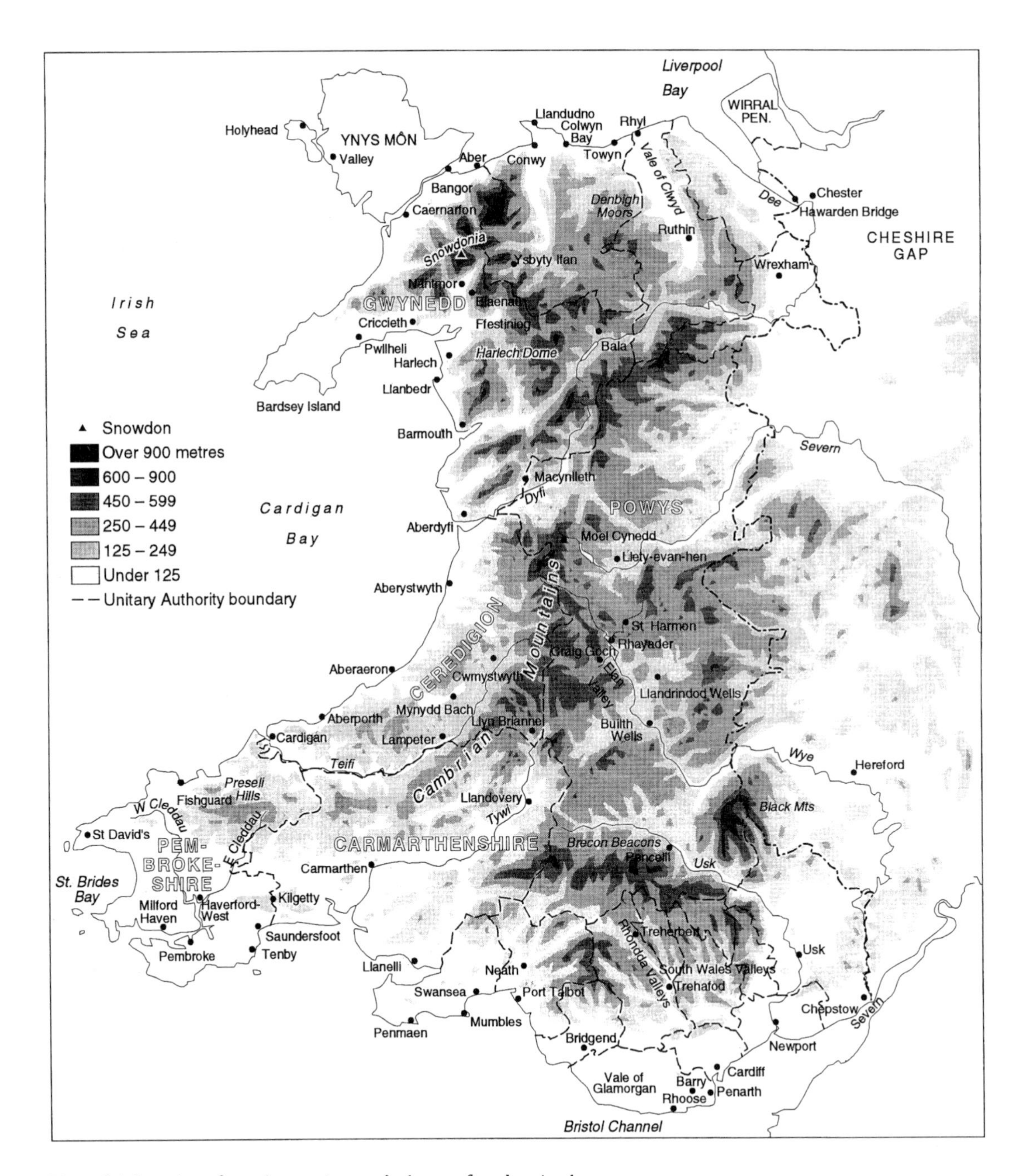

Figure 6.1 Location of weather stations and places referred to in the text

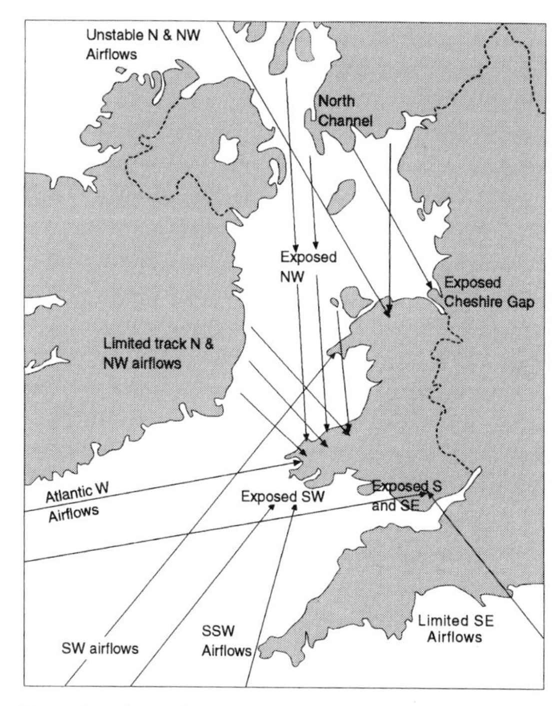

Figure 6.2 Critical airstream trajectories over the Irish Sea which affect Wales

unstable and showery, but showers may be directed through this gap on to particular parts of the coast of North Wales. The second, and major, opening exposes the south-west of Wales, Pembrokeshire and southern Carmarthenshire, and the extreme north-west, the Lleyn and Anglesey, to airflows between west and south which may possess an ocean fetch of two or three thousand kilometres. Such airflows are generally humid and stable, and these coasts, lacking the shelter afforded by the Irish landmass or other parts of Wales, are those most subject to sea fog.

The Welsh uplands are 'montane' rather than 'alpine', with low insolation, comparatively small seasonal ranges in temperature, high precipitation amounts and intensities, and frequent strong winds. In common with other temperate areas on western continental margins there is a rapid mean lapse of temperature with height, creating a climate which is at times far from equable or pleasant at higher elevations and away from the coasts. The upland massif does, however, occasionally, have its advantages. When anticyclones cover lowland England and the English border area to the east in the winter months, generally cloaking large areas in fog, the uplands and areas to their west are often sunny. Along the coast the additional maritime influence may enhance this effect, and, with overcast skies on the eastern flanks of higher ground, a weak föhn effect may also operate.

One of the consequences of Wales's upland nature is also that reliable climatological data are comparatively scarce in many areas. This is particularly so in the upland interior. Rain-gauges, for example, generally cluster at low elevations near to population centres, and even the few in upland regions are generally located in valleys. In 1930 only 12 per cent of climate stations were above 200 m and only seventeen appeared in the *Monthly Weather Report* (J. Mayes, personal communication, 1996). The situation has improved in recent years, and in particular with the introduction of automatic weather stations, so that by 1993 (the latest published at the time of writing) some sixty-three locations appeared in that publication, though still only six located at 200 m or above, and with few having a full 30-year availability (data from the 'Delta' AWS site in the Snowdonia area are cited in Table 6.3). The spatial distribution of weather parameters in the upland regions is thus known only in general terms, and maps of, for example, precipitation (Figure 6.5) must be regarded as only an approximation for large parts of the country. A further consequence has also been that some of the data included within this chapter are for periods other than 1961–90, and are sometimes for shorter durations.

TEMPERATURE

The three basic controls on Welsh temperatures are variations in air mass type, the proximity of the sea, and the changes introduced by the character of the local topography. The last two are ever-present, though their importance varies depending on the nature of the controlling air mass: its stability, the speed and direction of airflow, associated weather and cloud amount. While the more humid maritime (tropical or polar) air masses are the commonest, others from the north (Arctic), and particularly from the Continent (tropical and polar continental), may cause temperature conditions to vary significantly on a day-to-day basis or for longer weather spells.

The main ameliorating effect on temperature is that of proximity to the sea. Sea surface temperatures are, apart from inshore values at times of extreme cold, nearly always maintained at more than a minimum 5°C in winter and early spring, but are still less than 16.5°C by late summer and early autumn (Jones 1981), and mean values lie well within this range (Table 6.1). The maritime modification of land temperatures is evident around most of Wales's extensive coastline, exaggerated by the country's peninsular status, so that the mean seasonal march of temperature is everywhere comparatively small, and is particularly so for most coasts. A further important maritime impact on temperatures, because of the heat-conserving properties and mobility of deep water, is to delay the occurrence of the months of mean highest and lowest temperatures: from July to August and January to February over most parts of Wales. Sea surface temperature minima are similar in February and March, and maxima similar in August and September (Table 6.1). In January and July, conditions around the coast remain ameliorated by the sea surface temperatures, but by March and September the degree of land surface heating and cooling respectively is sufficient to overcome the sea-induced lag. The mean August maximum temperature exceeds that for July at two of the three coastal locations shown in Table 6.2. During late winter, minima at coastal locations are consistently lower in February than in January.

As a consequence of the strong maritime influence, plus the effect of altitude, the lowest mean winter temperatures are around 4 or 5°C for coasts, but near to 2 or 3°C at about 300 m in the uplands. Winter mean maxima are around 4 or 5°C at about 300 m, as opposed to 7 or 8°C in the most favoured coastal locations. Coastal amelioration is, however, most notable in night-time minima (Table 6.2). The most favoured locations in this respect are in the far

Table 6.1 Mean monthly sea surface temperatures (°C) for Bardsey Island for the period 1968–77

Jan.	*Feb.*	*Mar.*	*Apr.*	*May*	*June*	*July*	*Aug.*	*Sept.*	*Oct.*	*Nov.*	*Dec.*
8.5	7.5	7.7	8.6	10.2	12.7	14.1	15.2	14.9	13.7	11.4	9.8

Table 6.2 Mean monthly maximum and minimum temperatures (°C) for sites in Wales for the period 1961–90

Location	*Alt. (m ASL)*	*Jan.*	*Feb.*	*Mar.*	*Apr.*	*May*	*June*	*July*	*Aug.*	*Sept.*	*Oct.*	*Nov.*	*Dec.*	*Year*
Penmaen	85	7.1	7.0	9.2	11.8	14.9	17.6	19.5	19.3	17.2	14.1	10.2	8.3	13.0
		2.7	2.0	3.4	4.5	7.5	10.3	12.2	12.3	10.7	8.7	5.2	3.7	6.9
Moel Cynnedd	385	4.9	4.8	6.8	9.7	13.5	16.0	17.9	17.4	14.9	11.6	7.7	5.9	10.9
		−0.1	−1.0	0.3	1.0	3.6	6.6	8.8	8.6	6.8	4.7	2.0	0.8	3.5
Valley	10	7.6	7.6	9.3	11.4	14.4	16.9	18.4	18.5	16.7	14.2	10.6	8.7	12.9
		3.0	2.4	3.5	5.1	7.7	10.3	12.2	12.3	11.0	8.9	5.6	4.1	7.2
Rhoose	65	6.7	6.8	9.1	11.9	15.1	18.1	20.0	19.8	17.5	14.1	10.1	8.0	13.1
		1.7	1.5	2.6	4.3	7.1	10.1	12.0	12.1	10.5	8.5	4.5	2.8	6.4

Note: ASL = above sea level

west (for example, Valley and Penmaen). The slightly more continental nature of the climate of south-east Wales emerges in winter in its lower minimum temperatures. The lowland coasts of the Bristol Channel from Chepstow to Rhoose, including Cardiff and Newport, are sometimes afflicted by persistent low temperatures under cold anticyclonic spells in winter, when a temperature inversion maintains a large 'pool' of cold air over the Severn estuary, negating maritime influences. Along parts of the north coast, persistent cold air drainage off the uplands in the colder part of the year may have an impact on long-term average temperatures, particularly at night, producing average minima between 2 and 3°C.

While the maritime influence maintains comparatively high temperatures along the coasts throughout the colder half of the year, the reverse is true during summer days. Altitude is still a factor controlling summertime temperatures, but its effects are in part countered by increased opportunities for the daytime heating of the air over the warmed land surface, even in the uplands, and a reduction in values around the coast due to frequent sea breeze effects (see below). In July and August, mean temperatures are generally in the range 14 to 15°C around the coasts, rising to about 16°C in lowland inland areas, but are held back to 12 or 13°C at about 300 m. These values largely reflect the lower night-time minima for many high-level and inland valley sites (to which most data apply) and are a consequence of the combined effects of increased altitude and cold air drainage. Upland minima generally average 9 to 10°C in both July and August, while at lower levels inland minima are typically around 12°C. Coastal regions have minima averaging about 13°C. Around most coasts the ameliorating effect of the sea keeps average daytime maxima down to around 18°C. Similar maxima typically also apply to inland locations up to about 300 m. Here the increased opportunity for surface heating may produce potentially higher maxima on continuously sunny days, but this is often offset by the numerous occasions when afternoon temperatures are clipped back by significant convectional cloud development. The highest mean daytime temperatures are thus reached in inland lowland parts, such as Hawarden Bridge, where the mean July maxima approach 20°C. The capital city of Cardiff also experiences high summer temperatures which are close to those for lowland locations further inland, so that, taking summer and winter extremes together, this part of Wales experiences a more extreme seasonal range of temperature. Higher summertime maxima are also characteristic of some other southern coastal locations (compare Valley between May and September with Rhoose in Table 6.2), where local aspect favours more intense land surface heating, and with generally higher sea surface temperatures.

Secular Variations in Conditions

Against average, long-term, conditions must be also set considerable day-to-day variation. Wales's location at the margin of a large continent makes it particularly vulnerable to changes in temperature with changing air mass, with day-to-day changes in temperature of up to 10 degrees C, and with anomalously high or low temperatures under specific synoptic situations. In addition, it is not unknown for very high maxima to occur along the north coast in mid-winter because of a local föhn effect. Aber, on the north coast, reached 18.3°C on 10 January 1971 (the highest January temperature recorded anywhere in the United Kingdom) and 16.5°C on 5 February 1990. In the same season, temperatures in the south-west may exceed 10°C owing to the advection of extremely mild tropical maritime air, and even in the uplands maxima may reach 10°C under tropical maritime conditions.

By contrast, episodes of very cold continental air may yield subzero maxima over much of Wales away from the coast. Even around the western and south-western coasts subzero maxima have been recorded on a number of occasions. At these times, with an offshore wind, the close proximity of a relatively warm sea is not sufficient to prevent temperatures from plummeting. When the air mass is cold, dry and dense, clear conditions during the long winter nights will allow considerable out-radiation of heat. If the air is still, this cold air will stagnate and quite deep layers of subzero temperatures may accumulate near the surface. The whole of Wales suffered subzero night- *and* daytime temperatures during the week-long cold snap of January 1987, with exceptionally low maxima on 12 January, the temperature failing to rise above −5°C in all but parts of southern coastal Carmarthenshire and along the north Wales coast, the Lleyn and Anglesey (Brugge 1987). During the prolonged, very cold winter of 1962/63 the average January maximum temperature for Rhoose was 0.3°C and average minimum was −4.5°C (Richards 1989). These compare with 1961–90 mean maxima of 6.7°C and minima of 1.7°C.

During the warmer part of the year, too, averages disguise considerable day-to-day variation in conditions reflecting changes in air mass type. Even in upland areas daytime maxima can approach 27 or 28°C when hot tropical continental air invades from the east. In addition, in the absence of a sea breeze, there have been many occasions along the coasts of Wales when summer-time maxima have equalled those of sites inland. On 2 August 1995 Aberporth attained a maximum of 32.2°C, a figure matched or approached by a number of other sites along that part of the Cardigan Bay coast on that day. Valley reached 32.8°C on 29 July 1948 and 30.9°C on 21 August 1984; Port Talbot, 33.2°C on 2 July 1976; and Rhoose, 33.5°C on 3 August 1990 (Webb and Meaden 1993). These are among the highest values attained anywhere in Wales.

The Impact of Sea and Land Breezes

Temperatures at most coastal locations are moderated not merely by the simple proximity of the sea, but also, by day in the spring and summer, by the occurrence of sea breezes. These may be very marked around the west and south coasts, where, in general, elevation increases relatively slowly inland. For the Irish Sea the persistent upwelling of cold waters offshore may create pronounced subsidence within the sea breeze, so that the coast of Cardigan Bay may be particularly sunny while substantial cloud cover and shower activity persists inland, and particularly over the uplands. The impact of the sea breeze on daytime temperatures can be marked, with temperatures pegged back several degrees below inland maxima. The effect may extend many kilometres inland as the breeze invades. Within the sea breeze, however, near-continuous sunshine may appear to compensate for the fall in temperature. Figure 6.3 shows the air and dew point temperature traces for a spring day for Lampeter, some 20 km inland from the Cardigan Bay coast, with a sea breeze arriving just before 1200 GMT. The extent of the temperature fall as the breeze penetrates inland is proportional to the contrast in temperature between the inland and the sea surfaces (for Wales see Sumner 1977a), and sea breezes may be blocked by topography or assisted by anabatic flow up the major valleys (Sumner 1977b).

The converse of the sea breeze is the land breeze, which generally reaches its maximum intensity around the dawn period. The land breeze, funnelled by katabatic flow down river valleys opening on to the sea, may locally create unusually extreme nighttime coastal conditions. At times of major anticyclonic development or during very cold periods in the winter months, such flows may focus cold inland air down the larger valleys on to the coast. The strength of the flow may be such that the windchill factor considerably enhances what may appear to be only a moderately cold night. Worse, under some conditions it is possible for the down-valley flow to stagnate as it spreads out on leaving the confines of a river valley, particularly if the opening to the sea occupies a delta or more extensive low land. Under these circumstances very low temperatures may result. On 26 December 1995, for example, notably low temperatures occurred along parts of the Cardigan Bay coast, which, though higher than inland minima, were much lower than the normal expectation for these locations. Some coastal towns on the west coast of Wales, such as Aberdyfi or Aberaeron, are at the mouths of quite extensive river systems, and may as a result on some occasions experience unusually low minima.

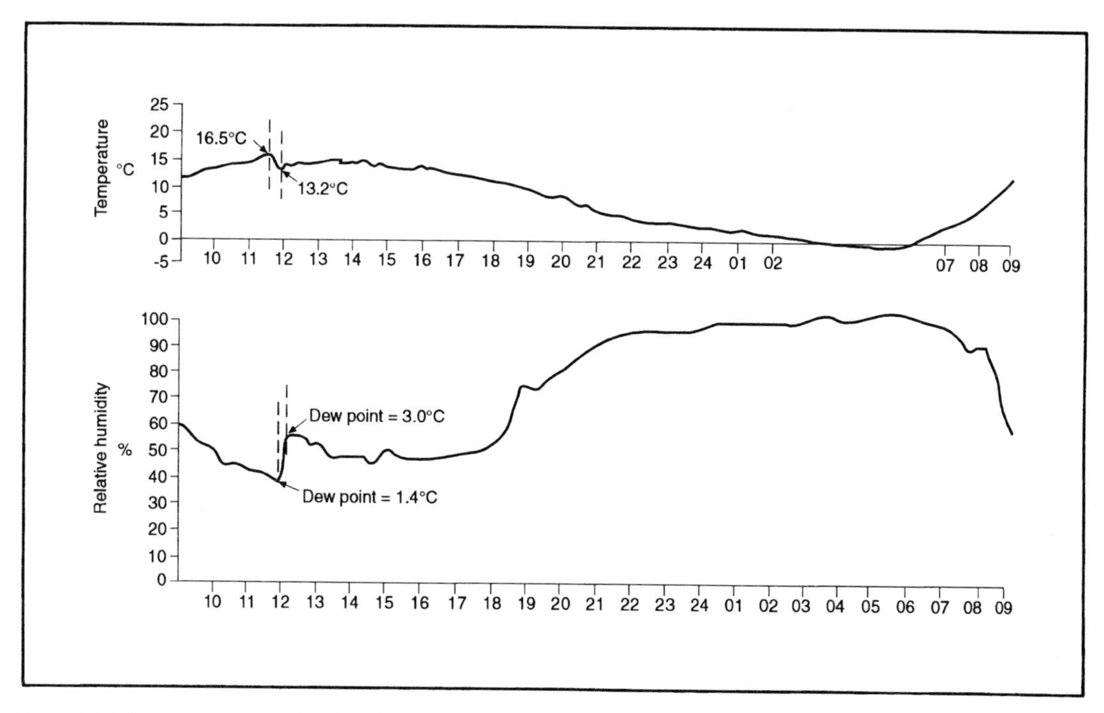

Figure 6.3 Temperature trace showing the arrival of the sea breeze at Lampeter, Ceredigion, on 23 April 1973

Topoclimates and Frost Incidence

The main controls on overall temperatures are distance from the sea and altitude, so that at the broadest scale there is the expected decrease of temperature with increasing elevation. However, at night-time, and particularly during the colder part of the year, a further complication may again be provided by cold air drainage into upland valley systems. Such locations are prone to a high frost incidence. Howe (1953) described conditions within a Ceredigion frost hollow near Aberystwyth during a winter's night, and showed that in spite of its being only a few kilometres from the sea, and at a comparatively low elevation, night-time minima may be remarkably low. In one case, while the minimum temperature at the coast was 0.5°C, it was −4.5°C in a frost hollow just 5 km inland. Sumner (1977a) contrasted frost incidence on a wind-swept hilltop site in mid-Wales at an altitude of 290 m (Llety-evan-hen; 47.6 air frosts per year), with Carmarthen, a valley site at an altitude of only 19 m (60.6 per year). Many inland valleys, both lowland and upland, have a notably severe night-time microclimate, and one of the former main climatological stations in Wales, St Harmon, situated in a deep upland valley at 279 m, with 500-m hills on either side, on many occasions recorded the coldest temperature anywhere in Wales, with air frosts registered for every month. During 1984 and 1985 this site was the coldest anywhere in the United Kingdom on 10 per cent and 18 per cent of nights respectively (Tout 1987a), belying Wales's general reputation for mildness. On 6 August 1987 the minimum was −1.1°C; on 19 September 1986, −5.5°C; on 25 April 1989, −7.8°C; and on 9 April 1990, −7.3°C. In the last two cases the minimum temperatures along the Cardigan Bay coast exemplified the importance of local topography and were much milder; between 0°C and +2°C.

The contrast between inland and coastal locations with respect to air frost frequency is well shown in

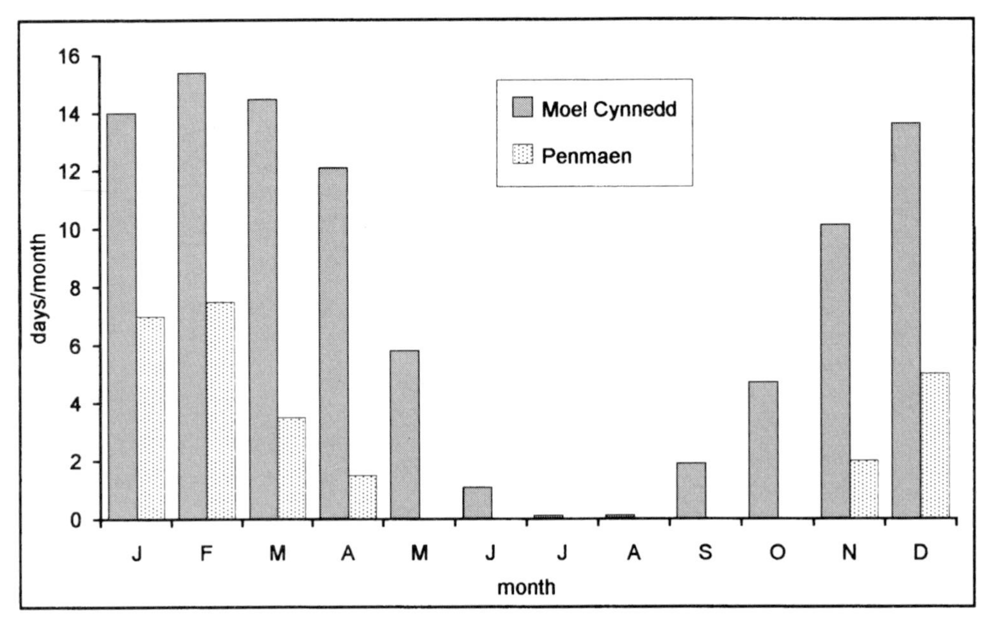

Figure 6.4 Mean monthly air frost frequency at Penmaen (West Glamorgan) and Moel Cynnedd (Powys)

the comparison in Figure 6.4 for Penmaen (near Swansea) and Moel Cynnedd (in the uplands of Powys; altitude 358 m). Frost frequency is generally much reduced near to coasts, reflecting long-term mean values of temperature; compare Penmaen's annual mean frequency of 26.0 air frosts with Moel Cynnedd's 93.4. In both instances, however, and reflecting a more general picture for maritime settings, February is the month most likely to witness air frosts. The 'average' condition is of course subject to huge year-to-year fluctuation, to the superimposition of local effects, as described above, and to occasional extreme events. Tout (1987b) gave the average annual incidence of air frosts at Rhoose (altitude 65 m) as 36.7, but reaching 85 in 1963. Subsequently the frequency has been as low as sixteen in 1989. In most western coastal locations the average incidence is less than twenty-five per year, but below ten per year on the western extremities of the Lleyn and Pembrokeshire. The incidence and mean times of occurrence of the last and first frosts of winter are also of considerable importance to agriculture, and shorten the growing season for specialist crops, particularly inland and in valleys. Mean dates for the first frost of winter vary from late November around the coasts (later still on the western extremities) to late September or early October well inland. The corresponding dates for the last frost range from mid-March to mid-May. The mean frost-free season inland may thus be restricted to about four and a half months, while in the most favoured areas it is about nine and a half months. The mean duration of the growing season also shortens considerably inland and at higher elevations. The growing season extends from mid-March to mid-December for the west and on coastal lowlands, but only from mid- or late April to late November in inland and upland areas (Jones and Thomasson 1985), although important local variations due to cold air drainage must be superimposed upon this general picture.

The climate of Wales, while subject to strong maritime influences, is thus perhaps only truly maritime around the coastal fringes and over the far western extremities. Winter-time minima in particular are highly variable in time and space. The

English borderland area and the south-east of Wales possess a more continental climate, with colder winter and night-time conditions, and warmer summer day temperatures. The general conservative control of sea surface temperatures, together with spring and summer sea breezes, maintains a comparatively small daily and seasonal range in temperature for most coastal locations for most of the time, depending on site and prevailing air mass. Mean temperature decreases with altitude, but during summer days the difference between coast and inland may be reduced by the chilling effects of sea breezes, and at night be exaggerated by cold air drainage into deep upland valleys.

PRECIPITATION

The dominant influence on the amount and distribution of precipitation over Wales is that of topography, so that the map of average annual precipitation (Figure 6.5) closely mimics the contours of higher ground. The highest annual totals follow the central upland spine between Snowdonia and the Brecon Beacons, and south-west towards the Preseli hills in Pembrokeshire, with annual amounts exceeding 1,500 mm over large areas, and exceeding 4,000 mm on the highest peaks in the north. The overall lapse rate of annual precipitation in Wales approximates 2.5 mm per metre of altitude (Barrowcliffe 1982; Roe 1984), but may reach 5 mm/m on exposed windward slopes, owing to orographic enhancement, and on leeward slopes, since the air dries out very rapidly on descent into the rain shadow area. Moisture-laden onshore winds may require uplift over higher ground to yield significant amounts of precipitation, so that even windward low-level coasts may be comparatively dry. Totals around the coasts and along the English border are appreciably lower than the adjacent uplands. Along some exposed parts of the south coast the annual average is around 1,100 mm, but shelter from prevailing precipitation-bearing winds provided by high ground along the north Wales and Ceredigion coasts may yield yet lower averages. Average annual amounts are less than 750 mm for the seaside resorts between Llandudno and Rhyl, and around 850 mm along parts of the west coast (Table 6.3). Values between 750 and 1,000 mm also occur along the English border through the middle and lower Wye and Severn catchments. The immediate coastal fringe in south-east Wales (incorporating Rhoose, Barry and Penarth) is somewhat drier than nearby Cardiff. Here shelter may sometimes be provided by Exmoor across the estuary, particularly in the warm sectors of depressions with south to south-west winds.

Estimates of annual evapotranspiration range from 400 to 800 mm, depending on location, local land use and vegetation. Jones and Thomasson (1985) give 1965–75 mean annual rates of 568 mm for Aberporth on the Cardigan Bay coast and 559 mm for Rhoose in the Vale of Glamorgan. Calder (1990) cites lysimeter records for an altitude of about 400 m in the upper Wye and Severn catchments east of Aberystwyth, indicating averages of 370 mm per year for the grassland-dominated Wye and 799 mm for the more forested Severn. Whichever figures are adopted, average annual precipitation exceeds average annual evapotranspiration in all parts of Wales, and handsomely so in the upland areas. This has meant that upland Wales has become a major source for water supply not only for the Principality but also for cities as far afield as Birmingham.

The distribution of mean annual precipitation is also reflected in mean monthly falls. These are shown for selected sites on the map in Figure 6.5 and in Table 6.3. Precipitation is most concentrated in the period between August and January. A relatively dry period occurs from February to July. The variation in seasonal precipitation amounts is greatest in the uplands, so that for some upland locations the mean of the wettest month is about twice that of the driest. More typically, however, this ratio is around 1.5. December totals (often on average the wettest for any month) exceed 400 mm in the highest parts of Snowdonia and widely reach 200 mm for elevations between 300 and 400 m. July totals are, except in the very highest parts, less than 100 mm,

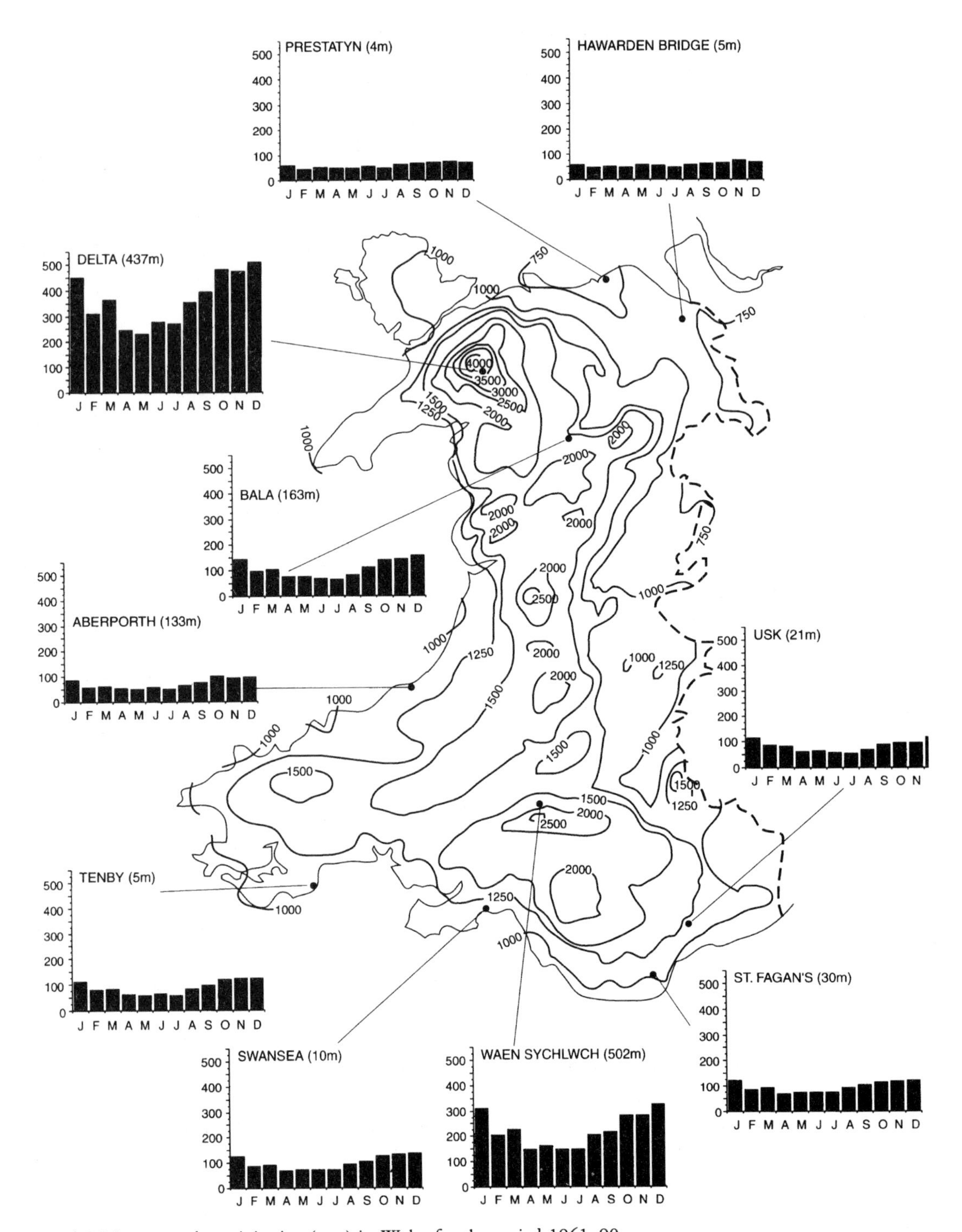

Figure 6.5 Mean annual precipitation (mm) in Wales for the period 1961–90

Table 6.3 Mean monthly and annual precipitation totals (mm) for the period 1961–90

Location	*Alt. (m ASL)*	*Jan.*	*Feb.*	*Mar.*	*Apr.*	*May*	*June*	*July*	*Aug.*	*Sept.*	*Oct.*	*Nov.*	*Dec.*	*Year*
Prestatyn	4	59	43	51	47	49	54	48	61	67	69	74	68	690
Valley	10	83	56	65	53	49	52	53	74	74	91	99	94	843
Bala	163	143	97	105	74	77	67	66	83	112	139	145	157	1,265
Aberporth	133	85	58	62	53	50	56	51	65	77	102	93	96	848
Swansea	8	126	86	91	67	73	73	71	94	105	125	132	135	1,178
Usk	21	115	86	83	60	66	58	54	70	90	96	96	118	992
St Fagan's	30	123	88	95	68	74	76	75	93	106	114	118	122	1,152
Waen Sychlwch	502	309	206	228	148	160	146	149	206	216	280	282	323	2,653
Abergorlech	76	182	118	131	93	94	90	91	120	140	172	174	185	1,590
Delta AWS	437	448	308	363	243	231	276	269	351	391	477	472	507	4,336
Alwen	335	136	95	104	81	80	73	74	98	115	131	148	152	1,287

and along parts of the north coast and in the Dee catchment's rain shadow area are less than 50 mm. As with all average data, however, monthly means may disguise a considerable degree of variation. High individual monthly totals exceeding 200 mm are not infrequent in coastal areas exposed to west and south-west airstreams during particularly westerly dominated and changeable months. The Meirionydd and north Pembrokeshire coasts and the Gower may at times be some of the wettest coastal locations in Britain.

Day-to-day Precipitation

Since Wales is often close to active weather systems, and given also the great volumes of moisture advected off the Atlantic and the country's topography, it is not surprising that high precipitation intensities and high daily, monthly and annual totals are not infrequent over Wales. Falls of over 50 mm occur at least once every two years in most areas (Rodda 1967), and Bleasdale (1963) has shown that daily falls of 200 mm or more may occur once every 100 years over parts of Snowdonia. In the northern uplands approximately 60 per cent of falls greater than 125 mm occur in the winter half of the year (October to March). For the southern hills the corresponding figure is 80 per cent. Bleasdale (1963) cited totals of up to 260 mm in a day for South Wales, and there have been suggestions, disputed by some, that extreme falls are becoming more frequent in the Welsh uplands (Howe *et al.* 1967; Perry and Howells 1982). Some of the more recent of these notable falls are listed in Table 6.4. Perhaps most important in terms of the overall perception of Welsh wetness, and certainly so in the context of flood potential, is the possibility of sustained high precipitation amounts spread over several days. At times of mobile west to south-west airflows, and when pressure is low, the orographic component will enhance existing long-duration, high-intensity precipitation, so that consecutive high daily totals occur. Of particular note are the high totals for Waen Sychlwch for March 1981 (Table 6.4).

The significant part that relief plays in the generation and enhancement over the long term is repeated in the short term, but it is important to recognise that wind direction at the time of precipitation will dictate where the highest falls will

Table 6.4 Notable precipitation totals in Wales (in 24 hours unless otherwise stated)

Date	*Location*	*Amount (mm)*	*Remarks*
18–19 June 1971	Above Llyn Brianne	82.4	Cyclonic rainfall (Jones 1977)
27 December 1979	South Wales	>75	Wide area of rainfall from cold front secondary system
26–28 December 1979	Above Trehafod	>150	See Perry (1980) and Browning (1980)
5–13 March 1981	Waen Sychlwch	897.4	
21–22 March 1981	Waen Sychlwch	206.6	
29 December 1986	Nantmor	104.9	
2 January 1991	Pencelli (Powys)	64.4	
9–10 August 1991	Blaenau Ffestiniog	89.0	
29 November–2 December 1992	Treherbert	250.0	Fell in 96 hours
11 June 1993	Aberporth	151.0	Active, near-stationary cold front
17–18 December 1993	Upper Rhondda valley	>100	
13 November 1994	Ysbyty Ifan (Gwynedd)	113.0	Mobile westerly
26–28 December 1994	Treherbert	164.0	Fell in 44 hours

occur (Bonell and Sumner 1992b; Sumner 1996). Because Wales is essentially a large peninsula, confined by the Bristol Channel to the south, the sweep of Cardigan Bay in the west and Liverpool Bay to the north, winds from all directions but east may pick up substantial amounts of moisture in their lowest layers. High totals are thus commonplace for many airflows from between north, through west to south-east on exposed slopes and at height. The dominant control on long-term precipitation is exposure to prevailing west to south-west winds, particularly when a mobile, changeable Lamb W or SW airflow dominates (Sweeney and O'Hare 1992; Sumner 1996), and it is significant that the central axis of high ground, from Snowdonia in the north to the Brecon Beacons and 'the Valleys' in the south, is almost exactly perpendicular to this prevailing flow. Significantly different patterns may result when precipitation is accompanied by airflows from other directions. These differences may be subtle. Precipitation distributions for airflows between south and west are broadly similar, but with progressively greater amounts over the southern uplands as the southerly component increases (Faulkener and Perry 1974; Sumner 1996). Patterns which are very different from the 'normal' will, on the other hand, be produced for north-westerly, northerly or easterly airflows. The highest daily totals are produced over the southern uplands or the south-east of Wales ahead of systems approaching from the south-west or west, since these systems are preceded by south to south-east winds. This latter situation is exemplified by the events of 29 October 1991 (Figure 6.6), when rain fell heavily over large areas of South Wales. Heavy rainfall over this area also often results from the passage of active cold front secondary depressions (Sumner 1996), and heavy snowfall may accrue in the winter months on the northern flank of depressions when they track to the south of, or over, Wales. Good examples are provided by the events of 17–18 January 1985, with heavy snow over South Wales, and by 9 February 1986 and 27 January 1990, both with heavy snowfall over mid-Wales.

For westerly and north-westerly showery situations high totals also accumulate along the central upland spine, but less so over the western extension through Mynydd Bach to the Preseli Hills (Figure 6.1). For showery north-westerly or northerly situations precipitation is more often concentrated over only northern hills, their exposed northern flanks, and the Cheshire Gap. The geographical restriction

Figure 6.6 Daily precipitation (mm) for 29 October 1991 resulting from an intense frontal depression preceded by south-easterly winds

of precipitation to the far north-east is also linked at these times to the passage of air mass showers through the North Channel and down the Irish Sea (Figure 6.2). Such is the association between surface airflow, synoptic situation and precipitation distribution, through topography, that it is possible to delineate distinct precipitation regions for Wales. The extent of these areas is closely controlled by major relief features, and the susceptibility of each to precipitation may often be linked to specific airflows (Bonell and Sumner 1992a, 1992b). As many as nine distinct spatial patterns of precipitation have been identified by Sumner (1996), each linked to different airflows or synoptic situations (see, for example, Box 6.1).

Mobile Westerly Conditions

Orographic enhancement over the uplands of Wales is particularly active under humid tropical maritime conditions and at times of significant frontal activity, when some exposed coastal areas may remain dry while substantial falls may occur inland. Daily precipitation over upland areas at these times may be fifteen to twenty times that on the coast (Sawyer 1956), and warm-sector rainfall is 'particularly productive' in upland parts (Lowndes 1968), a point reiterated by Pedgley (1970). Warm-sector rainfall is also highly subject to very localised orographic enhancement induced by aspect and the funnelling of winds along valleys (Pedgley 1970). When winds are strong, and valley funnelling yields marked local convergence, highly irregular upland rainfall distributions may be experienced. At the same time the massif itself causes a distinct 'barrier effect', sheltering areas downwind. Orographic enhancement is appreciably less marked, and the rain shadow less obvious, when warm sectors are accompanied by lighter winds (Pedgley 1971). While altitude and aspect determine where the highest precipitation intensities occur, wind speed determines by how much intensities increase (Browning and Hill 1981). At the highest elevations, a 40-knot wind produces a 2–4 mm/hour enhancement of intensity, but at 20 knots the enhancement is less than 1 mm/hour, while at 60 knots, by no means uncommon at higher elevations during the wetter part of the year, enhancement may reach 4 to 8 mm/hour. At times when several layers of cloud are present the main mechanism for this is the seeder–feeder mechanism, where lower-level clouds formed over hills are able to 'feed' on precipitation falling from above. Good illustrations of the power of orographic enhancement over Wales appear in Browning (1980) and Hill *et al.* (1981).

Proximity to the precipitation-generating disturbance dictates the overall level of potential precipitation development. Close to a very active front or low-pressure area, precipitation intensities will be naturally high, even away from the uplands (Sumner 1975). Precipitation resulting from the remains of Hurricane Charley on 26 August 1986 (Bank Holiday Monday) was general and extreme. Amounts were high over most parts of Wales, even away from higher ground (Shawyer 1987), with

Figure 6.7 The River Teifi at Cenarth Bridge, near Cardigan, 19 October 1987
Photo taken by Iolo Jones, kindly supplied by Elwyn Jones and reproduced by courtesy of the Institute of Hydrology

most of Wales experiencing more than 50 mm as a result of the storm. 'Charley' tracked across the south of the country and was particularly active, so that orographic effects were swamped by general uplift in the vicinity of the low: that is, over the whole of Wales.

A 'typical' significant wet-day precipitation distribution over Wales for a mobile westerly situation is shown in Figure 6.10. The surface synopsis indicates a strong west to south-westerly flow approaching Wales throughout 22 and 23 February 1991. During the morning of the 22nd an active warm front introduced extremely moist tropical maritime air. The following cold front approached, but did not cross Wales until the 24th, so that the feed of very moist air on strong to gale-force winds was maintained, and high daily totals resulted. Precipitation values exceeded 135 mm over Snowdonia and the northern part of the central spine. Over the Brecon Beacons amounts exceeded 50 mm. Isohyets follow closely the general trend of contours, both along the central spine and westwards along Mynydd Bach to Preseli in the south-west. The region of lower totals along the southern coast of Cardigan Bay, between Aberdyfi and Fishguard, is a common feature of the daily precipitation distribution under these conditions (Sumner 1996).

Showery Airstreams

Very different precipitation distributions may result when unstable, or conditionally unstable, polar or Arctic maritime air masses affect Wales. In general the greatest uplift will occur on the exposed north- or north-west-facing coasts and slopes, and widespread convection will occur inland when there is adequate insolation and surface heating. To this must be added the importance of the length of sea track across the Irish Sea, so that shower occurrence may be spatially limited to areas exposed to the

Box 6.1

THE HEAVY RAINFALL OF 17 AND 18 OCTOBER 1987

The 'Great Storm', the so-called hurricane, which afflicted southern and eastern England on the night of 15 and 16 October 1987 left Wales largely unscathed. The track of the centre of the depression kept the strongest winds well to the south. That freak Atlantic system, however, was one of many in an exceptionally wet month (Northcott 1988). Two days later on the weekend of 17 and 18 October, most of Wales, and particularly the west, experienced up to 36 hours of continuous and mostly heavy rain, subsequently producing severe flooding down the Tywi valley at Carmarthen (Lewis 1992). A cold front with secondaries moving along it remained nearly stationary over west Wales throughout most of the weekend. Figure 6.8 summarises the situation on the 18th. The distribution of cloud about the low pressure system is shown in the satellite image and casts a broad swathe across most of western Britain and Ireland. Figure 6.9 shows the distribution and intensity of rainfall in Wales over both days. Many locations experienced a total rainfall exceeding 100 mm from the system, and severe flooding resulted. Fishguard recorded 73 mm on the 17th (Frost and Jones 1988). A similar system, but further west, produced exceptionally heavy rainfall over Northern Ireland on the 21st.

Four people were drowned as a consequence of flooding produced by the heavy rainfall. A passenger train ran into the much swollen river Tywi between Llandeilo and Llandovery (Figure 6.1) after a section of a bridge crossing the river collapsed. The town of Carmarthen was virtually cut off, and parts of Haverfordwest were under nearly two metres of flood water. The *Western Mail* reported that the floods in west and north Wales were the worst in living memory. Twenty

Figure 6.8 Synoptic chart for 1200 GMT on 18 October 1987. Inset: infra-red image for 1415 GMT on the same day
Photo by courtesy of University of Dundee

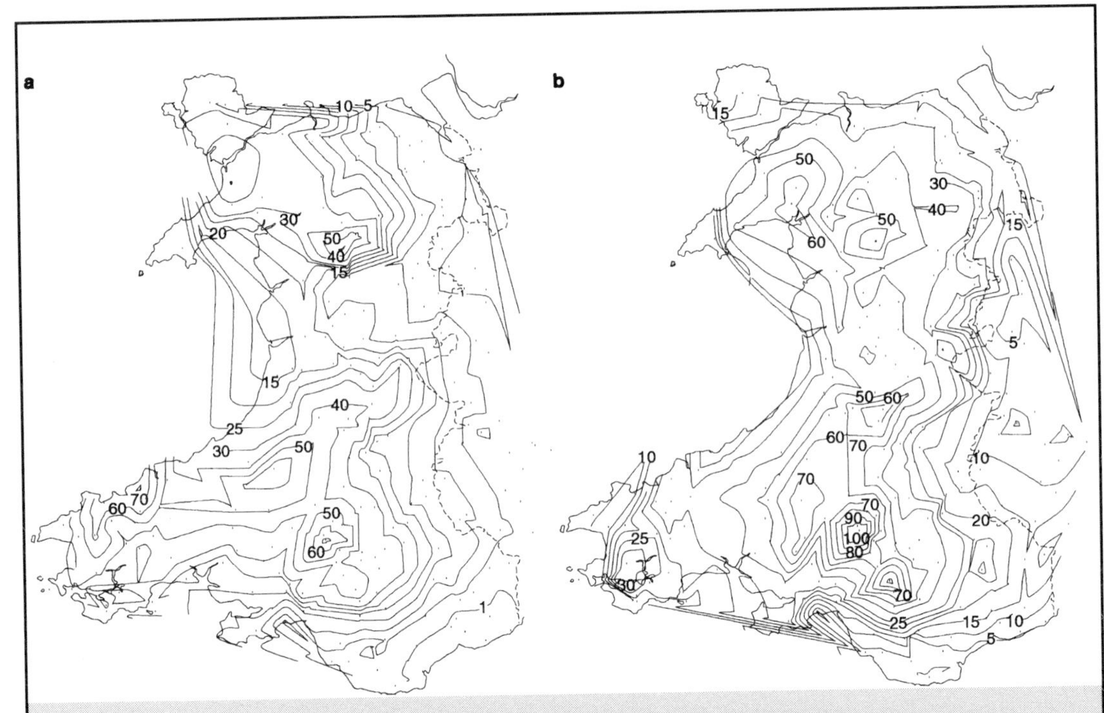

Figure 6.9 Daily rainfall totals (mm) for (a) 17 October and (b) 18 October 1987

thousand people had their drinking water contaminated near Cardigan, and thirty people were evacuated from their homes near Swansea as a hillside began to move. It is probable that these were also the worst floods on the river Tywi of the twentieth century (E. Jones, personal communication). The former county of Dyfed was the worst-affected administrative area, and particularly so in the far west of Pembrokeshire. There was severe damage to at least six road bridges in the county and an 'untold number' (*Western Mail*, 20 October 1987) of minor landslips on to roads.

longest sea tracks. Upwind lowland coasts, such as the Ceredigion coast, the Lleyn and Anglesey, may also experience far less shower activity, especially in spring and summer when the sea is cold, than at inland locations (Lowndes 1966a, 1966b; Oliver 1958). Again, the location and proximity of the depression controlling the airflow will be important (Lowndes 1966a). Orographically generated or orographically enhanced instability showers may also stream downwind from the mountain tops, destroying the 'neatness' of the traditional rain shadow (Pedgley 1971).

An important seasonal variation in shower occurrence can also be detected around the coasts. During the spring and summer, subsident conditions over the relatively cool Irish Sea, together with substantial inland uplift and convection, may combine to produce minimal coastal cloud and shower development, but often quite substantial development inland, and, in particular, over higher ground (Figure 6.11a). The occurrence of sea breezes at these times of year is thus important to precipitation distribution. Lines of showers may develop along sea breeze fronts or where two sea breeze systems

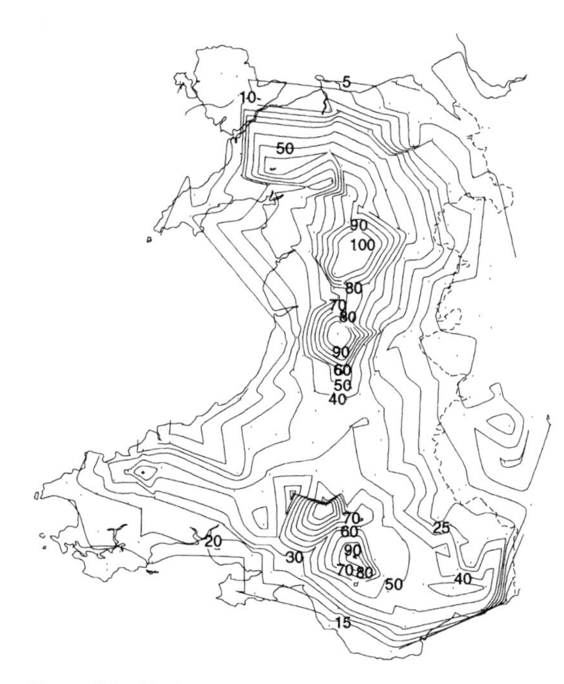

Figure 6.10 Daily precipitation (mm) for 22 February 1991 resulting from the passage of fronts embedded within a mobile westerly airflow

converge. Such conditions may sometimes produce severe storms over mainland Wales, often augmented by local orographic uplift. This situation occurred in late July 1973 over South Wales (Sumner 1977c) within a modified continental air mass. Violent thunderstorms may occur during the summer months under unstable airflows. On extreme occasions, both convectional and cold-front storm activity may trigger tornado development, though cases are rare, tracks are limited and the tornadoes are comparatively small. Nevertheless, they do occur, and a well-documented tornado caused chaos at Butlin's holiday camp in Pwllheli on 14 August 1989, and another affected Kilgetty, Carmarthenshire, on 3 March 1991.

In the autumn and winter, sea surface temperatures are often warmer than those inland and insolation is relatively weak, so that significant cloud and shower development may take place over the sea at times of unstable or conditionally unstable polar or Arctic maritime air masses. If the wind direction is right, lines of showers may be directed at 'favoured' points on the Welsh coasts, with the greatest degree of shower development where sea track is longest (Figures 6.11b and c). Figure 6.12 presents a good example of this form of activity, with a band of well-developed cumulus cloud extending down the Irish Sea. The same showers may continue to propagate inland and be augmented by orographic uplift, thereby constituting true 'air mass showers' relying little on the diurnal convection cycle. On rare occasions persistent lines of precipitation activity parallel to the airflow may occur, producing copious amounts of rainfall over limited areas (Lilley and Waters 1992; Figure 6.11d).

Snowfall

One consequence of increased altitude, topography and distance from the sea is a wide variation in the incidence of snowfall. High frequencies of falling or lying snow or sleet principally affect inland and, in particular, upland areas. Figure 6.13 shows the mean monthly incidence of snow or sleet falling and of snow lying at the standard observation hour of 0900 GMT for contrasting sites on the north-west coast (Valley) and in the upland interior (Moel Cynnedd). The incidence of snow or sleet falling at Moel Cynnedd is at least twice that at Valley during the winter months, and is much greater still in late spring and late autumn: snow has been observed at the higher site in all months except July, August and September. The maritime impact is, however, better reflected in the much greater contrast in mornings when snow is observed lying at 0900 GMT, and Valley's annual mean of 2.4 days hardly compares with Moel Cynnedd's 30.3 days. While snow or sleet may fall on the coast during the colder part of the year comparatively frequently, and quite reliably in most years, the higher overall temperatures mean that persistent snow cover is much less common in coastal areas than at height inland. Locations within a kilometre or so of the sea will rarely see snow lying for more than a few hours at a time, and even following heavy falls (Box 6.2) the cover has generally disappeared

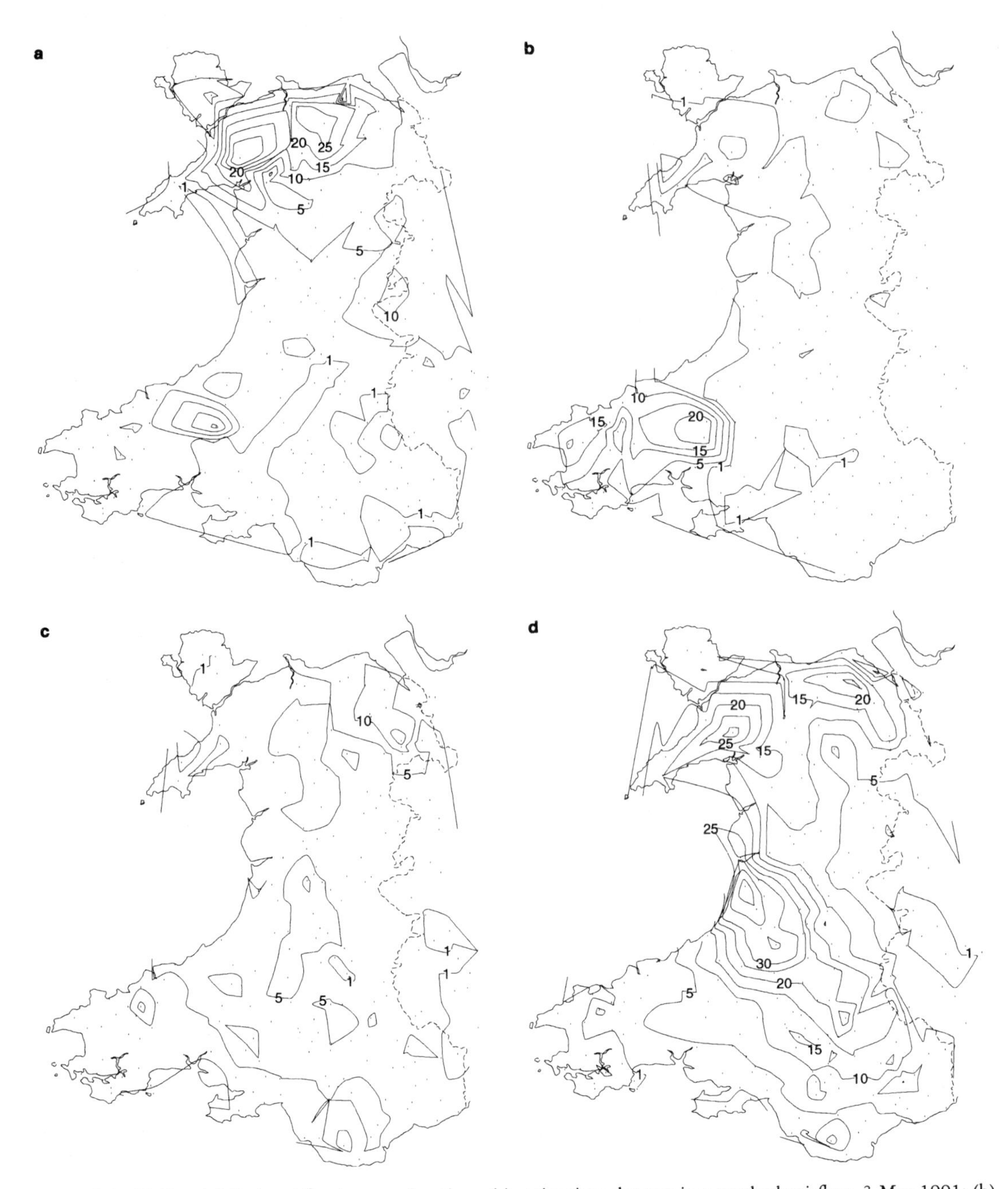

Figure 6.11 Daily rainfalls (mm) for showery situations: (a) springtime showers in a northerly airflow, 3 May 1991; (b) autumn showers streaming off the Irish Sea in a northerly airflow on 2 November 1990; (c) autumn air mass shower activity in a north-westerly airflow on 8 November 1991; and (d) high daily totals over north and mid-Wales in a northerly airflow on 20 November 1990

Figure 6.12 NOAA infra-red image taken at 1242 GMT on 2 November 1990 showing showers in a northerly airflow affecting Pembrokeshire (see also Figure 6.11b)
Photo by courtesy of University of Dundee

after a day or two. Snow cover may be persistent following even light falls at higher elevations in the depths of winter.

WIND

Exposed maritime climates are also windy climates, and Wales is no exception to this rule. Dominant winds are clearly from between south and west (Figure 6.17), and there is a relatively high frequency of strong winds, particularly in coastal areas when cyclonic conditions are dominant. While gales are not as frequent as in areas of Britain and Ireland much further north or north-west, gales have been recorded in every month of the year at Aberporth (Table 6.5) and Valley. Nevertheless, south to south-west winds may sometimes reach speeds as great in north and west Wales as on the west coast and north-west Highlands of Scotland (Prior and

Newman 1988). Bendelow and Hartnup (1980) give mean annual wind speeds of 13.4 knots for Valley, 12.8 knots for Aberporth, 10 knots for Port Talbot and 10.4 knots for Rhoose. Valley's and Aberporth's averages are the third and fourth highest anywhere in England and Wales, and are exceeded only by Great Dun Fell and Portland Bill. Aberporth also recorded the strongest gust anywhere in England and Wales for the Burns' Day (25 January) storm in 1990, with 89 knots at both 1400 and 1500GMT. Other recent notable recorded gusts are 83 knots at Rhoose (Cardiff–Wales Airport) on 1 April 1984; 81 knots at Milford Haven on 11 April 1989; 78.5 knots at Aberporth on 12 November 1991; 71 knots at Llanbedr (Gwynedd) on 3 February 1994; and 64 knots at Mumbles (Swansea) on

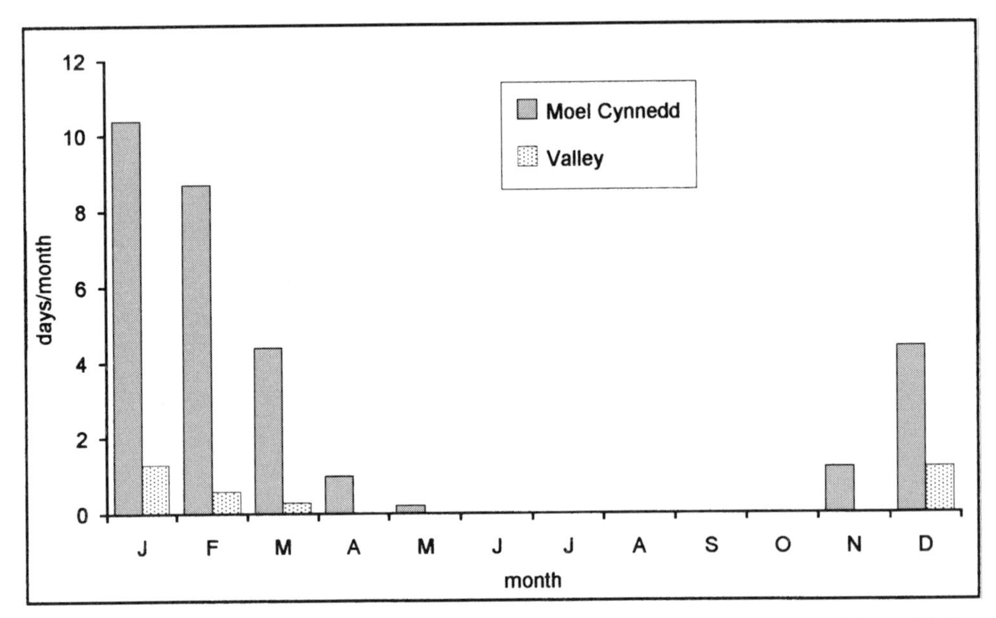

Figure 6.13 Mean monthly frequency of snow lying at 0900 GMT and days of snow falling at Moel Cynnedd, Powys, for the period 1961–90 and at Valley for the period 1981–95

Box 6.2

THE 'GREAT BLIZZARD' OF 8–10 JANUARY 1982

One of the worst blizzards for some years hit Wales between 8 and 10 January 1982. In many places the snow fell for up to 40 hours, commencing during the evening of the 8th, and not ending until early on the afternoon of the 10th. It was a typical 'battleground' situation. There had been a cold spell before Christmas, when wet snow brought down overhead cables over a wide area in south-west and South Wales, but thereafter the mild Atlantic air moved progressively, but slowly, northwards, adding heavy rain to now thawing snow. Over Christmas and the New Year the cold weather was restricted to Scotland, but during the first week in January a major thrust of very cold air moved south to cover all of Britain by the 7th. A depression then formed in the South-Western Approaches

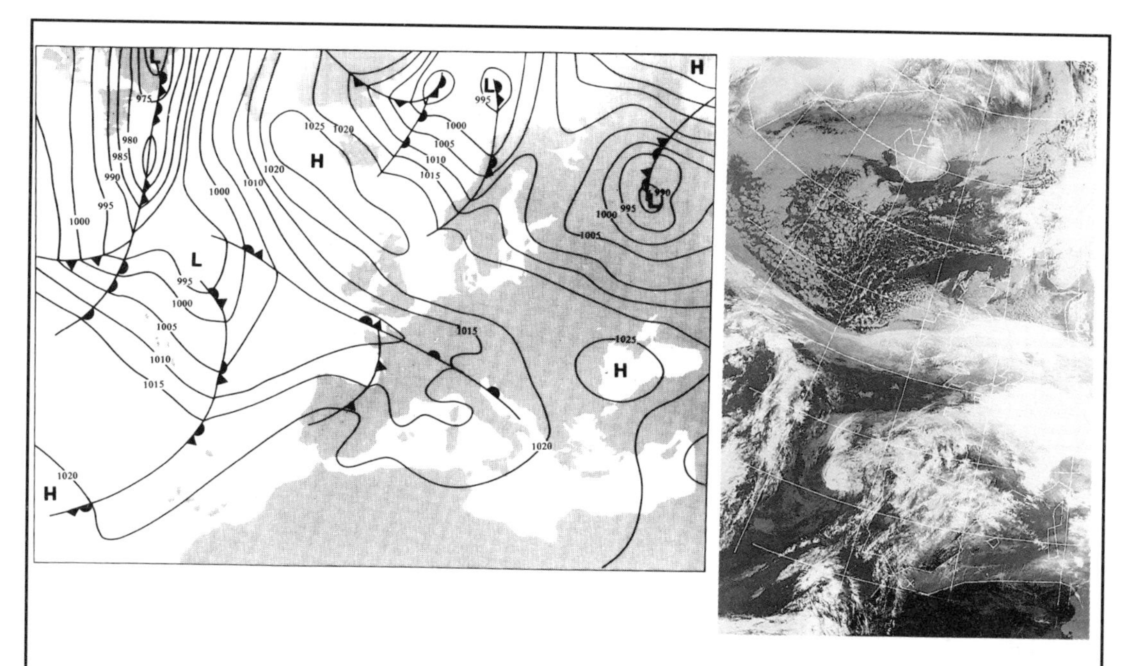

Figure 6.14 Synoptic chart for 1200 GMT on 9 January 1982. Inset: infra-red image for 1434 GMT on the same day showing a band of snow-bearing cloud across southern England and Wales
Photo by courtesy of University of Dundee

(see Figure 6.14). To the south of this was very mild and humid air, but temperatures were well below zero over much of Wales on its north-eastern edge. The satellite image inset in Figure 6.14 shows the band of snow-bearing cloud associated with the fronts of the weather system. Clearer skies to the north are associated with the anticyclone. The period of snow was accompanied at times by gale- or severe gale-force easterly winds, which led to serious drifting. Terry Wogan reported to his Radio 2 audience that 'the whole of Wales is cut off!'.

Wales was the worst-hit area in Britain. To the south the snow turned to rain, and floods resulted, but the system retreated south again during the 10th, and clear skies followed the snow over Wales until the 15th, yielding daytime maxima well below freezing, and night-time minima as low as −15°C in a number of places. Many parts of Wales suffered drifts in excess of 6 m (see Figure 6.15), and overland transport was impossible over most of the country; even Cardiff was cut off for a short time. Eight people were stranded for 19 hours in a train on the west coast near Tywyn. The M4 in west Wales was closed to all but four-wheel-drive vehicles. Seven hundred drivers spent five nights stranded at a community centre in Bridgend. In Cardiff the roof of the city's Sophia Gardens Pavilion collapsed under the weight of snow, and many thousands of homes were without electricity as the blizzard brought down power lines. The former county of Dyfed was the worst-hit area, with an estimated cost of clearing up put at the time at £2 million. The Territorial Army was called in to help the regular army to help Wales back on to its feet in *Operation Snowman*. Some more isolated communities remained cut off until 19 January.

Figure 6.15 The A48 in the town of Pyle, near Bridgend, after the blizzard of January 1982. Road traffic had been halted through south Wales with even major roads around Cardiff and the Vale of Glamorgan blocked by snow
Photo by courtesy of the South Wales Evening Post

6 April 1991. Comparable data for inland sites in Wales are not available. In general, wind speeds are significantly less for inland areas than they are on the coast, but stronger winds are always associated with higher altitudes. Local funnelling by valleys may also increase upland windiness.

Coastal locations are particularly vulnerable when winds are onshore, and all Welsh coasts may experience particularly strong winds from the direction to which they are most exposed. This is especially so for the southern and south-western coastal areas between the westernmost tip of Pembrokeshire and the Bristol Channel, but the major marine incursion at Towyn on the north Welsh coast, caused by a combination of very strong winds associated with the storm of 26 February 1990 (Starr 1990; Kay and Wilkinson 1992) and a high spring tide, illustrates well that coasts with very different orientations are also prone. One further consequence of the mixture of both rain and strong winds is that Wales is particularly subject to 'driving rain', which is a problem in both coastal and upland areas.

SUNSHINE AND FOG

The maritime location of Wales, its exposure to moist maritime airflows and its topography generally give the country a reputation for cloud as well as rain. Inland cloudiness relative to neighbouring

Figure 6.16 The scale of the snow clearance operation after the blizzard of January 1982 can be seen in this photograph of a road near Bridgend
Photo by courtesy of the South Wales Evening Post

coasts is most pronounced over the uplands and during the warmer part of the year. During the summer both convection and orographic uplift may lead to significant cloud development at a time when sunnier weather might be anticipated. Sunshine figures are at their highest around the coasts at all times of the year, with inland lowland locations (for example, Ruthin, Table 6.6) receiving about thirty minutes to one hour less per day than neighbouring coasts during the summer and about fifteen minutes per day less during the winter. In the uplands (Cwmystwyth is a good example), values are on average a further one hour below those of neighbouring lowlands in the summer, and a further fifteen minutes or so less during the winter. While these differences appear small on a day-to-day basis, their accumulation over a summer season holiday month, such as July or August, yields a potential 60-hour deficit in the uplands compared with the most favoured coastal locations such as Valley – a difference which is not lost on most tourists, but which is worthy of emphasis in the promotion of Welsh beaches!

Fog represents the reverse side of the climatic coin. Wales being a region surrounded on three sides by the Irish Sea (to the north and west) and Bristol Channel (to the south), sea fogs are relatively frequent and occur on average on around one day per

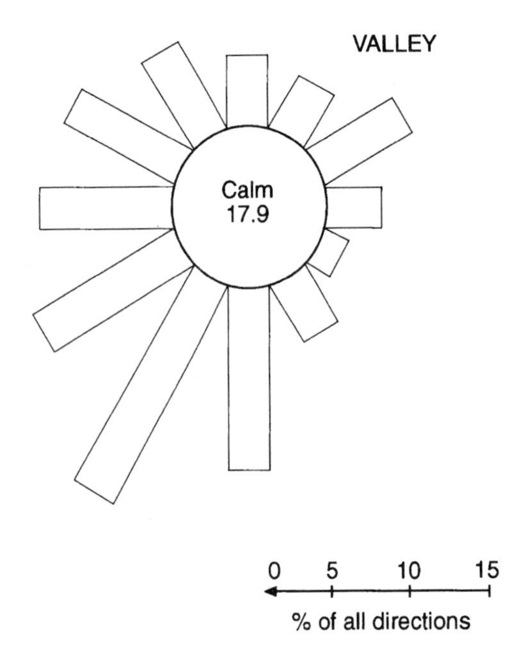

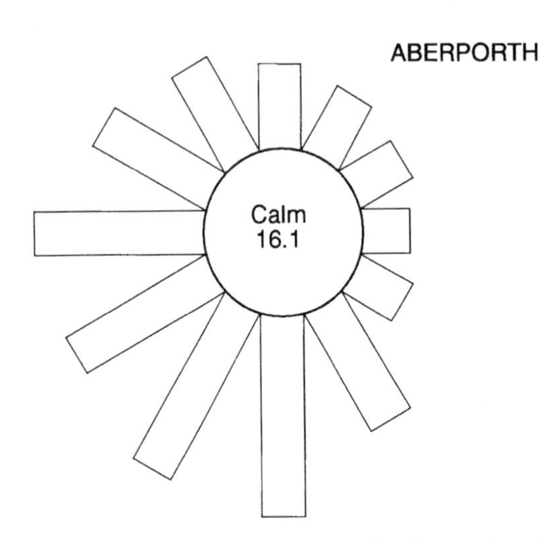

Figure 6.17 Median wind directions for Aberporth and Valley

month, but with slightly higher frequencies between May and August, and along the more exposed south coast (Figure 6.18). Mean annual days of fog at 0900 GMT are 11.1 at Rhoose but fall to 8.5 at Aberporth on the south side of Cardigan Bay. There are, however, variations from month to month, and the lowest frequencies of coastal fog occur between January and April. The plateau-like Vale of Glamorgan renders Cardiff–Wales Airport (Rhoose) highly prone to fog and low cloud at all times of the year (Mercer 1985), while daytime sea breezes in the summer months may draw sea fog on to any coastal margin from late morning onwards.

CONCLUSION

Wales thus possesses a variety of climates, much belying the common assumption that all of Wales is always wet. Exposure and topography together explain a large proportion of the detail of this variation. In common with other western, maritime, upland parts of Britain, Wales is in general milder, wetter, cloudier and windier than the lowlands further east. As is true elsewhere, though, its climate is inadequately summed up with reference only to 'average' conditions. There is considerable place-to-place and day-to-day variation in weather.

The central spine of high ground, between Snowdonia and the Brecon Beacons, plays an important role in separating the more maritime from the more continental districts, and in generating large precipitation amounts. To its east are areas nearly constantly in rain shadow, merging across the border into the English Midlands. To its west is the area most subject to oceanic influences, and where most orographic precipitation is generated. In the south there are the more rolling plateaux along Mynydd Bach into the Preseli Hills, which, with contrasting orientation, shelter the southern coastal fringe of Cardigan Bay, but offer increased exposure to rain-bearing winds along their southern flanks. Many upland locations experience a harsh climate, induced as much by local topography as by altitude, which bear comparison with the colder parts of the English Pennines or the Southern Uplands of Scotland. Many eastern lowland locations have climates indistinguishable from those of Shropshire and the English Midlands. Around the three main coastal orientations can be found quite extensive areas

Table 6.5 Average wind speed frequencies for Aberporth for the period 1950–59

	Jan.	*Feb.*	*Mar.*	*Apr.*	*May*	*June*	*July*	*Aug.*	*Sept.*	*Oct.*	*Nov.*	*Dec.*
Hours with winds in excess of gale force (33 knots)	12.7	2.0	4.5	0.7	0.7	0.0	0.7	1.5	1.4	6.7	2.9	12.7
Number of hours with gusts in excess of 43 knots	149	88	92	46	42	24	27	27	66	90	82	162
Number of hours with gusts in excess of 47 knots	23	8	7	2	3	0.3	0.7	2	5	9	11	28

Source: Meteorological Office (1968)
Note: Data are expressed as average hours of gale-force winds (first row) or as the average number of hours during which gusts in excess of 43 knots (gale force) or 47 knots (severe gale force) were experienced

Table 6.6 Mean monthly and annual sunshine totals for the period 1961–90

Location	*Alt. (m ASL)*	*Jan.*	*Feb.*	*Mar.*	*Apr.*	*May*	*June*	*July*	*Aug.*	*Sept.*	*Oct.*	*Nov.*	*Dec.*	*Year*
Penmaen	87	56	76	112	180	210	191	209	189	142	95	71	45	1,574
Valley	10	56	81	124	177	223	216	192	186	141	105	63	50	1,614
Ruthin	76	50	64	99	135	183	186	161	149	123	96	60	43	1,349
Cwmystwyth	301	31	56	81	126	152	156	136	130	102	81	45	31	1,127

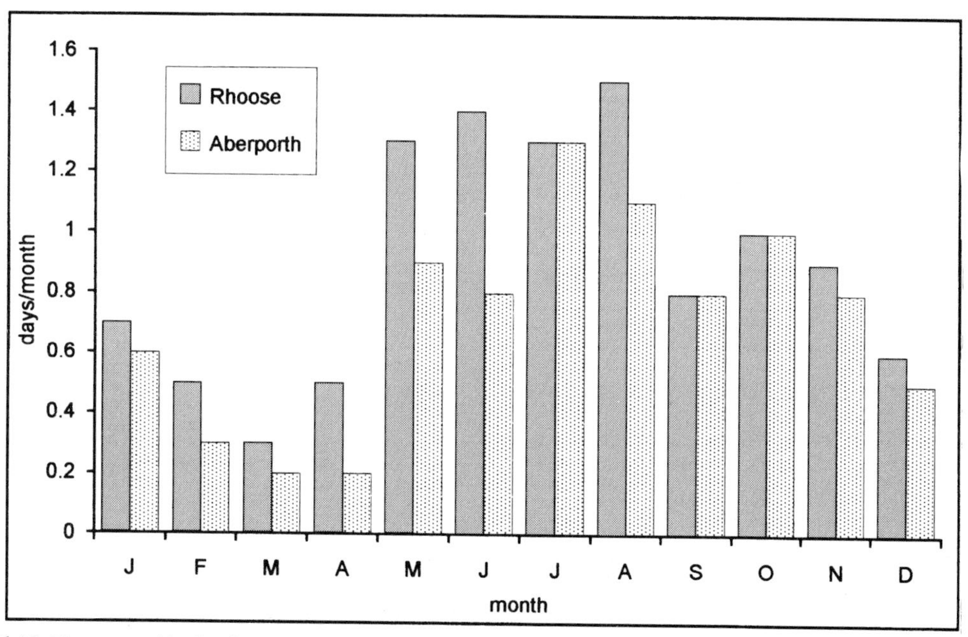

Figure 6.18 Mean monthly fog frequencies at 0900 GMT for the period 1961–90

with drier climates, and with greater winter warmth: lowland Lleyn and Anglesey and the Vale of Glamorgan in particular. The extent of such favoured climates is abruptly terminated by higher ground, and is restricted by the width of the coastal lowland. In the north-east the lowlands of the Dee estuary, extending into the Wirral and the Cheshire Gap in England, are areas subject to shelter for much of the time, but exposed to airflows through the North Channel and down the Irish Sea, and which may share temperatures more in common with lowland England.

Wales has been conveniently divided into five distinct climatic units (White and Perry 1989): the mild, but exposed, coastal lowlands of the Lleyn and Anglesey, together with the Vale of Glamorgan, with much in common with Cornwall and south Devon; the very exposed, wetter and mostly upland plateaux of Ceredigion, Carmarthenshire and Pembrokeshire, with maritime climates similar to those of north Devon and Dartmoor; the colder, wetter and cloudier central upland spine; the sheltered, rain shadowed eastern areas to the east of the upland spine; and the 'continental' lowlands in the far north-east and south-east, which share a climate similar to that of the English Midlands. Whether there is a welcome in the hillsides depends upon which hillside, the time of year and the synoptic situation.

REFERENCES

Barrowcliffe, R. (1982) 'The relationship between rainfall and height', Unpublished Agrometeorological Memorandum 976, London: MAFF.

Bendelow, V.C. and Hartnup, R. (1980) 'Climatic classification of England and Wales', *Tech. Monogr.* 15, Harpenden: Soil Survey.

Bleasdale, A. (1963) 'The distribution of exceptionally heavy daily falls of rain in the United Kingdom, 1863 to 1960', *Journ. Inst. Water Engrs*, 17: 45–55.

Bonell, M. and Sumner, G.N. (1992a) 'Autumn and winter daily precipitation areas in Wales, 1982/83 to 1986/87', *Int. J. Climatol.*, 12: 77–102.

—— (1992b) 'Atmospheric circulation and daily precipitation in Wales', *Theoretical and Applied Climatology*, 46: 3–25.

Browning, K.A. (1980) 'Structure, mechanisms and prediction of orographically enhanced rain in Britain', in R. Hide and P.W. White (eds) *Orographic Effects in Planetary Flows*, GARP Publication no. 23.

Browning, K.A. and Hill, F.F. (1981) 'Orographic rain', *Weather*, 36: 326–329.

Brugge, R. (1987) 'Low daytime temperatures over England and Wales, 12 January 1987', *Weather*, 42: 146–152.

Calder, I.R. (1990) *Evaporation in the Uplands*, Chichester: John Wiley.

Conrad, V. (1946) Usual formulas of continentality and their limits of validity. *Trans. Amer. Geophys. Union*, 27: 663–664.

Crowe, P.R. (1940) 'A new approach to the study of the seasonal incidence of British rainfall', *Q. J. R. Meteorol. Soc.*, 66: 285–316.

Faulkener, R. and Perry, A.H. (1974) 'A synoptic climatology of south Wales', *Cambria*, 1: 127–128.

Frost, J.R. and Jones, E.C. (1988) 'The October 1987 flood on the river Tywi', *Hydrological Data UK, 1987 Yearbook*, Wallingford: Institute of Hydrology.

Hill, F.F., Browning, K.A. and Bader, M.J. (1981) 'Radar and raingauge observations of orographic rain over south Wales', *Q. J. R. Meteorol. Soc.*, 107: 643–670.

Howe, G.M. (1953) 'Observations on local climatic conditions in the Aberystwyth area', *Meteorol. Mag.*, 82: 270–274.

Howe, G.M., Slaymaker, H.O. and Harding, D.M. (1967) 'Some aspects of the flood hydrology of the upper catchments of the Severn and Wye', *Trans. Inst. Br. Geogrs*, 41: 33–58.

Jones, A. (1977) 'An unusually large summer storm in central Wales', *Weather*, 32: 80–85.

Jones, R.J.A. and Thomasson, A.J. (1985) 'An agroclimatic databank for England and Wales', *Tech. Monogr. no. 16*, Harpenden: Soil Survey.

Jones, S.R. (1981) 'Ten years of measurement of coastal sea temperatures', *Weather*, 36: 48–55.

Kay, R. and Wilkinson, A. (1992) 'Lessons from the Towyn flooding', *The Planner*, 17 August.

Lewis, R.P.W. (1992) 'Flooding at Carmarthen in October 1987: historical precedents and statistical methods', *Weather*, 47: 82–89.

Lilley, R.B.E. and Waters, A.J. (1992) 'A persistent cloud band and heavy rainfall over west Wales', *Weather*, 47: 152–158.

Lowndes, C.A.S. (1966a) 'The forecasting of shower activity in airstreams from the northwest quarter over southwest England and south Wales in summertime', *Meteorol. Mag.*, 95: 1–15.

—— (1966b) 'The forecasting of shower activity in airstreams from the northwest quarter over northwest England in summertime', *Meteorol. Mag*, 95: 80–91.

—— (1968) 'Forecasting large 24-hour rainfall totals in the Dee and Clwyd River Authority area from September to February', *Meteorol. Mag.*, 97: 226–254.

Mackinder, H.J. (1925) *Britain and the British Seas*, Oxford: Oxford University Press.

Mercer, C.P. (1985) 'Rhoose (Cardiff–Wales) Airport', *Weather*, 40: 313–316.

Meteorological Office (1968) *Tables of Surface Wind Speed and Direction over the United Kingdom*, Bracknell: Meteorological Office.

Northcott, G.P. (1988) 'The rainfall of October 1987', *Weather*, 43: 338–340.

Oliver, J. (1958) 'The wetness of Wales: rainfall as a factor in the geography of Wales', *Geography*, 43: 151–163.

Pedgley, D.E. (1970) 'Heavy rainfalls over Snowdonia', *Weather*, 25: 340–350.

—— (1971) 'Some weather patterns in Snowdonia', *Weather*, 26: 412–444.

Perry, A.H. (1980) 'A note on the south Wales floods of late December 1979', *Weather*, 35: 106–110.

Perry, A.H. and Howells, K.A. (1982) 'Are large falls of rain in Wales becoming more frequent?', *Weather*, 37: 240–244.

Prior, M.J. and Newman, A.J. (1988) 'Driving rain: calculations and measurements for buildings'. *Weather*, 43: 146–156.

Richards, H.D. (1989) 'The winter of 1962/63 in south Wales remembered', *Weather*, 44: 473–475.

Rodda, J.C. (1967) 'A country-wide study of intense rainfall for the United Kingdom', *J. Hydrology*, 5: 58–69.

Roe, C.P (1984) 'The variation of rainfall with height in Wales', unpublished Agricultural Memorandum, 1001. London: MAFF.

Sawyer, J.S. (1956) 'The physical and dynamical problems of orographic rain', *Weather*, 11: 375–381.

Shawyer, M.S. (1987) 'Rainfall of 22–26 August 1986', *Weather*, 42: 114–117.

Starr, J.R. (1990) 'The storms of early 1990: mass media response to Met. Office warnings', *Weather*, 45: 365–369.

Sumner, G.N. (1975) 'Anatomy of the storm of 5th–6th August 1973 over Dyfed', *Cambria*, 2: 1–19.

—— (1977a) 'Climate and vegetation', in D. Thomas (ed.) *Wales – A New Study*, Newton Abbot: David & Charles.

—— (1977b) 'Sea breeze occurrence in hilly terrain', *Weather*, 32: 200–208.

—— (1977c) 'Sea breeze temperature and humidity contrasts at Lampeter, Dyfed', *Cambria*, 4: 187–198.

—— (1996) 'Daily precipitation patterns over Wales: towards a detailed precipitation climatology', *Trans. Inst. Br. Geogrs*, NS 21: 157–176.

Sweeney, J.C. and O'Hare, G. (1992) 'Geographical variations in precipitation yields and circulation types in Britain and Ireland', *Trans. Inst. Br. Geogrs*, 17: 448–463.

Tout, D.G. (1987a) 'Extremes of temperature in the United Kingdom, 1984/85', *Weather*, 42: 70–76.

—— (1987b) 'The variability of days of air frost in Great Britain 1957–83', *Weather*, 42: 268–273.

Webb, J.D.C. and Meaden, G.T. (1993) 'Britain's highest temperatures by county and by month', *Weather*, 48: 282–291.

White, E.J. and Perry, A.H. (1989) 'Classification of the climate of England and Wales based on agroclimatic data', *Int. J. Climatol.*, 9: 271–292.

7

NORTH-EAST ENGLAND AND YORKSHIRE

Dennis Wheeler

INTRODUCTION

North-east England and Yorkshire form for the most part an easily recognised region, with the Tweed as its northern boundary, the Humber estuary part of the southern limits, the Pennines watershed to the west and the North Sea to the east (Figure 7.1). Only to the south-west are the limits less clearly defined by natural features. The region thus comprises the counties (as they were defined until the reorganisation of 1996) of Northumberland, Tyne and Wear, Durham, Cleveland, North, West and South Yorkshire, North Humberside and small parts of Cumbria and Derbyshire.

The region's northerly location leaves it more frequently exposed to the attentions of active depressions and less subject to the advantages gained by exposure to the extensions of the Azores anticyclone. The relatively pronounced local relief and abiding coolness of the North Sea do much to modify these factors. In this respect the region's eastern and western boundaries are of climatic significance. The North Sea exercises a moderating control along the coast but is especially important in keeping summer conditions among the coolest in England. The influence of the Pennines is more far-ranging. In County Durham they attain their greatest altitude at Cross Fell (893 m above sea level (ASL)) and form an uninterrupted ridge over 700 m high between the Tyne Valley and the Stainmore Gap. The Yorkshire Pennines are less formidable, but at Whernside and Ingleborough they rise to 737 and 723 m respectively. These altitudes, though modest, are sufficient to create a harsh local environment of low temperatures, high rainfall and frequent cloud cover that extend north to south across the region. The Pennines also cast a rain shadow across the whole region through the shelter they afford from the dominant westerly winds. This shelter is, however, ephemeral and prevails only as long as winds remain in that critical quarter of the compass. Conversely, eastern and coastal districts of the region are notoriously exposed to winds from between north and south-east.

The region, while possessing great altitudinal range, is physiographically uncomplicated and consists, in its northern half, of a generally west-to-east sloping land surface. This surface is crossed by a number of eastwards-draining rivers, from the Tweed in the north to the Don in the south, forming the delightful Dales country. The general eastwards fall in the land surface is interrupted only by the North Yorkshire Moors and Yorkshire Wolds and to a lesser extent by the East Durham Plateau. The coastline is of similar simplicity, and the masters of vessels large and small will search in vain for the shelter provided by estuaries, bays and creeks that characterise southern and western regions. Over the years northerly gales in particular have exacted a heavy toll of lives from the local fishing and port communities.

The region's extensive moorlands have an inhospitable climate that restricts the activities of both human and plant communities, and even the less intimidating terrain of the Yorkshire Wolds presents few villages for the casual tourist to enjoy.

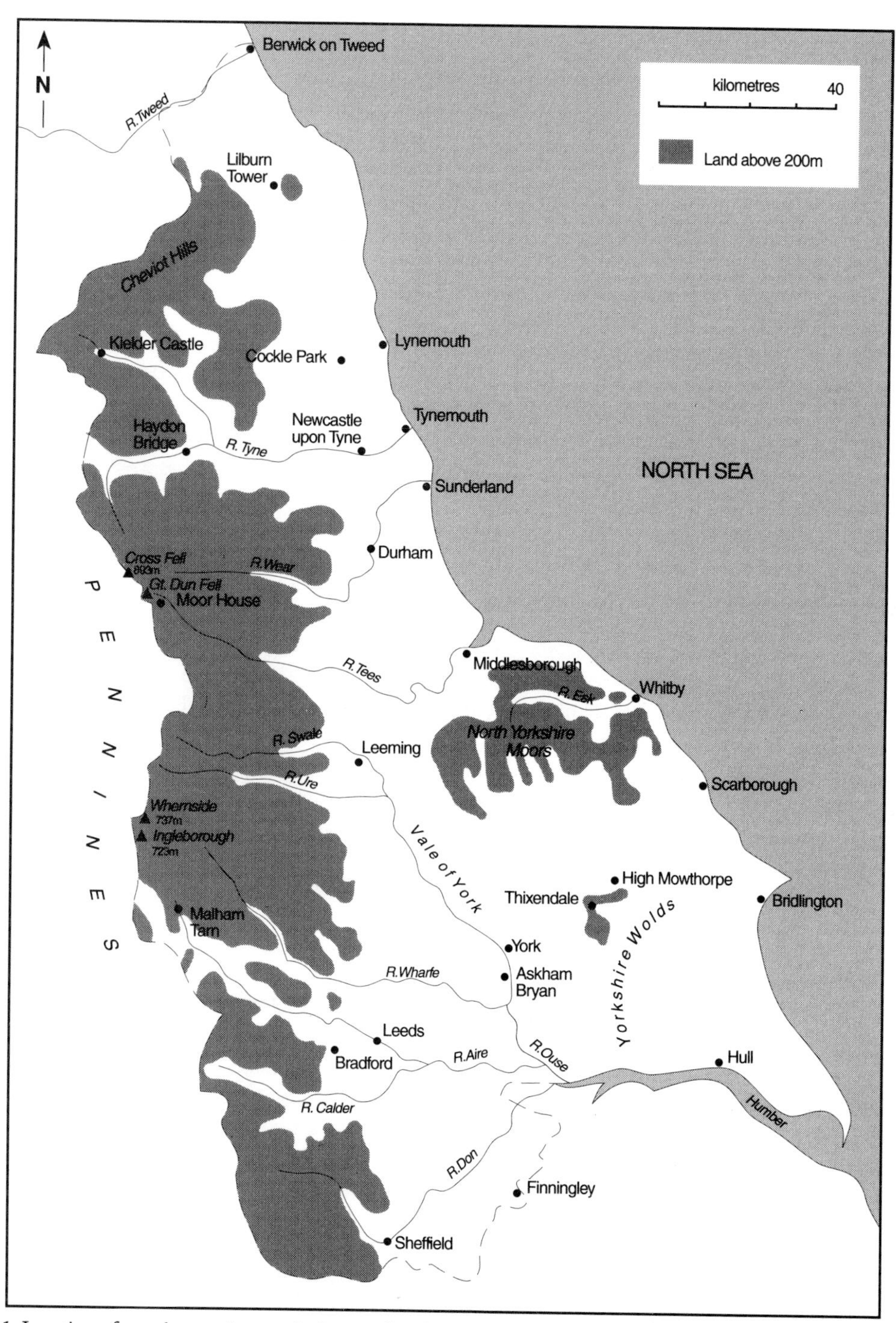

Figure 7.1 Location of weather stations and places referred to in the text

As testimony to the bleak but engaging isolation of these areas, three of the Britain's National Parks lie within the region. There are nevertheless many important towns and cities, most of which expanded from modest beginnings only in the Industrial Revolution. Some of them, such as Bradford and Huddersfield, are among the most elevated in the British Isles and suffer climatically as a result. The Yorkshire coast has fewer major settlements, though several picturesque fishing villages crouch for shelter from harsh northerly weather in small steep-sided coves. North of the Tees, however, the coastal coalfields of County Durham and south Northumberland have gathered to them the notable ports of Newcastle upon Tyne, Sunderland and Teesside. To the north of the Tyneside conurbation, the low-lying, sand-fringed coastline maintains its untouched splendour.

Weather Observations in the Region

The region contains many of England's largest cities, but, because it is dominated by sparsely populated moorlands, it cannot boast the same coverage of weather stations as is found in more populous regions. It is not, however, without a long history of weather studies, and Sunderland's Venerable Bede (*c.* 673–735) can be fairly claimed as the first Englishman to write a scientific treatise on climate (Wheeler 1993). Several centuries then elapsed until in 1611 the geographer John Speed described the region's weather, observing, 'the aire is sharpe and very piercing, and would be more, were it not that the vapours from the German Seas did helpe much to dissolve her ice and snow'. From that point onwards accounts become more commonplace. Christopher Sanderson's diary gives a good account of late seventeenth-century weather in Barnard Castle. A little later the diaries of William Elmsall describe the weather of the Dewsbury area for the period 1708 to 1740 while those of John Murgatroyd that of the Colne valley in the latter half of the eighteenth century (Tufnell 1987, 1991). Among the first instrumental records are those made daily in Newcastle by James Losh between 1803 and 1833. In Sunderland, Thomas Backhouse kept a detailed diary of all aspects of the weather between 1859 and 1915 (Wheeler 1984b). The most important observations are unquestionably those maintained by the Durham University Observatory. The observatory was established in 1841 and continues its work today as one of the country's longest-standing sites (Kenworthy 1994), and one which fortunately occupies a setting unchanged by any suburban development and consequent heat-island effects.

The majority of stations are on low ground but attention will be drawn to the important work carried out at the high-level sites of Moor House, Widdybank Fell and Great Dun Fell; the latter being England's highest station. Long periods of record are available from Askam Bryan (North Yorkshire) beginning 1936, Cockle Park (Northumberland) in 1898, Finningley (South Yorkshire) in 1943, Weston Park (Sheffield) in 1882 and South Shields in 1914. Some important weather stations are no longer active. The cessation of records from the Houghall College and Ushaw College sites, both near Durham City, is to be lamented, in the former case because the site is one of Britain's most notorious frost hollows.

TEMPERATURE

The region's north-easterly position underlies its prevailing coolness. This characteristic is given additional emphasis by the North Sea. The narrowness of the English Channel at its southern extremity and the shallowness of its wider northern margins limits the exchange with warmer Atlantic water. As a result, the coldest waters around Britain's shores are those of the western North Sea between Aberdeen and the Humber (Manley 1935). Jones (1981) has demonstrated that winter sea surface temperatures fall to 4 to 6°C (compared with 8 to 9°C off south-west England), and in summer they can be expected to rise to only between 12 and 14°C (16 to 18°C in the south-west). We should not, therefore, expect the same degree of moderation as is found elsewhere in the British Isles.

In common with all regions, the range of mean monthly temperatures increases with distance from the sea, which, though undeniably cold, usually sustains a winter surface minimum well above freezing point and keeps to an average of only twenty-five per year the number of air frosts in coastal areas. This frequency rises to over 125 in upper Teesdale. In this respect the steady increase in altitude towards the region's western limits exaggerates yet further its cool conditions. Although the winter coastal temperatures are higher than those further inland, they are, at around 4°C, among the lowest sea level winter monthly means in Britain. Even in coastal districts of north-west Scotland 5°C is more commonplace, while in south-west England 8°C is often approached. During summer the coastal temperatures are again subdued but not to the same extent as those of winter. Temperate but cool conditions are the key elements, and climatologists will search in vain for record-breaking high temperatures in this region.

Not only does the North Sea exercise a direct control on coastal temperatures but easterly winds and sea breezes during the summer months (see later section on wind regimes) can extend its influence several kilometres inland: However, the dominance of westerly winds would normally limit its landwards penetration. The region's physical geography is also important. Along most of the Durham and North Yorkshire coastlines the region presents a rugged and precipitous face to the sea. Only across the undulating plains of Holderness, along the Tees valley or over the low hills of eastern Northumberland can maritime influences pass easily westwards.

The Vale of York and its northern extension through County Durham constitute a further extensive area of low ground. The local topography is subdued, and its situation far enough from the North Sea not to be excessively influenced by its presence. The Vale of York enjoys additional shelter from the east afforded by the Yorkshire Wolds and North Yorkshire Moors. Thus Askham Bryan's July mean maximum (Table 7.1) is higher than Tynemouth's, where coastal influences are important, or Malham Tarn's, where altitude is the dominant element.

Extreme absolute maxima present a different picture in which prevailing synoptic conditions do not always place coastal sites at a disadvantage. A well-known example is offered by 3 and 4 August 1990, when temperatures reached 33°C over much of the region (Wheeler 1991). No less notable were the events of 30 September 1986 (Wheeler 1988) when much of eastern County Durham enjoyed maximum temperatures in excess of 25°C. This figure not only represents the highest recorded in the British Isles that day, but was also the highest in the whole of the month anywhere in Britain, and on that particular day Sunderland, with a maximum of 25.1°C, was warmer than such exotic locations as Alicante (17°C) and Casablanca (22°C).

The synoptic situation (Figure 7.2) that brought these conditions into being is a model for the warmest weather the region is likely to experience. High pressure lay to the south, advecting warm air northwards around its western margins from more southerly latitudes. This stable air retained much of its heat by contact with the warm land surface, and its trajectory towards north-east England did not bring it into prolonged contact with the sea. At the same time the strength of the anticyclonic circulation was sufficient to sustain light west to south-westerly winds and to inhibit cooling sea breeze circulations. Only under these conditions are high temperatures shared by coastal and inland sites alike; indeed, these are occasions when coastal temperatures might be the highest in the region, sometimes aided by a föhn effect as stable air moves down the Pennines' east-facing slopes.

These examples notwithstanding, high inland temperatures commonly generate sea breezes which frequently reduce daytime maxima by 3 or 4 degrees C and may be even more effective when they are accompanied by sea fogs. These circulations can also occur with bright sunny weather, leading to the 'bracing' weather for which east-coast resorts such as Bridlington and Scarborough are notorious (Lamb 1964)

Table 7.1 Mean monthly maximum and minimum temperatures (°C)

Location	*Alt. (m ASL)*	*Jan.*	*Feb.*	*Mar.*	*Apr.*	*May*	*June*	*July*	*Aug.*	*Sept.*	*Oct.*	*Nov.*	*Dec.*	*Year*
Kielder	201	4.5	4.7	7.0	10.1	13.7	16.8	18.2	17.8	15.1	11.9	7.3	5.3	10.9
Castle		−1.0	−1.4	−0.2	1.1	3.5	6.7	8.5	8.4	6.8	4.3	0.9	−0.6	3.1
Tynemouth	29	6.4	6.3	8.1	9.3	11.9	15.3	17.2	17.2	15.9	12.9	8.9	7.2	11.3
		1.7	1.8	2.8	4.4	6.9	9.9	11.7	11.9	10.2	7.9	4.3	2.5	6.3
Durham	102	5.8	6.0	8.4	11.0	14.3	17.5	19.2	19.1	16.8	13.2	8.7	6.7	12.2
		0.2	0.3	1.6	3.1	5.7	8.5	10.4	10.4	8.7	6.3	3.0	1.1	4.9
Great Dun	847	0.1	−0.9	0.7	3.5	7.8	11.2	11.8	11.8	10.2	7.5	2.3	0.9	5.5
Fell		−3.7	−4.3	−2.9	−1.2	1.9	5.1	6.4	6.8	5.5	3.2	−1.8	−3.3	1.0
Moor House	556	1.9	1.7	3.4	6.6	10.3	13.7	14.6	14.5	12.1	9.4	4.8	3.0	8.0
		−2.5	−3.0	−1.6	−0.2	2.2	5.4	7.0	7.3	5.9	4.0	−0.3	−2.0	1.9
High	175	4.7	4.7	7.2	9.8	13.2	16.6	18.6	18.8	16.3	12.4	7.7	5.5	11.3
Mowthorpe		−0.1	0.0	1.1	2.9	5.5	8.2	10.1	10.3	8.8	6.5	2.8	0.9	4.8
Askham	34	5.8	6.0	8.9	11.6	15.4	18.6	20.2	20.0	17.6	13.7	8.8	6.8	12.8
Bryan		0.3	0.4	1.8	3.4	6.0	8.8	10.7	10.6	8.9	6.3	2.8	1.0	5.1
Malham Tarn	395	3.7	3.4	5.9	8.7	12.3	15.1	16.5	16.2	13.8	10.6	6.6	4.5	9.8
		−0.7	−1.1	0.1	1.8	4.5	7.6	9.3	9.4	7.6	5.2	1.6	0.0	3.8
Sheffield	131	6.1	6.1	8.7	11.5	15.5	18.6	20.0	20.0	17.4	13.6	8.9	6.9	12.8
		1.4	1.2	2.4	4.2	6.9	10.1	12.1	11.9	10.0	7.4	3.9	2.3	6.1

Note: Data are calculated for the period 1961–90 for all sites except Great Dun Fell (1963–72) and Moor House (1961–80).
ASL = above sea level

The most inhospitable temperature regime is that to be found on the Pennines, where the paucity of weather stations allows us only to glimpse its true statistical character. Manley's (1936, 1938, 1942) early work at the Moor House and Dun Fell sites in upper Teesdale and Harrison's (1974) more recent studies identify the harshness of our upland climates and their sensitivity to altitude. Harding (1978) has shown that other factors such as season, synoptic state, soil and vegetation, and topography can also determine altitudinal temperature gradients. The Meteorological Office (1985) asserts that mean annual temperatures in the region decrease by 0.5 degrees C for every 100 m of altitude.

Such statistics, however, take no account of the general deterioration of climate with altitude that result from the combination of lower temperatures and increases in rainfall, cloudiness and wind speed. Vegetation responds to all these climatic elements, and in northern England the tree line stands at around 500 m ASL, above which the vegetation adopts a firmly sub-Arctic character – a curiosity that Manley (1952) relates to have prompted a visiting Swiss professor to declare 'this is the tundra' upon first seeing it. In this high-level zone the short growing season, usually restricted to the mid-April to October period, provides little opportunity for productive farming and land use. Rough grazing for hardy local sheep and plantations of equally hardy, if exotic, coniferous trees are widespread.

Finally, while the lowest mean monthly temperatures are associated with upland sites, some of which are among the coldest sites found in England, reference must be made to absolute minima. The region's coldest weather is most commonly associated with outbreaks of polar continental or Arctic air masses. Figure 7.3 shows the situation for one of the most recent outbreaks of polar continental air when even

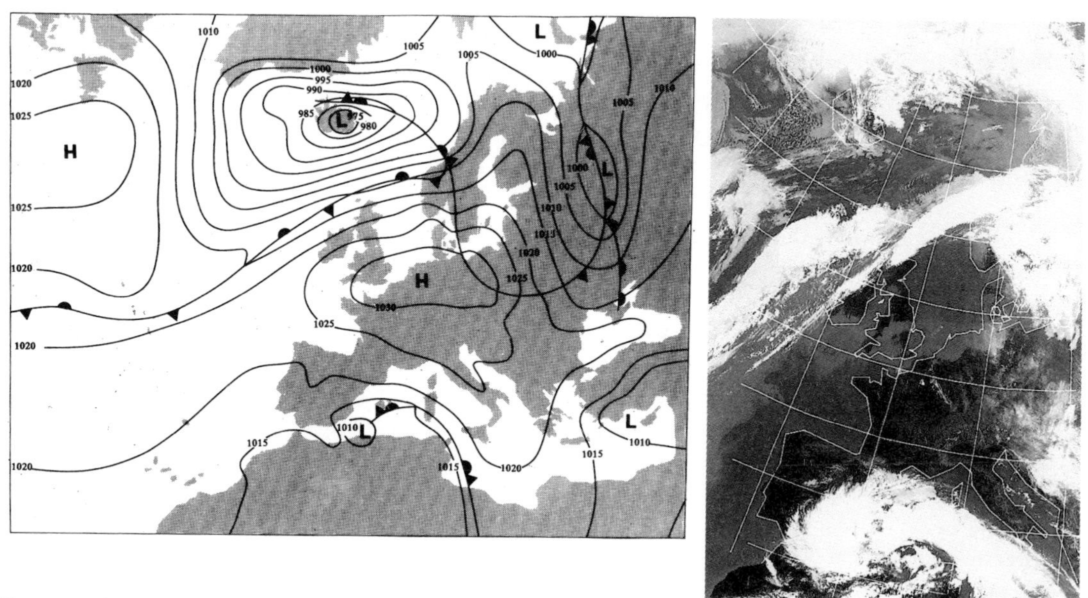

Figure 7.2 Synoptic chart for 1200 GMT on 30 September 1986. Inset: NOAA infra-red image for 1400 GMT on the same day

Photo by courtesy of University of Dundee

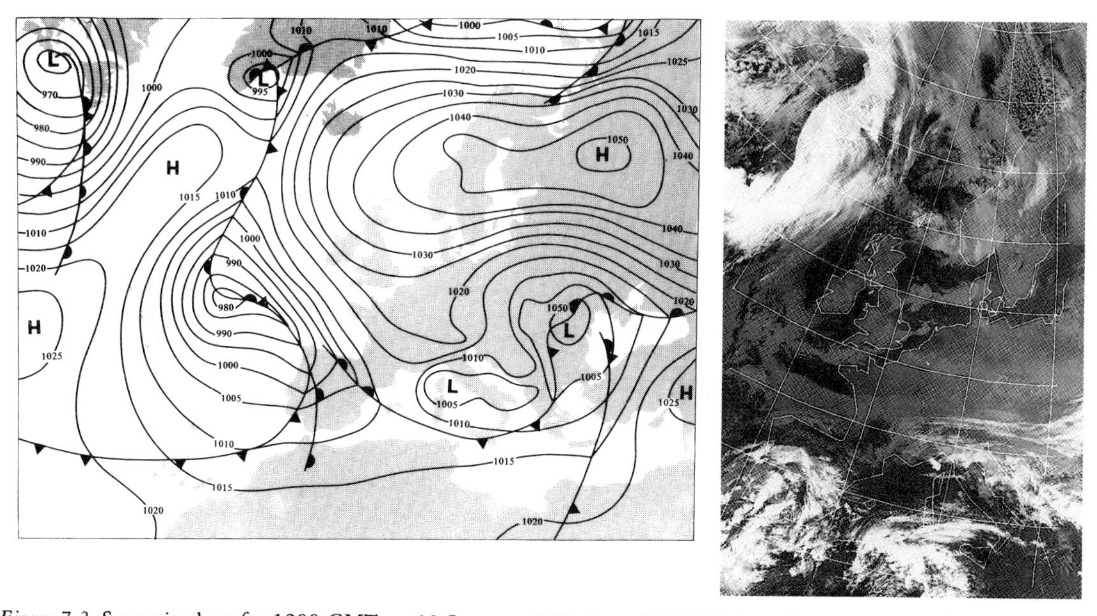

Figure 7.3 Synoptic chart for 1200 GMT on 12 January 1982. Inset: NOAA infra-red image for 1324 GMT on the same day

Photo by courtesy of University of Dundee

Box 7.1

HOUGHALL FROST HOLLOW – WHERE THE REGION'S LOWEST TEMPERATURES ARE RECORDED

No discussion of frost hollows in the north-east of England would be complete without reference to that of Houghall, one of the most pronounced and well-known examples of its type in the country. Houghall lies in the middle Wear less than 2 km from Durham City (Manley 1952). It is here, at an altitude of 50 m in an otherwise sheltered location in the floor of a deep valley, and not on the inhospitable slopes of Cross Fell, that the region's lowest absolute minima are consistently recorded. On 5 January 1941 the screen minimum fell to −21.1°C; this remains the county's lowest recorded minimum. During another cold spell of anticyclonic weather a minimum of −18.3°C was recorded on 16 February 1929, and on the 14th of the same month the maximum reached only −8.1°C as the pool of stagnant cold air lingered throughout the day. In contrast, Great Dun Fell's absolute minimum stands at −13.5°C. As in all such cases, the cooling effects were heightened by the existence of continuous snow cover that prevented heat from being conducted from the soil to replace that being lost by radiation at the surface. Houghall's extreme minimum regime is not apparent in its long-term daily means, which scarcely differ from those of nearby Durham University. Yet even here the movement of cold air into the valley floor is not a headlong rush but a slow drift of cold air downhill. Houghall, although an extreme case, is nevertheless representative of a wider class of topographically controlled temperature regimes of which the Dales valley floors are a member (Manley 1943).

coastal sites recorded temperatures as low as −12°C. The region is particularly vulnerable to polar continental air, which most often occurs during spells of weather dominated by high pressure centred over Scandinavia. If conditions are settled, cold air (katabatic) drainage is especially effective at these times when the ground is covered with snow and clear overnight skies encourage rapid radiant heat loss. Cold air drainage is a common feature across the region, giving rise to exceptional minima, and Stirling's (1982) list of annual absolute minima for English sites between 1939 and 1981 includes six citations for Houghall, seven for Moorhouse and four additional entries for other locations in Durham or Northumberland. The minimum of −23.3°C recorded at Haydon Bridge in the Tyne valley on 21 January 1881 was at the time the English record.

PRECIPITATION

Precipitation includes both rainfall and snowfall, and in this northern and hilly region the latter is a significant component of the annual totals. Whether in the form of rain or snow, precipitation is closely governed by altitude, and the relative simplicity of the region's relief, especially across the broad swathe of land between the Tees and Tweed, is reflected in an uncomplicated pattern of isohyets (Figure 7.4). Precipitation declines from high values on the Pennines ridges to minimum values along the east coast, though even the Pennines themselves lie in the rain shadow cast over them by the high ground yet further west in the Lake District and North Wales (Salter 1918). For this reason, high-level Pennine sites record annual totals between 500 and 1,000 mm less than those in the Lake District at similar altitudes. Comparisons can be made between Table 7.2 and Table 8.4, where Kendal's (36 m ASL) 1,323 mm far exceeds

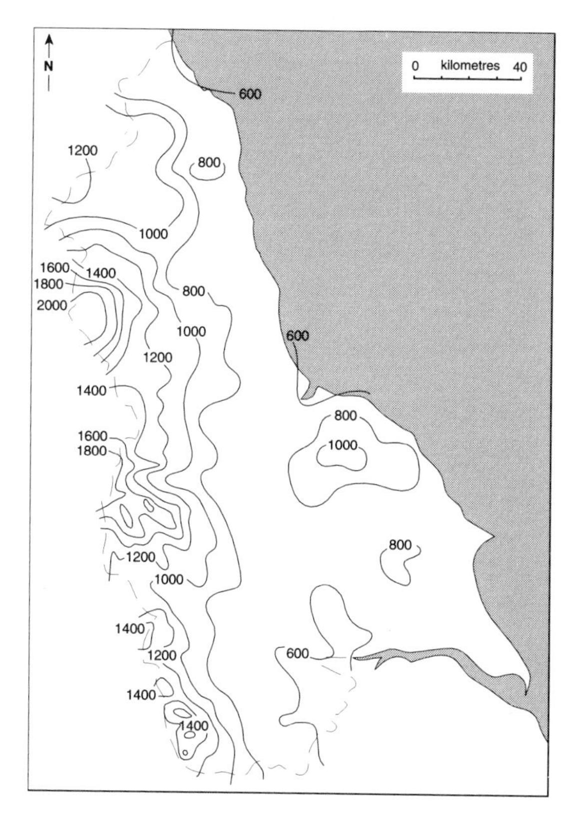

Figure 7.4 Mean annual precipitation (mm) for the period 1941–70

Askham Bryan's 622 mm, and Bolton's (107 m ASL) 1,186 mm is similarly greater than Durham's 652 mm.

Large areas receive annual precipitation of between 1,500 and 2,000 mm, but there are no data for the highest locations, where we can only speculate that average annual totals exceed 2,000 mm. Snaizeholme (at 579 m ASL in the upper Ure catchment) is, however, reported to have an annual total of 1,957 mm. High rainfall is also found in the far north of the region in the Cheviot Hills, where annual precipitation exceeds 1,100 mm over large areas. Rainfall of this order, although inimical to farming, can be turned to advantage in an age in which the demand for water continues to rise unabated. The North Tyne valley is the site of England's largest wholly artificial reservoir at Kielder, while the Yorkshire Pennines and Teesdale contain many smaller Victorian and later schemes.

The eastwards decrease in precipitation and surface elevation is marked, however, by occasional reversals. Overlooking the North Sea and to the east of the Vale of York the land rises again to over 400 m in the North Yorkshire Moors. This area, though not as wet as similarly elevated sites in the Pennines and receiving little more than 1,000 mm in an average year, supports large areas of moorland and blanket bog with its characteristic plant communities of sedges, cotton grasses and heather (Pearsall 1950). In all these upland areas precipitation exceeds evaporation and transpiration in all months of the year, giving scope for the accumulation of surface waters and permanently wet and acidic soils. Only on the rolling hills of the Yorkshire Wolds are these features absent. This distinctive 'downland' landscape is not, however, accounted for by any corresponding climatic contrast, and the hills are sufficiently high (rising to 250 m) to promote precipitation in excess of 750 mm in some parts of the Wolds north-western extremities close to the villages of Thixendale and Fridaythorpe. The explanation is geological, the Wolds being formed of permeable Cretaceous Chalk on which the absence of streams is testimony to its capacity to absorb most of the rainfall and snow melt, leaving little to sustain soil moisture levels.

Yet these are relatively minor variations in an otherwise dominant west-to-east decline in precipitation that is so marked that the east coast of northern England is one of the driest parts of the British Isles and annual precipitation along the Durham and Northumberland coasts is almost everywhere less than 650 mm, and in several places less than 600 mm. Similarly low totals can be found in the sheltered ground between the Pennines and the Yorkshire Wolds; around York and Leeming, for example. This precipitation gradient results from the rain shadow produced by the presence of the Pennines and the dominance of westerly winds. Frontal rainfall alone would diminish from west to east as the air masses slowly relinquish their

Table 7.2 Mean monthly and annual precipitation totals (mm) for the period 1961–90

Station	*Alt. (m ASL)*	*Jan.*	*Feb.*	*Mar.*	*Apr.*	*May*	*June*	*July*	*Aug.*	*Sept.*	*Oct.*	*Nov.*	*Dec.*	*Year*
Lilburn Tower	80	72	51	62	53	59	56	57	76	61	63	80	67	757
Cockle Park	99	64	48	56	48	55	50	60	76	66	55	71	62	711
Kielder Castle	201	134	87	114	74	83	77	92	109	110	122	136	138	1,276
Tynemouth	29	54	36	46	44	49	51	55	66	55	50	63	54	623
Durham	102	59	41	52	49	53	52	52	68	56	52	62	56	652
Askham Bryan	34	54	40	48	45	50	52	52	66	54	52	53	56	622
Bradford	134	87	69	74	64	63	63	59	72	74	76	84	88	873
Sheffield	131	83	66	70	63	62	62	54	64	63	67	79	83	816

moisture, but the effect of high ground is to exaggerate this loss. At the same time, as the land surface falls towards the east the important stimulus of altitude, which underlies so much British rainfall, also diminishes.

This effect is not constant; it differs significantly from month to month. Wheeler (1990) and Chuan and Lockwood (1974) have shown that the rain shadow's influence varies from a maximum in the winter months to a minimum in the summer. This variation reflects several factors. The effectiveness of the Pennines as a 'shield' against wet westerly weather depends largely on the frequency with which winds blow from that direction, but this itself varies from month to month. They are at their most persistent during the winter, but are less frequent and less vigorous during the summer when the northern hemisphere's westerly circulation is generally less active.

Summer is also the only season when ground-level heating will be sufficient to promote significant convective rainfall, and its frequency over low-lying areas such as the Vale of York and eastern Northumberland further diminishes the strength of association between precipitation and altitude. This point is emphasised in Table 7.2, from which it is clear that the low-level stations tend to have their wettest weather in August while more elevated sites have theirs in the late autumn and winter.

Wind direction and consequent exposure are evidently critical factors in determining the levels and the geographical distribution of precipitation. The region's shelter from westerly weather has already been noted. Conversely, easterly conditions can introduce prolonged spells of wet weather. Outbreaks of unstable northerly air can also lead to heavy showers along coastal districts and over the exposed North Yorkshire Moors. These points are illustrated in Table 7.3, where Sunderland's mean annual rainfall is disaggregated into those proportions that fall from winds of different directions. However, the picture is obscured by different durations for which those winds prevail. To overcome this, the average rainfall, based on a standard period of 1,000 hours of wind from each direction, is also given. These latter data more clearly demonstrate the vulnerability of east-coast sites to winds from between north and south-east and the relative dryness of those from between south-west to north-west.

Average monthly statistics are the clearest expression of any precipitation regime but individual extreme months and years can be equally informative. In 1989, for example, the north-east of England's unusual dryness was a consequence of the frequency of westerly weather and the enhanced rain shadow influence of the Pennines. Even long-established sites such as Durham recorded this as the

Table 7.3 Average annual rainfall (mm) from winds of different direction. Rainfall for a standard period of 1,000 hours of wind from each direction is also given

	N	*NE*	*E*	*SE*	*S*	*SW*	*W*	*NW*	*Calm*
Percentage of time for which wind prevails	9.0	4.3	4.9	10.2	7.5	23.5	22.7	5.6	12.3
Percentage of annual rainfall	14.8	8.7	9.1	17.6	8.9	14.2	9.5	9.3	8.0
Rainfall per 1,000 hours (mm)	104.6	128.5	117.9	109.4	75.0	38.2	26.8	104.2	41.3

Note: Data based on Sunderland for the period 1984–94

driest year on record. It was the most extreme of a number of such years between 1988 and 1992 whose effects were felt across much of eastern Britain (Marsh *et al.* 1994). The importance of the westerlies, in regard to mild as well as dry winters, has also been discussed by Cannell and Pitcairn (1993), who found that during the winter of 1988/89 westerly weather, as defined by Lamb's (1950) classification, was twice as frequent as usual. Yorkshire's mean precipitation in 1989 was estimated (National Rivers Authority 1990) to have been only 655 mm, 80 per cent of normal. The east coast was the driest area with many stations recording less than 400 mm. To emphasise the regional character of the rain shadow effect, the same conditions produced notable rainfall excesses in western England and Scotland as a result of greater exposure to humid maritime air masses.

Periods of prolonged and heavy rainfall in the eastern districts are usually associated with onshore winds on the northern side of depressions passing to the south of the region (Hay 1949; Sawyer 1956a; Finch 1972). One of the best examples of this situation occurred between 20 and 23 July 1930, when a depression became stationary in the southern North Sea and sustained a constant northerly airflow over the region. A total of 304 mm was recorded at Castleton in the upper Esk valley above Whitby, with daily totals of 68, 59, 145 and 32 mm. The 250 mm isohyet for the four days covered a large part of the surrounding moorland with the 100 mm isohyet extending across most of what is now the North Yorkshire Moors National Park (Meteorological Office 1930). A feature of this storm was its lack of intense rainfall but its notable persistence, with scarcely an interruption in the four days. More recent outbreaks of persistently wet easterly weather are not hard to find. On 15 July 1973 Sheffield and surrounding areas received over 100 mm, the detailed circumstances of which are discussed by Smithson (1974), and Wheeler (1985) has reviewed the unusually wet conditions of November 1984 in east Durham when on the 2nd and 3rd the rainfall was one of the heaviest of the century. Figure 7.5 shows the synoptic situation at the time, and the satellite image inset reveals the marked distribution of cloud over the North Sea and northern Britain.

In all such examples of wet, onshore weather there is a marked diminution in the otherwise well-established correlation between rainfall and altitude. This is a consequence of the vertical structure of the air masses, which, though easterly in their lower and middle layers, may still be dominated by circumpolar westerlies from the middle level upwards. Sawyer (1956b) suggested that a strong positive vertical wind profile is necessary for the orographic control to function efficiently; this would not be the case if low-level air flowed in a different direction compared with middle- and upper-level air. Jackson (1969) has also shown that depressions to the south of the region often produce rainfall patterns with a minimal orographic component.

A different picture is presented by the high ground close to the Pennines watershed. In these areas easterly winds have usually spent much of their

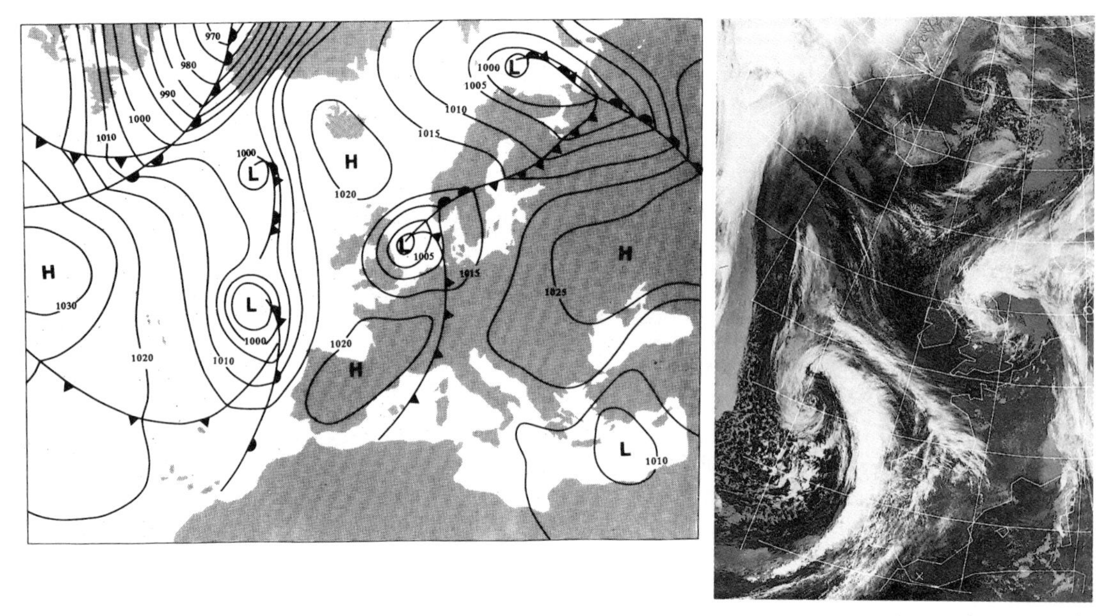

Figure 7.5 Synoptic chart for 1200 GMT on 3 November 1984. Inset: NOAA infra-red image for 1556 GMT on the same day. The low pressure-system over the North Sea is marked by the obscuring cloud
Photo by courtesy of University of Dundee

precipitation over low ground, and it is the westerly airstreams, especially the deep, moist, warm-sector air masses described by Douglas and Glasspoole (1947), that are the source of heavy and persistent rain. They are often the cause of floods in the rivers that drain these remote areas, particularly if the soils are already wet and unable to take up further moisture. This was the case in one of the most serious floods in recent years on the river Wear on 5 November 1967. Over 75 mm fell on a large part of the catchment in what had already been a wet month. The resulting runoff produced a record flow of 415 cubic metres per second on the Wear just above Durham (Archer 1992).

Not all the region's heaviest rainfalls are frontal or cyclonic in origin. Convective cells can produce short periods of intense rainfall. Because of the necessity of ground-level heating, convective activity is restricted to the summer months. Of all the intense falls noted by the Meteorological Office (1963) for the region, none has been earlier in the year than May, though several have been as late as October. Daily totals of more than 76 mm are not unusual, and more extreme events have been so marked as to be the source of considerable discussion of their accuracy. The Calderdale storm of 19 May 1989 has become something of a *cause célèbre* in this respect (Acreman 1989), when 193 mm was estimated to have fallen in two hours. Just over one year later on 24 August 1990 the County Durham and North Yorkshire areas experienced widespread heavy rainfall. Several sites recorded over 50 mm in two hours, but the most notable totals were 100.8 mm at Thornton Steward near Leyburn and 86.7 mm at Brignall near Barnard Castle (Archer and Wheeler 1991). Similarly severe storms that struck the West Yorkshire area in June 1982 have been usefully summarised by Chaplain (1982).

There is a general infrequency of the type of unstable conditions that more commonly prevail in, for example, south-east England, where summer ground-level temperatures are higher and invading masses of warm and humid air from the south are more effective vehicles for extreme rainfall. As a

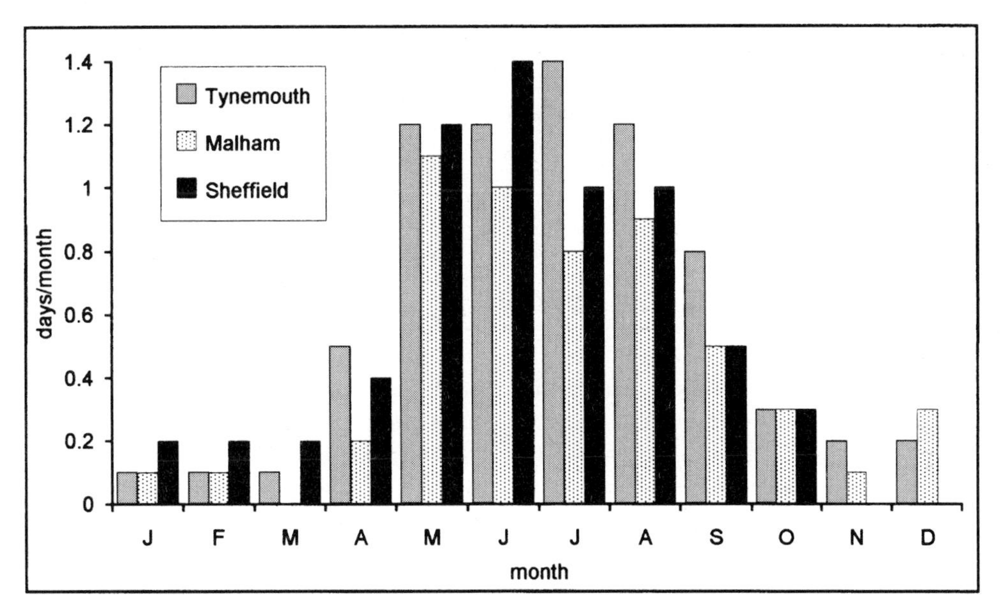

Figure 7.6 Mean monthly frequency of days with thunder for the period 1961–90

result, both hail and thunder tend to be less frequent in this region than to the south, and most locations (Figure 7.6) expect no more than five or six days of thunder each year. Summer is the most common season for thunder; winter thunder is infrequent. Thunder frequencies tend to vary little from one location to another, and Tynemouth's mean annual frequency of 7.3 days is similar to that of Sheffield at the other end of the region with 6.4 days.

Snowfall

Snowfall is an important part of the precipitation regime, more so than in any other region south of the Scottish border. It is important to recall also that the geography of snowfall in Britain differs significantly from that of rainfall, which is heaviest over western districts. Snowfall requires air temperatures of no more than 4°C at ground level, and while such temperatures can be introduced by westerly weather in winter, especially on high ground, they are more commonly associated with outbreaks of northerly air, particularly in the wake of eastwards-receding depressions. In these cases it is eastern and northern Britain that are most exposed and we find that the principal areas of snowfall have moved towards the east when compared with those of rainfall. Snow is a common sight on the region's moorland but not one necessarily restricted to the winter months. On Coronation Day on 3 June 1953 there was a notable snowfall on Cross Fell. More recently, and more remarkable because of the low altitude, on 2 June 1975 and again on 13 May 1995, coastal sites in County Durham saw some snowflakes, but these occurrences must be regarded as exceptional.

Exposure to northerly winds is critical, and the North Yorkshire Moors, which present a notable obstacle to these airstreams, can receive nearly as much snow as higher ground in the Pennines. Furthermore, northerly weather is typically unstable, comprising cold Arctic air, and the resulting snow is usually in the form of showers, some of which may be heavy. In such cases instability is initiated not over the cold land surface but over the North Sea, which though having low surface temperatures can be significantly warmer than Arctic air masses passing overhead. This temperature

Table 7.4 Mean number of annual days of snow falling and snow lying (at 0900 GMT and covering more than 50 per cent of the ground) for the period 1961–90

	Haydon Bridge/ Corbridge	Tynemouth	Moor House	Askham Bryan	Hull	Malham Tarn	Sheffield
Altitude (m ASL)	79	29	556	34	2	403	131
Snow falling	24.5	26.9	78.5	13.3	18.5	41.1	24.4
Snow lying	20.1	7.6	71.0	14.4	13.3	40.8	20.4

difference may be sufficient to produce small 'polar' lows embedded within the air masses. Lowndes (1971) suggests that as many as 84 per cent of falls of at least 7 mm result from polar lows or from warm and occluded fronts on southerly located depressions.

On the high Pennines, snowfall more commonly occurs with cold westerly and north-westerly streams of polar maritime air that, like their rain-bearing counterparts, have a diminishing effect as they move eastwards. The inevitable fall in temperature that accompanies altitude ensures that snow cover will here linger after the snow has fallen. At lower levels there is less opportunity for snow cover to last, and close to the coast the number of days on which snow falls is generally greater than those on which it lies. On high-level sites the two are more closely matched (Table 7.4). The cases of Tynemouth and Haydon Bridge, the latter lying in the middle Tyne valley over 50 km from the North Sea, are good examples of this contrast.

Periods of prolonged and heavy snowfall may, nevertheless, occur on the coast, where their explanation is found in synoptic situations similar to those that bring about persistent rainfall. As indicated above, the principal requirement is for a low-pressure system over the North Sea or slightly further south. To the north of this system moist air will circulate westwards towards the British Isles, possibly on the north side of a warm front. In late winter, high pressure over Scandinavia will not only contribute to the pressure gradient, but also add polar continental air to the northerly reaches of the circulation. The combined effect of these factors leads to blizzard-like conditions along much of eastern England and Scotland. An extreme example of this situation appeared in February 1941, which brought to the east of the region its heaviest snowfall, over 1 m deep in some areas, of the century (Wheeler 1991).

Snowfall can occur throughout much of the year at high level and is commonplace as late as May, though it is rarely seen on low ground before November. These averages disguise the year-by-year variations that result from the particular circumstances needed for snow to occur. Rainfall, by comparison, tends to be less variable. At sea level the number of days per year on which snow covers more than 50 per cent of the ground is approximately eight. This figure rises by ten for every additional 100 m of altitude gained, revealing again the sensitivity of precipitation and temperature to height changes (Figure 7.7). Great Dun Fell (847 m), for example, has an annual mean of 96.2 days of snow lying at 0900 GMT, at Malham Tarn (395 m) the figure falls to 40.8 while at Tynemouth (29 m) it is only 7.6. The latter is, of course, a coastal site where lower altitude and proximity to a relatively warm winter sea combine to minimise snow cover. The February peak of snow cover on Great Dun Fell coincides with the site's coldest month and again suggests the importance of outbreaks of easterly, polar continental air that often characterise that month.

Although snowfall is an important part of the region's precipitation regime, temperatures are rarely low for sufficiently long to support snow cover over a time span of several months; the famous winter of 1947 is the obvious exception. On low

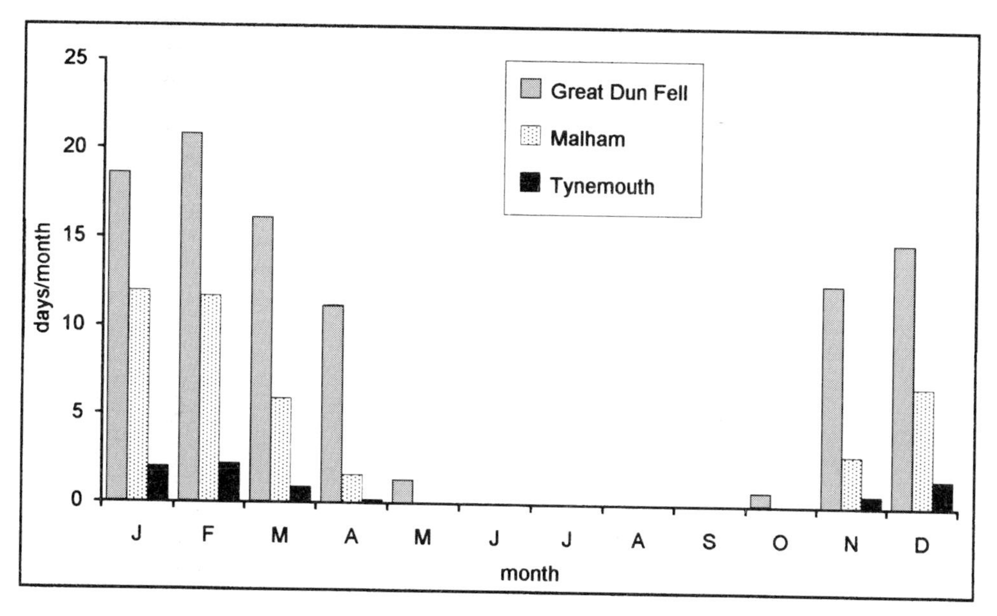

Figure 7.7 Mean monthly frequency of days with snow lying and covering more than 50 per cent of the ground at 0900 GMT for the period 1961–90, (Moorhouse 1961–80)

ground snow cover for periods of more than even ten consecutive days is exceptional (Lowndes 1971). Indeed, the frequent eruptions of mild tropical maritime air, especially when accompanied by rainfall, over large expanses of melting snow present a serious flood risk. It has been shown (Natural Environment Research Council 1975) that as many as 20 per cent of floods in the north of the region have a significant snowmelt element. The floods that followed the heavy snowfalls of 1941 and 1947, together with others of a similar origin, are described in Archer (1992).

WIND

The dominance of westerlies and south-westerlies (Table 7.5) is a feature common to all regions. These are also the directions for the majority of gales (Harris 1970), most of which are connected with active depressions passing eastwards to the north of the region (Figure 7.8). The truly regional aspects appear only in the detail provided by small-scale circulations and in relatively minor modifications to the prevailing airflows. The Pennines' ability to dissipate not only precipitation but also the vigour of the wind is again important, and westerly gales of the persistent ferocity and frequency familiar to Hebridean crofters or to Cornish fisherman are scarcely known beyond the highest Pennines ridges. Eastern and coastal districts are, on the other hand, exposed to northerly gales, which though less frequent can introduce unseasonal chilliness at any time of the year. More importantly, they have been known to wreak havoc far in excess of that of their westerly counterparts.

The latter point is illustrated by the storm surges that arise in the North Sea when winds and tides conspire to drive water into the shallow seas between mainland Europe and eastern Britain. An example of these destructive storms came at the end of January 1953. A deep low had moved south-eastwards into the North Sea area and at midnight on the 31st had deepened to 968 mb (Figure 7.9). Strong northerly winds followed, with gusts of between 90 and 100 knots. The combined effect of high tides, of water

Table 7.5 Summary of average wind directions for the period 1961–90

Location	N	NE	E	SE	S	SW	W	NW	Calm
Tynemouth	15.6	3.6	7.3	5.3	18.2	10.7	28.3	5.6	5.4
Finningley	14.1	5.5	5.6	7.3	17.8	14.3	18.5	12.2	4.7

Note: Data are expressed as percentages of all 0900 GMT observations

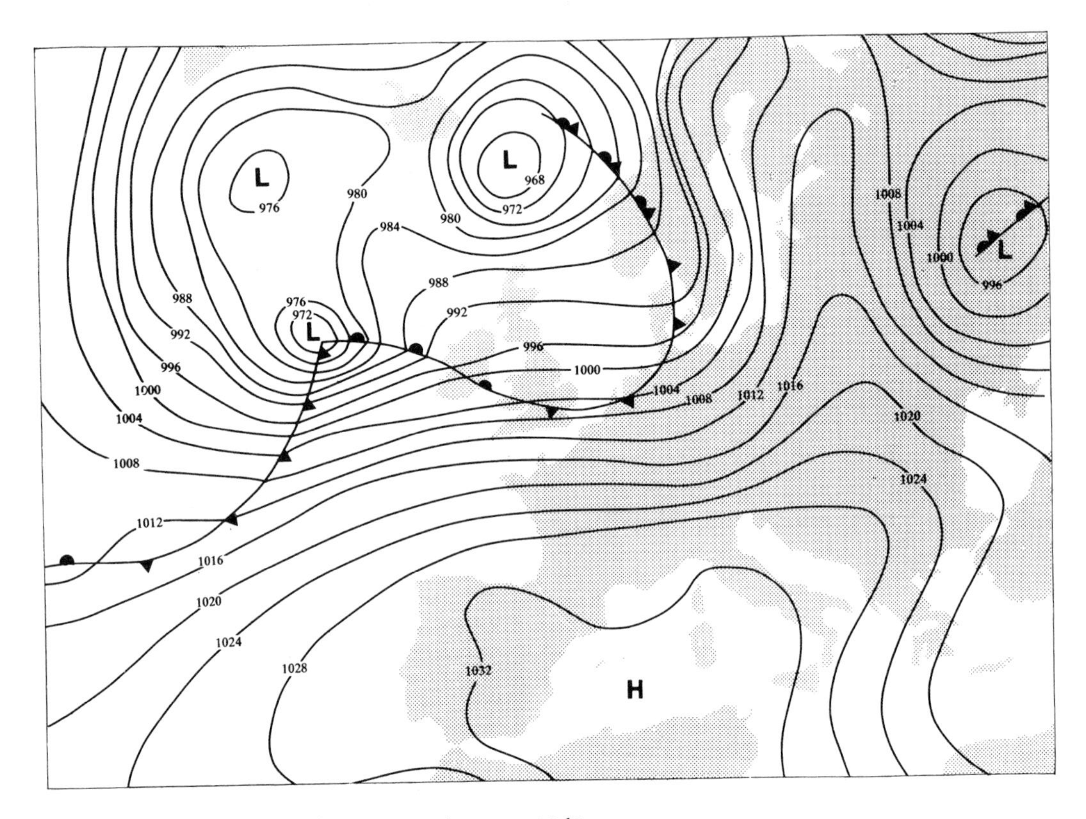

Figure 7.8 Synoptic chart for 1200 GMT on 14 January 1968

being driven by the winds into the ever narrower confines of the North Sea basin and the siphon effect produced by high pressure in the Atlantic but low pressure over the North Sea raised sea levels to catastrophic heights (Douglas 1953). Flooding in the region was confined to low-lying coasts of Holderness and the Teesside area, but in the Netherlands and the Thames estuary damage and loss of life was more widespread.

Lamb (1991) has shown that the events of 1953 were not unprecedented and that the North Sea area is peculiarly vulnerable to northerly winds. This is the only direction over which there can be any significant fetch, allowing the sea waves to build to heights more commonly found in open Atlantic waters. Even powerful sea defences have been known to give way to aggressive sea waves generated over hundreds of miles of open water. One part of the region is exceptionally vulnerable to this activity. The soft glacial sediments of East Yorkshire's Hol-

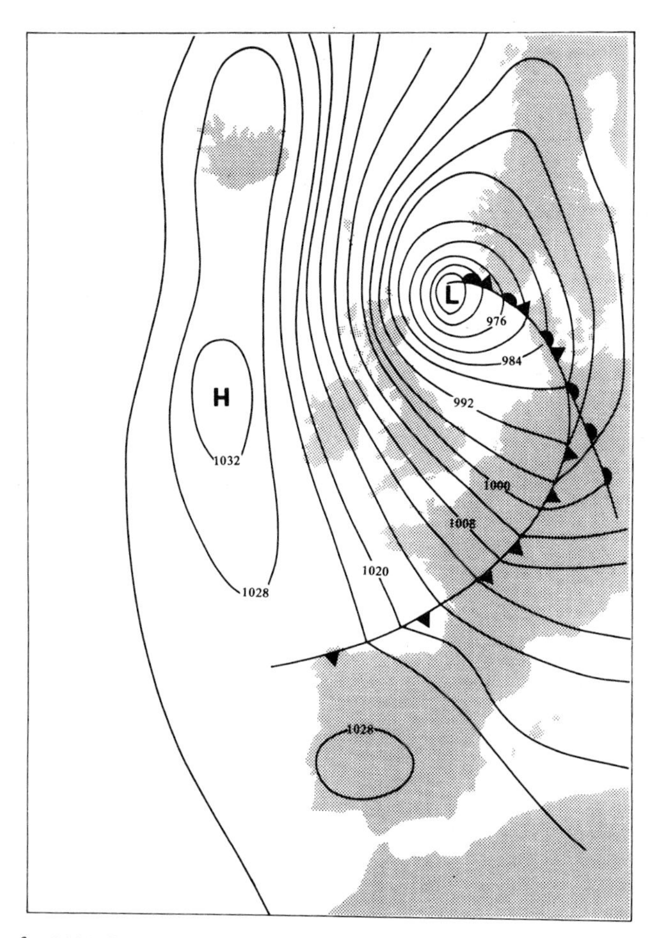

Figure 7.9 Synoptic chart for 1200 GMT on 31 January 1953

derness coast are notoriously unable to withstand the effects of wave action. Even under normal conditions the coastline retreats more than a metre a year but several metres can disappear in a few hours of north-easterly gales. Local legend claims that whole villages have disappeared overnight during such violent storms. Whether this is true or not, the list of East Yorkshire villages that have been swept away is impressive (Steers 1964).

Westerly gales are normally less hazardous, owing to their diminished vigour and to their offshore character, which is unlikely to encourage sea flooding. Nevertheless, they can occasionally prove destructive, and the severe gales of 16 February 1962 that left nearly two-thirds of all houses in Sheffield with some form of damage can scarcely go unnoted. Elsewhere in the region the combination of greater exposure and altitude can have a marked effect. On Great Dun Fell the region has an unrivalled location. On 15 January 1968 the site recorded England's highest gust of 116 knots. Not surprisingly, the direction of this gust was westerly. On the same day Durham recorded one of the region's highest gusts for a low-level site of 90 knots. Great Dun Fell also holds the record for the highest mean hourly speed, a remarkable 92 knots, recorded on 12 January 1974. The winds at the time were again from the south-west. Smithson (1987) has also described airflow at a high-level site in the southern Pennines at High Bradfield (395 m ASL). His analysis presents a similar picture of a propensity for gales at any season, but notes also that January and December are significantly stormier than June and July. South-westerly winds dominate the regime in this southern site also, although a secondary peak of north-easterly winds was indicated.

At lower levels winds of more than 100 knots have never been recorded, but speeds can be occasionally noteworthy. At South Gare on the Tees estuary a maximum low-level gust of 95 knots from the north was recorded on 2 June 1975. Gusts of over 90 knots have also been recorded at Lynemouth and Durham. The Newcastle University weather station recorded gusts on excess of 90 knots on 13 and 16 January 1984 (Sharp 1994). Spurn Point recorded 82 knots on 11 January 1978. The greatest low-level mean hourly speed was also noted at South Gare: 70 knots on 2 January 1976. Generally, at low and middle levels in the region maximum gusts are about 80 knots and maximum mean hourly speeds 50 knots, though in the Vale of York, which enjoys a degree of shelter from both east and west, RAF Leeming has a maximum hourly mean of only 45 knots.

These extreme values are perhaps less significant than the statistics for all gales. In this respect the region offers something very close to the two national extremes. The highest reaches of the Pennines are among the windiest locations in the country, though even this area scarcely matches parts of Highland Scotland (see Chapter 10), and Great Dun Fell has an average of more than 180 'gale days' per year. The east coast, less than 100 km distant, is one of the areas of northern England least troubled by

Box 7.2

THE HELM WIND OF CROSS FELL – NORTHUMBRIA'S MISTRAL

It is only within the past century that the Cross Fell area's ancient epithet 'Fiend's Fell' has passed into disuse. While the title certainly derives in part from the climatically hostile character of the area, it is further distinguished in suffering the occasional attentions of the Helm Wind, which is a local member of a family of similar winds that develop in mountainous areas in all parts of the world. This violent easterly wind, though not a Mistral type of airflow in the strict sense of the term (it is closer in origin to the Adriatic's Bora), is received with much the same sentiments of displeasure by local farmers as is its French counterpart. The Helm Wind's principal course rushes it across the Pennines ridge before it plunges with undiminished ferocity down its steep western-facing slope (Manley 1945). Most significant is the force with which this local wind can prevail; often it is of more than force 7 but sustains force 4 even on the low ground of the Vale of Eden. The principal requirements for its formation are a wind from north-north-east or east, i.e. within 30 degrees of the orientation of the main Pennines ridge, with a strength of at least force 4. More generally, for all winds of this type an uninterrupted hill crest is vital, and the Durham Pennines meet exactly this necessity. An inversion height above the hill crest also helps to concentrate and accelerate the low-level airflow. When fully developed, the Helm wind will be accompanied by the Helm itself, a stationary cloud over the highest part of the hills. Downwind, over the Vale of Eden, the Helm Wind is also replaced by a contrary, but lighter, south-westerly wind. This may also be accompanied by a bar of cloud at around 500 to 1,000 m altitude, this feature being explained by the rotor effect downstream of the ridge. The incidence of the Helm Wind is, however, highly variable and depends upon the frequency of easterly winds of the required force.

such events (Hardman *et al.* 1972), and Tynemouth's annual average is only fifteen.

Sea Breeze Circulations

At the other end of the wind scale sea breezes cannot be overlooked, not only because of their frequency but also because they exercise an important influence on coastal temperatures. They develop when there is a strong temperature contrast between a warm land and cool sea surface (see Chapter 1). They are therefore most frequent in late spring and summer. At this season the land surface will be warming quickly in response to the advancing year. Meanwhile the higher specific heat of the North Sea's waters allows only a slow response and it retains for longer its cooler winter conditions.

Even in summer the North Sea's surface temperatures rarely exceed 14°C. Onshore sea breezes introduce these cool conditions to coastal areas and, where topography admits, even further inland. However, low-lying coastal areas are extensive only in parts of Northumberland, the Tees valley and Holderness. The near-uninterrupted stretches of sea cliffs elsewhere diminish sea breezes' strength and their landwards penetration. They are scarcely ever able to negotiate such features as the East Durham Plateau, and locations such as Durham City only rarely feel their chilling touch despite being at no great distance from the North Sea.

The earliest date at which sea breezes develop is April, though they may be expected to occur as late as October. Along the region's coast they are from between north-east and south-east and can sustain speeds of 10 knots and gust up to 20 knots. Their morning onset, which can be as early as 0700 GMT, is typically sudden but their night-time decline is leisurely and vestiges of the circulation can be detected as late as 2200 GMT. They are, however, mere zephyrs compared with pressure gradient westerlies, which need attain only 10 knots to suppress completely their development.

SUNSHINE

Sunshine is determined by degree of cloud cover and responds, as does precipitation, to altitude. Nevertheless, synoptic weather types are also important and wet months need not necessarily be dull if characterised by showery weather while dry but overcast anticyclonic conditions can be notably gloomy. Latitude is also a significant factor, and southern English sites will more often fall under the influence of the Azores anticyclone and are less frequently exposed to fronts and associated cloud bands of depressions following their favoured routes to the north of Britain. Hence while coastal sites are the sunniest in the region they compare poorly with more southerly sites, but enjoy generally higher annual totals that those of eastern Scotland (Table 7.6).

Table 7.6 Mean monthly and annual sunshine totals (hours) for the period 1961–90 (Great Dun Fell for the period 1963–72 and Moor House 1961–80)

Location	*Jan.*	*Feb.*	*Mar.*	*Apr.*	*May*	*June*	*July*	*Aug.*	*Sept.*	*Oct.*	*Nov.*	*Dec.*	*Year*
Tynemouth	52	67	106	140	177	181	166	158	129	95	69	45	1,385
Durham	54	65	106	133	169	164	155	156	124	93	66	46	1,331
Great Dun Fell	21	36	60	86	109	130	104	93	75	54	28	28	824
Moor House	28	46	82	114	157	176	141	136	96	74	45	31	1,126
Askham Bryan	47	60	99	134	181	175	159	155	125	91	65	42	1,333
Sheffield	43	54	101	122	174	182	168	158	124	86	54	38	1,304

Although they are drier than their west-coast counterparts, there is no tendency for east-coast locations to be proportionally sunnier. Indeed, Blackpool and Carlisle are marginally more sunny than east-coast locations such as Hull and Tynemouth. This contrast results from seasonal factors. The shelter and consequently drier and brighter weather enjoyed by eastern districts under westerly conditions is most well-developed in winter and least evident in summer. As a result they are unable to take full advantage of the long hours of daylight offered by the latter season. Manley (1935) has reviewed these tendencies, though it is still difficult to interpret the data as sunshine is recorded at far fewer places than rainfall, and local factors of exposure and aspect can be extremely important.

It is easterly weather that most often brings dull conditions when large parts of the region might be shrouded in mist and low stratus cloud. On other occasions, especially in spring, the sea fogs and cloud might stop short of the Pennines watershed and cover only coastal areas. The role of the North Sea is vital and it encourages stratus cloud and fog in stable air masses from the east. Summer westerlies produce a different response, and in unstable polar maritime air masses convection resulting from ground-level heating will initiate cloud formation. Yet the same low sea surface temperatures that encourage fog in stable air masses will now be insufficient to sustain the air mass instability that soon breaks down, leading to the rapid dispersal of 'fair-weather' cumulus clouds. The onset of clear skies above the North Sea can be a notable feature of summer days (Hindley 1972), and in months dominated by westerly weather coastal stations may record as much as thirty hours' more sunshine than those even a few kilometres inland.

Winter presents a different set of responses, and the region's coastal sites can be among the sunniest in the country. At this time of year the often well-developed westerly circulations, with their attendant benefits of clearer skies for the eastern half of the region, are noticeable. The shortage of possible hours of daylight imposed by the region's latitude is then more than compensated by the temporary advantage to be gained from the prevailing synoptic conditions.

FOGS AND FRETS

Radiation fogs can occur at any season and are most abundant in low-lying areas of the region following a night of rapid heat loss or cold air drainage. Early-morning drivers through the Vale of York often have cause to proceed with care because of the limited visibility they might encounter. The hill fogs which often shroud the region's high ground, however, are also usually the result of low cloud, thereby helping to account for the astonishing fog frequencies noted for Great Dun Fell, where the annual mean of fog days (visibility less than 1,000 m at 0900 GMT) is 231.6. Elsewhere (Figure 7.10), fogs are less abundant but far from infrequent. The Moor House annual mean is 48.1, Askham Bryan's 25.1 and Tynemouth's 12.7. In common with what is found in so many other regions, winter (radiation) fogs dominate the statistics for inland sites such as Askham Bryan. Coastal fogs are less common but more evenly distributed through the year, with a tendency to late spring and summer maxima, when conditions are ideal for sea fogs (see Box 7.3). Figure 7.10, however, makes no attempt to distinguish between the origins or character of those fogs. It should also be recalled that the turbidity of the atmosphere even at distances of several kilometres from urban and industrialised areas can further reduce visibility. Long gone are the pre-industrial days when, as Defoe was informed in 1726, the Tyne's estuary could be seen from the Cheviot.

Humid tropical air masses of the type encountered along our western coasts are infrequent over the North Sea, where sea fogs result from the westwards advection of warm air from mainland Europe. The east coast's sea fogs are of such a striking character that they have been given a variety of local names. On the Yorkshire coast they might be known as 'sea rokes' or 'sea frets' and further north as 'haars' – a term of probable Norse origin. Other authors, notably Manley, have interpreted the term

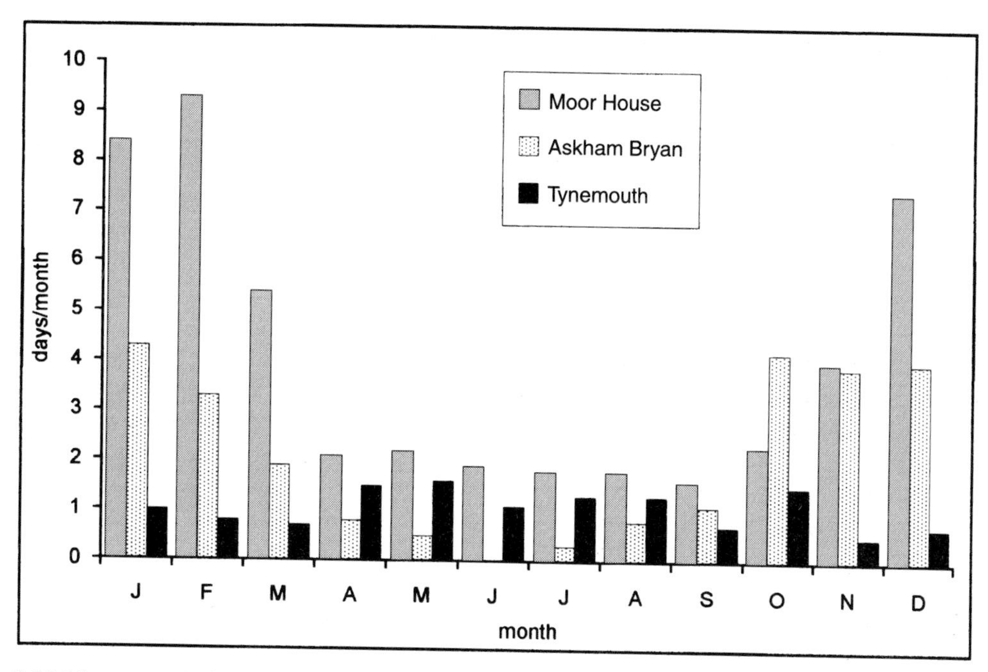

Figure 7.10 Mean monthly frequency of days with fog at 0900 GMT for the period 1961–90

more widely to include the banks of low stratus cloud that can cast a gloomy shade across the region in spells of easterly weather.

Dixon (1939) has already shown that sea fogs in eastern Scotland are highly seasonal with a maximum in spring and summer (see Box 9.3). The incidence of haars off north-east England is no different, and they occur when easterly airstreams are becoming warmer as they leave the coast of mainland Europe. They will subsequently gain moisture from the North Sea, which has yet to recover from its winter coolness, and, if the air masses are stable and therefore not subject to deep vertical mixing, will quickly become saturated in their lower layers (Roach 1995). These processes are more effective if the easterly winds are light and anticyclonic in origin. The resulting fogs need not be confined to the North Sea, and the airstreams may carry them a few kilometres inland (Box 7.3). Alternatively, on warm spring and early summer days, sea breezes may accomplish the same end. Not only do the haars reduce visibility to a few tens of metres or less, they also profoundly modify the local climate by blotting out the sunlight, raising the humidity and lowering the temperature. They are often not very deep, and it can be an unnerving experience to stand on the cliff tops at Flamborough Head in bright sunshine and peer down into a seeming cauldron of swirling sea mist.

Once having developed, sea fogs may lose heat by radiation from their upper surfaces (Douglas 1930). This initiates temperature inversions above them, which reduces the probability of their upwards dispersal. At the same time, it may set up instability within the fog layer, the consequent vertical mixing causing the fog base to rise, but often no higher than a few tens of metres. In such weather the tops of church steeples and high-rise blocks may be quite invisible while horizontal visibility might be several hundred metres. Under these circumstances the fog and low cloud can persist throughout a singularly dull and gloomy day.

Box 7.3

JULY'S HEATWAVE LEAVES THE NORTH-EAST COAST COLD

A good example of the manner in which the haar can modify local weather occurred in July 1983 (Wheeler 1984a). At that time most of the country was enjoying the spectacle of prolonged sunshine and high temperatures. While railway lines buckled in the heat and ice cream factories worked overtime to meet demand, the coastal districts of County Durham and Yorkshire languished in dull and cold conditions. On 9 July Durham University's maximum temperature was 21.4°C with 6.1 hours of sunshine. On the coast, however, no sunshine was recorded and maximum temperatures failed to rise above 15.4°C. On 11 July the contrast was yet more marked. Durham enjoyed 8.5 hours of sunshine and a maximum of 24.6°C while haars kept coastal maxima to no more than 17°C, with again no sunshine. Such behaviour is typical under haar conditions, though in recent years they have become less frequent – a consequence perhaps of the succession of warm winters that have not allowed the North Sea to cool to its usual levels. As a result it has lost a measure of its cooling capacity for the following spring's air masses. One further curiosity of the haar is the definition of its boundary, which is one not of gradual change but of an abrupt transformation in visibility, temperature and relative humidity – the former two declining, the latter increasing. So sudden can be the onset of a haar as it advances inland that haars can represent a hazard to aviation and have been the subject of much recent research (Findlater 1985). This feature can be seen in Figure 7.11, which shows the clear landwards limit of the haar on 11 July 1983. Readers should take note of the remarkable clarity of the skies over the rest of Britain. A further example of this aspect of east-coast weather is discussed in Box 9.4.

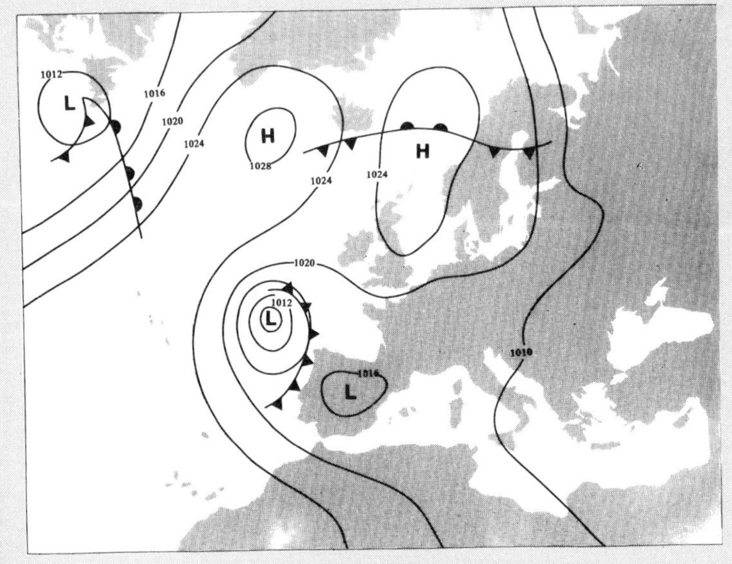

Figure 7.11 Synoptic chart for 1200 GMT on 11 July 1983. Inset: NOAA infra-red image for 1349 GMT on the same day. The seawards extent and marked coastal boundary of the haar are clearly seen
Photo by courtesy of University of Dundee

CONCLUSION

In common with all regions, north-east England and Yorkshire presents a picture of diversity, in this case principally between the Pennines and east-coast areas. If two aspects have to be identified to account for the region's overall character as well as internal diversity, they must surely be altitude and exposure. The former is highly variable across the region but fixed at any one point, where its influence may be dominant. The latter, however, depends almost entirely upon the synoptic state, which can vary from day to day as well as possessing seasonal characteristics. In a latitudinal zone where westerlies are the dominant wind direction, geographic factors ensure that they are, uncharacteristically, favourable to the region, bringing mild, reasonably dry and often sunny weather.

REFERENCES

Acreman, M.C. (1989) 'Extreme rainfall in Calderdale, 19 May 1989', *Weather*, 44: 438–446.

Archer, D. (1992) *Land of Singing Waters: Rivers and Great Floods of Northumbria*, Stocksfield: Spredden Press.

Archer, D. and Wheeler, D.A. (1991) 'Heavy rainfall in northeast England in August 1990 and some implications for calibration of rainfall radar', *Proc. Third National Hydrol. Symp., Univ. of Southampton.*

Cannell, M.G.R. and Pitcairn, C.E.R. (1993) *Impacts of the Mild Winters and the Hot Summers in the United Kingdom in 1988–1990*, London: HMSO.

Chaplain, H.R. (1982) 'Record rainfalls in Leeds and West Yorkshire during June 1982', *Weather*, 37: 282–286.

Chuan, G.K. and Lockwood, J.G. (1974) 'An assessment of topographical controls on the distribution of rainfall in the Central Pennines'. *Meteorol. Mag.*, 103: 275–287.

Dixon, F.E. (1939) 'Fog on the mainland and coasts of Scotland', *Meteorological Office Prof. Notes no. 88*, London: HMSO.

Douglas, C.K.M. (1930) 'Cold fogs over the sea', *Meteorol. Mag.*, 65: 133–135.

—— (1953) 'Gale of 31 January 1953'. *Meteorol. Mag.*, 82: 97–100.

Douglas, C.K.M. and Glasspoole, J. (1947) 'Meteorological conditions in heavy orographic rainfall in the British Isles'. *Q. J. R. Meteorol. Soc.*, 73: 11–38.

Finch, C.R. (1972) 'Some heavy rainfalls in Great Britain 1956–1971', *Weather*, 27: 364–377.

Findlater, J. (1985) 'Project haar', *Air Clues*. 350–353.

Harding, R.J. (1978) 'The temperature variation of the altitudinal gradient of temperature within the British Isles', *Geogr. Annr.*, 60A: 43–49.

Hardman, C.E., Helliwell, N.C. and Hopkins, J.S. (1972) 'Extreme winds speeds over the United Kingdom for periods ending 1971', *Meteorological Office Climatological Memo. no. 50A*, Bracknell: Meteorological Office.

Harris, R.O. (1970) 'Notable British gales of the past fifty years', *Weather*, 25: 57–68

Harrison, S.J. (1974) 'Problems in the measurement and evaluation of the climatic resources of upland Britain', in J. Taylor (ed.) *Climatic Resources and Economic Activity*, Newton Abbot: David & Charles.

Hay, R.F.M. (1949) 'Rainfall in east Scotland about the synoptic situation', *Meteorological Office Prof. Note no. 98*, London: HMSO.

Hindley, D.R. (1972) 'The importance of low sea surface temperatures in inhibiting convection along the North Sea coast in summer', *Meteorol. Mag.*, 101: 155–156.

Jackson, I.J. (1969) 'Pressure types and precipitation over north-east England', University of Newcastle upon Tyne, Department of Geography, Res. Series no. 5.

Jones, S.R. (1981) 'Ten years of measurement of coastal sea temperatures', *Weather*, 36: 48–55.

Kenworthy, J. (1994) 'The Durham University Observatory meteorological record: 150 years of Durham weather', in B.D. Giles and J.M. Kenworthy (eds) *Observatories and Climatological Research*. Occasional publication no. 29, Department of Geography, University of Durham.

Lamb, H.H. (1950) 'Types and spells of weather around the year in the British Isles: annual trends, seasonal structure of the year, singularities', *Q. J. R. Meteorol. Soc*, 76: 393–429.

—— (1964) *The English Climate*, London: Hutchinson.

—— (1991) *Historic Storms of the North Sea, British Isles and Northwest Europe*, Cambridge: Cambridge University Press.

Lowndes, C.A.S. (1971) 'Substantial snowfalls over the United Kingdom 1954–1969', *Meteorol. Mag.*, 100: 193–207.

Manley, G. (1935) 'Some notes on the climate of north-east England', *Q. J. R. Meteorol. Soc.*, 61: 405–410.

—— (1936) 'The climate of the northern Pennines: the coldest part of England', *Q. J. R. Meteorol. Soc.*, 62: 103–115.

—— (1938) 'High level records from the northern Pennines', *Meteorol. Mag.*, 73: 69–79.

—— (1942) 'Meteorological observations on Dun Fell, a mountain station in northern England', *Q. J. R. Meteorol. Soc.*, 68: 151–165.

—— (1943) 'Further climatological averages for the northern Pennines, with a note on topographical effects', *Q. J. R. Meteorol. Soc.*, 69: 251–261.

—— (1945) 'The Helm wind of Cross Fell, 1937–1939', *Q. J. R. Meteorol. Soc.*, 71: 197–219.

—— (1952) *Climate and the British Scene*, London: Collins.

Marsh, T.J., Monkhouse, R.A., Arnell, N.W., Lees, M.L. and Reynard, N.S. (1994) *The 1988–92 Drought*, Wallingford: Institute of Hydrology.

Meteorological Office (1930) *British Rainfall 1930*, London: HMSO.

—— (1963) 'Rainfall over the catchment areas of the Northumberland and Tyneside, and the Wear and Tees river boards', *Hydrological Memoranda no. 3*, Bracknell: Meteorological Office.

—— (1985) 'The climate of Great Britain: Pennines and Lake District', *Met. Office Climatological Memo. no. 128*, London: HMSO.

National Rivers Authority (1990) *Annual Rainfall Report 1989*, Leeds: NRA Yorkshire Region.

Natural Environment Research Council (1975) *Flood Studies Report*, vol. II: *Meteorological Studies*. London: NERC.

Pearsall, W.H. (1950) *Mountains and Moorlands*, London: Collins.

Roach, W.T. (1995) 'Back to basics: fog – part 3: the formation and dissipation of sea fog', *Weather*, 50: 81–84.

Salter, M. de C.S. (1918) 'The relation of rainfall to configuration', in *British Rainfall 1918*, London: British Rainfall Organisation.

Sawyer, J.S. (1956a) 'Rainfall of depressions which pass eastward over or near the British Isles', *Meteorological Office Prof. Note no. 118*, London: HMSO.

—— (1956b) 'The physical and dynamical problems of orographic rain', *Weather*, 11: 375–381.

Sharp, J.I. (1994) 'The climate of Newcastle upon Tyne: a statistical summary of the meteorological observations made by the Department of Geography of the University of Newcastle upon Tyne between March 1952 and September 1987'. University of Newcastle upon Tyne, Department of Geography Seminar paper no. 64.

Smithson, P. (1974) 'Heavy rainfall over central northern England', *Weather*, 29: 17–24.

—— (1987) 'An analysis of wind speed and direction at a high-altitude site in the southern Pennines', *Meteorol. Mag.*, 116: 74–85.

Steers, J.A. (1964) *The Coastline of England and Wales*, London: Cambridge University Press.

Stirling, R. (1982) *The Weather of Britain*, London: Faber & Faber.

Tufnell, L. (1987) 'Early weather observations in Northern England: part 1', *J. Meteorology (UK)*, 12: 79–82.

—— (1991) 'Early weather observations in Northern England: part 2', *J. Meteorology (UK)*, 16: 158–163.

Wheeler, D.A. (1984a) 'The July 1983 "heat wave" in north-east England', *Weather*, 39: 178–182.

—— (1984b) 'The work of Thomas Backhouse – Victorian meteorologist', *Weather*, 39: 240–246.

—— (1985) 'Heavy November rainfall in Sunderland', *Weather*, 40: 90–92.

—— (1988) 'Some observations on the remarkable weather of September 1986 in NE England', *Trans. Nat. Hist. Soc. Northumbria*, 55: 47–54.

—— (1990) 'Modelling long-term rainfall patterns in north-east England', *Meteorol. Mag.*, 119: 68–74.

—— (1991) 'The great north-eastern snowstorm of February 1941'. *Weather*, 46: 311–314, 319–320.

—— (1993) 'Meteorological studies in Sunderland and Monkwearmouth', in B.D. Giles and J.M. Kenworthy (eds) *Observatories and Climatological Research*, Occasional publication no. 29, Department of Geography, University of Durham.

8

NORTH-WEST ENGLAND AND THE ISLE OF MAN

Lance Tufnell

INTRODUCTION

Any account of climate in north-west England must emphasise its variety. Less than 250 km separates Cheshire from the Scottish border and half this distance lies between the Pennines and St Bees Head, where the region has its maximum east–west extent. Within these limits are both the coldest place in England (the summit of Cross Fell at 893 m above sea level (ASL)) and, at the head of Morecambe Bay, one of the areas of the country least affected by snow. The region also claims the wettest locality in England, on the fells around Seathwaite. Differences of this sort are primarily the result of topography and altitude, rather than distance from the sea. This is hardly surprising, as no English region possesses a greater range of altitudes.

The region's physiography consists of two main areas. One is a lowland, which extends southwards from narrow beginnings around Morecambe Bay to a more ample width in the Liverpool–Manchester and Cheshire areas. To the east and north is the second area, formed by the great inverted L shape of the Pennines and Lake District, joined by the Shap and Howgill Fells. In the north of the region the Eden valley separates the Lake District from the north Pennines and merges into the Solway lowlands. The region's western limits adjoin the Irish Sea, from which the main climatic characteristics are derived. More detailed topographical elements, such as the radial pattern of valleys in the Lake District and the Mid-Cheshire Ridge, also have noteworthy climatic effects. Standing apart from north-west England, the Isle of Man is essentially a mountain mass which rises to over 600 m, with a major area of lowland in the north.

To these topography–climate relationships must be added the contrasts which are partly caused by human activity. Though nature supplies the bracing winds of a Pennine moor or Lakeland felltop, it is people who create the fume-filled air of urban streets and road vehicles that have now taken over from the chimney in ensuring that the region's chief city, Manchester, retains one of the longest air pollution histories in the world.

Not unnaturally, this variety of weather and climate has provoked contrasting reactions. Dorothy Wordsworth was fascinated by the weather of rural Lakeland, as her *Grasmere Journals* testify, while L.S. Lowry found inspiration for his paintings among the smoking chimneys and polluted air of Manchester. Yet it was these same 'manufactorys' and 'clouds of smoke' that in 1830 had been so condemned by the young John Ruskin. Anyone who shares the late Gordon Manley's affection for the region will find that its climatic diversity generously repays careful study. The north-west of England certainly enjoys a fascinating history of weather observations (Box 8.1).

TEMPERATURE

The sea is an ever-present influence on weather in north-west England. The region is influenced not

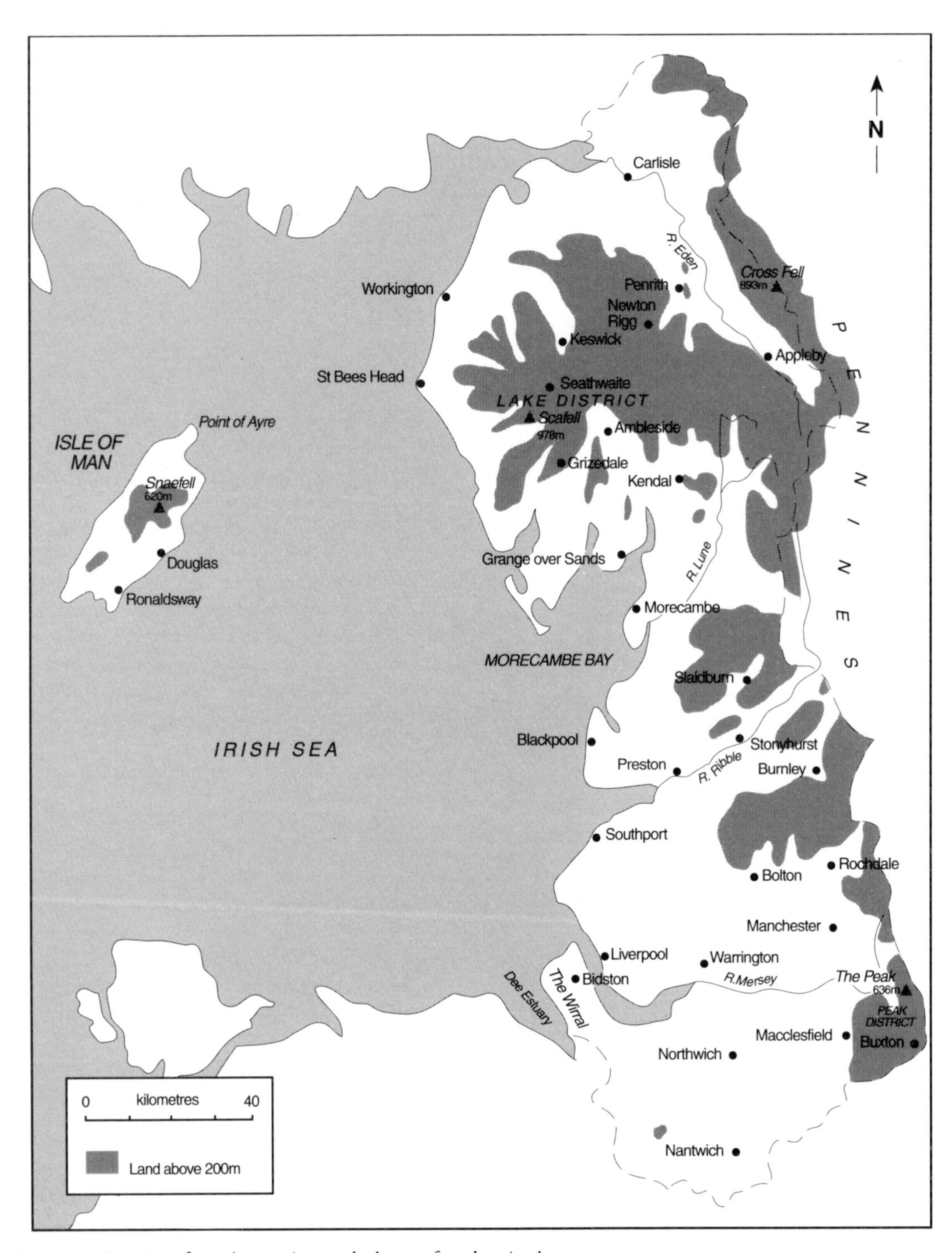

Figure 8.1 Location of weather stations and places referred to in the text

Box 8.1

OBSERVERS AND THEIR LEGACY

Observations from north-west England have played an honourable role in developing our knowledge of British climate. The first lengthy study of British rainfall was made near Burnley by Richard Towneley, who began his record in 1677 (Folland and Wales-Smith 1977). The famous Cumbrian scientist, John Dalton, published *Meteorological Observations and Essays* in 1793 and observed the weather of north-west England for more than fifty years (Manley 1968). During the second half of the nineteenth century, Robert Angus Smith did important work on air pollution in Manchester, while from the 1930s onwards the renowned Gordon Manley devoted much time to studying the region's climate. An equal contribution comes from the greater number of observers who are now largely forgotten. How many people can name, for example, those responsible for giving Carlisle the third longest continuous rainfall record in the UK (Jones 1983)?

The longevity of the region's climate record is impressive. Though there are few records from before 1600, more data exist for the years 1600–80 and 1680–1750 and there is a near-continuous *daily* record. Much of this is in diaries but includes instrumental data by Towneley (1694, 1700), Towneley and Dereham (1705) and Fleming (1689–93). Between 1753 and 1796 there was an upsurge of instrumental weather recording, many details of which appear in compilations by Garnett (1793, 1796), though descriptive sources remain very useful at this time (Winchester 1994). Over the next fifty years, however, instrumental records become increasingly dominant (Barnes 1830). Improved organisation and exchange of data is a feature of the region's weather studies between 1850 and 1900. As elsewhere, the appearance from 1860 onwards of *Symons's British Rainfall* was particularly helpful, and by 1880 it contained data from over 150 sites in north-west England. The twentieth century has been dominated by Gordon Manley, whose work in the region covered high-level studies in the north Pennines, climatic history and various overviews. Since 1984 the Manchester-based Acid Rain Information Centre (now the Atmospheric Research and Information Centre) has been an important development. Otherwise, of currently operating sites the oldest are Douglas (IoM), which opened in 1884, Newton Rigg (1904) and Wirral's Bidston Observatory (1908).

only by the comparatively warm Atlantic Ocean but also by the Irish Sea, which Crowe (1962) has described as an 'unresponsive mass of water'. Since the region's weather is therefore of maritime origins it has a typically modest temperature range, with essentially mild winters and cool summers. At times, however, there are brief interruptions of this dominance. In summer, invasions of tropical continental air from southern Europe may bring temperatures approaching 30°C, while in winter cold air may break out of the European continental interior to bring, even to this westerly region, some raw and cold conditions. The southward penetration of Arctic air can also introduce weather that is cold but typically clear as a result of the sheltering effects of the extensive Scottish mountains to the north (Manley 1974).

Differing altitudes, relief and exposure can be expected to increase further the variety of temperatures in north-west England. Yet Bilham (1938) perceived a 'peculiar' uniformity of temperatures over coastal and low-level inland sites, and pointed out how Southport and Manchester record very similar temperatures throughout the year. The existence of this characteristic is revealed in Table 8.1, which

Table 8.1 Mean monthly maximum and minimum temperatures (°C) for the period 1961–90

Location	*Alt. (m ASL)*	*Jan.*	*Feb.*	*Mar.*	*Apr.*	*May*	*June*	*July*	*Aug.*	*Sept.*	*Oct.*	*Nov.*	*Dec.*	*Year*
Douglas (IoM)	85	6.9	6.4	7.9	10.1	13.0	15.9	17.3	17.1	15.2	12.8	9.4	7.9	11.7
		3.0	2.5	3.3	4.7	7.2	9.8	11.5	11.7	10.3	8.6	5.2	3.9	6.8
Carlisle	26	6.3	6.7	8.7	11.3	14.9	17.6	19.0	18.8	16.4	13.6	8.8	6.9	12.4
		1.1	0.8	2.0	3.7	6.1	9.2	10.7	10.9	8.9	6.8	3.0	1.5	5.4
Grizedale	91	6.0	6.2	8.1	11.3	14.6	17.3	19.0	18.6	15.9	13.0	8.9	6.9	12.1
		−0.2	−0.2	0.7	2.2	4.6	7.6	9.4	9.6	7.9	5.2	2.0	0.5	4.1
Morecambe	7	6.1	6.5	8.4	11.3	15.0	17.6	19.0	18.8	16.6	13.6	9.1	7.1	12.5
		1.8	1.8	3.0	5.0	8.0	11.0	12.9	12.8	10.7	8.2	4.4	2.5	6.9
Slaidburn	192	5.0	5.2	7.4	10.3	14.1	16.9	18.1	17.8	15.4	12.2	7.9	5.9	11.4
		0.1	0.1	1.2	2.7	5.6	8.4	10.4	10.3	8.5	6.1	2.6	0.9	4.8
Manchester	75	6.3	6.5	8.8	11.5	15.2	18.2	19.6	19.4	16.9	13.6	9.0	7.1	12.7
Airport		1.2	1.1	2.4	4.3	7.2	10.1	12.0	11.8	9.9	7.4	3.5	1.9	6.1
Macclesfield	152	5.5	5.7	8.3	11.1	15.0	17.8	19.3	18.7	16.4	13.1	8.5	6.5	12.3
		0.7	0.6	2.1	3.8	6.8	9.6	11.5	11.3	9.5	6.9	3.0	1.5	5.7
Buxton	307	4.5	4.2	6.6	9.3	13.2	16.5	17.7	17.6	14.9	11.5	7.3	5.4	10.5
		−0.1	−0.4	1.1	2.6	6.5	8.7	10.5	10.3	8.3	5.8	2.5	0.9	4.5

Note: ASL = above sea level

contains Morecambe's and Manchester's current monthly average temperatures. On the other hand, pronounced, if short-lived, differences between coastal and inland temperatures can arise. Baxendell (1903) noted the propensity for sea breezes to develop across the expanses of the Lancashire plain, thereby subduing coastal maxima. In extreme cases coastal areas may be 10 degrees C cooler than inland locations. Substantial differences in temperatures are also a feature of localities at different altitudes. The Morecambe–Slaidburn and Macclesfield–Buxton transects suggest a fall in temperature with altitude of around 1 degree C per 100 m. Given this rate it is not surprising to find that the summit of Great Dun Fell in the north Pennines is colder than Reykjavik in Iceland.

January is normally the coldest month of the year on the mainland, but February lays claim to this distinction on the Isle of Man owing to the marked maritime influence on the island. The sea is also responsible for the mean daily maximum temperature on the Isle of Man being generally lower than at corresponding altitudes on the mainland and for extreme values being fewer than on the mainland (Meteorological Office 1983).

Though primarily of seasonal occurrence, frost can be recorded during any month, especially on higher ground. Inland localities in Cumbria at elevations below 140 m usually have between fifty and 100 air frosts annually. The total rises to 160 on Great Dun Fell and drops to about thirty over favoured coastal slopes around Grange-over-Sands (Manley 1973b). However, as few as fifteen to twenty-five air frosts occur at low elevations in the Isle of Man. Ground frosts are inevitably more frequent, numbering from between fifty-five and sixty on the Isle of Man, almost seventy for places like Stonyhurst and, on the coast, Southport, where sandy soils may have an important local effect (Gregory 1953), and over eighty at Manchester Airport and along the Pennine edge at towns such as Rochdale and Darwen (Crowe 1962). In the high Pennines there are over 175 ground frosts a year, though some annual totals exceed 200 (Rawes 1976, 1978). Local factors again figure prominently in accounting for contrasts such as those between Macclesfield and Grizedale. The latter, though at only 91 m ASL, contrives minima well below the former, which stands at 152 m. Grizedale, however, lies in a relatively narrow valley

Table 8.2 Record maxima and minima (°C) for north-west England by counties

County	*Maximum*	*Date*	*Location*	*Minimum*	*Date*	*Location*
Cumbria	33.3	20 July 1901	Newton Rigg	−21.1	21 Jan. 1940	Appleby
Lancashire	33.7	3 July 1976	Blackpool	−21.9	16 Jan. 1881	St Michaels-on-Wyre
Merseyside	34.5	2 Aug. 1990	Bidston	−14.0	12 Jan. 1981	Sefton Park
Greater Manchester	33.7	2 Aug. 1990	Manchester Airport	−17.8	21 Jan. 1940	Barton
Cheshire	34.6	3 Aug. 1990	Nantwich	−17.7	16 Jan. 1881	Hinderton
Isle of Man	28.9	12 July 1983	Ronaldsway	−11.1	17 Jan. 1881	Cronkbourne

Sources: Meaden (1983); Webb and Meaden (1983, 1993)
Note: cited temperatures are only for locations that fall within the region as defined in this text

which, typically for the Lake District, has a propensity at all seasons for cold air drainage of the surrounding high fells. Macclesfield by contrast occupies more open ground less subject to this influence.

As elsewhere, changeability is a fundamental characteristic, with occasional outbreaks of unseasonable weather. A minimum of −15°C was recorded at Newton Rigg long past mid-winter's day on 2 April 1917, while Liverpool enjoyed a maximum of 18.5°C on 1 November in 1984. Table 8.2 shows the region's highest values to be lower than in more southerly regions. It is also interesting to note that the outbreaks of cold, easterly weather of 1981 and 1982 that figure so frequently in the record minima for regions to the east and the south had less effect in this region to the west of the Pennines. The marked oceanicity of the Isle of Man, already seen in its mean temperatures, is again revealed in its extremes (Table 8.2).

It has often been low temperatures that have gained the interest of public and scientists. Much work is, however, needed before our understanding of the region's harshest winters can be satisfactorily fitted into the national picture given by Manley (1963) which shows that during the last 320 years the coldest winters were those of 1683/84, 1739/40, 1813/14 and 1962/63. Manley concluded that the first of these was the most severe, with a mean temperature 5 degrees C below the 1931–60 average. He suggested that west of the Pennines there was less snow, more clear weather and somewhat lower temperatures than further east. This view was based on reports of deeply frozen ground at Manchester, intense frost at Lancaster, and Towneley's record of low precipitation in December and January near Burnley (Manley 1975a). In his accounts of the 1739/40 winter, Manley (1958, 1975a) noted the freezing of Lake Windermere and of the rivers Eden and Lune. Fortunately, much more is known about the harsh winter of 1813/14. This is the first such winter in north-west England whose meteorology can be reconstructed from instrumental observations. Thus, temperatures recorded by William Pitt at Carlisle (Barnes 1830) and John Dalton at Manchester (Dalton 1819) show that from October 1813 to May 1814 all months except April were colder than the contemporary long-term average (Table 8.3). January was particularly extreme, with mean temperatures several degrees below normal, rivers and lakes frozen, and snow in abundance. By contrast, during the winter of 1962/63 much of north-west England was relatively snow-free and sunny (Manley 1963; Shellard 1968) and was characterised by easterly weather with high pressure persistent immediately north of the British Isles. Under these circumstances the normal west-to-east exposure and shelter regime exercised by the Pennines is reversed. The eruption of polar continental air made cold conditions widespread but while north-east England suffered heavy snowfalls and much cloud, Lancashire enjoyed twice the normal

Table 8.3 Temperatures recorded at Carlisle and Manchester, October 1813–May 1814

Location	*Period*	*Monthly mean temperatures (°C)*								
		Oct.	*Nov.*	*Dec.*	*Jan.*	*Feb.*	*Mar.*	*Apr.*	*May*	*Average*
Carlisle	1813–14	7.1	4.0	2.9	−4.1	1.7	3.4	9.3	8.4	4.1
	1801–24	9.1	5.4	3.1	2.4	3.7	4.8	7.4	11.0	5.9
Manchester	1813–14	8.5	4.0	2.6	−3.0	1.4	3.0	10.1	9.7	4.5
	1794–1818	9.3	5.3	2.7	2.1	3.6	5.7	8.0	11.1	6.0

Sources: Barnes (1830) for Carlisle, Dalton (1819) for Manchester

Figure 8.2 Skating on Derwentwater, 1925 (a postcard from the period)

sunshine (Stonyhurst noted its sunniest February since records began in 1887!) and less than half the usual precipitation.

Any assessment of low temperatures should incorporate the fact that Cumbria is the only English region well endowed with lakes. Unfortunately, data for the freezing of those lakes are not extensive. In the absence of methodically compiled observations, any attempt to study the history of lake freezing in Cumbria has to rely on a wide range of sources. These can be used to identify cold spells as distant as 1571, when Derwentwater froze, or in the winter of 1607/08, when Ullswater became a thoroughfare (there was even a 'Boone fire builded on the Ise'!). In the winter of 1813/14 Derwentwater was recorded to have frozen to a depth of 254 mm.

When, on the other hand, lakes are unfrozen they can have a noticeable influence on temperatures around their shores. This is most likely to occur during winter and on cooler nights in summer, when anticyclonic conditions prevail and wind and cloud amounts are low. Hence, night minima at Keswick, which is near Derwentwater and Bassenthwaite, can be several degrees higher than at a site like Appleby, where lakes are absent (Manley 1973b). The Eden valley floor, with its drumlin landscape, is also a good example of how small-scale relief variations can influence temperatures by assisting the development of frost hollows (Manley 1975b). At night, relatively low temperatures may also be recorded where soils contain a high percentage of air. This happens, for example, over the sandy terrain at Blackpool and Southport. On the other hand, temperatures can be raised by the effects of urbanisation, as Manley (1944) has demonstrated for Greater Manchester.

PRECIPITATION

Exposure to the dominant westerly winds and the importance of polar and tropical maritime air masses is a theme in any examination of the region's precipitation regime, but is given variety by the shelter afforded by the Pennines when easterly conditions prevail and by the localised shelter provided by the diverse relief within the Lake District. As early as 1777, Matthew Dobson of Liverpool pointed out that the amount of rain which falls during a year is 'a very uncertain test of the moisture or dryness of any particular season, situation, or climate'. Nowhere is this more strongly confirmed than in the Lake District, which has some of Britain's wettest localities, but whose average relative humidity (at least in the valleys) is similar to that for the rest of north-west England (Manley 1973b).

Given the region's exposure to polar and tropical maritime air masses, it should come as no surprise to find that the Lake District has some of the wettest towns and villages in the country: Ambleside and Windermere (Table 8.4) both feature in a recent list of the twelve wettest towns in Britain (Eden 1995 and Box 8.2). Places on the edge of the mountains do not fare much better: Keswick averages nearly 1,500 mm a year and Kendal gets 1,323 mm (Manley 1974) – figures typical of upland Wales. Precipitation totals on the high fells are matched only by those found in limited areas of north-west Scotland and North Wales. By contrast, the reputedly wet city of Manchester has a mean annual total of only 859 mm (Beck 1992), while some places in Cheshire average less than 750 mm (Rowsell 1970).

Rainfall gradients are noticeably steep along the

Table 8.4 Mean monthly and annual precipitation totals (mm) for the period 1961–90

Location	*Alt. (m ASL)*	*Jan.*	*Feb.*	*Mar.*	*Apr.*	*May*	*June*	*July*	*Aug.*	*Sept.*	*Oct.*	*Nov.*	*Dec.*	*Year*
Douglas (IoM)	85	120	81	91	71	63	70	66	92	104	125	125	123	1,131
Sprinkling Tarn	600	393	268	320	224	207	255	284	339	382	422	429	430	3,953
Windermere	175	179	115	141	91	93	95	108	133	158	177	175	180	1,645
Ambleside	49	219	139	172	105	101	100	115	151	184	223	220	230	1,959
Langdale	108	296	184	218	133	129	132	147	196	237	291	275	306	2,544
Kendal	36	131	89	112	77	76	84	94	112	130	144	139	135	1,323
Morecambe	7	93	61	77	57	63	66	74	96	102	111	102	100	1,002
Bolton	107	115	74	93	70	77	89	83	109	109	127	120	120	1,186
Bidston	60	60	44	51	49	55	54	57	71	72	77	74	69	733
Manchester Airport	75	69	50	61	51	61	66	65	79	74	77	77	78	808
Macclesfield	201	91	62	79	69	71	79	79	91	88	91	99	97	996
Buxton	307	136	97	115	90	87	80	84	106	106	122	131	139	1,293

Box 8.2

SEATHWAITE – THE WETTEST PLACE IN ENGLAND?

Anyone interested in British climate will have heard of Seathwaite in Borrowdale and few will hesitate in accepting the claim that this is the wettest place in England. Manley (1952) has mentioned that Seathwaite hamlet gets a mean annual rainfall of 3,100 mm, while according to J. Glasspoole the nearby fell below Great End receives about 4,700 mm (Manley 1946). Such comments raise the question of where precisely is the spot that receives the highest rainfall in England.

Some of the earliest attempts to measure the rainfall of the Lake District were made at Kendal and Keswick. Dalton's work at the former locality in the 1780s and 1790s established it as the wettest place known. In 1836, rainfall at Esthwaite was found to exceed that for Kendal and the heavier rainfall of the mountains was confirmed in 1843 with data from Grasmere. This prompted John Fletcher Miller of Whitehaven to establish a gauge at Seathwaite in the following year. In 1845 it recorded 3,861 mm and he was told that even greater totals might occur nearby. As a result, he extended his work in the area to cover localities at higher altitudes including the siting of a gauge near the top of Scafell. The practical difficulties of such an undertaking were enormous, and Miller's assistants had to endure both hardship and injury. Yet he carried on these observations to within three years of his death in 1856. Since then the work has been continued by J. Dixon, I. Fletcher, J. Maitland and others up to the present day (Symons 1868, 1896, 1897; Manley 1946; Simpson 1995). The results obtained are important not only because they relate to the wettest area of England, but also because they are a tribute to the inquisitiveness, fortitude and even the philanthropy of our Victorian forefathers (Miller, Fletcher and Maitland all bore the cost of their Borrowdale observations). This begs the question of how well present efforts to understand Lakeland rainfall compare with those of the Victorian era. Since mean annual rainfall apparently varies by up to 1,600 mm in the short distance between Seathwaite hamlet and the nearby fells, and with yearly totals at Seathwaite able to differ by almost 2,500 mm, we are still not sure exactly which is the wettest spot in England. Clearly, Seathwaite is only the wettest *inhabited* place in England.

western flanks of the Pennines as topography is generally more abrupt than on the eastern sides of the range (Barrett 1966; Chuan and Lockwood 1974). Variations in rainfall gradients also exist in the Lake District, where the profiles radiate outwards from a central area of high precipitation around the Scafell–Bowfell group. In the sixteen or so kilometres from this centre to the coast at Ravenglass, mean annual rainfall declines by more than 3,500 mm. Local variations in shelter are also indicated in Table 8.4, which shows that Ambleside, despite lying at only 49 m above sea level (ASL), receives nearly 300 mm a year more rain than nearby Windermere at 175 m.

It has long been recognised that there exists a seasonal pattern in the region's rainfall. Data from Merseyside, Greater Manchester and Stonyhurst indicate that the period February to June tends to be drier than July to January (Gregory 1953; Crowe 1962). Similarly, in Cumbria the period March to June inclusive yields only about one-quarter of the year's total while the three months November to January account for over one-third (Manley 1974). Table 8.4 emphasises that while April and May are the driest months in the north of the region, Feb-

Figure 8.3 The atmosphere of Derwentwater
Photo by courtesy of Friends of the Lake District

ruary more commonly claims this title in the south. The period October to December is universally the wettest.

The airstreams which cross north-west England are modified primarily by topography, the classic example of this being in the Lake District. Here, orographic uplift and the radial pattern of valleys combine to produce the convergence of rising air and enhanced precipitation. Orographic uplift also occurs at a number of other places, including some where relief is distinctly modest, the Liverpool and Prescot uplands being good examples (Gregory 1953). Equally, the rain shadow effects of higher ground are also present. Thus, under prevailing west or south-westerly airstreams, the Eden Valley will be in the rain shadow of the Lake District. As a result, towns like Appleby and Penrith get substantially less rain than, for example, Kendal, which is more exposed to influences particularly from the south-west. In the same way, the Cheshire plain and Merseyside have fairly low rainfall, owing to the influence of the north Welsh mountains. This is not, however, the same in all places, and the Liverpool Observatory at Bidston, for example, does not experience as strong a rain shadow effect as do less exposed places around the head of the Dee estuary (Reynolds 1953).

Any general picture of rainfall in north-west England inevitably masks the details of extreme events. Yet such events are often important, so it is not surprising that attempts to measure them began well over a hundred years ago. A questionable rate of 170 mm per hour was reported for Bolton on 3 July 1892, and a more startling 381 mm per hour for Preston on 10 August 1893. More recently and reliably, Bolton experienced a downpour on 18 July 1964, when 56 mm fell in just 15 minutes. Another example was from the Dunsop valley, where 117 mm was recorded in 90 minutes on 8 August 1967 (Webb 1988). The synoptic conditions of the

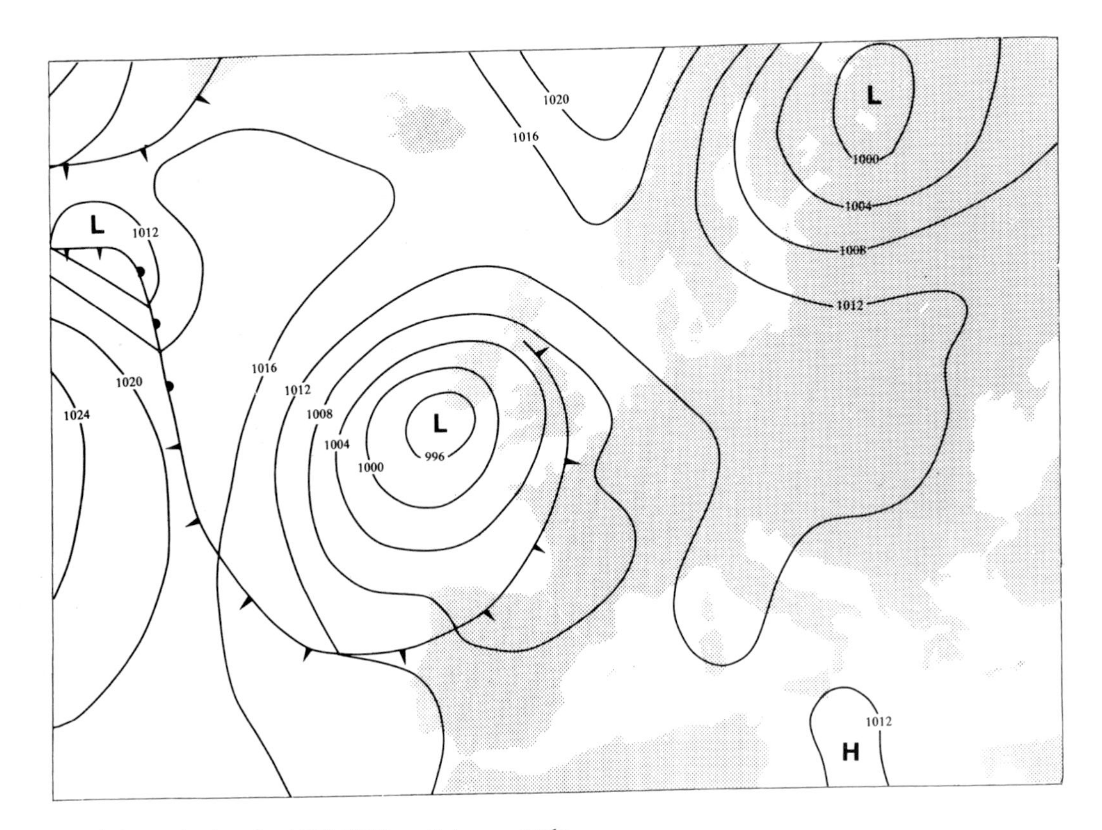

Figure 8.4 Synoptic chart for 1200 GMT on 8 August 1967

day are outlined in Figure 8.4. This event was typically confined within a short space of time and resulted from convective and instability processes.

The rainfall intensities decline when heavy daily falls are considered and the emphasis shifts to include rainfall from the fronts of slow-moving depressions. In 1911 a table of localities in Britain which had recorded over 101.6 mm (4 inches) of rain in a single day was prefaced by the remark that the number of times falls exceeded this value in Cumbria was far greater than for other counties of Britain. More recently, a table published by Webb (1987) for the years 1860–1985 has underlined how Seathwaite and Cumbria dominate any list of heavy daily rainfalls in north-west England. The former, for example, recorded 204 mm on 12 November 1897. Most of the Cumbrian entries are for the autumn, while those for the other counties of north-west England have tended to occur in the summer. Webb (1987) has pointed out that lengthy orographically intensified frontal rain in a mountainous area such as the Lake District is most likely in the autumn, owing to the relatively high sea temperatures and the predominance of south-westerly winds. Convectionally driven storms help to account for the remainder.

Autumn has also registered very high monthly rainfall totals, some of which have approached, even exceeded, 700 mm. December 1986 is a good recent example, when westerly cyclonic weather dominated over most of the month and, although all of the British Isles suffered above average rainfall, few areas were wetter than the Lake District. In this month Coniston recorded over 500 mm, Ambleside 365 mm and even coastal sites such as Workington had 174 mm. When Atlantic fronts and depressions are unusually persistent, high monthly totals become juxtaposed to give an extended period of very wet weather. This situation is commonly associated with high zonal indices in the upper westerly circulations and can lead to some notable seasonal totals; for example, Cutforth (1987) has shown the wettest October to December periods at Coniston to have yielded 1,465 mm (1954), 1,388 mm (1929) and 1,318 mm (1986).

Nonetheless, dry spells are far from unknown, and the summer of 1995 was so dry that by late September a drought order was in force at Seathwaite. Such droughts tend to occur when east and north-east winds prevail, something that is most likely between April and June. It means that the Lake District becomes sheltered by the north Pennines and the mountains of southern Scotland, and there have been months such as February 1932 at Seathwaite and February 1986 along the south Cumbrian coast when gauges have recorded no rain at all (Manley 1974; Cutforth 1987).

Snowfall and Snow Cover

The Lake District's frequent snow cover colours our impression of its region-wide incidence. The reality is that over much of the region, despite its propensity to rainfall, snow is not as common as is often imagined. On average, the likelihood of snow within the region increases with latitude and altitude, though distance from the sea and aspect are also important. This means that any snow cover over low ground in the more southerly parts of the region is normally restricted to the period December to March, while at similar altitudes in Cumbria this period extends from November to April. In the uplands the period of possible snow cover is increased further from October to May. The average number of days with snow lying in north-west England can vary from five or less in parts of the Isle of Man to about 100 in the north Pennines. This last area is, in fact, decidedly snowier than the Lake District mountains, because the heavier falls in these regions largely come from the north-east and east (Manley 1973b), and the latter's greater exposure to the snow-bearing Arctic air masses outweighs the fact that its altitude is less than that of the Lake District. The different exposures on the eastern and western sides of the Pennines to these snow-bearing air masses is registered in other statistics. Thus whereas the average snow cover increases by eight days per year in north-west England, east of the Pennines watershed the rate is ten days per 100 m. At sea level the regionally averaged number of days with snow cover is yet more marked, being 4.0 and 7.7 respectively.

Table 8.5 shows that snowfall is not frequent on low ground and snow usually lies for only a short time before melting. Only on high ground or in the inland fastnesses of Lake District valleys do the days of snow lying even approach those of snow falling. Mean annual and monthly snow cover is variable across the region and heavily dependent upon local factors. Figure 8.5 gives some indication of this variability of snow cover, which can be encountered as late as May or as early as October even in Buxton, where the annual days of over 50 per cent snow cover are 30.2. The figure for Slaidburn is 16.8 days, for Carlisle 8.4 days but only 4.6 for Morecambe and 3.6

Table 8.5 Mean number of annual days of snow falling and snow lying (at 0900 GMT and covering more than 50 per cent of the ground) for the period 1961–90

	Douglas (IoM)	*Carlisle*	*Grizedale*	*Morecambe*	*Manchester Airport*	*Buxton*
Altitude (m ASL)	85	26	91	7	75	307
Snow falling	12.7	29.4	19.3	7.0	30.8	35.9
Snow lying	3.6	8.4	19.2	4.6	7.6	30.2

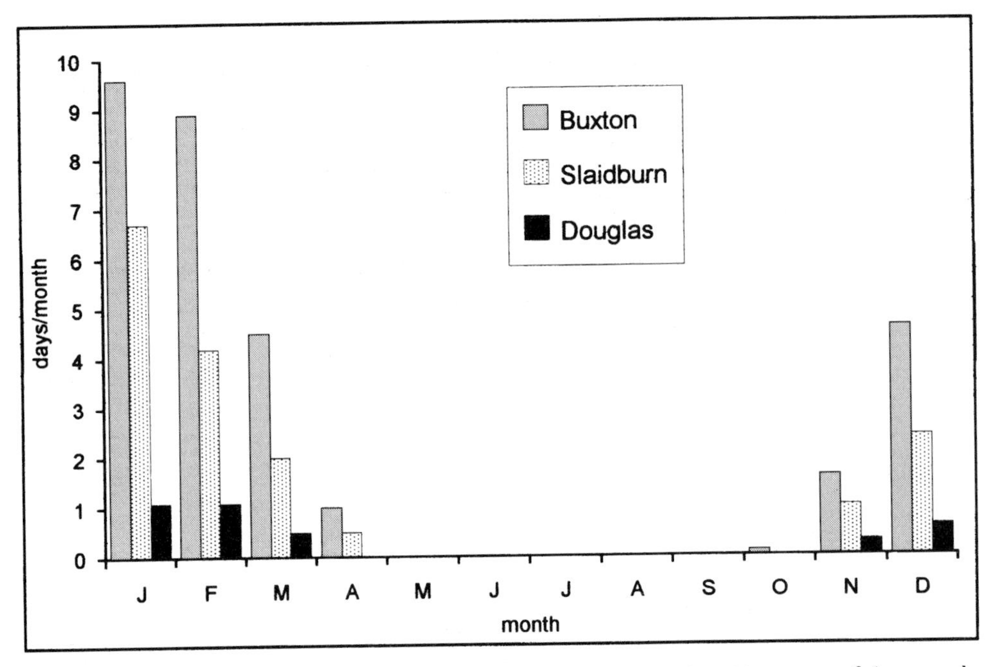

Figure 8.5 Mean monthly number of days with snow lying and covering more than 50 per cent of the ground at 0900 GMT for the period 1961–90

for Douglas. Great Dun Fell's figure is a remarkable 96.2 days.

Remarks about 'average' snow in north-west England must be treated with care, as this is a highly variable weather element. In some years the first snows of the winter occur in September, as they did on Great Dun Fell in 1995, whereas in others January is snow-free. Equally, there have been times when snow has either fallen or been lying in June – the events around Kendal in 1869 are part of local folklore – or, more rarely, in July or even August. The more recent events of 2 June 1975 have passed into sporting history when heavy snow in Buxton brought the county cricket match between Derbyshire and Lancashire to an early conclusion (Figure 8.6).

Variability also characterises the depth of snow in north-west England. This is especially true at altitudes of over 460 m, where drifting of snow by wind is universal (Manley 1943). It means that the area covered may fall below the 50 per cent needed to record a day with snow lying. Manley (1942, 1943) has suggested that without this factor the mean annual number of days with snow lying on the summit of Great Dun Fell might be recorded as 140. Aspect, as noted above, also helps determine patterns of snow cover by influencing the amounts received and lost. Thus, in the north Pennines, Wildboar Scar (550 m) on the west-facing escarpment overlooking the Eden valley registers less snow than does Moor House field station (558 m), which is on the eastern dip slope. Manley (1973b) has pointed to similar differences occurring in the Lake District.

Altitude is evidently an important factor in snow cover, but low-lying areas, though demonstrating less frequent snow cover, do not escape occasional notable falls. Clark (1986) has identified eighteen snowfalls of 15 cm or more since 1880 in Manchester, and has shown that these resulted from either fronts and frontal depressions (mainly from between south and west) and polar lows and troughs (chiefly from the north-west). Clark also found that such heavy

Figure 8.6 Snow stops play in Buxton, 2 June 1975
Photo by courtesy of Derbyshire Times Newspapers

falls were more frequent in December, suggesting the importance of evaporation from the still warm Irish Sea and Atlantic, but were less common later in the winter when cooler oceanic conditions prevail.

Thunder and Hail

Crowe (1962) has maintained that thunderstorms 'play no significant role in the Lancastrian scene'. Yet monthly reports published during the past twenty years in the *Journal of Meteorology* contain hundreds of references to storms in the region. On the other hand, thunderstorms tend to be less frequent in much of the north-west region than in many counties to the south and east. Indeed, observers in Cumbria and the Isle of Man can expect to hear thunder on somewhat under half the number of days that it is recorded in south-east England. They are also less likely to hear it than are their colleagues further south in the region, where unequal heating of the Pennine slopes can act as a trigger to storm development. The importance of ground-level heating is amply displayed by the summer peak of activity revealed in Figure 8.7. This was the case,

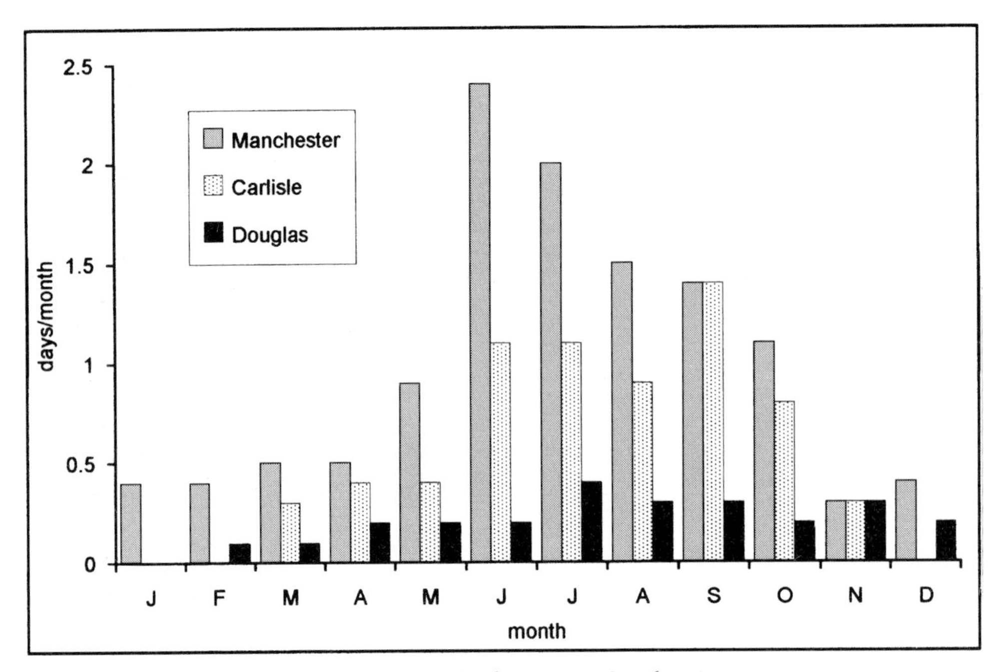

Figure 8.7 Mean monthly number of days with thunder for the period 1961–90

for example, with an isolated thunderstorm which occurred in the Manchester area on 23 June 1956 (Prichard 1985). With the help of topographic influences, some places in north-west England manage to average close to ten days a year with thunder. Thus Manchester Airport's annual average is 12.0 days of thunder, those for Macclesfield and Buxton are 8.5 and 6.4 respectively, but Douglas (IoM) averages only 2.8 days, as does Grizedale in the Lake District. The plains of the lower Eden around Carlisle, where the annual average is 6.7 days, are conducive to summer heating and instability leading to thunder. On the other hand, coastal Morecambe averages less than two days annually and must rank as one of the least thunder-prone areas of Britain.

Annual thunderstorm frequency may vary considerably. Thus, in 1983 Blackpool recorded twenty-six days with thunder, whereas its 1961–80 average was only nine. On average, however, the seasonal pattern is marked, and thunder is most likely to occur from May to August (Figure 8.7). Even so, there is no insurance against such storms developing in mid-winter, usually along active cold fronts.

Sometimes, thunderstorm precipitation is in the form of hail. Although this tends to be less important in north-west England than at many places further south and east, it is still a noteworthy element of the region's climate. Webb (1988) has pointed out that Lancashire experiences a high frequency of severe hail, because of the influence of the Pennines, which often allows the conditional instability of polar maritime air masses to be realised. In 1983, large hailstones caused much damage in the Wirral on 14 May and were followed by an even worse series of storms on 7 June, which particularly affected Greater Manchester and some areas of Cheshire. Other notable outbreaks were on 1 and 2 July 1968 in the Burnley, Greater Manchester and Northwich areas, and on 8 July 1893 in the Eden valley (Webb 1993). Observations of hail from around sixty localities have been tabulated for the June 1983 event and a detailed analysis of the synoptic situation has proved that it was due to a rare combination of favourable factors. Indeed,

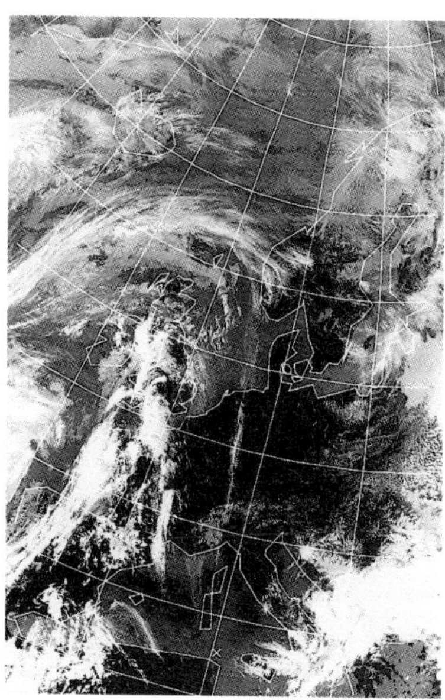

Figure 8.8 Synoptic chart for 1200 GMT on 7 July 1983. NOAA infra-red image for the same day at 1401 GMT
By courtesy of University of Dundee

many of these, such as strong vertical wind shear, low-level south-east winds and upper south-west flow are typical of major self-propagating storms everywhere in Britain. The hail produced was up to 70–75 mm in diameter and caused much damage to greenhouses and windows, outdoor crops and road vehicles (Dent and Monk 1984). Figure 8.8 shows that the conditions on the day were exactly those required for the classic 'Spanish plume' advecting warm and moist air northwards ahead of a cold front. Cloud cover was particularly well-developed over north-west England (Figure 8.8 inset). The main events took place during the late afternoon and early evening when ground-level heating over the Lancashire plains encouraged notable atmospheric instability.

WIND

Table 8.6 summarises the distribution of wind for two sites in the region. The anticipated concentration of observations into the west and south-west sectors is notable at Point of Ayre on the northerly tip of the Isle of Man but is less marked at Manchester Airport, where subregional topographic elements account for the abundance of southerlies. Given that most of the severe gales that we experience in Britain have a strong westerly component, the region's exposure to this direction provides a notable number of such events, some of which are summarised in Table 8.7. The high ground of the Pennines and Lake District is another reason why strong winds are much more common in north-west England. Thus on the summit of Great Dun Fell the mean annual wind speed is close to 30 knots, though it puts matters into perspective to note that values for the top of Ben Nevis have been suggested to exceed 40 knots (Manley 1952, 1974). Similarly, the 100 days of gale each year on Great Dun Fell represent less than half the number recorded on the singularly exposed island site of Snaefell (Meteorological Office 1983, 1985). One way of gauging the region's overall windiness is to compare it with south-east England, where average winds speeds are markedly less, coastal gales dis-

Table 8.6 Summary of average wind directions for the period 1961–90

Location	*N*	*NE*	*E*	*SE*	*S*	*SW*	*W*	*NW*	*Calm*
Point of Ayre IoM	6.3	10.0	8.2	13.3	12.2	15.5	22.0	9.4	3.2
Manchester Airport	5.5	7.5	9.8	7.0	25.2	14.4	16.8	8.9	4.9

Note: Data are expressed as percentages of all 0900 GMT observations

Table 8.7 Examples of wind damage in north-west England

Date	*Locality*	*Nature of damage*
January 1839	Lowther Castle, near Penrith, and the Liverpool area	Woodland extensively damaged and loss of life on land and at sea
February 1903	Leven Viaduct, near Ulverston	Train blown over
March 1907	Morecambe	Extensive damage in the town (see Figure 8.9)
August 1944	Ribble estuary	Aeroplane takes off and encounters a tornado – 59 people killed
January 1962	Egremont area	Many buildings damaged by a tornado
December 1964	Greater Manchester	Tornadoes damage 200 houses at Salford and 80 houses at Cheetham Hill
January 1976	Liverpool	Recently completed landing stage for Mersey ferry sunk
January 1984	Fiddler's Ferry, near Warrington	Power station cooling tower collapses

tinctly fewer and calms less infrequent, or with north-east England, which frequently enjoys the shelter of the Pennines.

Winds with a westerly component are not as dominant in the region as might be expected (Table 8.6). Thus over the Cheshire plain and the Manchester area, surface winds are often from the south and result from channelling between the Pennines and the mountains of North Wales (Crowe 1962). Likewise, pronounced channelling occurs along the Eden valley, giving an important south-easterly component to the winds at Carlisle (Meteorological Office 1985). In the Lake District also, winds are often channelled along the numerous valleys. The curious 'bottom wind' of Derwentwater may also owe something to topography and may be the result of gusts descending from the surrounding hills. A good early description of this phenomenon was given by William Gell in 1797 (Rollinson 1968). Descending air also characterises the well-known Helm Wind of the Cross Fell escarpment (see Chapter 7), which has less impressive parallels in the Lake District, for example in the Grasmere area.

A study of gales at Southport has indicated that these are far more common during winter than in summer. Such winds lose their strength at low level as they pass inland, with the result that according to Crowe (1962), Southport gales have five times the duration of those recorded at Manchester Airport.

The hazard of tornadoes is less common in north-west England than in eastern or southern counties (Elsom 1985). Despite this, reports in the *Journal of Meteorology* prove that tornadoes have occurred during every month of the year. The damage caused has at times been substantial and there has occasionally been loss of life. On 17 January 1962 a severe tornado damaged many buildings in the Egremont area (Rowe 1985). Nor has the region wholly

Figure 8.9 Storm damage at Sandylands, Morecambe, 24 March 1907 (from a postcard of the period)

escaped multiple tornado damage, and on 23 October 1964 tornadoes wrought much destruction throughout Greater Manchester with 200 houses damaged at Salford, another 80 at Cheetham Hill (Meaden 1980). On 23 November 1981 Lancashire, Merseyside, Greater Manchester and Cheshire were also part of what was later described as Europe's biggest tornado outbreak, when 105 individual examples were recorded in England and Wales (Rowe 1985; Meaden and Rowe 1985).

VISIBILITY, AIR POLLUTION AND SUNSHINE

Visibility relates to the way light is scattered by water droplets and solid particles in the atmosphere. The former have been a major influence on visibility in Cumbria, while the latter have been more important in the industrial areas of Merseyside and Greater Manchester. This contrast has, however, narrowed in recent decades, as smokestack industries have declined and as the internal combustion engine has increasingly penetrated the whole of north-west England.

Fog in the North-west of England

In Cumbria the visibility problems caused by valley fogs are usually avoided, because of a relatively clean and dynamic atmosphere (Manley 1974). At greater altitudes, however, visibility becomes more of a problem, owing to low cloud. This may be illustrated by comparing the summit of Great Dun Fell (847 m), which averages over 230 days of fog, with Carlisle (25 m), with only 6.0 days a year. Especially in winter, Great Dun Fell often records months when nearly all days have fog at 0900 GMT (Rawes 1973). Buxton is also sufficiently high to record an annual average of 44.8 fog days. Fortunately, the hill fogs of Cumbria are often accompanied by strong winds, so they are rarely as dense as

Figure 8.10 'Beautiful Oldham', *c.* 1910 (from a postcard of the period)

the radiation fogs which can plague inland sites on lower ground in the more southerly parts. These two types of fog have a broadly similar annual pattern with winter being the most likely time for them to occur (see the example of Buxton in Figure 8.11). Most lowland sites are more vulnerable in winter when net heat loss,. leading to radiation fogs, is greatest and relative humidities in maritime air masses, leading to advection fogs, at their highest. The pattern is different again at a coastal site like Douglas (Figure 8.11), where the spring and early summer fog maximum is attributable to advection fogs forming over the cool Irish Sea before summer heating has had full effect. This distinction between the annual patterns of fog in coastal and upland locations has already been seen in many other regions.

Visibility and Pollution

For over two hundred years air pollution has combined with natural causes to reduce visibility in north-west England. As early as 1785 a foreign visitor, François Rochefoucauld, made a complaint that became depressingly familiar in succeeding centuries: 'What a fog, what a dense black and offensive smoke infects the whole atmosphere' (quoted in Scarfe 1995). The place he described was Liverpool, but it could easily have been any of the industrialising towns in north-west England. Over a hundred and fifty years later, Bilham (1938) similarly complained that England's lowest sunshine totals were all in the north-west of the country in heavily polluted towns, such as Manchester, Bolton and Burnley (see also Box 8.3). Further north in Cumbria the situation has been better, though even here there was never complete immunity from the airborne drift of pollutants. Thus, Cohen and Ruston (1925) have published a photograph of soot deposits near the water's edge at Coniston and remarked: 'In the Lake District such deposits . . . are frequently observed.' The consequences of this pollution are still not fully understood. For example, Gilbert (1970) and Ratcliffe

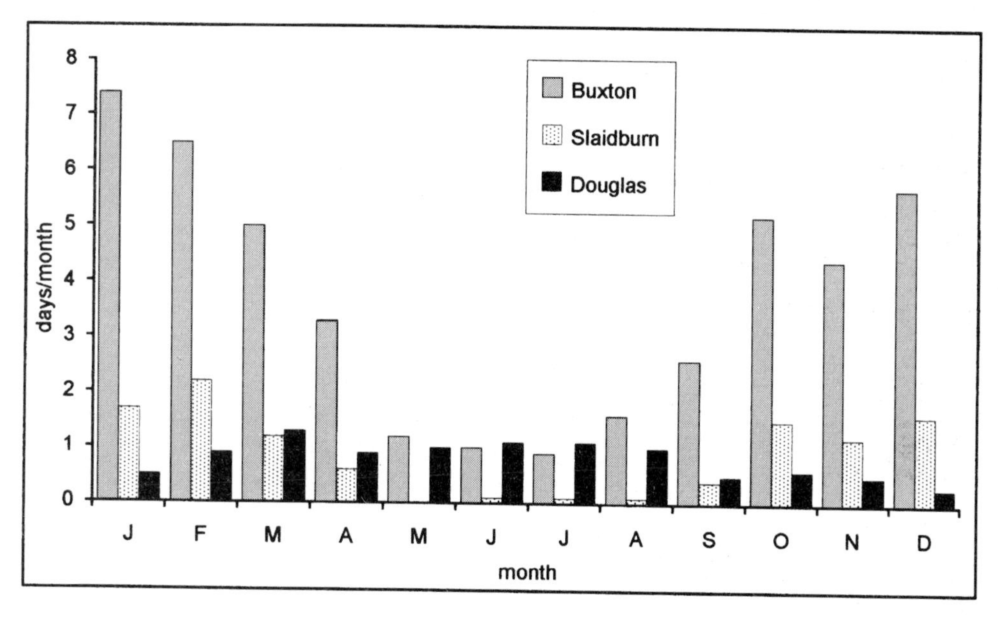

Figure 8.11 Mean monthly number of days of fog at 0900 GMT for the period 1961–90

(1973) differ over the extent to which tree lichens in the Lake District have been depleted by imported air pollution. On the other hand, Batterbee's (1988) study of sediment cores from Cumbrian lake beds such as Scots Tarn, Low Tarn and Greendale Tarn has provided evidence of acid deposition dating from the mid-eighteenth century to the present day.

One major consequence of fog and air pollution has been to reduce the intensity and duration of sunlight. This is especially the case when conditions were anticyclonic, settled and when the sun is low over the horizon in winter. Other seasons are not immune. For example, the summer of 1906 was particularly good, yet at Manchester sunshine was mainly recorded during the early morning, before industry was in full swing (Graham 1914). Around the turn of the century, smoke was intercepting half the effective sunshine in central Manchester (Simon and Fitzgerald 1922). In 1913 the light received between 10 am and 4 pm at the city centre was less than half that some 8 km to the west-south-west at Davyhulme (Shaw and Owens 1925). As recently as 1930, the larger urban areas of north-west England were losing around 35–40 per cent of bright winter sunshine because of pollution (Tout 1979).

During the past fifty years, progress in reducing amounts of smoke and sulphur dioxide has led to improved visibility and sunshine levels in many areas. As early as the fog of November 1955 it was observed that visibility was better in those parts of central Manchester which had become a smokeless zone; even so, poor air quality during this fog killed between 60 and 90 people (Medical Officer of Health 1955). From 29 November to 1 December 1977, meteorological conditions in the area again favoured the build-up of dense fog. Around the city centre levels of smoke and sulphur dioxide approached 700 $\mu g/m^3$, values which were, however, much less than would have occurred fifteen or more years earlier. Despite this, visibility was reduced to under 100 m (Elsom 1978).

A gratifying result of the Clean Air Acts of 1956 and 1968 has been that improvements in the levels of air pollution and visibility (Figure 8.12) have often extended beyond the industrial and urban areas. In 1974 it was reported that pollution-sensitive lichens

Box 8.3

MANCHESTER'S BLACK SMOKE TAX

Manchester was the world's first industrial city and has a long history of air pollution. Among the early records is one from 1800, when the local Commissioners of Police set up a committee to examine smoke pollution and other nuisances. Despite this and much subsequent action, Manchester's smoke problem persisted throughout the nineteenth century. In 1848 it became the first place to record the dark form of the peppered moth, while in 1859 L.H. Grindon made it the first locality where disappearance of lichens was ascribed to air pollution (Richardson 1975). Between 1850 and 1915 R.A. Smith and others connected with the town's Literary and Philosophical Society investigated the problem scientifically (Smith 1852; Knecht 1905), but responses of this sort did little to reduce the environmental impact of a rapidly expanding population. By the late nineteenth century the better off were escaping at an increasing rate to the suburbs and nearby countryside. Local parks, which were created to improve the environment, had by now been so contaminated that many of their trees had died. Even where vegetation did survive, the growing season was shortened and leaves fell early. Arsenic, soot and acids were all blamed for this situation (Royal Commission on Arsenical Poisoning 1903; Cohen and Ruston 1925). One estimate claimed that during ten days in 1902 more than 300 tonnes of soot fell on the city from Manchester's polluted air (Irwin 1902). This was also a time of keen debate over whether or not the area's fogs were getting worse (Graham 1914; Jenkins 1915). Because air pollution degraded the environment, caused health problems and affected people's financial situation, for example by increasing household washing bills, Mancunians thought of it as imposing 'the black smoke tax' (Graham 1914; Manchester City Council Air Pollution Advisory Board 1920). These disadvantages having been recognised, the obvious step was to try to remedy them. In facing this challenge, the city was to earn itself a reputation as 'the home of smoke abatement'. It was here that Charles Gandy proposed the concept of the 'smokeless zone', an idea first put into practice in 1952 when 42 ha in Manchester city centre was so designated. By 1972 Salford had completed its smoke control programme, while that for Manchester was finished in 1985. Between 1956 and 1988 mean annual concentrations of smoke declined by 90 per cent and those for sulphur dioxide fell by 85 per cent. More importantly, deaths from bronchitis were reduced in line with these trends (Manchester Area Council for Clean Air and Noise Control 1983; Manchester Area Pollution Advisory Council 1988). As early as 1969 it became possible for a market gardener to grow *white* chrysanthemums in Wythenshawe (Medical Officer of Health 1969)!

had reappeared in the Buxton area for the first time this century (Hawksworth 1974).

It is indeed ironic that as smoke and sulphur dioxide were being markedly reduced, the levels of other pollutants (e.g. nitrogen oxides and airborne lead) began to increase. In Greater Manchester the concentrations of smoke and sulphur dioxide, though much lower than formerly, are still able to damage materials and affect health (Manchester Area Pollution Advisory Council 1988). Another element of the pollution cocktail, acid precipitation, has been especially well studied throughout the conurbation since the formation of the Acid Rain Information Centre in 1984. Its work has demonstrated the importance of acid precipitation in Greater Manchester and has found that both local and continental European sources are responsible for the problem (Longhurst *et al.* 1990). With the number of road

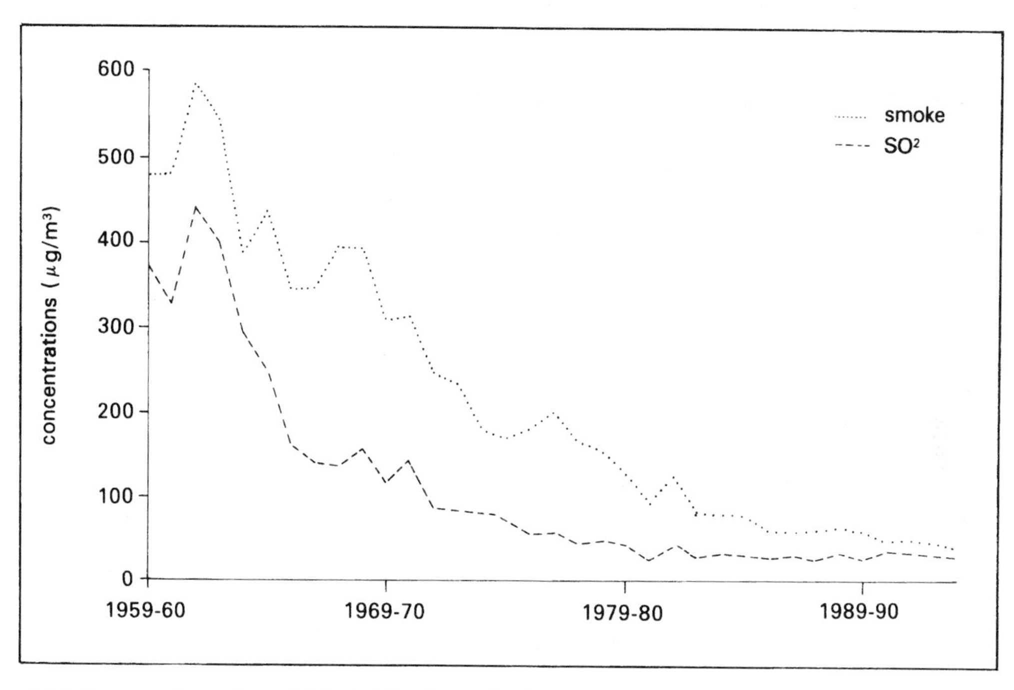

Figure 8.12 Decreases in smoke and SO_2 in Manchester for the period 1959–94

vehicles continually increasing, nitrogen dioxide has become a more important hazard. Surveys in 1986–88 found that levels of the pollutant throughout Greater Manchester were within the EC limit of 200 μg/m, though the 50 μg/m guide value was being breached along main roads, in Manchester's commercial centre and around some chemical works. Surveys have also revealed that heavy metal concentrations in Manchester's atmosphere are at least ten times higher than at a rural site like Windermere. It is fortunate that levels of potentially hazardous lead have recently declined as changes to vehicle fuel have been introduced (Manchester Area Pollution Advisory Council 1988).

Sunshine

Despite a history of poor air quality and reduced visibility it would be wrong to ignore the positive side of things. Thus, even in the 1920s and 1930s Southport recorded far more days of good to excellent visibility than of slight to very dense fog (Baxendell 1936). The town has always enjoyed a relatively high amount of sunshine, especially in spring and summer, owing to the shelter afforded by the Welsh mountains (Table 8.8). Reductions in air pollution and fogs have also led to improvements in the duration and intensity of sunshine: again, these have been greatest in central Manchester, though they have also caused mean annual sunshine totals at the airport to improve from 1,263 hours between 1931 and 1960 to 1,346 hours from 1961 to 1970. These improvements have to be seen against the longer-term increase of annual sunshine in Greater Manchester during the present century (Wood 1973; Tout 1978) and are revealed in Table 8.9.

Present-day average sunshine hours over the region also vary by altitude (Table 8.8), and sites such as Slaidburn have significantly lower totals than more favourably located low-level sites. Douglas claims one of the sunniest sites in the region,

Table 8.8 Mean monthly and annual sunshine totals (hours) for the period 1961–90

Location	*Alt. (m ASL)*	*Jan.*	*Feb.*	*Mar.*	*Apr.*	*May*	*June*	*July*	*Aug.*	*Sept.*	*Oct.*	*Nov.*	*Dec.*	*Year*
Douglas (IoM)	85	52	70	113	171	215	210	196	179	129	93	64	44	1,536
Carlisle	26	47	74	103	150	190	187	176	164	117	92	63	44	1,409
Morecambe	7	49	73	105	153	198	194	182	171	123	93	62	42	1,446
Slaidburn	192	40	57	93	136	180	175	162	153	115	82	55	36	1,283
Manchester Airport	75	48	67	100	138	186	181	169	163	123	97	61	45	1,378

Table 8.9 Mean annual duration of sunshine (hours) at Manchester and Bolton for the period 1901–60

Years	*Manchester (Whitworth Park)*	*Bolton*
1901–10	1,009	981
1931–40	1,022	1,104
1951–60	1,072	1,288

Source: Wood (1973)

and west-coast sites generally, despite their high degree of exposure to moist westerlies, compare well with east-coast locations. At either end of the Tyne–Solway gap Carlisle has indeed longer hours of sunshine than Tynemouth (see also Table 7.5). As Manley (1974) has commented, sunshine amounts in north-west England are 'quite good for the latitude'.

CONCLUSION

It is now around three hundred years since Richard Towneley and Sir Daniel Fleming observed the weather in north-west England, and two hundred years have passed since John Gough, John Dalton and Peter Crosthwaite made their instrumental observations at Kendal and Keswick. Despite the progress since then some aspects of climate in north-west England remain to be researched. Indeed, the history of research is itself a largely unwritten story excepting always the inestimable contribution of Gordon Manley. As a result, many of the observations gathered by earlier workers await collation and analysis, especially those in the numerous volumes of Symons's *British Rainfall* and its successors.

The region's variable relief, from the rugged fastness of the Lake District to the undulating Lancashire and Cheshire plains, and its exposure to the dominant westerlies of our latitudes are thrown into sharp focus as the key climatological themes. Environmental gradients can be steep as a consequence, a feature readily acknowledged by Gordon Manley (1973a, p. 11), who observed:

> witness the quite extraordinary change in the vegetation in what will thrive and what will fail, that accompanies a walk of 16 kilometres, a climb of only 300 metres, a fall of a mere 2°C, a doubling of rainfall, between the shores of Morecambe Bay and the Lake District or Bowland Fells

Thus do we account for Blackpool's persistent appeal as a holiday resort, for Buxton's more stringent allure and for the enduring appeal of harshness of the Pennines moorlands. A yet finer grain exists within the Lake District, for here the fells rise abruptly from the Irish Sea. From the centre, deep valleys radiate in every direction to form a kaleidoscope of local effects that almost defy scientific description, leaving instead a preference for artistic licence.

REFERENCES

Barnes, T. (1830) 'Remarks explanatory, and tabular results of a meteorological journal kept at Carlisle by the late Mr William Pitt during twenty-four years', *Trans. R. Soc. Edinburgh*, 11: 418–432.

Barrett, E.C. (1966) 'Regional variations of rainfall trends in northern England, 1900–59', *Trans. Inst. Br. Geogrs*, 38: 41–58.

Batterbee, R.W. (1988) *Lake Acidification in the United Kingdom 1800–1986: Evidence from the Analysis of Lake Sediments*, London: HMSO.

Baxendell, J. (1903) 'The meteorology of Southport', in *Southport: A Handbook of the Town and Surrounding District*, Southport: Fortune & Chant.

—— (1935) *Southport Auxiliary Observatory (The Fernley Observatory of the Corporation of Southport): Annual Report, and Results of Meteorological Observations, for the Year 1935*, Southport: Robert Johnson.

Beck, R.A. (1992) 'Manchester's rainy reputation reviewed', *J. Meteorology (UK)*, 17: 256–257.

Bilham, E.G. (1938) *The Climate of the British Isles*, London: Macmillan.

Chuan, G.K. and Lockwood, J.G. (1974) 'An assessment of topographical controls on the distribution of rainfall in the central Pennines', *Meteorol. Mag.*, 103: 275–287.

Clark, J.B. (1986) 'Major snowfalls in Manchester since 1880', *Weather*, 41: 278–282.

Cohen, J.B. and Ruston, A.G. (1925) *Smoke: A Study of Town Air*, London: Arnold.

Crowe, P.R. (1962) 'Climate', in *Manchester and Its Region*, Manchester: Manchester University Press.

Cutforth, P.R. (1987) 'Wet and dry in Lakeland', *J. Meteorology (UK)*, 12: 130–131.

Dalton, J. (1793) *Meteorological Observations and Essays*, London: T. Ostell.

—— (1819) 'Observations on the barometer, thermometer, and rain, at Manchester from 1794 to 1818 inclusive', *Mem. Manchester Literary and Philosophical Soc.*, 3: 483–509.

Dent, L. and Monk, G. (1984) 'Large hail over north-west England, 7 June 1983', *Meteorol. Mag.*, 113: 249–263.

Dobson, M. (1777) 'Observations on the annual evaporation at Liverpool in Lancashire; and on evaporation considered as a test of the moisture or dryness of the atmosphere', *Phil. Trans. R. Soc.*, 67: 244–259.

Eden, P. (1995) *Weatherwise*, London: Macmillan.

Elsom, D.M. (1978) 'The changing nature of a meteorological hazard', *J. Meteorology (UK)*, 3: 297–299.

—— (1985) 'Tornadoes in Britain: where, when and how often', *J. Meteorology (UK)*, 10: 203–211.

Fleming, D. (1689–93) *Meteorological Observations at Rydal*, Kendal Record Office. WD/Ry/HMC no. 4602a.

Folland, C.K. and Wales-Smith, B.G. (1977) 'Richard Towneley and 300 years of regular rainfall measurement', *Weather*, 32: 438–445.

Garnett, T. (ed.) (1793) 'Meteorological observations made on different parts of the western coast of Great Britain', *Mem. Literary and Philosophical Soc. of Manchester*, 4: 234–272.

—— (1796) 'Meteorological observations', *Mem. Literary and Philosophical Soc. of Manchester*, 4: 517–641.

Gilbert, O.L. (1970) 'Lichens', in G.A.K. Hervey and J.A.G. Barnes (eds) *Natural History of the Lake District*, London: Frederick Warne.

Graham, J.W. (1914) 'Coal smoke: its causes, consequences and cures', *Trans. Manchester Statistical Society*, 39–62.

Gregory, S. (1953) 'Weather and climate', in W. Smith (ed.) *A Scientific Survey of Merseyside*, Liverpool: University Press of Liverpool.

Hawksworth, D.L. (1974) *Report on the Lichen Flora of the Peak District National Park*, Shrewsbury: Nature Conservancy Council (Midlands Region).

Irwin, W. (1902) 'The soot deposited on Manchester snow', *Journal Soc. Chem. Industry*, 21: 533–534.

Jenkins, W.C. (1915) 'Note on foggy days in Manchester', *Mem. Manchester Literary and Philosophical Soc.*, 59: 1–4.

Jones, P.D. (1983) 'Further composite rainfall records for the United Kingdom', *Meteorol. Mag.*, 112: 19–27.

Knecht, E. (1905) 'On some constituents of Manchester soot', *Mem. Manchester Literary and Philosophical Soc.*, 49: 1–10.

Longhurst, J.W.S. *et al.* (1990) *Annual Report 1989/90*, Manchester: Acid Rain Information Centre.

Manchester Area Council for Clean Air and Noise Control (1983) *Twenty Five Year Review.*

Manchester Area Pollution Advisory Council (1988) *Air Pollution Monitoring 1988.*

Manchester City Council Air Pollution Advisory Board (1920) *The Black Smoke Tax: An Account of Damage Done by Smoke, with an Inquiry into the Comparative Cost of Family Washing in Manchester and Harrogate*, Manchester: Henry Blacklock.

Manley, G. (1942) 'Meteorological observations on Dun Fell, a mountain station in northern England', *Q. J. R. Meteorol. Soc.*, 68: 151–165.

—— (1943) 'Further climatological averages for the northern Pennines, with a note on topographical effects', *Q. J. R. Meteorol. Soc.*, 69: 251–261.

—— (1944) 'Topographical features and the climate of Britain: a review of some outstanding effects', *Geogrl J.*, 103: 241–263.

—— (1946) 'The centenary of rainfall observations at Seathwaite', *Weather*, 1: 163–168.

—— (1952) *Climate and the British Scene*, London: Collins.

—— (1958) 'The great winter of 1740', *Weather*, 13: 11–17.

—— (1963) 'The climatologist's view', in *The Long Winter 1962–3*, Manchester: Manchester Guardian.

—— (1968) 'Dalton's accomplishment in meteorology', in D.S.L. Cardwell (ed.) *John Dalton and the Progress of Science*, Manchester: Manchester University Press.

—— (1973a) 'Climate in Britain over 10,000 years', in A.R.H. Baker and J.B. Harley (eds) *Man Made the Land: Essays in English Historical Geography*, Newton Abbot: David & Charles.

—— (1973b) 'Climate', in W.H. Pearsall and W. Pennington (eds) *The Lake District*, London: Collins.

—— (1974) *Enjoy Cumbria's Climate*, Kendal: Cumbria Tourist Board.

—— (1975a) '1684: the coldest winter in the English instrumental record', *Weather*, 30: 382–388.

—— (1975b) 'Weather and climate of the Lake counties', in *The Lake District National Park*, London: HMSO.

Meaden, G.T. (1980) 'British tornadoes and waterspouts of the 1960s, Part 2: 1963–1964', *J. Meteorology (UK)*, 5: 173–184.

—— (1983) 'Britain's highest temperatures: the county records', *J Meteorology (UK)*, 201–203.

Meaden, G.T. and Rowe, M.W. (1985) 'The great tornado outbreak of 23 November 1981', *J. Meteorology (UK)*, 10: 295–300.

Medical Officer of Health (1955) *Fog in Manchester, November 1955*. Mimeographed.

—— (1969) *Report on the Health of the City of Manchester for 1969*, Manchester: Health Department.

Meteorological Office (1983) *The Climate of Great Britain: Lancashire and Cheshire and Isle of Man*, Climatological Memorandum 130, Bracknell: Meteorological Office.

—— (1985) *The Climate of Great Britain: Pennines and Lake District*, Climatological Memorandum 128, Bracknell: Meteorological Office.

Prichard, R. (1985) 'The spatial and temporal distribution of British thunderstorms', *J. Meteorology (UK)*, 10: 227–230.

Ratcliffe, D.A. (1973) 'The habitats: the native woodlands', in W.H. Pearsall and W. Pennington (eds) *The Lake District*, London: Collins.

Rawes, M. (1973) *Moor House, 14th Annual Report*, Grange-over-Sands: Nature Conservancy Council.

—— (1976) *Moor House, 17th Annual Report*, Grange-over-Sands: Nature Conservancy Council.

—— (1978) *Moor House, 19th Annual Report*, Grange-over-Sands: Nature Conservancy Council.

Reynolds, G. (1953) 'Rainfall at Bidston, 1867–1951', *Q. J. R. Meteorol. Soc.*, 79: 137–149.

Richardson, D. (1975) *The Vanishing Lichens*, Newton Abbot: David & Charles.

Rollinson, W. (ed.) (1968) *A Tour in the Lakes Made in 1797 by William Gell*, Newcastle upon Tyne: Frank Graham.

Rowe, M. (1985) 'Britain's greatest tornadoes and tornado outbreaks', *J. Meteorology (UK)*, 10: 212–220.

Rowsell, H. (1970) *Rainfall over the Areas of the Cheshire, Mersey, Lancashire and Cumberland River Boards 1916–1950*, Hydrological Memorandum 11, Bracknell: Meteorological Office.

Royal Commission on Arsenical Poisoning (1903) *Minutes of Evidence and Appendices*: vol. 1, *Evidence Received in 1901*, London: HMSO.

Scarfe, N. (1995) *Innocent Espionage: The La Rochefoucauld Brothers' Tour of England in 1785*, Woodbridge: Boydell Press.

Shaw, N. and Owens, J.S. (1925) *The Smoke Problem of Great Cities*, London: Constable.

Shellard, H.C. (1968) 'The winter of 1962–63 in the United Kingdom: a climatological survey', *Meteorol. Mag.*, 97: 129–141.

Simon, E.D. and Fitzgerald, M. (1922) *The Smokeless City*, London: Longman.

Simpson, C.J. (1995) 'The raining king of Lakeland weather', *Cumbria*, 45: 35–37.

Smith, R.A. (1852) 'On the air and rain of Manchester', *Mem. Manchester Literary and Philosophical Soc.*, 10: 207–217.

Symons, G.J. (1868) 'Origin, progress, and present state of our knowledge of the rainfall in the Lake District', *British Rainfall 1867*, 13–21.

—— (1896) 'Seathwaite's jubilee, 1845–94', *British Rainfall 1895*, 15–25.

—— (1897) 'The most rainy part of England', *British Rainfall 1896*, 16–26.

Tout, D. (1978) 'A remarkable increase in November sunshine in parts of the United Kingdom', *J. Meteorology (UK)*, 3: 129–131.

—— (1979) 'The improvement in winter sunshine totals in city centres', *Weather*, 34: 67–71.

Towneley, R. (1694) 'A letter from Richard Towneley, of Townley in Lancashire, Esq. containing observations on the quantity of rain falling monthly, for several years successively', *Phil. Trans. R. Soc.*, 18: 51–58.

—— (1700) 'An account of what rain fell at Townley in Lancashire, in the years 1697, and 1698, with some other observations on the weather; being part of a letter of the 12th of Jan. 1698 from Richard Towneley Esq; to Mr. William Dereham', *Phil. Trans. R. Soc.*, 21: 47–48.

Towneley, R. and Dereham, W. (1705) 'Prospect of the Weather, Winds and Height of the Mercury in the Barometer, on the first day of the Month; and of the whole Rain in every Month in the Year 1703, and the beginning of 1704: Observed at Towneley in Lancashire', *Phil. Trans. R. Soc.*, 24: 1877–1881.

Webb, J.D.C. (1987) 'Britain's highest daily rainfalls: the county and monthly records', *J. Meteorology (UK)*, 12: 263–266.

—— (1988) 'Hailstorms and intense local rainfalls in the British Isles', *J. Meteorology (UK)*, 13: 166–182.

—— (1993) 'Britain's severest hailstorms and "hailstorm outbreaks" 1893–1992', *J. Meteorology (UK)*, 18: 313–327.

Webb, J.D.C. and Meaden, G.T. (1983) 'Britain's lowest temperatures: the county records', *J. Meteorology (UK)*, 8: 269–372.

—— (1993) 'Britain's highest temperatures by county and by month', *Weather*, 48: 282–291.

Winchester, A.J.L. (ed.) (1994) 'The diary of Isaac Fletcher of Underwood, Cumberland, 1756–1781', *Cumberland and Westmorland Antiquarian and Archaeological Soc.*, Extra Series XXVII.

Wood, C.M. (1973) 'Visibility and sunshine in Greater Manchester', *Clean Air*, 3: 15–24.

9

CENTRAL AND SOUTHERN SCOTLAND

John Harrison

INTRODUCTION

Central and southern Scotland can be defined as lying roughly between the Highland Boundary Fault, which runs north-eastwards from the Firth of Clyde to the North Sea coast south of Aberdeen, and the border between Scotland and England, which is itself defined by the course of the river Tweed and the Cheviot Hills. Such a large area, which has a maximum west-to-east extent of 210 km and a north–south extent of 260 km, inevitably contains a complex array of local climates and it is no easy task to identify a broad subregional zonation. However, the identification of the principal climatic controls opens the door, if only a little, to an understanding of the most noteworthy features of a climate which ranges from a cold north-east coast, where Robert Louis Stevenson described Edinburgh as having 'one of the vilest climates under heaven' (Paton 1951), to the mild south-west, where the climates of Wigtownshire and Dumfriesshire have been variously described as 'genial' (Sutherland 1925) and 'amenable' (Wood 1965).

Another major geological fault provides a primary topographic division within the region. The Southern Upland Fault, which extends north-eastwards from near Ballantrae on the Ayrshire coast to Dunbar in East Lothian, separates the older grits and shales of the Southern Uplands from the younger sediments of the Midland Valley. However, the presence of resistant lavas, conglomerates and grits to the north of the fault tends to blur the boundary and there are substantial volcanic uplands within the Midland Valley, such as the Ochil Hills. The dissected Southern Uplands rise to more than 500 m in the Lammermuir Hills to the south of Edinburgh but reach heights in excess of 800 m from Broad Law (840 m) across to Merrick (843 m) in Galloway (Figure 9.1). There is a steep altitudinal deterioration in climate, which is typical of northern Britain and which results in the very marked contrast between the farms of the lowlands and the open moorland pastures dominant above altitudes as low as 200 m.

The region has coasts to both west and east, providing distinctly different influences on climate. Not only is the North Sea generally cooler than the Irish Sea but it also experiences slightly delayed seasonal extremes of temperature. There are four main estuarine areas, these being the Clyde, Tay, Forth and Solway, which generate particular local climatic variations that are related to the morphology of the coastline and the spatial extent of the intertidal zone (Harrison 1987). The tidal movement of water across extensive mud and sand flats results in complex temperature and moisture regimes within the sediment and in the near-surface air layer. Estuaries also provide routeways for the inland penetration of weather systems.

There are two major conurbations, in Glasgow and Edinburgh, and the city of Dundee, plus a large number of substantial towns, many of which are located on flood plain sites. The industrial development of the Glasgow region has resulted in a marked urban climate and historically high air pollution loadings. Of particular relevance is the

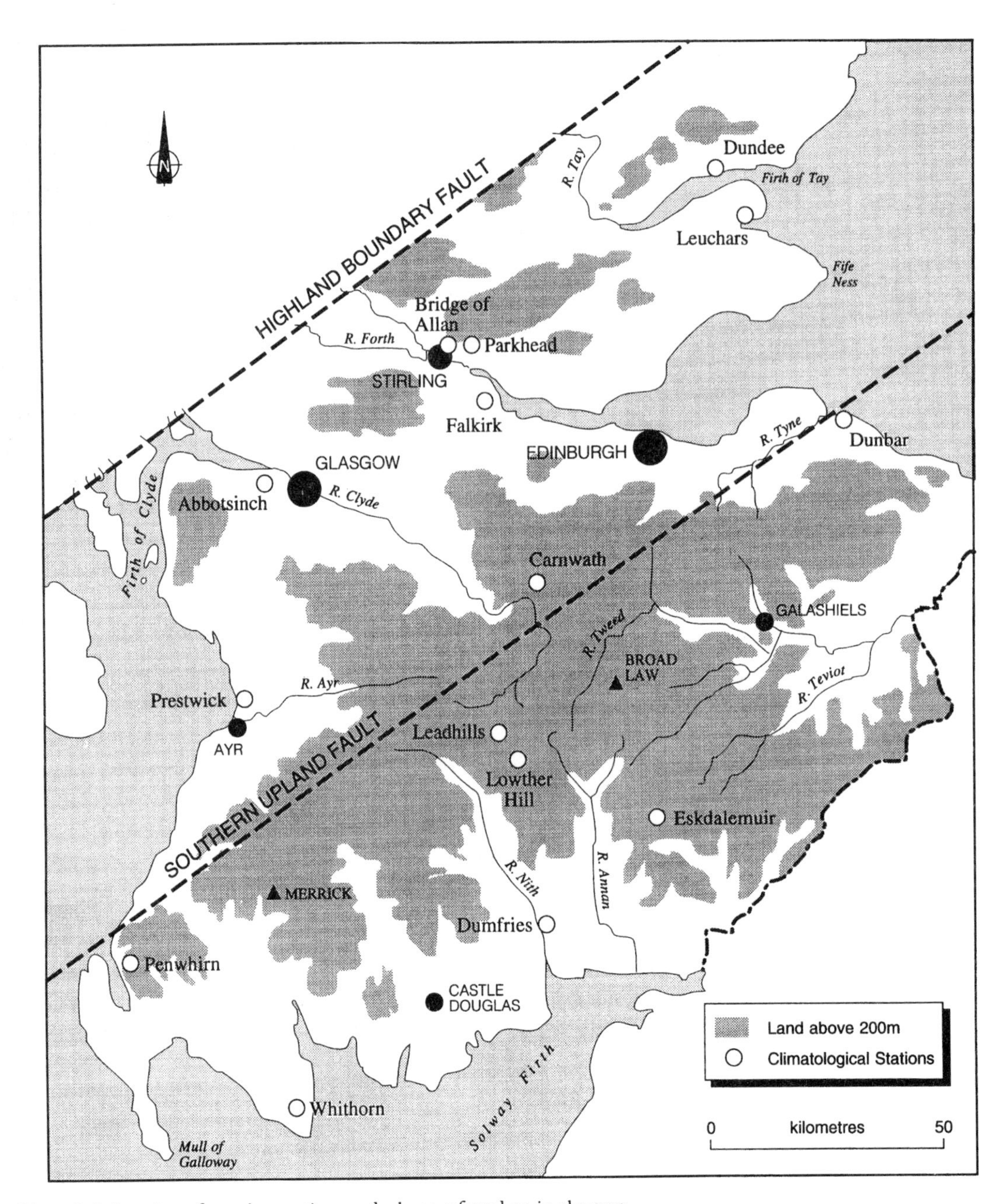

Figure 9.1 Location of weather stations and places referred to in the text

role of the region as the gateway into Scotland. Major south-to-north trunk roads, such as the A/M74, A1 and A68, and both the West Coast and East Coast Main Lines all pass through the Southern Uplands, which renders them vulnerable to closure or restriction by high winds, driving rain or snow accumulation.

Although parts of the Midland Valley are moderately well served in the distribution of climatological stations, there are important gaps elsewhere in the region. There has, for example, been a shortage of observations for Galloway and for much of the Southern Uplands (Halstead 1956). The latter is a familiar problem in upland areas, and there are only six climatological stations above 250 m in the region currently reporting to the Meteorological Office Monthly Weather Report. Of these the highest is at Leadhills (388 m) in the Strathclyde region, which is officially listed as making returns since 1953, but records were kept at the Scots Mining Company's counting house as early as 1816 (Wood 1965). Undoubtedly the most widely known upland observatory in the region is that at Eskdalemuir (242 m) in the Dumfries and Galloway region, which, because of the longevity of its record, has been used as a representative station for southern Scotland in analyses of temporal changes in climate. The observatory was established in 1908 on the west side of the White Esk valley, but over the past twenty years the landscape surrounding the site has changed progressively from open moorland to coniferous forest (Young 1985), which will have affected the homogeneity of the climatological record. The records for Dundee began in 1865 but some of the earliest observations were made in Edinburgh, and have been documented by Robert Mossman (1896, 1897, 1902). Of the lowland stations, the majority are of recent origin, but particularly long records exist for Glenlee in Dumfries and Galloway (1887), Dumfries (1906) and in the Borders region in West Linton (1907). In view of this, use has been made of short records, for the period 1981 to 1990, which although not climatological normals in the strictest sense, help to establish climatic contrasts within the region. In more recent years, the arrival of the Weather Watchers organisation in Laurieston, near Castle Douglas (Dumfries and Galloway), has served to raise the profile of weather observing in the region (Chaplain 1985).

TEMPERATURE

With airflow dominantly from west-south-west, and the primary long-distance source of advected heat energy being from between south and west, it is reasonable to expect this to be reflected in the geographical distribution of air temperature. Historically, Gauld (1922), Sutherland (1925) and Halstead (1956) have highlighted the effect of the warmth of south-west Scotland on plant growth, whether this be of exotic species in Galloway or early potatoes in Ayrshire. The available climatological data do not, however, reveal particularly stark contrasts between south-west and north-east coasts. For example, a comparison between two coastal stations at Prestwick and Dunbar reveals only slightly higher maxima at the former and lower minima at the latter (Table 9.1). An analysis by the Meteorological Office (1989) has indicated that the extreme south-west of Scotland may have January maximum temperatures of the order of 1 degree C higher than the east coast, with similar differences for minimum temperatures. This contrast is borne out, in part, by differences in the frequency of air frosts between the south-west and north-east coasts. In order to reduce the influence of site-specific bias in minimum temperatures, total frost frequencies over the period 1981–90 have been derived for pairs of stations, viz. Penwhirn and Whithorn in the south-west and Leuchars and Dunbar in the north-east. Comparison between these two pairs indicates that the former experience fewer frosts during the core winter months of December and January. However, within the region the most significant variation in air temperature arises from the combined effects of proximity to the coast, topography, and urban and industrial development.

The effect of a coastal location is to ameliorate winter air temperatures and depress summer temperatures. Smith (1985), for example, compared temperatures along the Ayrshire coast with those recorded at Abbotsinch in the upper Clyde Estuary some 30 km from the open sea. During the winter months Abbotsinch, which is in a moderately sheltered site, registers approximately 50 per cent more

Table 9.1 Mean maximum and minimum air temperatures (°C) for the period 1961–90

Location	*Alt. (m ASL)*	*Jan.*	*Feb.*	*Mar.*	*Apr.*	*May*	*June*	*July*	*Aug.*	*Sept.*	*Oct.*	*Nov.*	*Dec.*	*Year*
Leuchars	10	6.0	6.3	8.6	10.8	13.4	16.8	18.5	18.3	16.1	12.9	8.6	6.7	11.9
		0.2	0.1	1.5	2.9	5.4	8.3	10.2	10.0	8.3	5.9	2.1	0.8	4.6
Dunbar	23	6.5	6.5	8.4	10.1	13.0	16.1	17.8	17.7	15.9	13.1	9.1	7.3	11.8
		1.6	1.4	2.6	3.9	6.4	9.3	10.9	11.0	9.5	7.3	3.7	2.4	5.8
Abbotsinch	5	6.3	6.6	8.7	11.6	14.8	17.6	18.8	18.5	15.9	12.9	8.8	6.9	12.2
		0.4	0.3	1.6	3.1	5.9	8.8	10.5	10.2	8.4	6.1	2.0	0.9	4.8
Prestwick	16	6.6	6.8	8.6	11.3	14.5	16.9	18.1	18.0	15.7	13.1	9.1	7.4	12.1
		1.0	0.8	2.0	3.4	5.9	8.8	10.6	10.5	9.0	6.7	2.8	1.7	5.2
Dumfries	49	6.0	6.2	8.3	11.1	14.3	17.2	18.5	18.2	15.7	12.9	8.6	6.8	11.9
		0.7	0.6	1.8	3.3	5.8	8.8	10.5	10.4	8.6	6.3	2.6	1.3	5.0
Eskdalemuir	242	4.5	4.7	6.0	10.0	13.4	16.3	17.6	17.2	14.6	11.6	7.2	5.3	10.7
		−1.0	−1.2	0.1	1.4	4.0	7.1	8.8	8.7	6.9	4.7	1.0	−0.2	3.3

Note: ASL = above sea level

Table 9.2 Record absolute maximum and minimum temperatures (°C) for central and southern Scotland by administrative regions

Region	*Maximum*	*Date*	*Location*	*Minimum*	*Date*	*Location*
Fife	30.8	2 Aug 1990	Kirkcaldy	−18.1	11 Jan. 1982	Cupar
Lothian	32.4	4 Aug. 1975	Turnhouse	−21.1	8 Jan. 1982	Livingstone
Borders	32.4	2 July 1976	Wauchope	−26.6	11 Jan. 1982	Bowhill (Selkirk)
Strathclyde*	32.2	29 July 1948	Prestwick and Kilmarnock	−24.8	11 Jan. 1982	Carnwath
Central	31.7	12 July 1911	Buchlyvie	−17.2	11 Jan. 1982	Stirling (Parkhead)
Dumfries and Galloway	32.8	2 July 1908	Dumfries	−23.9	10 Feb. 1895	Drumlanrig

Sources: Meaden (1983); Webb and Meaden (1983, 1993)
* The Strathclyde region embraces also parts of Highland Scotland (as defined in this text). The data apply only to sites within the area covered by this chapter

air frosts than the coastal plain, whilst during the summer cool sea breezes tend to suppress daytime maximum temperatures along the coast. Because the annual temperature cycle in the North Sea is slightly delayed relative to that in the Irish Sea, lower coastal water temperatures have a tendency to delay spring warming along the east coast, but a later maximum tends to lead to more favourable autumnal temperatures. Along this coast, daily and seasonal temperature variation increases quite rapidly inland, away from the ameliorating influence of coastal waters. This is very much in evidence in a comparison between Dunbar and Leuchars, the latter being only 5 km inland from the Fife coast (Table 9.1). The average daily range is 1.3 degrees C greater and the annual range of mean temperature 1.0 degree C greater at the latter.

Topography provides several important thermal controls, the principal ones being aspect, exposure to the winds and altitude. During the summer, sheltered inland straths and glens can become extremely warm, especially if aspect is favourable and there is some degree of shelter from the wind. The whole question of solar radiation receipts in areas of sloping terrain has been reviewed by Varley *et al.* (1996), who stress the significance of season as well as aspect and slope angle.

It is, however, the Forth lowlands that experience some of the highest summer daytime temperatures in Scotland (Meteorological Office 1989). During the spell of hot weather in late June 1995 the absolute maximum temperature reached 29.2°C at Parkhead in the Forth valley, whilst it reached only 24°C near the Fife coast at Leuchars. Table 9.2 reveals some further, more notable, record maxima for the region that stand comparison with those for some more southerly districts of the British Isles. The warm spell of early August 1975 stands out clearly in this respect, and although Glasgow's 31.2°C (also registered on 4 August) was not a regional record it was the highest in a composite series that goes back to 1868. This period was marked by anticyclonic conditions with a southerly airflow that placed central Scotland in a highly favourable position to the lee of the Southern Uplands where long-standing maximum records held intact. On the other hand, from September through to April the combined effects of shading and katabatic drainage of cold air can make many of the same locations very cold, especially at greater distances from the ameliorating influence of coastal waters. For example, during the exceptionally cold spell of January 1982, the absolute minimum temperature fell to −17.2°C at Parkhead, but only −13.7°C at Leuchars. Falkirk (3 m above sea level) in the low-lying Forth valley, though only a few kilometres inland, experiences a higher frequency of air frosts (with an annual average of fifty-one) than coastal stations. Dunbar, on the other hand, has one of the lowest frost frequencies on the east coast with an annual average of just thirty-one (Figure 9.2).

The effect of elevation on air temperature is particularly marked in Scotland, especially inland from the west coast, where the cooling effect of increasing altitude combines with a decreasing coastal influence

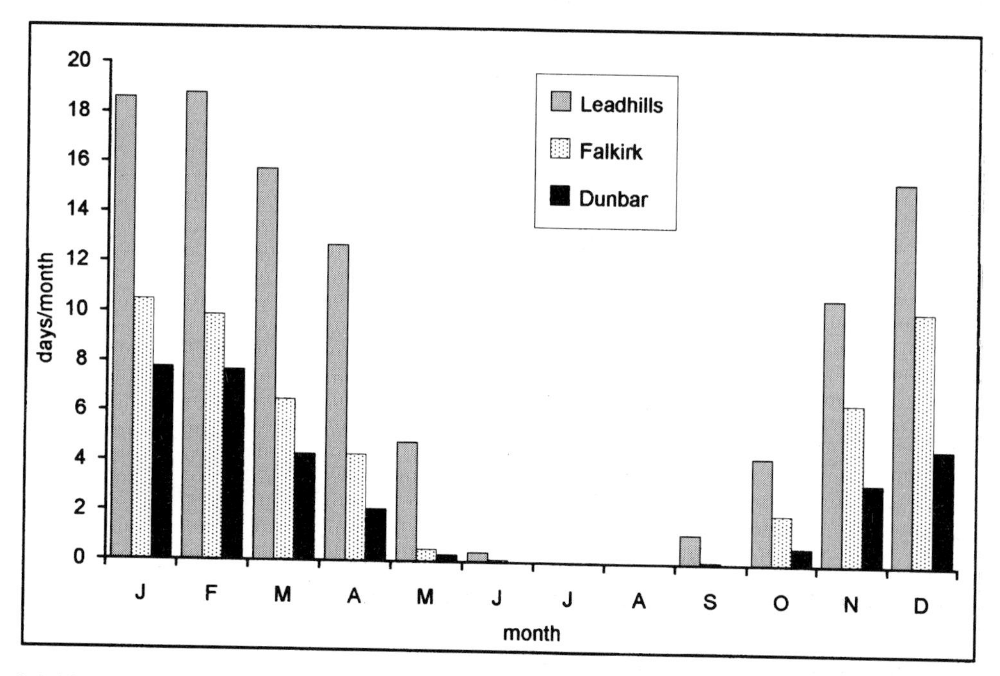

Figure 9.2 Mean monthly frequency of air frosts for the period 1981–90

during the winter and spring. Values of lapse rates in maximum temperature in excess of 13.0 degrees C/1,000 m have been measured in a similar setting in mid-Wales (Harrison 1973), which is exceptionally steep. However, away from coastal influences, lapse rates are more typically 10.0 to 10.5 degrees C/1,000 m for maximum temperatures and 4.0 to 4.5 degrees C/1,000 m for minimum temperatures (Harding 1978; Johnson 1985; Harrison 1994). Derivations of lapse rates in minimum temperature are always subject to potential error when comparing records from paired stations (Harrison 1994). The particular location of Eskdalemuir, for example, makes it unwise to use observations from there in the derivation of temperature–altitude relationships. However, minimum-temperature observations from Leadhills provide an indication of the very marked effect of altitude on the incidence of air frosts (Figure 9.2). The mean annual frequency of frosts at Leadhills (1981–90) is 103, in contrast to the general average of forty-four along south-west and north-east coasts, and the frost-free season may be as little as twelve weeks.

The katabatic drainage of radiatively cooled air creates inversions of temperature in sheltered hollows and valleys when regional pressure gradients are relatively weak (Box 9.1). Many climatological stations in southern Scotland, of which Carnwath (Strathclyde) is the most notable example, have registered among the lowest minimum temperatures in the British Isles (Lyall 1981). On the night of 12 January 1979 the air temperature at Carnwath fell to −24.6°C, which at the time was the lowest recorded anywhere in the United Kingdom since 1955 – although even this minimum was to be exceeded within a few years (Table 9.2). More

Box 9.1

FORTH VALLEY FROST HOLLOW

A comparison has been made between frost frequencies near Bridge of Allan (10 m) located towards the centre of the Forth valley, and the University of Stirling's climatological station at Parkhead (35 m), which is on the north side of the valley (Table 9.3). This reveals a much greater probability of frost and a shorter frost-free season at the former. Between 1980 and 1991 temperature data were available from a recording station at 332 m in the Ochil Hills, from which it was possible to examine the seasonal variation of temperature inversions in the Forth valley (Harrison 1990). Inversions were steepest and most frequent during October but relatively rare during May and June. The mechanism of frost occurrence in the valley is relatively complex and results in a double minimum of temperature, which usually falls steadily during the evening but can rise again between midnight and 0300 hours before resuming its fall towards a post-dawn minimum. The reason for this not been fully established but is most probably linked to the onset of turbulent mixing in the mountain–valley wind system which is developed in the Forth valley.

During December 1981 an intensive study of the Forth valley frost hollow was undertaken (Harrison and Wallace 1982). By the late afternoon of 11 December a ridge of high pressure extended over much of the British Isles. The air was generally calm and skies cloud-free for much of the following night, which resulted in a rapid radiative heat loss from the already cold ground surface. Air temperatures were recorded every 3 hours using a whirling psychrometer along a north–south transect across the Forth valley, between Thornhill and Kippen. By 1800 hours on the 11th the frost hollow effect was already evident but during the course of the night it intensified as temperatures in the middle of the valley fell to −14.6°C, whilst those on the valley sides remained above −10.0°C (Figure 9.3).

Table 9.3 Monthly mean air frost frequencies at Bridge of Allan (Westerlea) and Parkhead for the period 1988–95

Location	*Alt. (m ASL)*	*Jan.*	*Feb.*	*Mar.*	*Apr.*	*May*	*June*	*July*	*Aug.*	*Sept.*	*Oct.*	*Nov.*	*Dec.*	*Year*
Parkhead	35	9.5	10.5	6.6	2.9	0.9	0.0	0.0	0.0	0.1	1.9	6.8	11.6	50.6
Bridge of Allan	10	11.8	10.0	8.4	5.8	2.1	0.1	0.0	0.0	1.0	4.4	10.0	13.8	67.3

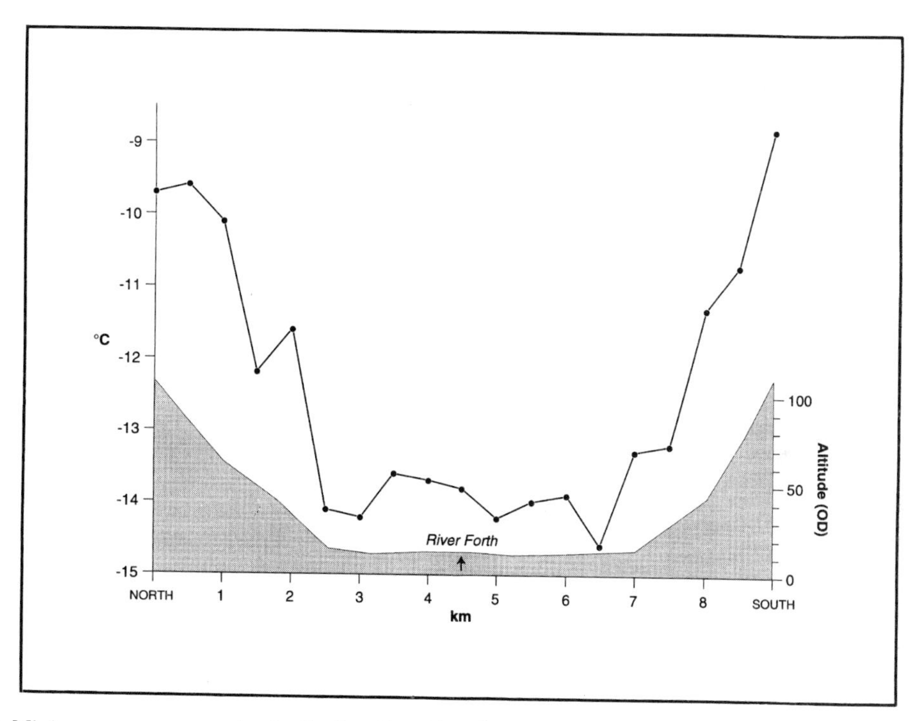

Figure 9.3 Minimum temperatures in the Forth valley, 11 and 12 December 1981

recently, a minimum of −9.9°C was recorded here on 17 October 1993, which was the lowest October minimum on record in the United Kingdom. Although 180 m lower in altitude, Carnwath experienced, on average, 100 frosts per year over the period 1981–90 in comparison to 103 at Leadhills (Figure 9.2). It should be noted also that air frosts occur regularly in all twelve months in Carnwath but are far less frequent in the summer months at Leadhills. The greater relief of the Southern Uplands provides ample opportunity for minima not far removed from those of the Highlands of Scotland (Chapter 10). In addition to Carnwath's −24.8°C, temperatures around Kelso and Selkirk have fallen to below −26°C. On the other hand, on the more subdued topography of the Midland Valley −21°C may represent a lower limit at present. Outbreaks of easterly weather (see Box 1.7, for example) account for many of these record minima such as those for 11 January 1982. This same outbreak of polar continental air set record minima further afield in north-east England (see Figure 7.3).

Little is known of the effects of either urban areas

or industrial complexes on temperature distribution within the region. Hartley (1977) undertook a short-term study of the urban heat-island effect in Glasgow and found an absolute maximum urban–rural temperature difference of 8.3 degrees C. The Grangemouth industrial complex on the Forth estuary undoubtedly does raise local air temperatures, evidence for which is found in the much reduced incidence of road icing and a relatively lower incidence of air frost as compared with nearby Falkirk (Harrison 1993a). However, it is difficult to disaggregate the potential industrial effect on temperature from the ameliorating effect of the waters of the Forth estuary.

The combined effect of latitude and dominant wind direction on the thermally defined growing season means that in the south-west of Scotland this is in excess of 260 days while in the north-east it falls below 240 days. Inland, away from coastal effects and as altitude increases, the season decreases very rapidly to less than 140 days in the highest parts of the Southern Uplands. The rate of decrease has been variously estimated as eighteen days for every 100 m rise (Halstead 1958) and nearly seventeen days (Harrison and Harrison 1988).

PRECIPITATION

Rainfall is derived primarily from eastwards-moving Atlantic depressions and their associated frontal systems. In the absence of any other controlling factors this brings about a west-to-east decrease in total falls. This is indicated in the annual totals for lowland sites in Table 9.4, Abbotsinch, Prestwick and Dumfries receiving approximately 60 per cent more rainfall than Leuchars, Edinburgh and Dunbar. The dominance of cyclonic rainfall is embodied in local weather lore throughout the region, the most common warning of rainfall being a general lowering of the cloud base. Thus in the Borders region:

> When Cheviot ye see put on his cap
> Of rain ye'll have a wee drap
>
> (quoted in Inwards 1994)

and in the Forth valley

> When the Castle of Stirling gets a hat
> The carse of Cornton pays for that.
>
> (quoted in Harrison 1980)

The frequency of mid-latitude depressions tends to be greatest during the winter half-year but, unlike southern England, Scotland tends to remain under their influence during much of the summer. Thus, although there is a marked seasonality in average monthly rainfall, heavy falls can, and do, occur throughout the summer half-year. Indeed, on average, August is the wettest month at Dunbar, but the period September through to January contains the highest average monthly totals in the majority of cases. Rainfall decreases sharply after January, and the months February through to April are, on average, the driest, April being consistently

Table 9.4 Mean monthly and annual precipitation totals (mm) for the period 1961–90

Location	*Alt. (m ASL)*	*Jan.*	*Feb.*	*Mar.*	*Apr.*	*May*	*June*	*July*	*Aug.*	*Sept.*	*Oct.*	*Nov.*	*Dec.*	*Year*
Dundee	45	69	46	51	45	55	55	57	67	68	66	60	62	701
Leuchars	10	66	44	47	41	54	50	52	62	61	62	57	59	655
Dunbar	23	49	30	39	36	48	45	49	64	55	55	54	48	572
Abbotsinch	5	109	74	86	50	67	60	62	85	111	115	110	106	1,035
Prestwick	16	89	56	71	45	56	59	69	91	110	109	98	89	942
Edinburgh (Bot. Gardens)	26	54	40	47	39	49	50	59	63	66	63	56	52	638
Dumfries	49	110	76	81	53	72	63	71	93	104	117	100	107	1,047
Penicuik	185	83	56	75	52	64	62	68	82	83	90	87	76	878
Eskdalemuir	242	166	110	137	83	95	94	98	125	149	164	152	165	1,538

the driest month in most places. A particular feature of the seasonal distribution of rainfall across the region is that, as a general rule, the amplitude of variation between wettest and driest months decreases from west to east (Table 9.4). Thus differences in rainfall between western and eastern coasts are greatest in autumn and winter when westerly airflow is more frequent and least in spring when it is less common.

Not all rainfall is cyclonic and frontal in origin, and localised convection can generate some remarkable falls. The latter can be triggered by the uplands and can occur at almost any time of year, particularly in unstable polar airstreams and as very active cold fronts move south-eastwards across the region. Thunderstorms in warm air travelling northwards from continental Europe occasionally reach the south-west of Scotland, where they tend to move inland from the Solway coast along the line of the river valleys. Gauld (1922) observed that 'thunderclouds are often seen travelling up the Dee from seawards to the vicinity of Loch Ken and then west into the hill country, later returning across the Ken valley and down the water of Urr to the Castle Douglas plain'. Further north, in the Forth valley, thunderstorms are relatively rare, there being usually fewer than eight per year, but an analysis of published accounts of exceptional storms during the period 1870–1919 (Harrison 1980) revealed a very clear late-summer maximum. These summer storms can produce some remarkable falls, none more so than that described by Morris (1899). A heavy thunderstorm occurred to the west of Stirling in August 1897, shortly after which

> one of the shepherds walking on the hill about five in the afternoon saw what he supposed was a sheep lying on its back. . . . He thought that the sheep might have been killed by lightning but on approaching he discovered that it was a block of ice . . . he estimated it to weigh about a hundredweight (51 kg).
>
> (Morris 1899)

Clearly, such a large lump of ice had not fallen from the sky, but the possibility is that there was a sufficient depth of hail on the steeply sloping ground for a roller to form (Harrison 1982).

Cyclonic mechanisms provide the basic canvas on which the distribution of rainfall is painted while the effect of topography provides the detailed brush-strokes of the smaller, and more complex, variations across the landscape. Orographic uplift over upland areas increases the intensity and duration of precipitation, which is demonstrated in the relatively higher mean annual fall at Eskdalemuir (Table 9.4). The rate at which annual rainfall increases with altitude tends to be steepest in the west of Scotland and is considerably less steep in the rain shadow area to the east. Typical rates of increase are 2.5 mm/m in the former and 1.0 mm/m in the latter (Weston and Roy 1994). Halstead (1958) derived the average elevation of fixed isohyets, which indicated gradients of 2.07 mm/m for the Southern Uplands, but he did not differentiate between western and eastern slopes, while Harrison (1986) derived values of 1.18 mm/m for paired rain-gauges on a north-east-facing slope above the Forth valley. On this basis, the mean annual rainfall for the highest parts of the Southern Uplands, which are the headwaters for several steeply graded rivers, can be assumed to be of the order of 3,000 mm. In the shorter term, the altitudinal increase is dependent on the weather system within which orographic enhancement is taking place (Smithson 1969), and there is a well-marked seasonal variation with maximum steepness occurring during the winter half-year. The value of altitudinal gradients is, however, dependent upon the absolute amount of rainfall, but once this variance has been removed, the underlying seasonality of monthly rainfall gradients above the Forth valley (Harrison 1986) varies from 0.68 mm/10 m of altitude in February to 1.26 mm/10 m in October.

The rain shadow effect of the Southern Uplands and the hills of central Scotland is in evidence in the geographical distribution of rainfall. The former enhances the west-to-east decrease in rainfall, the east coast being subject to occasional water shortages, and also contributes to the low rainfall around Ayr, which is sheltered from the south and south-west by the hills of Galloway and Carrick. In the Forth valley the uplands to the west result in a steep rate of decrease in annual rainfall over short eastwards distances (Harrison 1993a).

Whilst 24-hour rainfalls in excess of 100 mm are not infrequent in the Highlands to the north and Cumbria to the south (Reynolds 1985), these are relatively rare in the southern half of Scotland, where the elevation is lower. However, there were notable occurrences such as the 150 mm which fell in parts of the Tweed basin on 12 December 1948, bringing flooding to several Borders rivers (Learmonth 1950). This destroyed seven mainline railway bridges and covered farmland with gravel deposits. Flooding is a particular problem in the south of Scotland (Sargent 1989) as a high proportion of rainfall goes to short-term runoff (Learmonth 1950; Werritty and Acreman 1985). To some extent this can be attributed to steeply graded catchments on the northern side of the Southern Uplands, but much of the flooding problem has been generated by flood plain development. In Haddington (Lothians), a long history of damaging floods on the Tyne has led to the installation of a fully automated flood warning scheme by the Forth River Purification Board, whilst in Dumfries, on the river Nith, engineering solutions to flood protection have not been adopted because of local amenity considerations (Werritty and Acreman 1985). Another example of flooding is discussed in Box 9.2.

Heavy rainstorms in rural areas, apart from causing flooding, contribute to a growing soil erosion problem throughout the region, particularly where the soil has been cultivated (Davidson and Harrison 1995). Although it was not under cultivation, there

Box 9.2

THE STRATHCLYDE FLOODS OF DECEMBER 1994

Urbanisation of catchments involves not only the waterproofing of the ground surface but also the entrainment of flow in culverts, some of which may be unable to accommodate extremes of discharge. The floods which affected parts of the Strathclyde region in December 1994 caused damage approaching £100 million. Rain entered the region by 2200 hours on Friday 9 December as a warm front approached from the south-west (Figure 9.4). By 0600 hours on the 10th, heavy, continuous rain was falling in a freshening south-westerly wind, which persisted unabated until the early hours of the 12th. The rain was orographically enhanced and there was additional topographical convergence over the upper reaches of the Firth of Clyde (M.G. Roy, personal communication). Picketlaw, in the White Cart catchment, registered a 24-hour rainfall of 100.8 mm on the 10th with a further 57 mm on the 11th, and the intensity approached 8 mm/hour towards noon on the 10th (Figure 9.5) (Black and Bennett 1995). The band of rain remained almost stationary over Strathclyde and was sustained by a supply of warm and very moist air from the south-west. The rain shadow effect of the hills to the north of Glasgow was very marked, with the Forth valley receiving 50 per cent less rainfall. The rain fell on an area with a long history of flooding (e.g. Sweeney 1977), and flood warnings were issued by Strathclyde police late on the 10th. By the early hours of the 11th, the Cart, Kelvin and Irvine had overflowed their banks and there was widespread flooding in Kirkintilloch, Irvine, Johnstone and Beith. Many residents were forced to seek temporary accommodation and more than 600 were homeless for at least two nights. Parts of the tunnel carrying ScotRail's Argyle Line beneath Glasgow contained water to a depth of more than 3 m and an early-morning commuter train at Glasgow Central Low Level had to be abandoned as the flood waters rose.

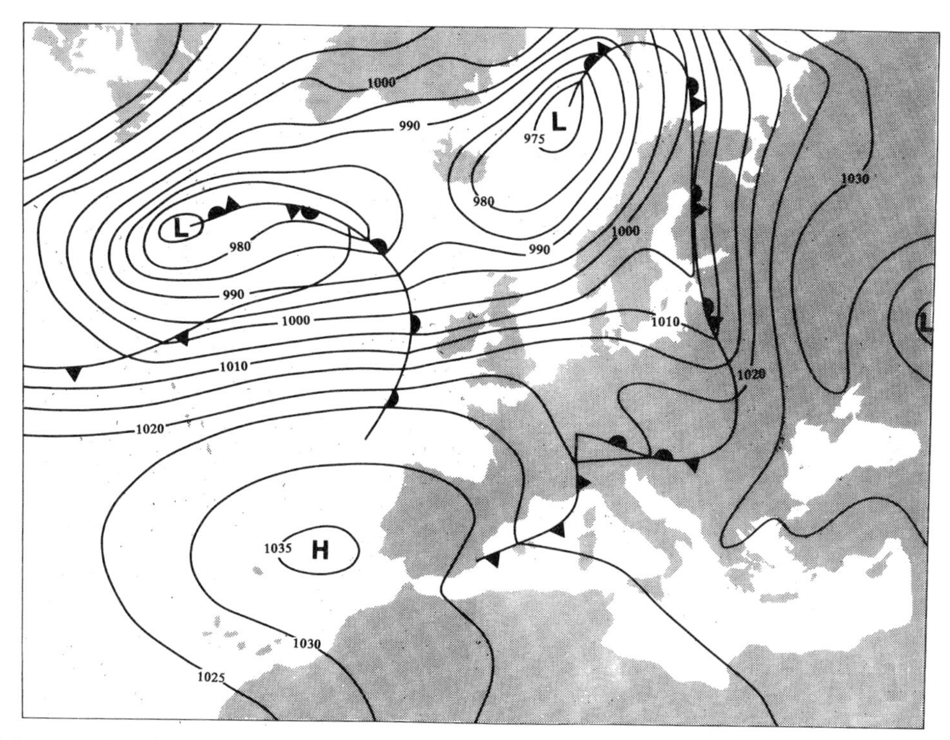

Figure 9.4 Synoptic chart for 1200 GMT on 10 December 1994

was a massive slope failure above Menstrie, to the north-east of Stirling, in November 1984. Over the 24 hours to 0900 GMT on 4 November 68.3 mm was registered at nearby Parkhead climatological station as an occluded front lingered over Scotland and rainfall intensities reached a maximum of 9 mm per hour. At Menstrie, on the south-facing slope of the Ochil Hills, a resulting slope failure released an estimated 350 m^3 of debris downslope which inundated domestic properties (Jenkins *et al.* 1988).

In contrast, in parts of the eastern coastal plain, annual potential evapotranspiration exceeds 450 mm, approximately 75 per cent of the mean annual rainfall, which means that summer droughts are a real possibility. During the months June to October in the years 1971 to 1973 severe restrictions had to be placed on the use of water in Fife after annual rainfall fell to less than 75 per cent of average (Tricker and Miller 1977).

Snowfall

The geographical distribution of average snow depth and the frequency of days with snow lying are determined by a complex interaction between local topography and distance inland from the coast, and the direction of movement of snow-bearing weather systems. Coastal areas record, on average, fewer than six days per year with snow lying at 0900 GMT, but snow is least frequent along the extreme south-west coast of the Dumfries and Galloway region. Sutherland (1925) observed an almost total absence of snow along the shoreline, and the Reverend Nicholson, parish minister of Whithorn in 1839, commented that 'frost and snow have been of such rare occurrence and such short continuance, that the children in Whithorn run the risk of becoming as incredulous about the effect of cold upon water as the Emperor of China' (quoted in Dick 1916). Snow rarely accumulates to any great

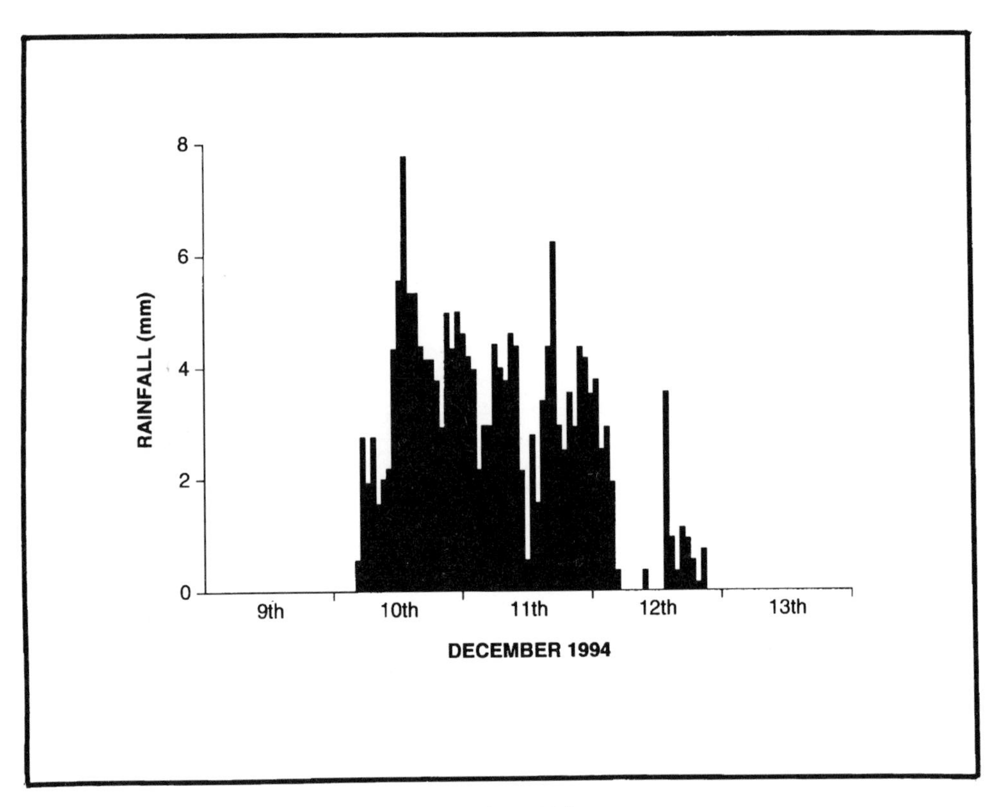

Figure 9.5 Hourly rainfall (mm) at Picketlaw, 9–13 December 1994

depth in this area (Figure 9.6), and Lowndes (1971) has recorded that Prestwick, on the Ayrshire coast, received fewer substantial snowfalls than the Isles of Scilly over the period 1954–69.

Snow is more common from Arctic continental and polar continental air arriving from the east over the North Sea. Under such conditions air temperatures can be very low and snow can accumulate on ground surfaces down to sea level, often on beaches down to the water's edge. Accumulations may be substantial along the east coast, and Fife suffers particularly badly from drifting snow in strong easterly winds, while the west coast may receive very little. Snow also falls in cold Arctic maritime and polar maritime air from between south-west and north, when it may be associated with fronts crossing slowly eastwards across Scotland, but also in the subsequent and cooler north-westerly or northerly airstreams. The intensity of fall can be exceptionally high, leading to rapid accumulations on higher ground whilst the lowlands may escape with sleet. In this case greatest falls usually occur in the west whilst in areas sheltered from the prevailing westerly winds, such as the lee of hills and the east coast, falls may be relatively small.

At the lowest altitudes, snowfall tends to be restricted to the months November to April with January being, on average, the snowiest month. Snow does fall at other times of the year but rarely lies on the ground. Because of the large number of factors that determine whether snow lies after it has fallen, it is perhaps not surprising that the apparent differences in snow-lie between the north-east and the south-west coasts are not always statistically significant. However, during January the generally lower air temperature along the east coast tends to

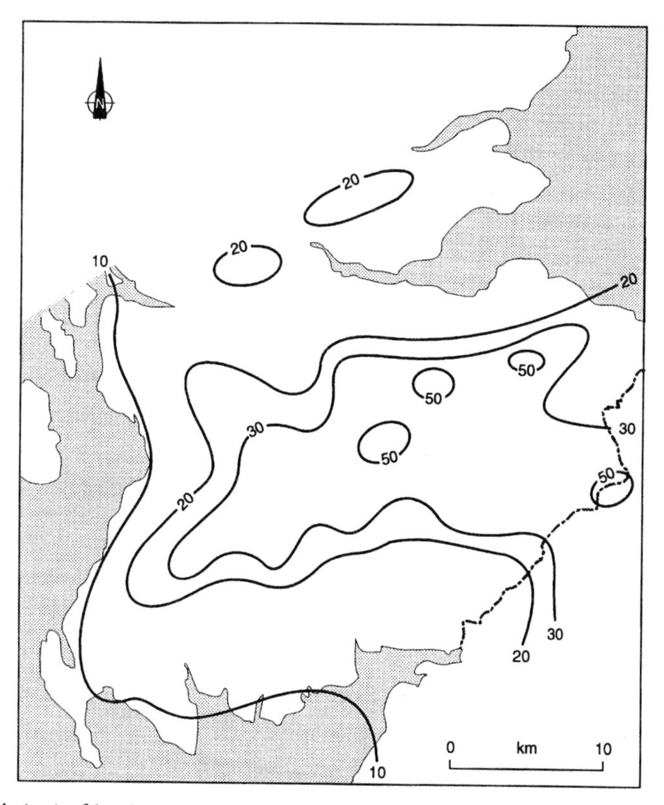

Figure 9.6 Isolines of depth (cm) of level snow reached or exceeded in ten winters per century

result in a high frequency of days when snow lies on the ground in this part of the region even when compared with much more elevated sites in the west. Figure 9.7 shows that if altitude is taken into consideration, east-coast Leuchars, at 10 m ASL, has a January snow cover not much less than that of Eskdalemuir, over 200 m higher at 242 m ASL. The corresponding difference between west-coast Tiree (January mean 2.3 days of snow cover) and high-level Balmoral (15.2 days) in the Highland region (Figure 10.11) is much greater. The snow season lengthens as altitude increases, and at Leadhills snow frequently lies during both October and May. Above 700 m in the Southern Uplands there may be in excess of 100 days with snow lying during the cold 'easterly' type of winter, but fewer than fifty during the milder 'westerly' type (Harrison 1993b).

Because of the effect of topographic exposure and drifting, patterns of snow accumulation are spatially very complex and there is inevitably a large amount of scatter in the relationship between duration and depth of snow cover and altitude. However, Manley (1971) estimated that the duration of snow cover increases by one day for every 5 to 8 m increase in altitude, and in the Southern Uplands estimated rates of change based on Leadhills data appear to lie within this range, with the gradients being steeper on the more western hills. Snow depth increases with altitude as precipitation intensity and duration increase and air temperatures decrease, which is indicated in Figure 9.6. However, higher wind speeds in hill areas means that drifting is an increasingly important factor, the more exposed sites being swept clear of snow. In contrast, in the more sheltered sites, accumulations may be substantial with deep drifts several metres deep. Pearson (1975), for

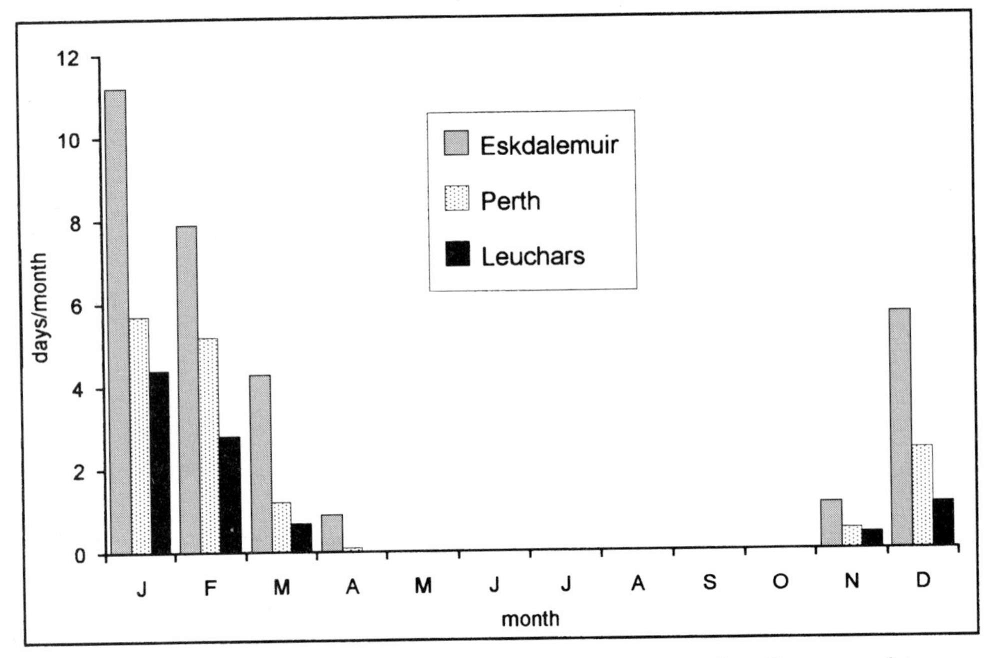

Figure 9.7 Mean monthly frequency of days with snow lying and covering more than 50 per cent of the ground at 0900 GMT for the period 1972–92 (Perth 1951–79)

example, has described a snowstorm in southern Scotland in March 1827 during which accumulations in excess of 150 cm occurred between Moffat and Dumfries, with deeper drifts in places. More recently, heavy snow in January 1978 (Perry and Symons 1980), January 1984 (Perry *et al.* 1984) and February 1996 have resulted in the closure of road and rail links between Scotland and England. During the last of those events the snow, which fell as an occluded front moved slowly eastwards across the British Isles, left more than 1,000 vehicles stranded on the A/M74 and a state of emergency was declared in the Dumfries and Galloway region.

WIND

Because of the nature of the interaction between airflow and spatially complex terrestrial surfaces, wind speed and wind direction are probably the most variable descriptors of climate. Point measurements are subject to local site exposure and it is unwise to read too much into the limited number of available wind observations. However, in a mid-latitude maritime climate it is not surprising that the dominant wind direction is broadly westerly or south-westerly in southern Scotland. The strongest, and most persistent, winds are, therefore, experienced along the west coast and on hill slopes with open exposure to the west. The coast of south-west Scotland is, however, afforded some degree of shelter from Ireland and the Kintyre peninsula, and there is a restricted fetch over the North Channel and the Firth of Clyde. Thus Prestwick, for example, may experience only five or six gales per year whilst Tiree, 120 km to the north, is fully exposed to Atlantic storms and experiences on average more than thirty gales per year. The strength and persistence of the westerly winds can, however, be seen in the very marked deformation of tree crowns along the west coast (Sutherland 1926). The low-lying nature of much of the Central Valley of Scotland also leaves it open to stormy westerly conditions – as the events of January 1968 testify (Box 9.3).

Box 9.3

THE GLASGOW STORM OF JANUARY 1968

There have been several notable westerly winter storms in the west of Scotland. On 28 January 1927 wind speeds reached 87 knots in Paisley and there was widespread damage in the Glasgow area. On 31 January 1953 the ferry *Princess Victoria* foundered in heavy seas after leaving Stranraer with the loss of 133 lives (Holford 1976). However, the most notable was what has become to be known as 'the Glasgow Storm', which hit central Scotland in the early hours of January 1968. An Atlantic depression deepened as it moved eastwards, and by midnight on the 14th its central pressure was 957 mb over Ross and Cromarty (Figure 9.8). At Glasgow Airport the storm-force WSW winds averaged 52 knots with gusts in excess of 87 knots, reaching their strongest between 0200 and 0300 hours. By 0600 the centre of the depression had moved out over the North Sea and the winds slackened and veered to a blustery north-westerly. Over a little less than eighteen hours, the centre of the depression had moved eastwards at a remarkable speed in excess of 45 knots. In the few short hours that the storm lasted, more than 340,000 buildings were damaged and there was widespread windthrow in the woodlands. The overall cost has been estimated at £30 million (Perry 1981). Much of the damage was caused by the funnelling of the wind into the Clyde valley, by tornadoes set off by the storm (Holford 1976), and by mechanical turbulence created by buildings and local topography.

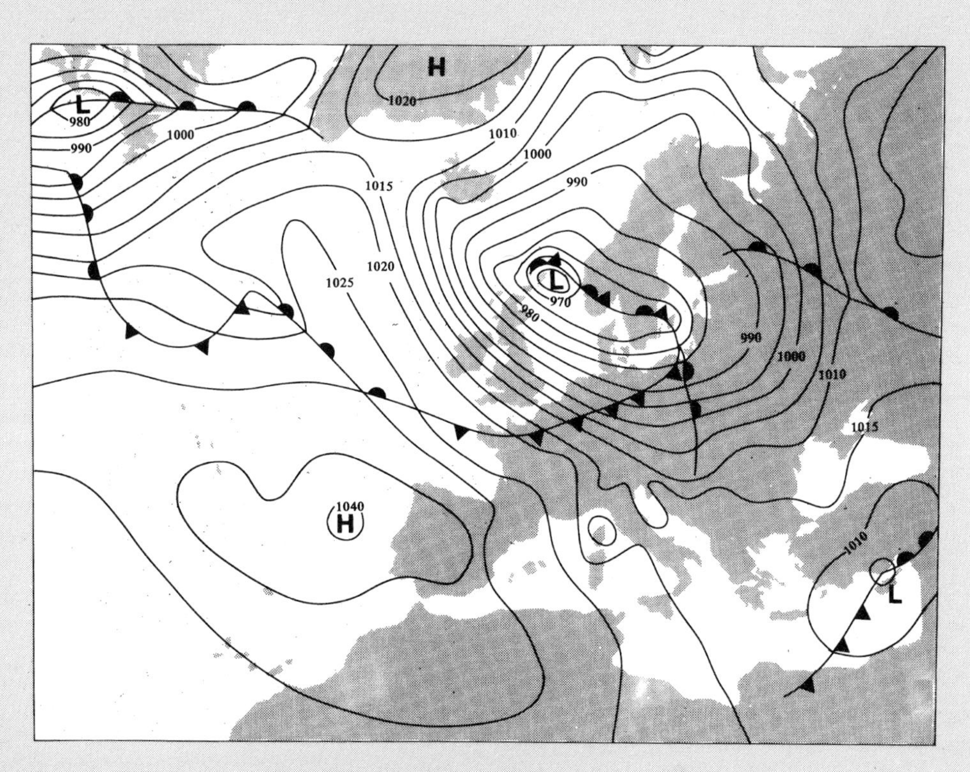

Figure 9.8 Synoptic chart for 1200 GMT on 14 January 1968

The penetration of westerly winds across Scotland is controlled, to a large extent, by topography. The Forth–Clyde lowlands provide ready access eastwards, and lateral topographic constriction can cause local acceleration of flow. The Tay Bridge disaster of December 1879 was due, in part, to a moderate westerly gale blowing down the Tay estuary, but has been attributed to structural deficiencies in the bridge. Leuchars, which has a moderately open exposure, is affected by westerly gales, whilst Dundee, only 12 km to the north, is afforded shelter from the west and experiences relatively few gales (Table 9.5). During the 'Burns' Day' storm of 25 January 1990 (McCallum 1990), gusts in the strong south-westerly wind exceeded 68 knots in south-west Scotland but failed to reach 49 knots in the more sheltered eastern location of Dunbar. The funnelling effect of topography is very much in evidence in the Forth valley, where there is a marked dominance of winds along the south-west to north-east axis (Harrison 1987).

Easterly winds occasionally reach gale force, especially when there is a strong anticyclone over Scandinavia. These are relatively infrequent but can be associated with severe wind-chill along the east coast, especially in late winter and spring, when the North Sea is at its coldest. Robert Louis Stevenson referred to the climate of Edinburgh in spring as 'downright meteorological purgatory' (Paton 1951). Easterly gales tend to blow inland up the Forth and Tay estuaries but have usually lost much of their energy by the time they reach the west coast.

Average wind speeds exhibit a well-marked seasonality with the highest speeds occurring between November and March when cyclonic activity is at its most intense, and the lowest in August (Smith 1983). The number of days with gusts in excess of 40 and 50 knots (Table 9.6) at anemometer stations throughout southern Scotland also indicates this strong seasonality (Meteorological Office 1968). This concentration of stronger winds into the winter season is of course a common theme

Table 9.5 Mean seasonal frequency of gale days for the period 1981–90

Location	*Alt. (m ASL)*	*Jan.*	*Feb.*	*Mar.*	*Apr.*	*May*	*June*	*July*	*Aug.*	*Sept.*	*Oct.*	*Nov.*	*Dec.*	*Year*
Dundee	45	0.2	0.1	0.5	0.0	0.0	0.0	0.0	0.0	0.1	0.1	0.2	0.1	1.3
Leuchars	10	1.6	1.0	0.9	0.3	0.1	0.0	0.1	0.0	0.3	0.5	0.6	0.7	6.1
Leadhills	388	4.6	3.5	3.5	0.8	1.0	0.3	0.2	1.1	2.1	2.6	2.8	2.6	25.1
Prestwick	16	1.6	0.9	0.7	0.0	0.1	0.0	0.0	0.0	0.2	0.4	0.5	0.4	4.8
Dumfries	49	1.3	0.4	0.7	0.1	0.1	0.0	0.1	0.0	0.2	0.9	0.9	0.9	5.7

Table 9.6 Mean frequency of days with gusts exceeding fixed wind speed thresholds for the period 1950–59 (after Meteorological Office 1968)

Location	*Jan.*	*Feb.*	*Mar.*	*Apr.*	*May*	*June*	*July*	*Aug.*	*Sept.*	*Oct.*	*Nov.*	*Dec.*	*Year*
Exceeding 40 knots													
Leuchars	3.9	3.0	1.5	1.2	1.0	0.7	0.5	1.3	1.2	2.3	3.2	5.2	25.0
Prestwick	5.4	2.9	1.2	1.2	0.9	0.5	0.4	1.0	2.9	2.6	3.5	7.8	30.3
Eskdalemuir	5.8	3.8	2.7	3.6	3.2	1.7	0.8	1.4	3.7	4.6	4.4	9.0	44.7
Exceeding 50 knots													
Leuchars	0.7	0.7	0.1	0.3	0.0	0.0	0.0	0.1	0.1	0.4	0.6	1.3	4.3
Prestwick	1.7	0.6	0.1	0.3	0.1	0.0	0.0	0.2	0.6	0.2	1.0	2.0	6.8
Eskdalemuir	2.1	0.9	0.6	0.6	0.3	0.1	0.1	0.7	0.8	0.5	1.4	2.7	10.8

through British climates and reflects the intensification of mid-latitude circulations at that time of the year.

There is a general increase in wind speed with altitude and consequent distance from the frictional effects of the land surface, although the air can be relatively calm in a well-sheltered upland hollow while blowing strongly at sea level over an exposed coast. Measurements of wind speed made over several years by students of the University of Stirling during early afternoons on the south-facing slope of the Ochil Hills have established that the effect of altitude on wind speed is highly variable (Harrison 1993a). The long-term average difference between altitudes of 30 and 250 m was measured at 4.2 m/s (8 knots), which is equivalent to a rate of increase of the order of 2 m/s (4 knots) per 100 m rise. Smith (1985) refers to Lowther Hill (754 m), for which wind speed records were published in the *Monthly Weather Report* up to December 1991, as being 'amongst the windiest places in Britain', and the frequency of gales at Leadhills is more than four times greater than coastal lowland sites (Table 9.5). A further problem in hill areas is their effect on airflow over adjacent lower-lying areas. For example, in the Forth valley to the east of Stirling, the Ochil Hills can create turbulent downdraughts which cause minor damage to structures such as garden sheds and greenhouses. The Ochils, and the Gargunnock Hills to the south of the valley, also generate lee waves in appropriate stability conditions (Fyfe 1953), which not only create ripple effects (Figure 9.9) and lenticular cloudforms but also provide good soaring conditions for birds over the centre of the Forth valley.

Strong winds on higher ground, and along exposed coasts, have everyday consequences and cause operational problems for electrified railway lines due to an increase in the risk of fouling of overhead wires by pantographs. Wind sensors and alarm systems have, for example, been installed at Carstairs, Elvanfoot and Cranberry on the West Coast Main Line as it passes through the Southern Uplands (Crawford 1989). The problem at Saltcoats, on the Ayrshire coast, is heavy sea-spray being driven inland during westerly gales.

Figure 9.9 Rippling effect in clouds over the Forth valley created by air flowing over the Ochil Hills. Photograph taken in July 1983 looking westwards from the University of Stirling. Winds, moderate north-easterly
Photo by courtesty of Mr J. McArthur, University of Stirling

Wind speeds, particularly at lowland sites during the summer months, exhibit diurnal variation, with highest speeds in the early afternoon and lowest in the post-dawn period. This is often accompanied by directional changes associated with the development of land and sea breezes. The latter usually develop during the morning, reach a maximum strength in the early afternoon and die away by early evening. Weak and intermittent sea breezes occur in the Clyde estuary but these develop more strongly along the Ayrshire coast. Smith (1986) described the formation of a sea breeze along this coast in June 1985. As the breeze arrived at Prestwick, wind direction changed from east-south-east to west with an attendant increase in speed of 12 knots. The air temperature fell from 25 to 20°C and the relative humidity increased from 34 to 68 per cent. Over the west Galloway coast the breeze tends to blow from the north-west and its dying away in the evening is expressed in a local proverb: 'An honest man and the north-west wind [sea-breeze] go to sleep together' (quoted in Gauld 1922).

On the east coast, sea breezes blow from between south-east and north-east, depending on local sea and coastal conditions (Lamb 1943), and can penetrate considerable distances up the Forth and Tay estuaries. In the Forth valley, on warm and relatively calm days from May through to September, a sea breeze front may reach Stirling by noon, its arrival being marked by a freshening of the wind and a distinct drop in temperature in the cooler North Sea air. This is particularly welcome during hot summers such as that of 1995, when the arrival of the sea breeze can significantly suppress maximum daytime temperatures in the same manner as they do in north-east England.

As a sea breeze dies away during the evening it may be replaced by an offshore land breeze which is strengthened by the katabatic drainage of cold air from inland. This nocturnal breeze is considerably less vigorous than the sea breeze but is readily detected in, for example, changes in the direction of smoke plumes from sources near to the ground such as bonfires. A particular feature of the Forth valley is an interaction between coastal and slope wind systems. A marked mountain–valley wind circulation develops between the Highlands to the west and the Forth estuary, which reinforces the onshore sea breeze during the afternoon and the land breezes during the early morning, although no specific research has been undertaken on the relative strengths of the interacting components of flow.

SUNSHINE

Although higher latitude results in longer day length during the summer months, the greater cloudiness of the Scottish climate means that average daily hours of bright sunshine are generally lower than for southern England, for example. A comparison between the Forth valley and the Isle of Wight (Harrison 1993a) reveals that this may amount to an average 1.5 hours per day in high summer. Greater cloudiness on the west coast than on the east means that these areas receive as much as 24 per cent fewer hours of bright sunshine during the winter months, but the presence of east-coast haar tends to offset this difference during May and June (Table 9.7). Weather patterns, however, exert a considerable influence on the geographical distribution of bright sunshine. The summer of 1995 provides a very clear indication of the importance of surface pressure patterns. High pressure lay to the west of the British Isles during June and August, and cloud amounts were lowest along the west coast. The respective total hours of bright sunshine for these months were 247 and 242 at Leuchars (east coast) and 280 and 254 for Abbotsinch (west coast). In contrast, high pressure lay south of the British Isles in July, which favoured the east coast, and monthly totals for Leuchars and Abbotsinch were 212 and 200 respectively.

Away from the coast, local topography exerts a strong influence on the amount of solar radiation received. During the winter months, when the solar elevation can be less than 10 degrees, deep glens and north-facing slopes can be in shade for lengthy periods. At higher altitudes, increased cloudiness reduces the duration of bright sunshine, indicated in the average values for Eskdalemuir (Table 9.7). The rate at which average annual hours of bright

Table 9.7 Mean monthly and annual sunshine totals (hours) for the period 1961–90

Location	*Alt. (m ASL)*	*Jan.*	*Feb.*	*Mar.*	*Apr.*	*May*	*June*	*July*	*Aug.*	*Sept.*	*Oct.*	*Nov.*	*Dec.*	*Year*
Leuchars	10	58	75	118	155	185	190	184	164	126	100	76	50	1,481
Dunbar	23	56	78	117	153	186	192	190	169	129	104	72	46	1,492
Abbotsinch	5	39	64	97	147	182	182	170	148	110	82	54	34	1,309
Prestwick	16	47	76	99	161	202	196	178	159	114	88	60	41	1,421
Dumfries	49	45	69	98	145	180	177	162	157	111	90	62	38	1,334
Eskdalemuir	242	39	63	86	132	161	158	151	135	95	74	56	39	1,188

sunshine decrease with altitude in Scotland has been estimated by Gloyne (1966) to be 58 hours per 100 m.

VISIBILITY AND FOG

A high frequency of polar maritime air tends to make visibility throughout Lowland (and Highland) Scotland greater than, for example, in south-east England. However, moist airstreams from the south-west tend to bring a low cloud base, particularly in the west of Scotland, which means that visibility in upland areas can be quite poor. This is indicated in the higher frequency of fogs at Leadhills (Figure 9.10), which approaches an average of twenty per year. The Scots weather saying 'When the mist taks the hills guid weather spills' refers to a low cloud base being indicative of an advancing warm front with its rain. However, the second part of this saying continues, 'When the mist taks the howes, guid weather grows', referring to a rise in pressure with more stable air bringing fogs to low-lying ground. The frequency of fogs over low ground varies very markedly over the region, from more than eight days per year along the coasts of East Lothian and the Solway Firth to a virtual absence at Prestwick on the west coast, this being a major factor in the decision to site a major international airport there.

Fogs on low ground fall into two broad types, radiation fogs and sea fogs. The katabatic drainage of cold air to low-lying ground can result in the air reaching its dew point, following which fogs begin to develop from the ground upwards, particularly over moist ground. In the Forth valley, for example, the Wallace Monument may stand, sentinel-like, above the sea of fog which surrounds it (Figure 9.11). Long nocturnal hours of cooling in damp Scottish glens during the winter months leads to the formation of these valley fogs, which may develop to depths in excess of 100 m.

Nearer coasts, fogs are more frequently sea fogs, which affect both the west and the east coast of

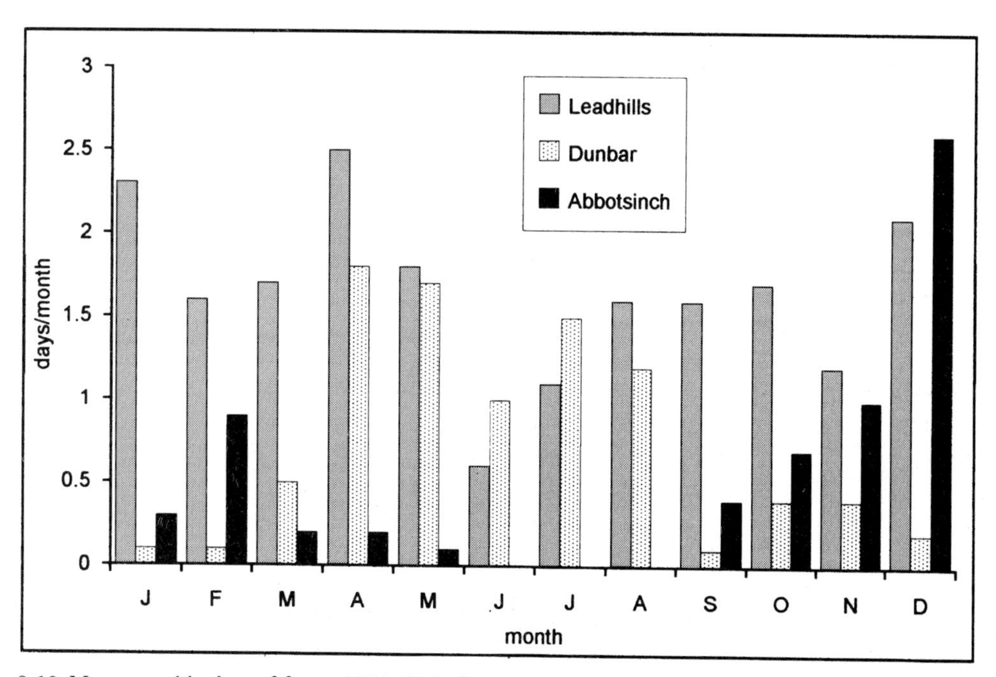

Figure 9.10 Mean monthly days of fog at 0900 GMT for the period 1981–90

Box 9.4

THE EAST COAST HAAR

The east-coast sea fog, or *haar*, which makes the seas around the Firth of Forth so foggy (see also Box 7.3 for an example from north-east England), occurs as a result of warm-air advection over a cold sea and, if associated with an easterly airflow, can penetrate considerable distances inland, particularly through the Forth–Clyde lowlands. Paton (1951) described the haar as 'the most deplorable characteristic of east coast weather'. As the haar moves inland, near-surface mechanical and thermal turbulence may cause it to lift from the surface to become a sheet of stratus cloud which envelops the tops of taller buildings and higher ground. At the coast, the landward limit of the haar may be marked by the water's edge, thus introducing a tidal influence such as that observed by Alexander (1964) in the Eden estuary north of St Andrews in Fife.

A detailed analysis of visibility observations from Inchkeith lighthouse in the Forth estuary (Harrison and Phizacklea 1985) revealed close associations between fog occurrence and both wind direction and state of tide. During the summer months fogs were associated almost exclusively with a north-easterly airflow, which drives the haar towards the coast. However, during the winter months airflow was most frequently south-westerly. In this case fogs were primarily radiation fogs formed over the coastal plain which were drifting out to sea in a land breeze or, in some cases, were sea-smoke caused by very cold air advecting over much warmer coastal waters. At the 1000 hours observation, fogs were observed most frequently on a falling tide, taking fog offshore, while at 1600 hours fogs occurred most frequently on an incoming tide, driving the haar towards the coast.

Scotland (see Box 9.4), although the height of the observation above sea level has a considerable bearing on recorded fog frequencies. Fogs observed at the top of cliffs and hills may be due to low cloud rather than sea fog. Leadhills at 388 m in the Southern Uplands reveals this tendency throughout the year (Figure 9.10). Whithorn also has less well marked seasonal variation in fog occurrence as it not only experiences radiation fogs but also is subject to sea fogs during spring and summer. Abbotsinch, where radiation fogs dominate, has a marked maximum in December with scarcely any such activity during the summer months. The seasonality of sea fogs or haars (see Box 9.4) is, however, most marked on the east coast, where Dunbar, for example, has a very distinct spring and summer maximum. Dixon (1939), in his analysis of sea fogs around the Scottish coast, calculated average annual hours of fog for sea areas. The Firth of Forth with 443 hours was the foggiest, with the south-west coast experiencing only 129 hours. The Firth of Clyde was relatively less prone to fog, with eighty-nine hours, whilst the Ayrshire coast experienced less than fifty hours.

Significant long-term reduction in visibility due to the presence of particulates in the atmosphere has been restricted to the major urban areas of Edinburgh and Glasgow. Edinburgh's old image as 'Auld Reekie' owes much to the entrainment of smoke and other emissions from the many breweries into the low stratus of the haar. The heavy industrial base in Glasgow and Clydeside produced significant levels of particulate pollution. The general murkiness of the air over the city was such that Halstead (1958) was led to comment, 'the clarity of the atmosphere during Glasgow Fair fortnight (industrial holidays) is very striking'. In recent years, with the closure of much of the heavy industry, visibility has improved, and there was a 93 per cent improvement in median visibility at Glasgow Airport

Figure 9.11 Valley fog surrounding the Wallace Monument near Stirling. Photograph taken at 0630 GMT in January 1984 after heavy snow. View looking eastwards
Photo by courtesy of the Department of Environmental Science, University of Stirling

between the 1960s and the early 1980s (Harris and Smith 1982; Lee 1985). However, as particulate concentrations have decreased, the improved transparency of the atmosphere has led to increases in photochemical pollution, which is related to the increasing volume of traffic flows in central Scotland (Sweeney 1981).

CONCLUSION

The very many travellers who pass through the region, either northwards towards the Highlands or westwards towards the ferry at Stranraer, will undoubtedly experience a wide range of climatic conditions, from the mildness of the lowlands around the Solway Firth to the generally cooler, wetter and windier Southern Uplands, and from the comparative shelter of the glens to the raw exposure along the North Sea coast. No description of the climate within a region can ever hope to embrace all the subtle, but no less significant, local variations, and will be limited inevitably by the availability of appropriate data. In the longer term, climates are, of course, constantly in a state of flux, so any description must be seen as nothing more than a snapshot at some fixed point in time. Of particular relevance in the context of the Scottish climate is the observed increase in westerly weather types since the mid-1970s (Mayes 1991), which will have far-reaching consequences for the climate of southern Scotland. Rainfall in western Scotland, particularly during the winter half-year, has been increasing over the past two decades (Smith 1995), which has already resulted in measurable increases in the annual mean discharges of rivers such as the Clyde, Nith and Teviot (Smith and Bennett 1994). Changes in the character of the Scottish winter will lead to very different contrasts between west and east coasts, not only in terms of rainfall, but also in terms of the nature and frequency of snowfall, average hours of bright sunshine, the incidence of frost and the frequency of strong winds.

REFERENCES

Alexander, L.L. (1964) 'Tidal effects on the dissipation of haar', *Meteorol. Mag.*, 93: 379–380.

Black, A.R. and Bennett, A.M. (1995) 'Regional flooding in Strathclyde December 1994', in *Hydrological Data UK 1994 Yearbook*, Wallingford: Institute of Hydrology.

Chaplain, R. (1985) 'Weather Watchers Network in Scotland', *J. Meteorology (UK)*, 10: 91–92.

Crawford, C.L. (1989) 'The effect of severe weather on the Scottish rail system', in S.J. Harrison and K. Smith (eds) *Weather Sensitivity and Services in Scotland*, Edinburgh: Scottish Academic Press.

Davidson, D.A. and Harrison, D.J. (1995) 'Water erosion on arable land in Scotland: results of an erosion survey', *Soil Use and Management*, 11: 63–68.

Dick, C.H. (1916) *Highways and Byways in Galloway and Carrick*, Wigtown: GC Books.

Dixon, F.E. (1939) 'Fog on the mainland and coasts of Scotland', *Meteorological Office Prof. Notes no. 88*, London: HMSO.

Fyfe, A.J. (1953) 'Lee waves of the Ochil Hills', *Weather*, 7: 137–139.

Gauld, W. (1922) 'Galloway: an introductory study', *Scot. Geogr Mag.*, 38: 22–39.

Gloyne, R.W. (1966) 'Some features of the climate of the immediate area served by the Forth and Tay Road Bridges', *Scot. Geogr. Mag.*, 82: 110–118.

Halstead, C.A. (1956) 'Climatic observations in Scotland', *Scot. Geogr. Mag.*, 72: 21–23.

—— (1958) 'The climate of the Glasgow region', in R. Miller and J. Tivy (eds) *The Glasgow Region*, London: British Association for the Advancement of Science.

Harding, R.J. (1978) 'The variation of the altitudinal gradient of temperature within the British Isles', *Geogr. Annlr*, 60A: 43–49.

Harris, B.D. and Smith, K. (1982) 'Cleaner air improves visibility in Glasgow', *Geography*, 67: 137–142.

Harrison, S.J. (1973) 'An ecoclimatic gradient in north Cardiganshire', unpublished PhD thesis, University of Wales.

—— (1980) 'Rainfall in the Stirling area', *Forth Naturalist and Historian*, 5: 23–34.

—— (1982) 'A nineteenth century hail roller?', *J. Meteorology (UK)*, 7: 77–78.

—— (1986) 'Spatial and temporal variation in the precipitation–elevation relationship in the maritime uplands of Scotland', in M. Erpicum (ed.) *Topoclimatology and Its Applications*, Liège: University of Liège Press.

—— (1987) 'Climatic conditions over the estuary and Firth of Forth', *Proceedings of the Royal Society of Edinburgh*, 93B: 245–258.

—— (1990) 'The frequency of lower atmosphere temperature inversions in the Stirling area', *Report CHU/01/90*, University of Stirling.

—— (1993a) 'Climate', in L. Corbett (ed.) *Central Scotland*, Stirling: Forth Naturalist and Historian.

—— (1993b) 'Differences in the duration of snow cover on Scottish ski slopes between mild and cold winters', *Scot. Geogr. Mag.*, 109: 37–44.

—— (1994) 'Air temperatures in the Ochil Hills, Scotland: problems with paired stations', *Weather*, 49: 209–215.

Harrison, S.J. and Harrison, D.J. (1988) 'The effect of elevation on the climatically determined growing season in the Ochil Hills', *Scot. Geogr. Mag.*, 104: 108–115.

Harrison, S.J. and Phizacklea, A.P. (1985) 'Tide and the climatology of fog occurrence in the Forth estuary', *Scot. Geogr. Mag.*, 101: 28–36.

Harrison, S.J. and Wallace, R.W. (1982) 'Frost in the Forth valley', *J. Meteorology (UK)*, 7: 84–86.

Hartley, M. (1977) 'Glasgow as an urban heat island', *Scot. Geogr. Mag.*, 93: 80–89.

Holford, I. (1976) *British Weather Disasters*, Newton Abbot: David & Charles.

Inwards, R. (1994) *Weather Lore*, London: Senate.

Jenkins, A., Ashworth, P.J., Ferguson, R.I., Grieve, I.C., Rowling, P. and Stott, T.A. (1988) 'Slope failures in the Ochil Hills, Scotland, November 1984', *Earth Surface Processes and Landforms*, 13: 69–76.

Johnson, R.C. (1985) 'Mountain and glen contrasts at Balquhidder', *J. Meteorology (UK)*, 10: 105–108.

Lamb, H.H. (1943) 'Haars or North Sea fogs on the coast of Great Britain', *Met Office M.O.504*, Bracknell, Meteorological Office.

Learmonth, A.T.A. (1950) 'The floods of 12 August 1948 in south-east Scotland', *Scot. Geogr. Mag.*, 66: 147–153.

Lee, D.O. (1985) 'A preliminary analysis of long-term visibility trends in central Scotland', *J. Climatology*, 5: 673–680.

Lowndes, C.A.S. (1971) 'Substantial snowfalls over the United Kingdom 1954–69', *Meteorol. Mag.*, 100: 193–207.

Lyall, I.T. (1981) 'The occurrence of severe frost in Scotland 1969–1975', *J. Meteorology (UK)*, 6: 247–249.

McCallum, E. (1990) 'The Burns Day storm 25 January 1990', *Weather*, 45: 166–173.

Manley, G. (1971) 'The mountain snows of Britain', *Weather*, 26: 192–200.

Mayes, J.C. (1991) 'Regional airflow patterns in the British Isles', *Int. J. Climatol.*, 11: 473–491.

Meaden, G.T. (1983) 'Britain's highest temperatures: the county records', *J. Meteorology (UK)*, 8: 201–203.

Meteorological Office (1968) *Tables of Surface Wind Speed and Direction over the United Kingdom*, Met. O.792, London: HMSO.

—— (1989) *The Climate of Scotland: Some Facts and Figures*, London: HMSO.

Morris, D. (1899) 'Large hailstone at Sheibrae', *Trans. Stirling Nat. Hist. and Archaeol. Soc.*, 21: 175.
Mossman, R.C. (1896) 'The meteorology of Edinburgh', *Trans. R. Soc. Edinburgh*, 38: 681–755.
—— (1897) 'The meteorology of Edinburgh, part 2', *Trans. R. Soc. Edinburgh*, 39: 63–207.
—— (1902) 'The meteorology of Edinburgh, part 3', *Trans. R. Soc. Edinburgh*, 40: 469–509.
Paton, J. (1951) 'Weather and climate', in A.G. Ogilvie (ed.) *Scientific Survey of South Eastern Scotland*, London: British Association for the Advancement of Science.
Pearson, M.G. (1975) 'Never had it so bad', *Weather*, 30: 14–21.
Perry, A.H. (1981) *Environmental Hazards in the British Isles*, London: Allen & Unwin.
Perry, A.H. and Symons, L. (1980) 'Economic and social disruption arising from the snowfall hazard in Scotland: the example of January 1978', *Scot. Geogr. Mag.*, 96: 20–25.
Perry, A.H., Symons, L. and Williams, P. (1984) 'Snow depth and snowfall disruption in Scotland in January 1984', *J. Meteorology (UK)*, 9: 133–135.
Reynolds, G. (1985) 'Extreme rainfall events in Scotland', in S.J. Harrison (ed.) *Climatic Hazards in Scotland*, Norwich: GeoBooks.
Sargent, R.J. (1989) 'Water resource management and flooding', in S.J. Harrison and K. Smith, *Weather Sensitivity and Services in Scotland*, Edinburgh: Scottish Academic Press.
Smith, K. (1985) 'The physical environment', in J. Butt and G. Gordon (eds) *Strathclyde: Changing Horizons*, Edinburgh: Scottish Academic Press.
—— (1986) 'The climate of the estuary and the Firth of Clyde', *Proc. R. Soc. Edinburgh*, 90B: 43–54.
—— (1995) 'Precipitation over Scotland 1757–1992: some aspects of temporal variability', *Int. J. Climatol.*, 15: 543–556.
Smith, K. and Bennett, A.M. (1994) 'Recently increased river discharge in Scotland: effects on flow hydrology and some implications for water management', *Applied Geography*, 14: 123–133.
Smith, S.G. (1983) 'The seasonal variation of wind speed in the United Kingdom', *Weather*, 38: 98–103.
Smithson, P.A. (1969) 'Effects of altitude on rainfall in Scotland', *Weather*, 24: 370–376.
Sutherland, D. (1925) 'The shore vegetation of Wigtownshire', *Scot. Geogr. Mag.*, 41: 1–23.
—— (1926) 'The vegetation of the Cumbrae Islands and of south Bute', *Scot. Geogr. Mag.*, 42: 272–286.
Sweeney, J.C. (1977) 'The Glasgow rainstorm of 28 September 1976', *Scot. Geogr. Mag.*, 93: 52–55.
—— (1981) 'Photochemical air pollution comes to central Scotland', *Scot. Geogr. Mag.*, 97: 50–55.
Tricker, A.S. and Miller, K.M. (1977) 'Three consecutive dry years in Fife 1971–1973', *Scot. Geogr. Mag.*, 93: 55–60.
Varley, M.J., Bevan, K.J. and Oliver, H. (1996) 'Modelling solar radiation on steeply sloping terrain', *Int. J. Climatol.*, 16: 93–104.
Webb, J.D.C. and Meaden, G.T. (1983) 'Britain's lowest temperatures: the county records', *J. Meteorology (UK)*, 8: 269–272.
—— (1993) 'Britain's highest temperatures by county and by month', *Weather*, 48: 282–291.
Werritty, A. and Acreman, M.C. (1985) 'The flood hazard in Scotland', in S.J. Harrison (ed.) *Climatic Hazards in Scotland*, Norwich: GeoBooks.
Weston, K.J. and Roy, M.G. (1994) 'The directional dependence of the enhancement of rainfall over complex orography', *Meteorol. Appl.*, 1: 267–275.
Wood, D.J. (1965) 'The complicity of climate in the 1816 depression in Dumfriesshire', *Scot. Geogr. Mag.*, 81: 5–17.
Young, W.K. (1985) 'Instrumentation at Eskdalemuir Observatory', *Meteorol. Mag.*, 114: 202–211.

10

THE HIGHLANDS AND ISLANDS OF SCOTLAND

Marjory Roy

INTRODUCTION

The region considered in this chapter (Figure 10.1) has been defined as the area of Scotland lying to the north-west of the Highland Line, the geological boundary which marks the northern limit of the Central Valley of Scotland. The Highland Line can be traced from the Mull of Kintyre in the south-west, through the southern end of Loch Lomond, north-west of Stirling and Perth to reach the sea at Stonehaven. Although the area of mainland Scotland which lies to the north-west of the Highland Line is generally referred to as 'the Highlands of Scotland', it includes terrain which is far from mountainous, such as around the Moray Firth and in north-east Aberdeenshire.

There is also a marked contrast in the nature of the mountainous terrain between the west and east of the Highlands. The western mountains rise steeply from the glens and the fjord-like sea-lochs whereas the eastern Grampians are generally much more rounded, although they contain some steep-sided corries and deep valleys. A notable feature of the mountains of the Western Highlands is the long north–south effective barrier which they present to Atlantic winds from directions between south-west and north-west. There is a break in the northern part of the mountain chain, to the north of Ullapool, where the mountains (Stac Polly, Cul Mor, Suilven, Canisp) are isolated peaks rising from undulating moorland, and this break is mirrored in the pattern of average rainfall, which shows very strong orographic features.

With the exceptions of the Dee and the Don in Aberdeenshire, most of the major rivers of the Highlands of Scotland (Tay, Spey, Ness, Conin, Oykel) rise in the mountains of the west and then flow east to the North Sea. Flooding of their lower reaches is generally due to heavy orographic rain falling well to the west. For example, flooding of the river Ness at Inverness in February 1989 was due to prolonged heavy rainfall in the area near to Loch Quoich, which lies close to the west coast.

The islands of Scotland comprise the Inner Hebrides (Islay, Jura, Colonsay, Mull, Tiree, Coll, the Small Isles to the south of Skye and Skye itself), the Outer Hebrides, often referred to as the Western Isles or the Long Island (Barra, South Uist, Benbecula, North Uist, Harris and Lewis), and the Northern Isles of Orkney and Shetland. These islands are very diverse in character and topography, ranging from mountainous in Rhum (one of the Small Isles), Harris, Mull and Skye to low-lying in the island of Tiree, at one time the 'granary of the Isles'. Even within individual islands there are striking contrasts, for example on the island of Skye between the barren Red and Black Cuillin in the centre of the island and the lush vegetation of Sleat in the south.

Most of the islands which make up Orkney are low-lying, with fertile soils. Shetland, which lies on the same latitude (60° N) as Bergen, is generally bleaker, although there are sheltered fertile areas,

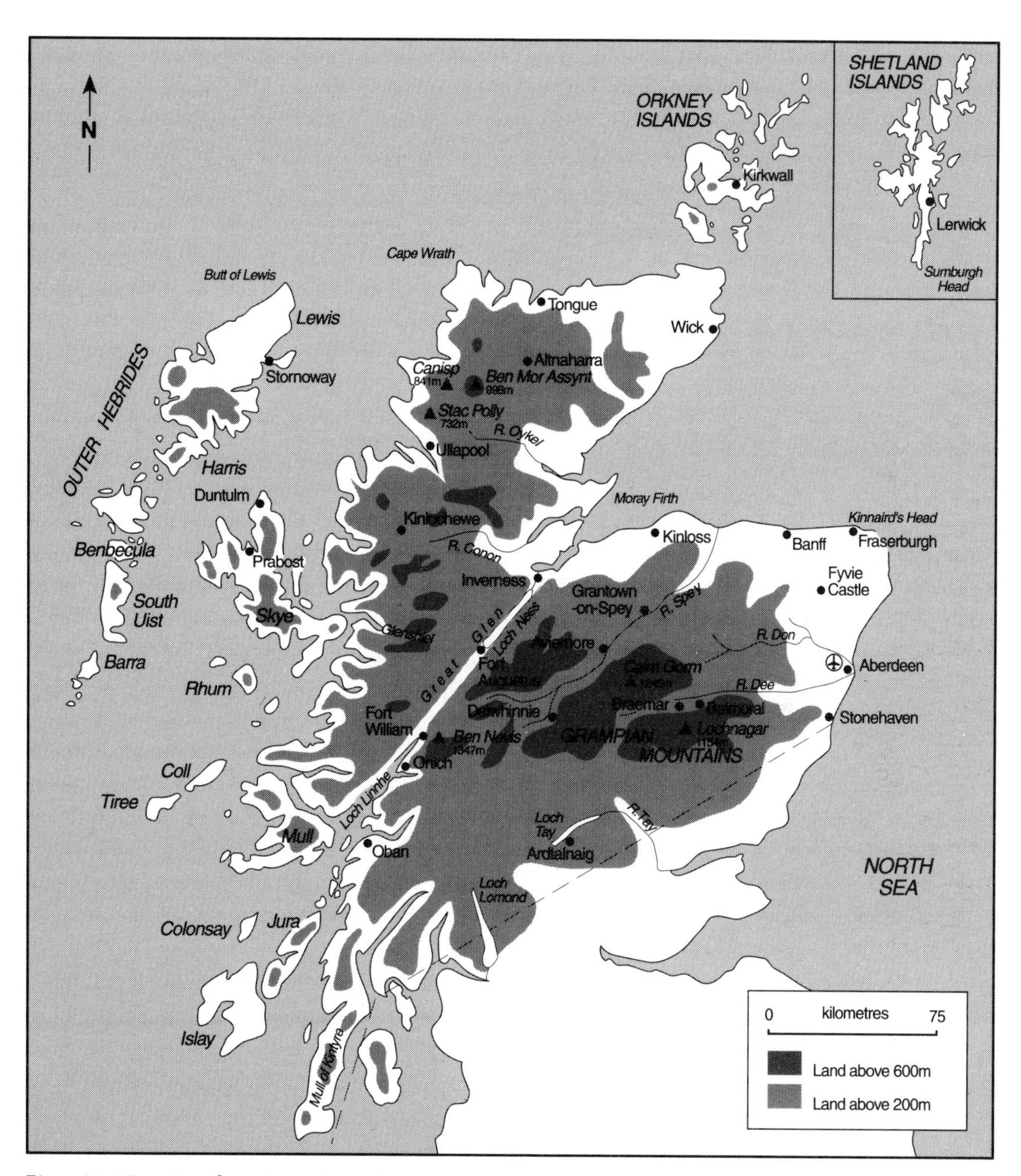

Figure 10.1 Location of weather stations and places referred to in the text

and considerable parts of Shetland, especially on Yell, are covered in a thick blanket of peat.

The complex topography of this region leads to great diversity of climate on both the local and the large scale. On the local scale the effects of altitude, slope, aspect and the immediately surrounding topography can lead to very rapid changes in temperature over short distances and somewhat smaller changes in precipitation. Large-scale effects are related to proximity to the sea and, depending on the low-level airflow at any particular moment, upwind shelter from areas of high ground. The vertical and horizontal scales of the mountains of the Highlands of Scotland are such that topographic shelter plays a dominant role in many aspects of the region's climate.

Weather Observations in Highland Scotland

With the setting up of the Scottish Meteorological Society in 1855 a network of climatological stations was established, with standard observing instruments and methods of observation. This network included a number of sites at which observations were already being made, for example at Culloden House near Inverness, where activity commenced in 1840. Monthly summary data for 1856 for the new network were published by the Scottish Meteorological Society, and the Edinburgh Meteorological Office archives hold the actual returns of daily data from January 1857 onwards. By 1860 there were twenty-one climatological observing sites in the Highlands and Islands of Scotland. Responsibility for the running of the climatological network in Scotland was finally taken over by the Meteorological Office in 1920, under an agreement with the Scottish Meteorological Society.

The site with the longest period of near-continuous observations at the same location is at Braemar (339 m), on the upper reaches of the river Dee in Aberdeenshire, where an observing station was set up by Prince Albert in 1855. The site is not an ideal one, since there are nearby trees which have grown over the period since it was set up, but there are no 'urban heat-island' problems. Although long-period observations have been made in the Aberdeen, Stornoway, Wick and Lerwick areas, in each case there have been major changes of site that affect the homogeneity of the records.

Most of the sites at which climatological observations have been made are at relatively low altitudes, with the exception of the climatological site near to the car park on Cairngorm (663 m) and the automatic weather stations at the ski centres of Cairngorm, Aonach Mor and Glenshee. The longest-running mountain automatic weather station (AWS) site is that at the summit of Cairngorm (1,245 m), where Heriot-Watt University installed a prototype instrument in 1976. Data from this instrument have been archived by Heriot-Watt University and (from 1985) by the Meteorological Office. In 1990 a Meteorological Office 'Severe Icing Environment Synoptic Automatic Weather Station' (SIESAWS) came into operational use on the same site. This instrument makes continuous measurements of wind, temperature and dew point.

The most complete set of hourly high-level climatological observations in the United Kingdom was that made at the Ben Nevis Observatory (1,344 m) from 1 November 1883 to 1 October 1904 (Paton 1954; Roy 1983). This observatory was set up by the Scottish Meteorological Society but eventually had to close owing to the fact that government financial support was insufficient to cover the costs of its operation (Figure 10.2). The hourly data were published *in extenso* in the *Transactions of the Royal Society of Edinburgh*, together with averages and other statistics. A low-level observatory was set up at Fort William in 1890, providing a valuable nearby sea-level comparison.

Figure 10.2 The Ben Nevis Observatory as it appeared at the start of the twentieth century

TEMPERATURE

Within this region the range of temperature regimes is possibly greater than that which occurs anywhere else in the British Isles, ranging from mild oceanic in the islands off the west coast to semi-continental in the Highland valleys which lie many kilometres from any coastal influence. To this must be added rapid changes of temperature with height. Some of the main features of these temperature regimes can be seen in the monthly and annual values of average daily maximum and minimum temperatures (Table 10.1).

Geographic Factors Influencing Highland Temperatures

The moderating influence of the sea is most apparent in the data for the island sites, where the differences between the warmest and coldest months are only of the order of 9 degrees C. The winters are mild, though considerably less so than those of south-west England. Because of the moderating influence of the sea, on average August is the warmest month and February the coldest. The difference between the average maximum and minimum temperatures is a function of both the diurnal heating by the sun and the day-to-day variations in temperature due to the passage of weather systems. The average difference is least in the winter and is greatest in April, May and June before declining in July.

Table 10.1 Mean monthly maximum and minimum temperatures (°C) for the period 1961–90 (Ben Nevis October 1885–September 1904)

Location	*Alt. (m ASL)*	*Jan.*	*Feb.*	*Mar.*	*Apr.*	*May*	*June*	*July*	*Aug.*	*Sept.*	*Oct.*	*Nov.*	*Dec.*	*Year*
Lerwick	82	5.3	5.1	6.0	7.7	10.0	12.5	13.7	14.1	12.3	10.2	7.2	6.0	9.1
		1.2	1.1	1.6	2.7	5.2	7.5	9.1	9.5	8.0	6.3	3.1	1.8	4.7
Kirkwall	26	5.8	5.8	7.0	8.9	11.4	14.0	15.1	15.3	13.5	11.3	7.9	6.5	10.2
		1.5	1.3	2.0	3.0	5.4	7.9	9.5	9.7	8.3	6.6	3.4	2.2	5.0
Stornoway	15	6.7	6.8	7.9	9.6	12.1	14.4	15.5	15.7	13.9	11.9	8.6	7.4	10.8
		1.7	1.4	2.3	3.3	5.9	8.3	9.9	9.8	8.4	6.4	3.1	2.3	5.2
Tiree	9	7.3	7.1	8.3	10.1	12.6	14.7	15.8	16.0	14.5	12.5	9.5	8.1	11.3
		2.9	2.6	3.3	4.4	6.7	9.1	10.8	10.9	9.8	8.1	4.8	3.8	6.4
Onich	15	6.6	7.0	8.6	11.5	14.8	16.9	17.7	17.7	15.3	12.8	8.8	7.3	12.0
		0.9	0.7	1.8	3.0	5.6	8.2	9.8	9.7	8.2	6.2	2.6	1.6	4.8
Kinloss	5	6.2	6.6	8.6	10.7	13.8	16.8	18.1	17.9	15.6	12.9	8.5	6.8	11.8
		0.2	−0.2	1.6	2.9	5.7	8.7	10.4	10.2	8.5	5.9	2.3	0.9	4.7
Braemar	339	3.8	3.9	6.0	9.2	12.8	16.2	17.5	16.9	14.0	10.8	6.3	4.6	10.1
		−2.2	−2.7	−0.7	0.6	3.4	6.5	8.4	8.0	6.3	3.8	0.1	−1.0	2.5
Aberdeen Airport	65	5.7	6.0	8.0	10.3	13.0	16.3	17.8	17.7	15.5	12.5	8.2	6.5	11.4
		−0.3	−0.2	1.1	2.4	5.1	8.0	9.8	9.6	7.9	5.5	1.9	0.6	4.2
Ardtalnaig	130	5.2	5.3	7.7	11.1	14.6	17.6	18.6	18.2	15.1	11.8	7.6	6.1	11.5
		0.4	0.1	1.3	2.7	5.4	8.4	10.0	9.7	8.0	6.0	2.4	1.3	4.6
Ben Nevis	1,344	−2.5	−2.8	−2.6	−0.5	2.6	6.5	7.1	6.4	5.3	1.4	0.1	−2.0	1.6
		−6.3	−6.4	−6.3	−4.3	−1.6	2.2	3.1	2.8	1.4	−2.1	−3.5	−5.5	−2.2

Note: ASL = above sea level

This decline has been ascribed by Lamb (1972) to the 'European monsoon' – a period during which the focus of cyclonic activity shifts southwards, bringing Scotland more firmly into a regime of westerly, often wet and cloudy, circulations.

By comparison with inland areas of Scotland very low minimum temperatures do not occur on the islands, and frosts, though not uncommon, are rarely severe. For example, at Prabost on Skye, in the period from 1959 to 1989, the lowest air minimum temperature was −8.9°C in February 1960, but there was an annual average of 42 days of air frost and 121 days of ground frost. Nevertheless, Irvine (1969b) reported that Shetland experienced temperatures of −5.7°C as late as 6 April in 1968 during an eruption of northerly air. Earlier that winter the sea was reported to have frozen sufficiently around Fair Isle to have supported the weight of adventurous pedestrians (Irvine 1969a). Conversely, on the larger islands, such as Skye and Mull, maximum temperatures in summer have exceeded 28°C, but at Lerwick the highest maximum was only 23.4°C, which was achieved on 5 July 1991.

Although Onich lies on a western sea-loch, it is surrounded by mountains and the oceanic influence is less than in the islands. Consequently the average differences between the warmest (July and August) and coldest (January and February) months are greater, as are the differences between the average maximum and minimum temperatures. Owing to the sheltered character of the site very high temperatures have been recorded, and a record maximum of 32.1°C was observed on 1 August 1995.

Kinloss on the Moray Firth has an east-coast setting, and the moderating influence of the sea is largely confined to periods of north-easterly winds or when a sea breeze is set up. These conditions are most frequent in spring and early summer, when there is the highest frequency of anticyclones over or close to the north of Scotland. This is also the time of year when sea fogs, or 'haars', are most prevalent in the Northern Isles and along the east coast of Scotland (Findlater *et al.* 1989), and these too can lower daytime maximum temperatures by several degrees (Boxes 7.3 and 9.4).

In comparison with Onich, Kinloss is colder in April and May, when cold easterly winds and haars are more prevalent, but warmer in July and August, the months in which there tends to be a return to a more westerly regime. Its highest maximum, of 30.7°C, was recorded on 31 July 1995. If the wind direction lies between southerly and north-westerly the Moray Firth area lies to the lee of very considerable ranges of mountains, and strong föhn effects are frequently observed.

Lying about 9 km from the North Sea at a height of 65 m, Aberdeen Airport's temperature regime is more typical of an inland than a coastal site, except when winds are from an easterly direction. Like Kinloss, Aberdeen lies in an area where very strong föhn effects have been observed. An example of this (Figure 10.3) occurred from 1 to 2 January 1992, when there was a very mild, moist, warm sector over Scotland, with strong west-south-westerly winds (Figure 10.3a). Aberdeen was within the warm sector from 2000 GMT on 1 January until 2200 GMT on 2 January, during which period the increase in temperature difference between Benbecula (on the west coast) and Aberdeen was about 3 degrees C, and the temperature at Aberdeen reached a maximum of 15.1°C on 2 January, the highest recorded January temperature at that site. Figures 10.3b and c draw attention to the notable temperature contrast between west-coast Benbecula and east-coast Aberdeen and to the different humidities at the two sites. During the height of the föhn the dew point and dry bulb (air) temperatures at Benbecula were very similar, indicating conditions close to saturation point, while at Aberdeen the dew point stood up to 4 degrees C lower, with consequently much lower relative humidities.

Braemar and Ardtalnaig both lie in deep valleys, well within the main mountainous area of Scotland, but Ardtalnaig lies mid-way along the banks of Loch Tay, where the loch is deep enough for winter freezing never to occur. (The average temperature of the water throughout its depth must fall below 4°C, its temperature of maximum density, for overturn-

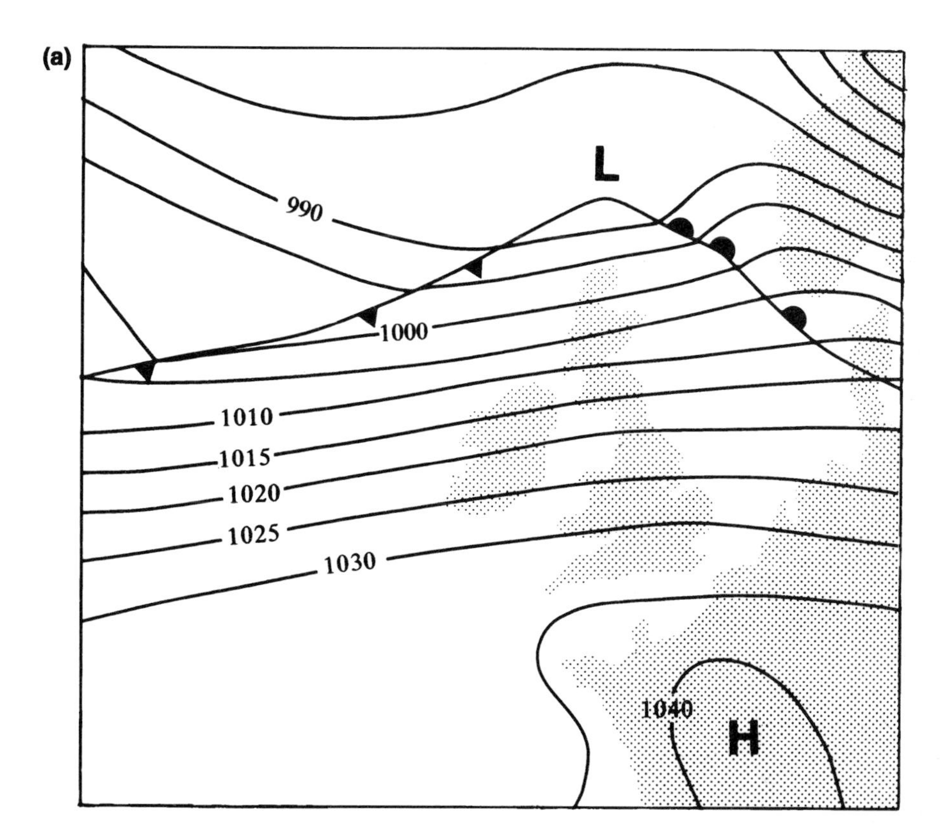

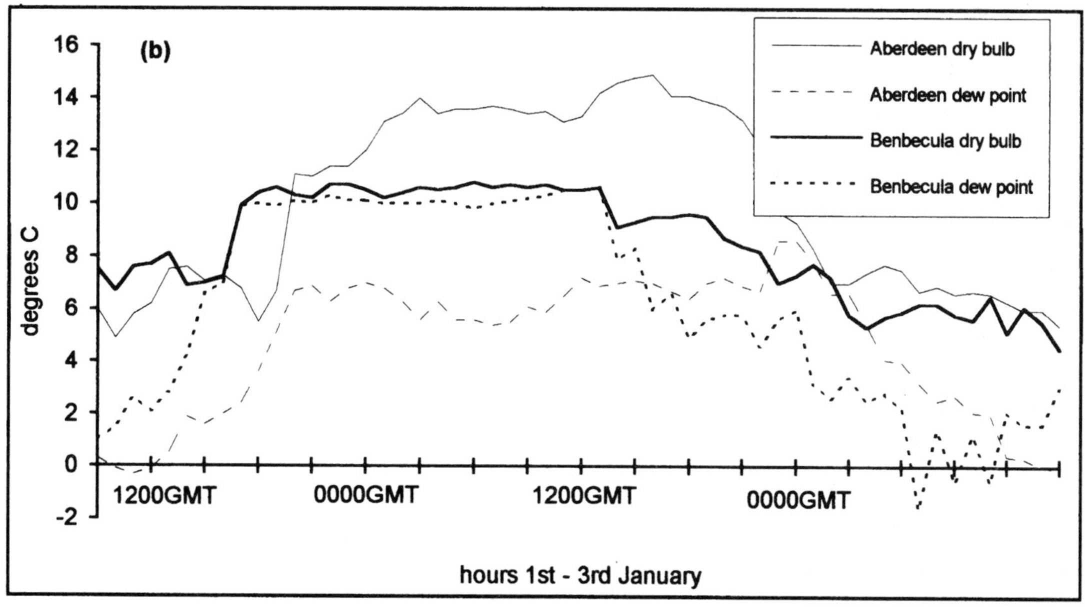

Figure 10.3 (a) Synoptic chart for 0000 GMT on 2 January 1992. Föhn effect 1–3 January 1992 over Highland Scotland revealed by data for Benbecula and Aberdeen in respect of (b) air and dew point temperatures, and (c) relative humidity

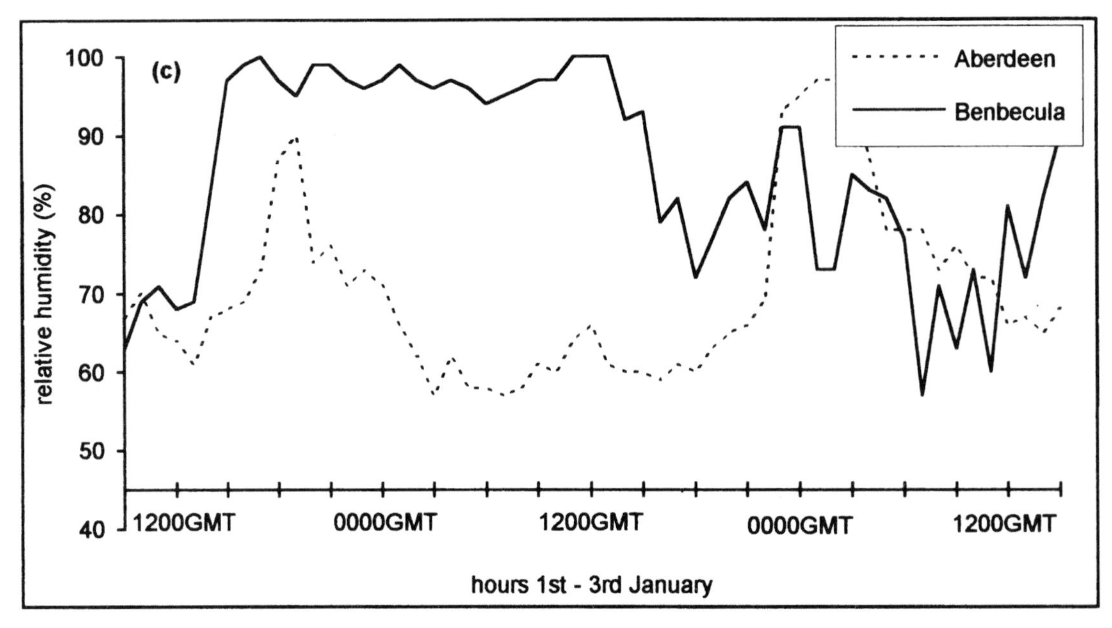

Figure 10.3 (Continued)

ing, which brings warmer water to the surface, to cease, allowing the surface layers to freeze.) As a consequence, the differences between the Braemar and Ardtalnaig average temperatures in winter are greater than would be expected from their respective altitudes (339 m and 130 m), and very low minima are not observed at Ardtalnaig, the lowest being −13.4°C on 30 December 1995. These ameliorating effects are confined to areas close to the loch (Tabony 1987). Similar effects occur close to the long sea-lochs which penetrate deeply into the Western Highlands. Shallower lochs do, however, freeze over during prolonged cold spells.

Temperature Extremes in Highland and Island Scotland

The official record extremes for the region are summarised in Table 10.2, in which the extreme maritime character of the island settings, notwithstanding some of the peculiarities noted in earlier sections, stands in marked contrast to the more continental character of the inland locations.

The weather situation that gives rise to very low temperatures is usually one in which northerly winds bring very cold, dry air from the Arctic, followed by a rise of pressure as an anticyclone becomes established over Scotland. During the winter months, with clear skies and a snow cover (not necessarily deep) on the ground, the loss of heat over a 24-hour period by long-wave radiation from the ground is greater than the gain of heat from solar radiation and from long-wave radiation from the atmosphere. Unless winds introduce warmer air, over a period of days the average daily temperature will fall steadily. Roach and Brownscombe (1984) studied the cold spells of December 1981 and January 1982, from which they concluded that snow cover, light winds and clear skies all contributed to extreme minima. They also confirmed that local and large-scale topography play a part in determining which sites experience the lowest temperatures – a point reiterated by Tabony (1987).

Braemar lies in a natural bowl in the mountains, into which, in clear, calm winter weather, cold air drains from the surrounding heights. The only escape for this air lies down the valley of the Dee, so that ponding effects are especially strong, and it is

Table 10.2 Record maxima and minima (degrees C) for Highland and Island Scotland by administrative regions

Region	*Maximum*	*Date*	*Location*	*Minimum*	*Date*	*Location*
Shetland	25.0	12 July 1926 2 July 1958	Baltasound	−10.0	21 Jan. 1958	Baltasound
Orkney	26.5	16 July 1972	Stenness	−12.8	12 Jan. 1963	Stenness
Western Isles	27.2	30 July 1948	Benbecula	−11.1	20 Jan. 1960	Stornoway
Highlands	32.1	1 Aug. 1995	Onich	−27.0	30 Dec.1995	Altnaharra
Grampian	31.6	1 July 1976	Lossiemouth	−27.2	11 Feb. 1895 10 Jan. 1982	Braemar

Sources: Annual summaries of the *Monthly Weather Report* 1920–93; Edinburgh Meteorological Archives for 1993–7 data
Note: Strathclyde and Tayside do not lie wholly within Highland Scotland (as defined in this text) and are not included in the table

not surprising that the lowest recorded minimum temperature in the United Kingdom (−27.2°C) has twice been recorded there (Table 10.2). In late December 1995 this record came close to being broken when a temperature of −27.0°C was recorded by the automatic weather station at Altnaharra in Sutherland, and temperatures below −20°C have been recorded on a number of occasions in Highland valleys as slack pressure gradients, clear skies and an extensive snow cover combined to provide optimum conditions for intense night-time heat loss. Such minima are impressive in an essentially maritime part of the globe, but McClatchey *et al.* (1987) have offered persuasive evidence that temperatures in the Spey valley, especially around Grantown-on-Spey, may occasionally fall to −30°C or below.

Whereas the south-east of England generally experiences its lowest temperatures with cold, dry air which has been brought in on easterly winds across the short sea track from continental Europe, airstreams from the east have had a long track over the North Sea before they reach the Highlands of Scotland and have been considerably modified, becoming in winter both warmer and more moist in their lower layers. Usually the temperature at the surface is just above freezing point when the air reaches the east coast, but on 12 January 1987 the air was exceptionally cold through a considerable depth and the coastal temperature was about −2.5°C. At Aberdeen Airport, the increase in height and cooling over snow-covered ground had reduced the temperature to −5.5°C and on the summit of Cairngorm (1,245 m) the temperature reached the lowest value recorded in the eighteen years of measurements at that site, −16.5°C. At Ben Nevis Observatory the lowest recorded temperature was −17.4°C on 6 January 1894, and this also occurred when easterly winds brought very cold air from the continent to the British Isles. From these examples it can be seen that the lowest temperatures occur not at the summits of the mountains but in the valleys.

Although average summer temperatures in the Highlands of Scotland are considerably lower than those in the south of England, high temperatures are not infrequent. Temperatures of around 30°C have been recorded at several inland valley sites and also on the coast when offshore winds have been strong enough to prevent a sea breeze developing. The record maxima throughout the region are listed in Table 10.2.

Climatic Variations with Altitude: the Example of Fort William to Ben Nevis

At the summit of Ben Nevis the differences between the average temperatures for the coldest month (February) and warmest month (July) are about 9.5 degrees C for both maximum and minimum temperatures. The differences between the average daily maxima and minima are very similar throughout the year, ranging from 3.5 degrees C in October and December to 4.0 degrees C in July. Since hourly observations were made at the observatory it is possible to produce graphs showing the true average diurnal variation during each month (some of which are plotted in Figure 10.4). The diurnal variation due to solar heating is extremely small in the winter (when it is less than 0.5 degrees Celsius) and increases in the summer months to over 2 degrees C. At this site the difference between the average maximum and minimum temperatures is almost entirely due to the varying temperature of the air masses that cross the region. This is in contrast to the large plateau of the Cairngorms, which can provide a major heat source in the summer months. Ben Nevis and the surrounding mountains have relatively small summit areas and are divided by deep valleys, so that high-level heating by incoming radiation is very limited.

Comparison of the Ben Nevis observations with those from Fort William Observatory during the period 1891 to 1903 can provide values for the average change with height of both maximum and minimum temperature (Table 10.3).

Because the Fort William Observatory was situated close to the shore of Loch Linnhe there are no local 'frost hollow' complications to be taken into account when comparing the minimum temperatures at the two sites, and the differences show little variation in the course of the year. By contrast, the differences in maximum temperature are about 2

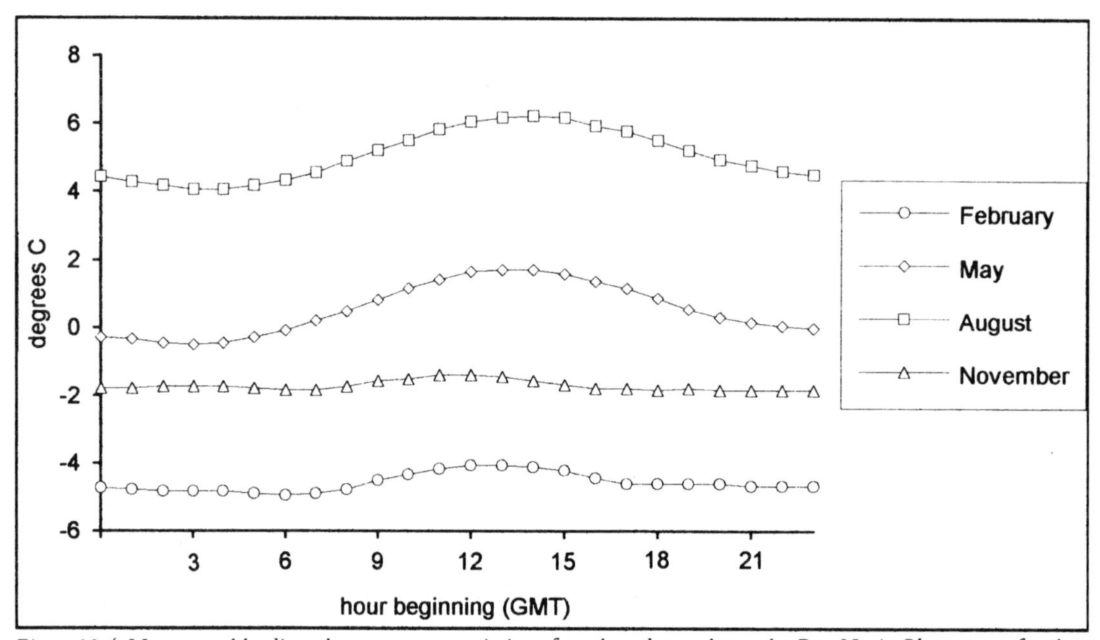

Figure 10.4 Mean monthly diurnal temperature variations for selected months at the Ben Nevis Observatory for the period 1884–1903

Table 10.3 Average fall of maximum and minimum temperatures (°C) in 1,000 m calculated from Fort William and Ben Nevis data for 1891–1903

	Jan.	*Feb.*	*Mar.*	*Apr.*	*May*	*June*	*July*	*Aug.*	*Sept.*	*Oct.*	*Nov.*	*Dec.*	*Year*
Maximum	6.6	6.5	7.5	8.3	8.4	8.0	7.7	7.7	7.4	7.1	6.3	6.3	7.3
Minimum	6.1	5.6	6.0	5.8	5.6	5.2	5.6	5.6	5.5	5.6	5.6	5.6	5.6

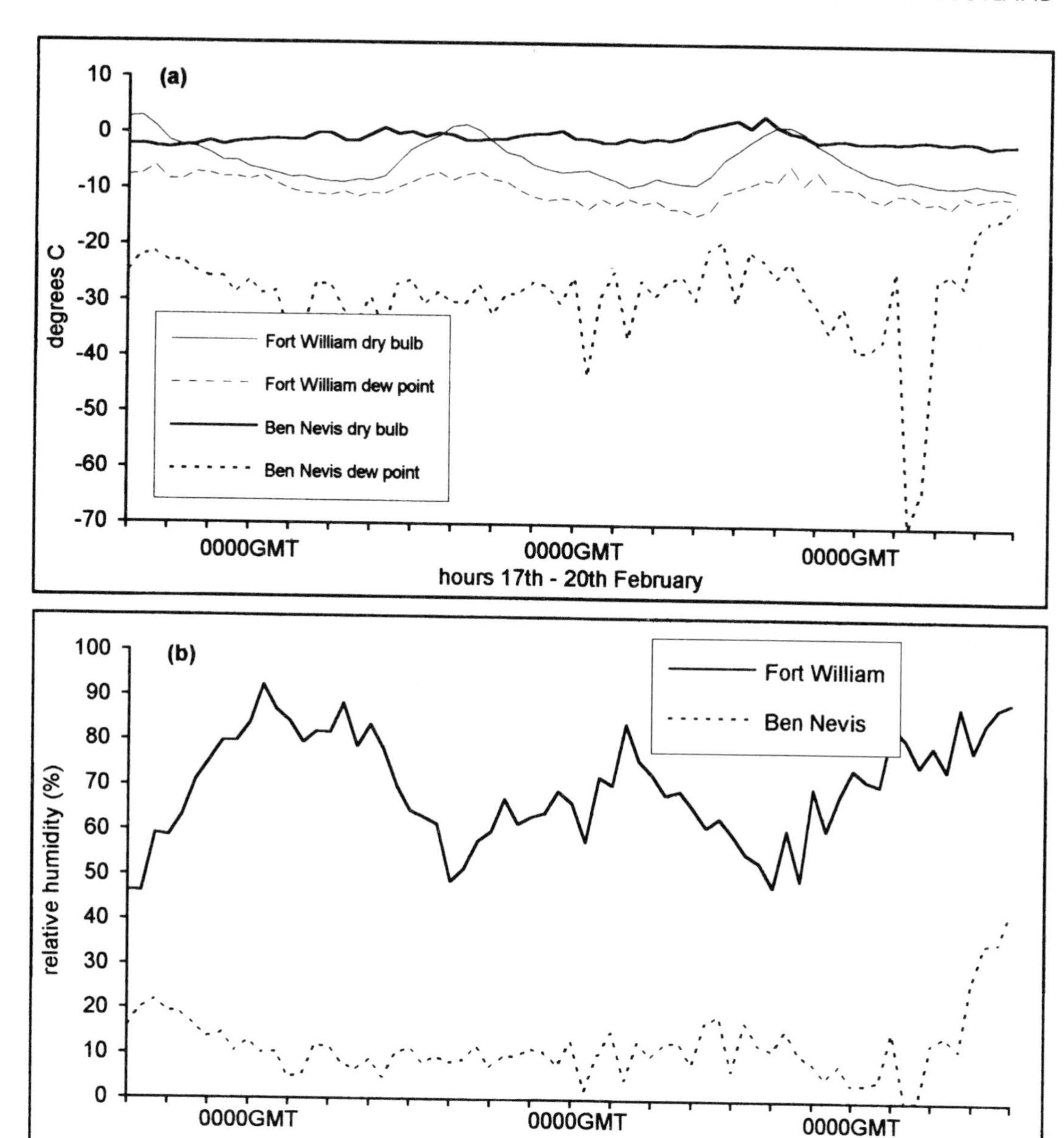

Figure 10.5 (a) Air and dew point temperatures (°C), and (b) relative humidities (percentages) at Fort William and Ben Nevis for February 1895

degrees C greater in April and May than those in November and December.

Although on average temperature falls with height, this is by no means always the case. Reference has already been made to the much lower temperatures in valleys on cold, clear nights, when there is a low-level temperature inversion. Other situations in which a temperature inversion can occur are ahead of an approaching warm front or occlusion, when the lower boundary of the warm air aloft gradually descends towards the surface, and in large anticyclones. Anticyclones are areas in which the air aloft is descending slowly, undergoing adiabatic warming, and a situation may be reached in which the resulting inversion level is below the summits of the higher peaks. As a consequence of the warming the relative humidity of the air above the inversion can become extremely low.

An example of such an occurrence is shown in Figures 10.5a and b. The graphs show Fort William Observatory and Ben Nevis Observatory values of dry bulb (air) temperature, dew point and relative humidity for the period from 1500 GMT on 17 February to 0900 GMT on 20 February 1895. The largest increase of temperature between the low- and the high-level sites was 9.8 degrees C at 0900 GMT on 19 February. The relative humidity on Ben Nevis was below 20 per cent from 1800 GMT on the 17th until 0600 GMT on the 20th, and for several hours during that period it was less than 10 per cent. Green (1967) has also reviewed the peculiarities of humidity over Ben Nevis, drawing similar attention to the tendency to occasional very low values. During such periods low cloud is frequently trapped below the inversion, giving damp, miserable conditions in the valleys while the peaks are bathed in warm sunshine.

PRECIPITATION

Annual precipitation is made up of both rain and snow, and over the higher mountains in this region the latter can make a substantial contribution to the total, although amounts vary widely from year to year. Snow has been observed to fall on the high tops in every month of the year but, conversely, winter precipitation is by no means always of snow.

Over much of this topographically complex

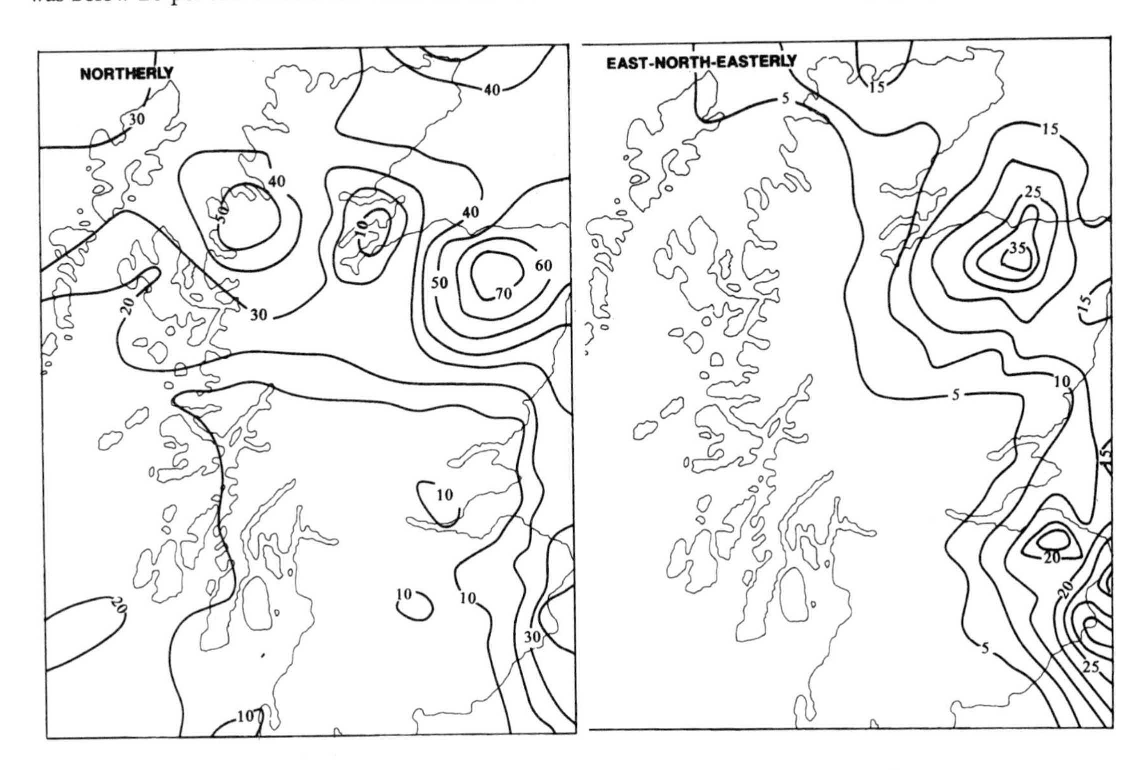

Figure 10.6 Distribution of rainfall under different airflow types (after Weston and Roy 1994)

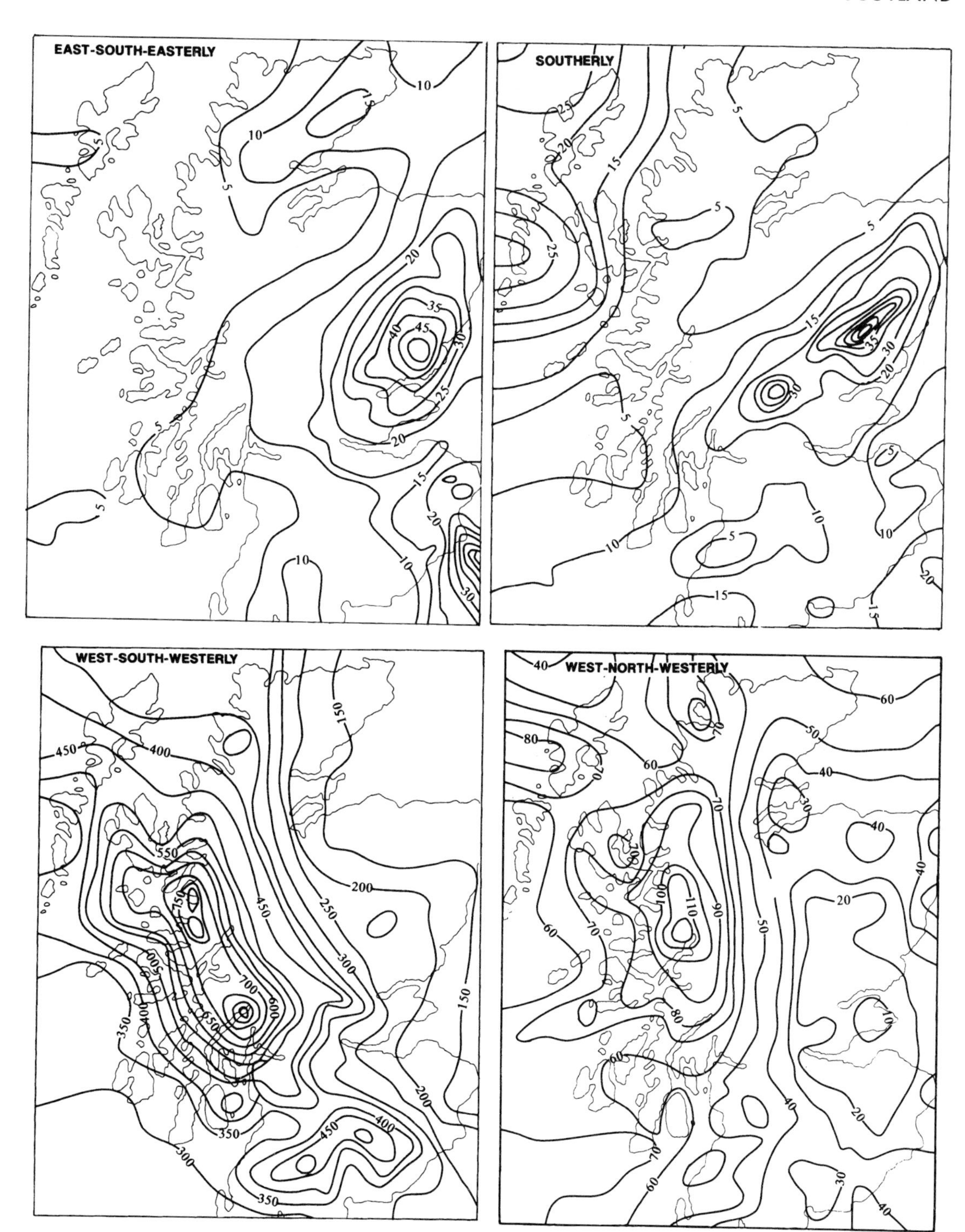

Figure 10.6 (Continued)

region the rain-gauge network is too sparse for the detail of even the average annual rainfall distribution to be mapped with confidence. However, relatively dense networks of gauges around some of the major hydro-electric and water supply schemes and the Ben Nevis observations provide useful data on the variation of rainfall with height. Close to the west coast, where the sea-level value of annual average rainfall is 1,600 mm, the rate of increase with height is about 240 mm/100 m, whereas over the Cairngorms the increase is about 120 mm/100 m. In areas, such as the Western Highlands, where the mountains are intersected by very deep glens, the effective altitude of low-level sites is much greater than the true altitude, owing to wind-drift effects. Various methods of interpolation have been used (e.g. Bleasdale and Chan 1972), and a complex fitting technique was developed by Spackman (1994) to produce the Meteorological Office's maps of the 1961–90 averages of rainfall.

Factors Influencing Precipitation and Precipitation Regimes

Since much of the rain which falls over Scotland is associated with active weather systems, the effects of orographic enhancement over the mountains and downwind shelter in their lee play a major part in determining rainfall patterns. The dominance of the westerlies means that the most exposed areas are the Western Isles and the west coast of Scotland. Along much of the west coast of the mainland the mountains rise steeply from the sea, and it is not surprising that the highest annual average rainfall totals are found in these areas, with the wettest being around Loch Quoich, to the north-west of Fort William, where the annual total exceeds 4,000 mm. Over the lower-lying islands of the Hebrides the average annual rainfall is about 1,200 mm and to the lee of the mountains, around the Moray Firth, it is less than 650 mm. Another notable feature is the relatively low total (~2,000 mm) over the Cairngorms.

Orographic effects are not confined to events with westerly winds, as was shown by an analysis by Weston and Roy (1994). Figure 10.6 shows the total rainfall amounts recorded on the days which could be allocated to distinct airstream categories. In all cases the rainfall distribution was largely determined by the interaction of the airstream and the topography. With the predominant west-south-westerly trajectory there was a more rapid west-to-east decrease in rainfall over the Highlands than was evident in the annual rainfall distribution, and it can be seen that much of the rainfall over Aberdeenshire occurred with northerly and easterly winds. The distributions also showed that the area around the Moray Firth benefited from orographic shelter not only with winds from the west, but also with winds from all directions except east-north-easterly. The west of Scotland benefited from orographic shelter when winds were from an easterly direction. A notable example of this latter influence was reported by Green (1963), who described the extreme dryness of north-west Scotland during the winter of 1962/63 when easterly winds associated with high pressure to the north of Britain dominated the climatic scene. Green noted extreme desiccation of plants and soils, and estimated the potential evapotranspiration for the January–February period to be 18 mm at Prabost and 43 mm at Stilligarry (South Uist) – normally it is zero in those areas. Not surprisingly, the same conditions brought record sunshine to Stornoway.

Pedgley (1967), studying the heavy rainfall of mid-December 1966, found strong evidence of orographic enhancement within a persistent and moist westerly airstream moving at high speed. On this occasion at least, the contribution from low-level frontal convergence was small. Another factor contributing to precipitation is the steepness and orientation of the terrain, since the vertical velocities induced in the airstreams passing over the mountains depend on both the horizontal wind speed and the topographic gradient at right angles to the wind direction. The Cairngorms are more rounded than the western mountains and they also lie to the lee of other (lower) mountain ranges for all wind directions, not just westerly, hence their relatively low annual precipitation totals compared with other high-altitude sites to the west.

Table 10.4 Mean monthly and annual precipitation totals (mm) for the period 1961–90 (Ben Nevis 1885–1903)

Location	*Alt. (m ASL)*	*Jan.*	*Feb.*	*Mar.*	*Apr.*	*May*	*June*	*July*	*Aug.*	*Sept.*	*Oct.*	*Nov.*	*Dec.*	*Year*
Lerwick	82	133	93	115	73	62	62	64	77	118	136	143	144	1,220
Kirkwall	26	103	73	84	56	50	52	56	78	102	116	123	113	1,009
Stornoway	15	122	85	105	65	61	63	73	84	116	138	133	128	1,173
Tiree	9	127	79	96	59	59	61	78	95	129	140	122	120	1,165
Onich	15	225	153	197	99	98	114	134	160	216	233	231	236	2,096
Kinloss	5	53	39	46	36	46	49	55	71	58	56	62	52	623
Braemar	339	106	62	72	48	66	58	54	71	81	93	87	91	889
Aberdeen	65	81	51	58	53	59	53	60	75	68	77	75	73	783
Ardtalnaig	130	159	102	116	59	80	64	69	82	116	138	127	140	1,252
Ben Nevis	1,344	466	344	387	215	201	191	274	339	400	392	390	484	4,084

Data for a number of locations are listed in Table 10.4. At all of the western and northern sites the wettest part of the year is the autumn and early winter, with the five months from September to January accounting for about 55 per cent of the total annual rainfall. Ardtalnaig shows a similar pattern, but the percentage decreases to 48 per cent at Aberdeen Airport and 45 per cent at Kinloss. This is the season of the year when westerly types are of most frequent occurrence and are at their most vigorous. During late winter and early spring there is an increasing tendency for anticyclonic and easterly spells to occur, which leads to lower rainfall totals in February and March.

During late spring and early summer the frequency of winds from a westerly direction is at its lowest, and this is the time when there is the highest frequency of anticyclones situated over or close to Scotland. Consequently rainfall totals are at their lowest during the months of April, May and June in western and northern areas. July frequently sees the European monsoon and increased westerlies, but these are much less vigorous than during the winter, and, although it is more cloudy, there is little increase in rainfall. August sees a further increase in average rainfall, but the major change does not come until September. In the low-rainfall area around the Moray Firth the seasonal variation in rainfall is less than that in the rest of this region, with the wettest month being August. This is also the only part of the region where the total rainfall in the summer half of the year exceeds that in the winter half, and then only marginally.

Aberdeen is often in a sheltered position, but ahead of eastward-moving warm fronts and occlusions the surface wind direction is frequently southerly or south-south-easterly, and this means that even with 'westerly' types of weather system, there may be considerable rainfall in this area. As a result, despite its location on the eastern side of Scotland, rainfall amounts in the autumn and winter are not as low as those around the Moray Firth.

If any particular month is dominated by one weather type, either westerly or easterly, a map of the percentage of average rainfall for this region shows a marked split between the west and the east, with one part having well above average rainfall and the other well below average rainfall. For example, in January 1989 total rainfall over much of the Western Highlands of Scotland was more than two and a half times the average, whereas parts of Aberdeenshire had less than 10 per cent of the average. At Achnangart in Glenshiel the monthly total was 855 mm, whereas at Fyvie Castle to the north-west of Aberdeen it was only 4 mm.

By contrast, in September 1995 persistent north-easterly winds throughout the first third of the month brought heavy rainfall to the east of Scotland. The heaviest falls were in the area from Aberdeen northwards, with Kinloss on the Moray Firth recording an 11-day total of 271 mm, 43 per cent of the average annual rainfall. The month's rainfall in this area was over 500 per cent of the average, whereas to the south-west of a line from Harris to Loch Lomond it was below average. In circumstances like these, a mean monthly rainfall value for the whole of the Highlands of Scotland serves little useful purpose and can be very misleading.

Although there have been a number of notable flash floods caused by summer thunderstorms which have been set off over high ground, most flooding in the Highland area is due to prolonged heavy rainfall which continues for many hours and affects a wide area. The large high-level plateau of the Cairngorms is a preferred area for the initiation of thunderstorms, and the potential for cars to become trapped at the Cairngorm car park, as happened in August 1978 (Werrity and Acreman 1985), is one that summer visitors to the area should take into consideration.

In general, thunderstorms are much rarer events in the Highlands and Islands than they are in the south of England. Along the west coast and in the Western Isles they are an autumn rather than a summer phenomenon and are associated with heavy showers which are set off over the relatively warm sea. Since they are normally borne along by a brisk wind they are often 'one-clap' thunderstorms, which may be missed unless an observer is very alert.

Notable rainfall events, such as that described in Box 10.1, tend to be periods of prolonged, often orographically enhanced, activity. Floods which are due to heavy rainfall in the west are almost invariably winter events, when heavy rainfall may be accompanied by a rapid snow thaw. By contrast, flooding of rivers which rise in the eastern Grampians normally occur with easterly winds in late summer and early autumn (Werrity and Acreman 1985). The most notable of these events was the Moray floods of 1829. The synoptic weather situations which give rise to such flooding are those where a mature (occluded) depression becomes slow-moving to the south or east of the area, giving persistent winds from an easterly direction. If the winds are strong, orographic enhancement of the rainfall can be considerable. In winter and early spring the precipitation over the hills is of snow and the flooding risk is considerably diminished.

Although over much of the Highlands of Scotland average rainfall in each month exceeds the potential transpiration, periods of low rainfall can lead to drought conditions. This is most likely to occur in the spring, but in most years summer rains bring the soil back to field capacity. Summer droughts, such as that of 1984, do also occur. This drought affected a greater area of the Highlands of Scotland than that of 1976. During the period from 1 April to 31 August 1984 a large part of the Highlands (Figure 10.9), to the east of a line from near Tongue in Sutherland to Loch Lomond, had less than 50 per cent of the average rainfall for those months, and around Inverness rainfall was close to 30 per cent of average. In late summer and autumn 1993 a lack of westerly weather types meant that the normally reliable rains of the Western Highlands failed to materialise and Fort William had its driest autumn since 1915.

Snowfall

Statistics of days with snow falling are less reliable than those of days with snow lying at 0900 GMT. The latter show great variability both within the region and from year to year, and the popular image of the whole Highland and Island area of Scotland being covered in snow throughout the winter is very far from the truth. At Stornoway on the east side of Lewis the average number of days with snow lying is only about ten. Slightly higher values are found around the eastern and northern coasts of the mainland, especially in areas which are exposed to northerly winds, where coastal values are between fifteen and twenty days.

The frequency of snow lying increases with distance from the coast and with height, and at the summit of the Drumochter Pass (460 m) the average number of days with snow lying is about seventy. At Braemar (339 m), where ponding of cold air and

Box 10.1

THE EAST HIGHLAND FLOODS OF FEBRUARY 1989

Although the floods occurred mainly over the lower reaches of the Spey, Ness, Conon and Oykel in the Eastern Highlands, they had their origin in prolonged heavy rainfall which fell in the Western Highlands, far from the area in which the worst flooding occurred. In this part of Scotland the watershed between eastward- and westward-flowing rivers lies close to the west coast, and in the 48–hour period ending at 0900 GMT on 7 February 1989 the total rainfall exceeded 200 mm over much of their headwaters (Figure 10.8). A warm front introduced a strong, moist, west-south-westerly airstream over the area during the evening of 5 February. Among the mountains rainfall was of moderate intensity ahead of the warm front, but it became very heavy within the warm sector. At Kinlochewe, just to the west of the main watershed, rainfall accumulated to a total of 174 mm in the 24 hours preceding 0900 GMT on the morning of the 6th. A cold front moved south-east across the Hebrides during the morning of 6 February, but areas to the south-east of a line from Tiree to near Banff remained within the warm sector and heavy orographic rainfall continued to fall until the front cleared southwards on 7 February. At Kinloch Hourn the two-day total of 306.1 mm set a new Scottish record. The flood waters from the west travelled down Glen Garry and Glen Moriston into the Great Glen, then eastwards down Loch Ness to the short river Ness and the sea. On the final stage they swept away the railway bridge, which had stood since 1862.

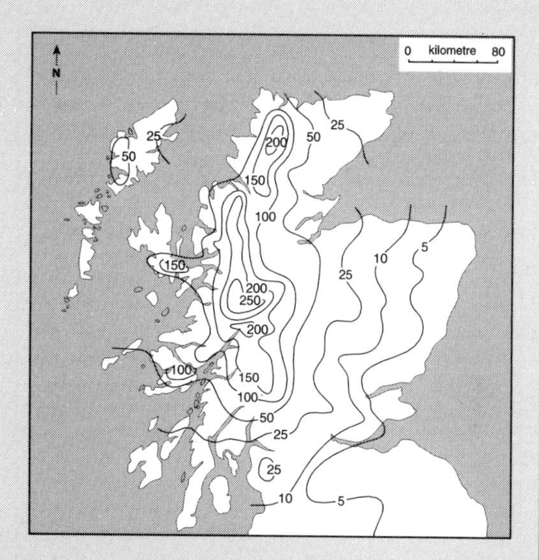

Figure 10.7 Total precipitaion in Highland Scotland for the 48 hours ending 0900 GMT on 7 February 1989

restriction of sunshine in winter by the surrounding mountains is likely to delay the thaw of snow, it is about sixty days, with the highest value being 102 days in the winter of 1950/51 and the lowest fourteen days in 1956/57. There are few centres of habitation in the Highlands of Scotland at a higher altitude than Braemar, which can be taken as representing the extreme value for towns and villages. By contrast, the average of about fifteen days for Fort Augustus at 21 m (Figure 10.10) reflects the degree of shelter it enjoys by virtue of its location in the Great Glen. In the winters of 1963/64 and 1971/72 only one day was recorded as having snow lying at 0900 GMT. At Tiree (at 9 m on the west coast) the annual mean number of days of snow cover is only 3.9. Few sites give better expression to the mildness of the west coast's coastal climate.

In a windy climate, such as that of Scotland, snow cover over the mountains is strongly affected by the terrain, since snow is blown off sharp ridges and into corries. Over the higher mountains snow can fall with winds from any direction (Davison 1987). In most years snow cover is permanent from about mid-November to May, although on the New Year's Day of the exceptionally mild winter of 1988/89

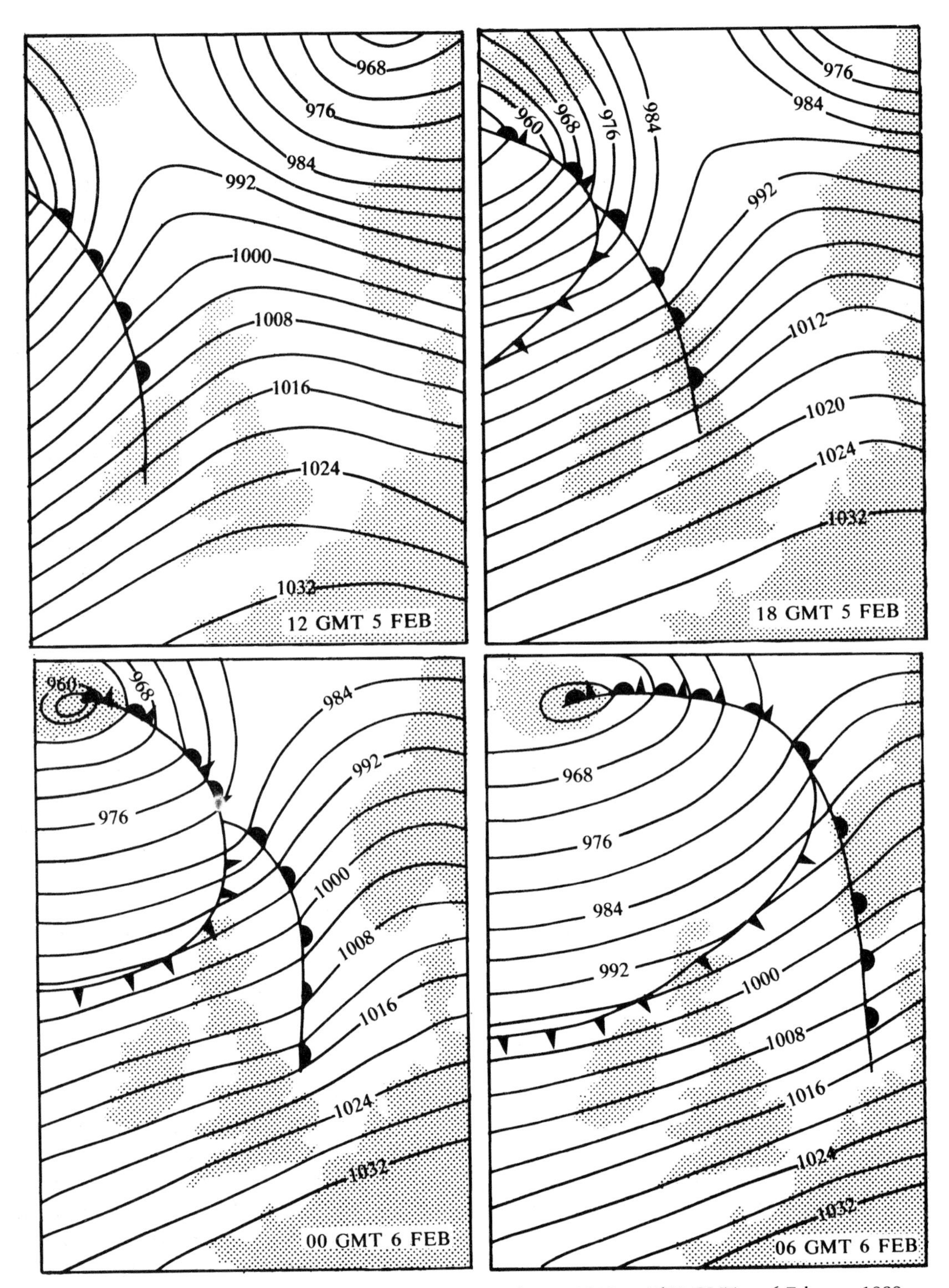

Figure 10.8 Synoptic charts for the period 1200 GMT on 5 February 1989 to 0600 GMT on 6 February 1989

Figure 10.9 Rainfall (mm) over Scotland for the period 1 April–31 August 1984 as a percentage of the April–August average for 1941–70

there was no snow at the summits of the Scottish mountains.

On low ground, snow most frequently falls with northerly or easterly winds. In showery situations with northerly winds, snowfall is largely confined to the Western and Northern Isles, the Northern Highlands and the area to the south of the Moray Firth. Other parts of the Highlands are very well sheltered from northerly winds and enjoy days of clear skies and brilliant visibility. Easterly winds may give only light snow flurries along the east coast or there may be frequent snow showers, depending on the depth of the unstable layer. However, there may be prolonged periods of heavy snow if a frontal boundary between mild air to the south and much colder air to the north lies just to the south of or across the Highlands.

Another situation which can give prolonged snowfall is one in which a mature (occluded) depression becomes slow-moving to the south-east of the area, giving north-easterly or northerly winds over the Highlands and Islands. This was the scenario which gave rise to the blizzard of 28/29 January 1978, which affected the north of Scotland (Figure 10.11).

In showery airstreams with south-westerly or westerly winds from the Atlantic the air is not normally cold enough to give snow showers over low ground, but notable exceptions to this occurred in January 1984 and in January 1993, when snow fell at all levels. In the January 1993 event heavy snow showers were swept by a strong south-westerly wind across central Scotland, giving large accumulations as far east as the eastern Grampians. When the cold, showery airstream was followed by very mild warm-sector air, there was a rapid thaw over the whole area of the Tay catchment, including the eastern tributaries of the Isla and the Ericht. Meanwhile, heavy orographic rain was falling in the west of the catchment, and this combined with the melting snow to give floods which inundated low-lying parts of Perth.

WIND

On average the west and north of this region are the windiest parts of the British Isles, since they are fully exposed to the Atlantic and lie closest to the deep depressions of the winter months. Winds of the strength which caused major damage in the south of England during the storm of October 1987 occur on average in one year in two over the Hebrides and the Northern Isles. In Shetland there are on average over 45 days with gale (mean wind speeds over a 10-minute period exceeding 33 knots) per year. In general, wind speeds are much lower during the period from April to August than from September to March, but gales can occur in all months of the year. Inland from a windward coast the roughness of the land causes a decrease in the mean wind speed compared with that which occurs at sea, but gusts may still be of the same strength.

The flow of wind over mountains of steep slopes, such as those of Western Highlands of Scotland, is very complicated, sometimes setting up reverse circulations in the valleys or very strong downslope

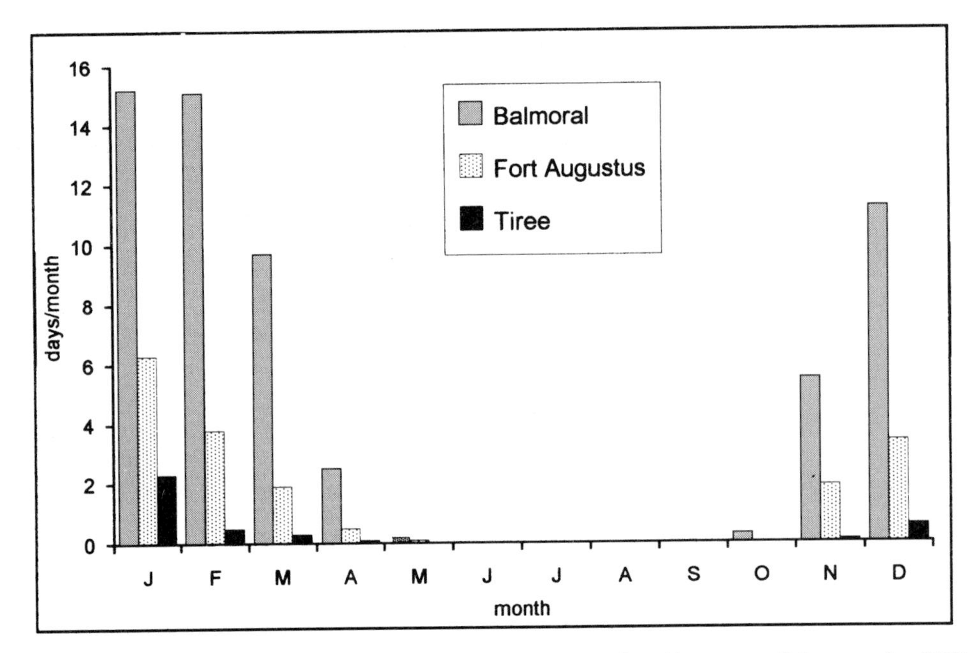

Figure 10.10 Mean frequency of days with snow lying and covering more than 50 per cent of the ground at 0900 GMT for the period 1961–90 (Tiree 1973–92)

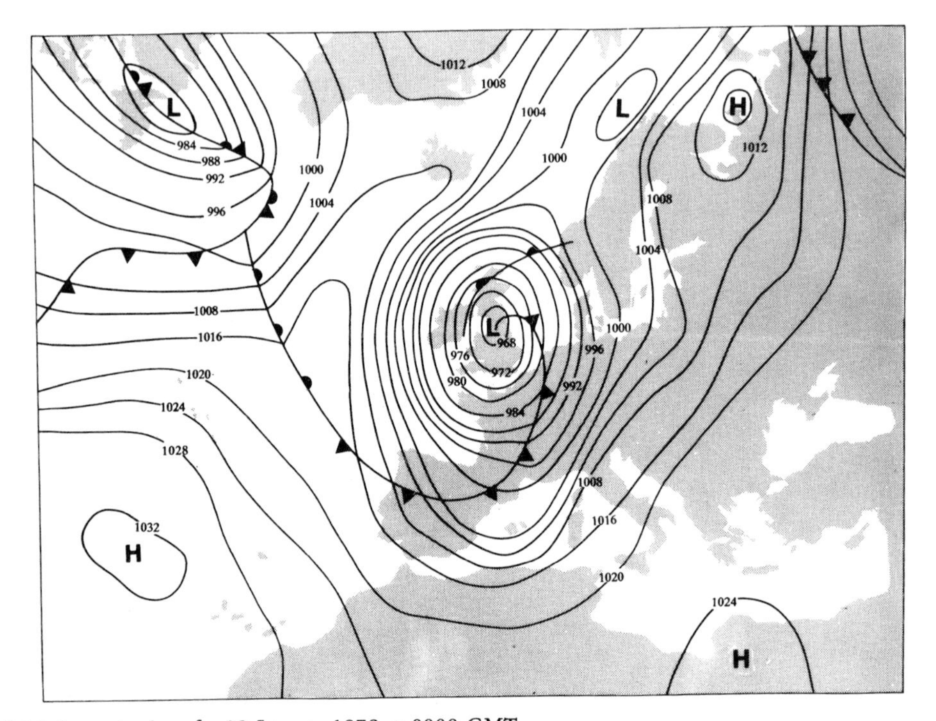

Figure 10.11 Synoptic chart for 28 January 1978 at 0000 GMT

winds. Topographic influences can be profound in areas of abrupt relief such as Highland Scotland, and if the direction of the geostrophic wind approximates to that of a valley, funnelling effects may give rise to locally intensified flow. The degree to which geostrophic airflow may be distorted has been examined by Gunn and Furmage (1976) for sites in the region. They found that airflow distortion depended upon the relative orientations of the geostrophic wind field and the grain of the surface topography, and on other aspects such as the degree of atmospheric stability. Airflow is most seriously disrupted, for example, when geostrophic airflow is perpendicular to valley orientation. Wind speeds can also be significantly reduced from the geostrophic levels.

Over smooth, rounded summits, such as the top of Cairngorm, on the other hand, wind speeds often approximate to or exceed those of the geostrophic wind, whereas the strongest winds on Ben Nevis came from a south-easterly direction and were estimated to reach speeds in excess of 113 knots. The highest gust measured by the Heriot-Watt AWS at the summit of Cairngorm was 150 knots at 0048 GMT on 20 March 1986. Gusts of this order have also been recorded on a number of occasions by the Meteorological Office AWS. Such high values, combined with the effect of below-zero temperatures, give an indication of the severity of the conditions which can occur at the summits of the Scottish mountains. Spectacularly high values have also been recorded at lowland sites (see Box 10.2).

On the low-lying island of Tiree, local topographic effects are minimal, and the annual wind rose (Figure 10.12) for the period 1970–86 is a good indicator of the distribution of wind directions over the open sea off the west coast of Scotland. A notable feature is the high frequency of winds from south-south-easterly and southerly directions. These include a substantial number of gales and normally occur ahead of eastward-moving fronts. It is likely that the barrier provided by the mountains of the Western Highlands leads to a deflection of the winds to a southerly direction and increases their strength. The concept of a 'dominant south-westerly' is not supported by this windrose.

At the summit of Cairngorm wind speeds are much stronger than on Tiree, and higher wind speed boundaries were used in constructing wind roses for 3-month periods (Figure 10.13) for eight years of data from 1985 to 1992 inclusive. The grouping of months shows that the late spring and early summer period had a greater frequency of winds from an easterly direction than high summer and early autumn. The mean wind speed exceeded 40 knots for 27 per cent of the time in the winter half of the year but only for 11 per cent of the time in the summer half of the year. The high frequency of winds from the west-north-westerly direction at all times of the year may be due to local topographic effects.

SUNSHINE

In winter the length of the period between sunrise and sunset is considerably shorter than it is in the south of England, restricting the potential number of hours of sunshine, but in summer it is much longer in the North. Despite the longer summer day length in the North, nowhere in this region do the average hours of sunshine for the summer achieve the values recorded in the south of England, although low-lying islands off the west coast, such as Tiree and Colonsay, have a well-deserved reputation for high sunshine totals during the late spring and early summer. Average monthly sunshine totals are given in Table 10.5.

At all of the sites the sunniest months of the year are May and June. July, despite having longer day lengths than May, is less sunny than the latter. The decrease in hours of sunshine in July is greatest in the west and the Northern Isles. This fall in sunshine in July in the west is another consequence of the European monsoon, which brings cloudier skies to these areas.

For the year as a whole, Tiree is only slightly more sunny than Aberdeen Airport, but the distribution through the twelve months is very different,

Box 10.2

THE STORM WHICH GAVE THE HIGHEST RECORDED GUST AT A LOWLAND SITE IN THE UNITED KINGDOM

The record maximum gust of 123 knots at a low-level site in the United Kingdom was recorded on 13 February 1989 at Kinnaird's Head Lighthouse, near Fraserborough on the east coast of Scotland, during an event which was notable for a rapid increase in wind speed over the northern half of Scotland. During the afternoon and evening of 13 February a deep depression moved east at about 35 knots, just to the north of Scotland, towards southern Norway, and the associated occlusion cleared the east coast soon after midday. Behind the occluded front winds were south-westerly, with mean speeds around 15 knots but over 25 knots in exposed western areas. Within the cold air a trough moved east behind which pressure rose quickly by as much as 19 mb in three hours.

The wind then veered to the north-west and there was a rapid increase in wind speed. At exposed sites such as Butt of Lewis and Benbecula the mean hourly wind speed increased to 60 knots. At Fraserburgh the mean speed in the hour beginning 2000 GMT was 68 knots, i.e. hurricane force. Speeds at inland sites were lower, in the range 45 to 55 knots (severe gale to storm force on the Beaufort Scale). Maximum gusts of 100 knots were recorded at Fair Isle and 93 knots at Benbecula. The north-westerly wind direction meant that upwind from Fraserburgh there was a long sea track across the Moray Firth. In the hour beginning 2000 GMT gusts of 112 and 123 knots were registered, the latter setting a new record for a low-level UK site.

Conditions subsided within four hours but not before a great deal of damage had been done across northern Scotland. Over 75,000 people in this sparsely populated area lost their electricity supply. On the Isle of Skye wooden posts carrying power lines had been broken by the force of the wind. It was estimated that speeds of 130 knots would have been necessary to accomplish this. Timely warning had, however, been given of this storm and there was no loss of life.

with Aberdeen being sunnier during the more westerly period of autumn, winter and early spring and Tiree being much sunnier during the late spring and early summer, when high pressure is most likely to lie over or near to Scotland. The dryness of the area around the Moray Firth might lead to the expectation that this would be an area with high monthly sunshine averages, but the Kinloss figures are lower than those at Aberdeen.

The mountainous character of much of this region leads to a greater frequency of cloudy skies over inland areas than occurs over the sea. In areas where there are narrow valleys surrounded by steep-sided mountains the surrounding terrain may cut off the sunshine for a considerable part of the day, with very different amounts reaching north- and south-facing slopes. This is especially important during the winter, when the Sun's elevation is low.

Across Scotland the cloudiness of the skies shows a similar response to exposure and shelter to that observed with rainfall; with predominantly westerly winds the west is much cloudier than the east and vice versa. Over the western half of the Highlands, March 1994 was the wettest March since records began, being a month of unremitting westerly winds, but the total rainfall at Aberdeen Airport was only half the monthly average. Sunshine showed a similar east–west split: Onich recorded only forty-four hours of sunshine and Aberdeen Airport 165 hours. By contrast, in September 1993 winds

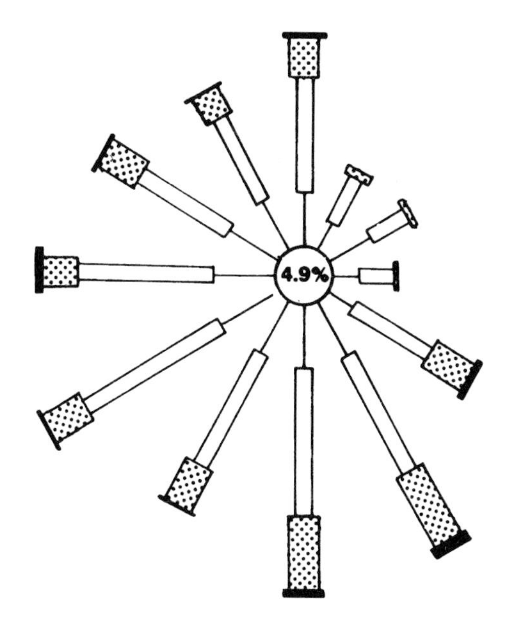

Calm or light & variable

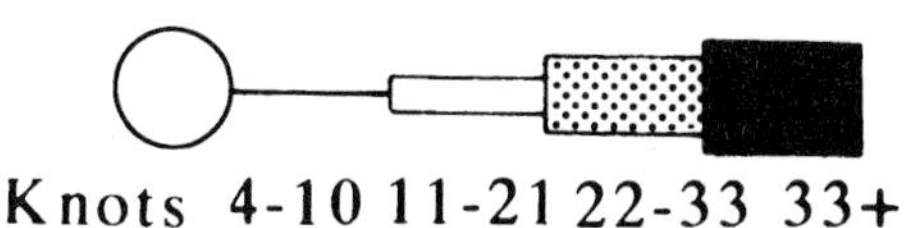

Figure 10.12 Distribution of wind speed and direction at Tiree for the period 1970–86 (after Roy 1993)

were predominantly easterly, with the western Highlands and islands recording well below average rainfall and above average sunshine while the east coast had near average rainfall and below average sunshine. At Tiree the total sunshine for the month amounted to 161 hours, whereas at Aberdeen Airport it was only 100 hours.

Summers such as 1976, which as a result of the northwards extension of the Azores anticyclone are exceptionally sunny over England and southern and eastern Scotland, are frequently cloudy over the Western and Northern Isles of Scotland, which often remain within a moist, cloudy, south-westerly airstream. By contrast, sunny summers in the west of Scotland are frequently those in which much of England experiences dull, wet and miserable weather, since the predominant pressure pattern is one of high pressure over Scotland and low pressure over the south of England. A notable example of this was the summer of 1968, which was the sunniest on record in Skye, with an average of seven hours of sunshine per day for the months of June, July and August, while the south of England had one of its cloudiest summers on record. In the west of Scotland most records for average hours of sunshine in any calendar month have been set in May, with Tiree recording 329 hours in May 1946 and again in 1975. At Prabost on Skye the sunniest month was again May 1975 with 300 hours. This was a month when air pressure was high to the north-west of Scotland, reversing the normal climatic gradient across the British Isles.

FOG AND VISIBILITY

With a lack of major sources of pollution this part of Scotland enjoys remarkably good visibility, particularly in polar maritime air masses, and it is possible on very clear days for the observers on Ben Nevis to see the Hills of Antrim, almost 200 km away. The three stations cited in Figure 10.14 therefore have annual mean frequencies of fog days far below those of most English regions: Duntulm (Skye), 4.1; Braemar, 5.3; and Kinloss, 2.1. Radiation fogs are generally short-lived, and the major sources of poor visibility are sea fog, which may creep in over the coast, and hill fog, where low cloud covers all high ground. The poetic name for the Isle of Skye, Eilean a' Cheo (Island of Mist), refers to low cloud swirling about the mountains and not prosaic lowland mist or fog. At Prabost on Skye the average frequency of visibility of less than 1,000 m at 0900 GMT was less than one in any month of the year.

A study made by Buchan (1902) of lighthouse observations of fog showed that the greatest frequency occurred at sites, such as Barra Head (208 m), Cape Wrath (122 m) and Sumburgh Head (91 m), where the lighthouses were situated above high cliffs and frequently lay above the cloud

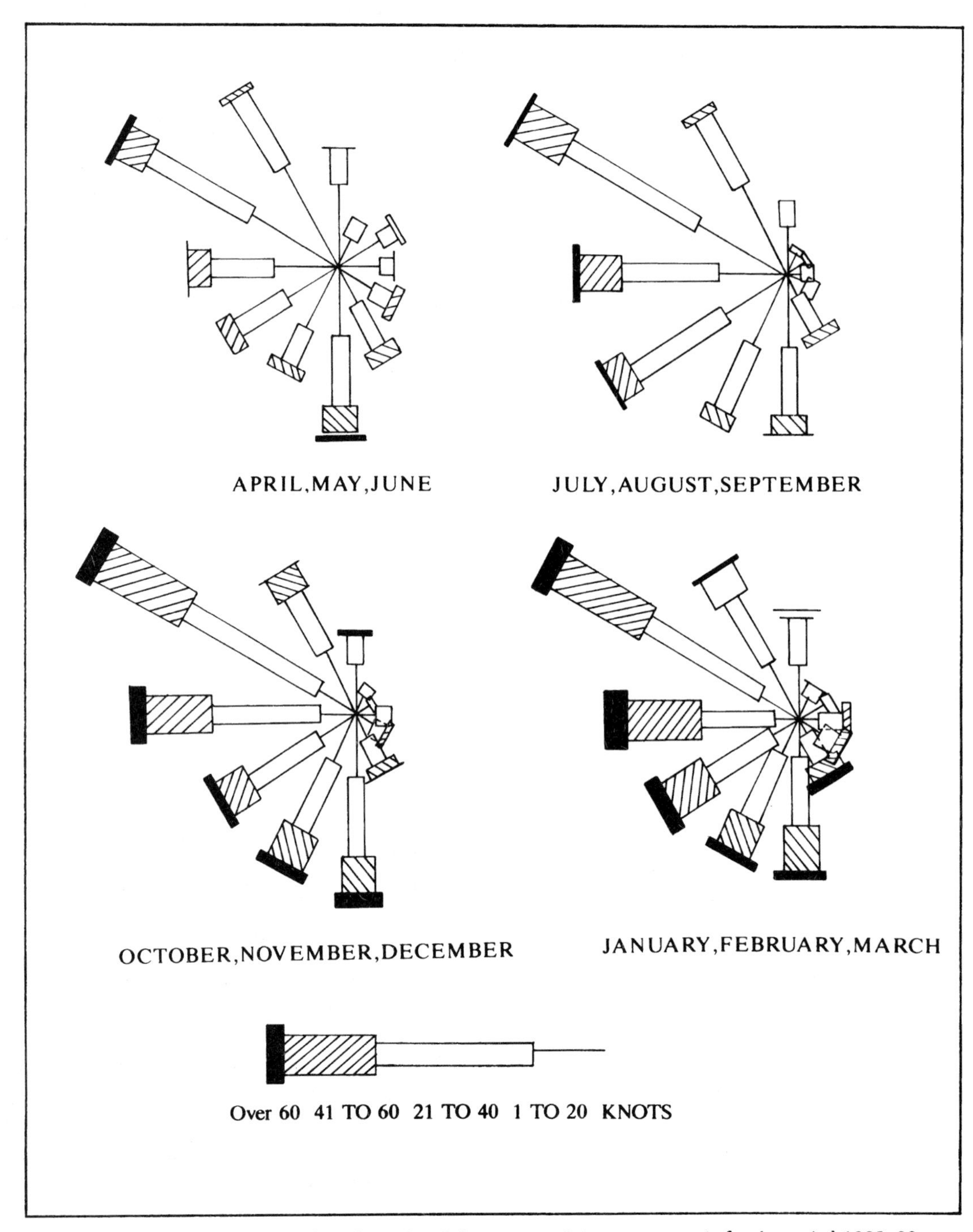

Figure 10.13 Seasonal distribution of wind speed and direction on Cairngorm summit for the period 1985–92

Table 10.5 Mean monthly and annual sunshine totals (hours) for the period 1961–90 (Ben Nevis 1884–1903)

Location	*Alt. (m ASL)*	*Jan.*	*Feb.*	*Mar.*	*Apr.*	*May*	*June*	*July*	*Aug.*	*Sept.*	*Oct.*	*Nov.*	*Dec.*	*Year*
Lerwick	82	22	52	85	132	147	151	123	124	94	61	31	13	1,037
Kirkwall	26	29	60	97	135	161	156	129	129	96	73	36	20	1,119
Stornoway	15	34	68	106	155	184	169	129	133	102	80	44	25	1,229
Tiree	9	39	66	108	170	214	195	159	156	111	78	48	30	1,375
Onich	15	24	58	81	129	156	145	119	117	78	60	33	17	1,022
Kinloss	5	42	76	105	136	167	163	147	135	105	84	50	35	1,244
Braemar	339	25	55	95	139	162	169	158	147	103	69	36	19	1,177
Aberdeen	65	52	73	112	145	176	173	157	151	116	93	62	43	1,353
Ardtalnaig	130	22	54	91	136	164	163	154	134	95	63	33	13	1,122
Ben Nevis	1,344	22	42	55	80	116	127	85	58	62	42	28	18	736

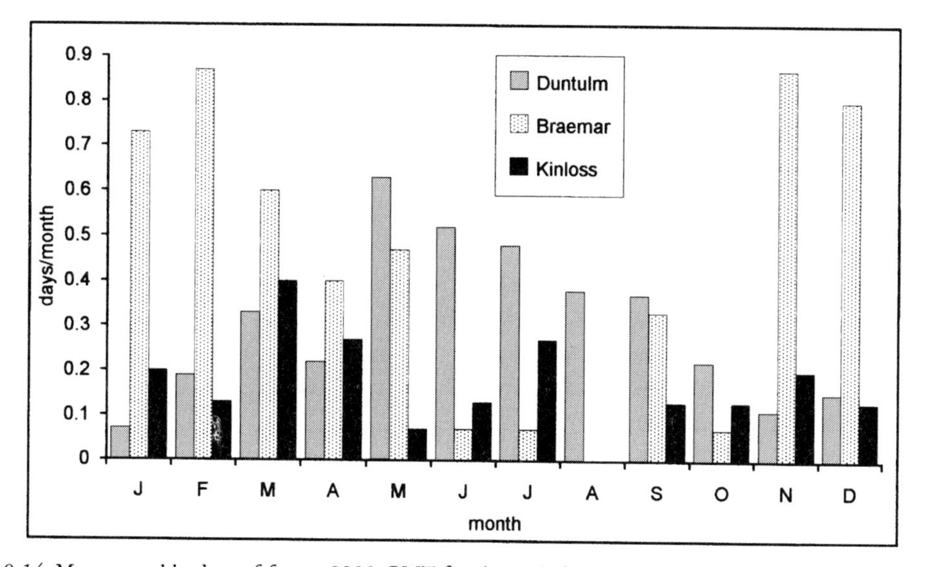

Figure 10.14 Mean monthly days of fog at 0900 GMT for the period 1978–90 (Duntulm 1946–72)

base. For most west-coast sites the greatest frequency of fog was during the summer months (Duntulm (Figure 10.14) is a good example). In Orkney and Shetland May and July were equally prone to fog. Buchan also showed that along the east coast the fogs occurred almost entirely with south-easterly and easterly winds and along the west coast with winds from a westerly direction, but that in Orkney and Shetland they occurred with winds from all directions. The greater frequency of fogs in July along the west coast as compared with the east coast is another feature of the European monsoon.

Findlater *et al.* (1989) have described the haar (sea fog) of the Moray Firth. These fogs are most common in spring and early summer (see Kinloss in Figure 10.14 for a typical example). Once established under anticyclonic conditions, such haars can persist almost indefinitely over the sea, with their movement being determined by the local winds, especially sea and land breezes. During particular events local circulations of this type can lead to some areas suffering from persistent haar, while others show daytime clearances.

The other major cause of poor visibility in this region is hill fog. In very moist warm sectors the cloud base along the west coast may be as low as 100 m, with all the high ground being shrouded in hill fog. To the lee of high ground, cloud bases are much higher, and very marked changes can be observed over short distances. Showery westerly airstreams may also be moist enough for cloud to cover the summits of the mountains in the west, and at Ben Nevis Observatory the percentage of time that the summit was in fog ranged from about 78 per cent between November and January to about 55 per cent in May and June.

CONCLUSION

The region shows great diversity of climate, a result of both its location in the north-east Atlantic and the nature of the topography. One consequence of the region's location is the windiness of the winter half of the year. This, more than temperature, sets the altitudinal limits at which trees will grow and makes local shelter of great importance. Given shelter, the mildness of the winters along the west coast can be exploited, as in the world-famous gardens at Inverewe and elsewhere.

The climate reflects the interaction between the weather systems and the topography. The vertical and horizontal scales of the mountains of the Highlands of Scotland are large enough for very considerable shelter to be experienced by areas which at any particular time lie to their lee. Because of the overall predominance of westerly systems the long-period averages show very high rainfall totals and cloudiness over the mountains just inland from the west coast and much drier and less cloudy conditions in the east, but in months when the normal synoptic pattern is reversed, so also are the rainfall and cloudiness patterns.

Since on average the temperature decreases with height by about 6 degrees C for every 1,000 m, the contrast in temperature between the low ground and the summits of the mountains is very much greater than that experienced at sea level between southern Argyllshire and the most northerly point in Shetland. Similarly, the periods of winter snow cover are much greater over the mountains than in the Northern Isles. When the high wind speeds and periods of intermittent drought during anticyclonic conditions are also taken into consideration, it is not surprising that the plant and animal life of the high tops is similar to that of the Arctic tundra.

REFERENCES

Bleasdale, A. and Chan, Y.K. (1972) 'Orographic influences on the distribution of precipitation', in *Distribution of Precipitation in Mountainous Areas*; Geilo Symposium, Norway, WMO Report no. 326, Part II, 322–333.

Buchan, A. (1902) 'Fogs on the coasts of Scotland', *J. Scot. Meteorol. Soc.*, 12: 3–12.

Davison, R.W. (1987) 'The supply of snow in the eastern Highlands of Scotland 1954–5 to 1983–4', *Weather*, 42: 42–50.

Findlater, J., Roach, W.T. and McHugh, B.C. (1989) 'The haar of north-east Scotland', *Q. J. R. Meteorol. Soc.*, 115: 581–608.

Green, F.H.W. (1963) 'Features of the weather in the north of Scotland in January and February 1963', *Weather*, 18: 364–365.

—— (1967) 'Air humidity on Ben Nevis', *Weather*, 22: 174–184.

Gunn, D.M. and Furmage, D.F. (1976) 'The effect of topography on surface wind', *Meteorol. Mag.*, 105: 8–23.

Irvine, S.G. (1969a) 'The Shetland blizzard 1968', *Weather*, 24: 74.

—— (1969b) '1968 – Shetland's record breaking weather year', *Weather*, 24: 234–235.

Lamb, H.H. (1972) *Climate Past, Present and Future*, vol. 1: *The Fundamentals of Climate Now*, London: Methuen.

McClatchey, J., Runacres, A.M.E. and Collier, P. (1987) 'Satellite images of the distribution of extremely low temperatures in the Scottish Highlands', *Meteorol. Mag.*, 116: 376–386.

Meaden, G.T. (1983) 'Britain's highest temperatures: the county records', *J. Meteorology (UK)*, 8: 201–203.

Paton, J. (1954) 'Ben Nevis Observatory 1883–1904', *Weather*, 9: 291–308.

Pedgley, D.E. (1967) 'Why so much rain?', *Weather*, 22: 478–482.

Roach, W.T. and Brownscombe, J.L. (1984) 'Possible causes of the extreme cold during winter 1981–82', *Weather*, 39, 362–372

Roy, M.G. (1983) 'Ben Nevis Meteorological Observatory 1883–

1904, Part 1: Historical background, methods of observation and published data', *Meteorol. Mag.*, 112: 318–329.

—— (1993) 'Climate', in G. Robertson (ed.) *Isle of Skye Data Atlas*, Portree: Skye Forum.

Spackman, E.A. (1994) 'Calculation and mapping of rainfall averages 1961–90', personal communication based on paper presented at BHS National Meeting, University of Salford, December 1993.

Tabony, R.C. (1987) 'The estimation of extreme minimum temperature in the United Kingdom', *Meteorol. Mag.*, 116: 285–290.

Webb, J.D.C. and Meaden, G.T. (1983) 'Britain's lowest temperatures: the county records', *J. Meteorology (UK)*, 8: 269–272.

Werrity, A. and Acreman, M.C. (1985) 'The flood hazard in Scotland', in S.J. Harrison (ed.) *Climatic Hazards in Scotland: Proceedings of the Joint Royal Scottish Geographical Society and Royal Meteorological Society Symposium, University of Stirling 1984*, Norwich: GeoBooks.

Weston, K.J. and Roy, M.G. (1994) 'The directional-dependence of the enhancement of rainfall over complex topography', *Meteorological Applications* 1: 267–275.

11

IRELAND

John Sweeney

INTRODUCTION

Ireland functions as a meteorological sentry post for much of north-western Europe, and it is in its vicinity that the skirmishes between air masses which determine the climatic fingerprints of much of the continent are often first observed, and their sting removed. It is here that the harbingers of weather for areas further east may be first assessed and the knowledge used to provide early warning of imminent weather events. The shelter effect of Ireland constitutes one of the principal controls on climate for much of Britain, and it is in the vicinity of Ireland that any significant changes in oceanic circulation associated with global warming in Europe will be first detected.

It is important to realise that Ireland possesses a rich and varied climatic mosaic of its own within its 84,000 km^2. About 5 per cent of its area lies over 300 m above sea level (ASL), and the sometimes complex interplay between airstreams and relief ensures that the stereotypical view of the island's climate as a relatively homogeneous maritime one is only partly correct.

Unlike most islands of its size, Ireland possesses a mountainous perimeter of hard ancient rocks and a relatively flat interior composed of softer, younger rocks. This can be seen in Figure 11.1, where the almost unbroken chain of north-west- to south-east-trending mountains of Caledonian age form much of the western and eastern coasts. Significant gaps in the western uplands at Sligo Bay and Galway Bay, and along the Shannon estuary, provide easier access routes for maritime influences to penetrate inland, and the interdigitation of land and sea influences gives western Ireland a complex climatic character, some parts being quite sheltered and others exposed to the full rigours of the Atlantic. To the south, the east–west trending folds of the Hercynian orogeny along the Cork and Kerry coastline provide a partial bulwark against southerly airstreams, whilst even in the extreme north-

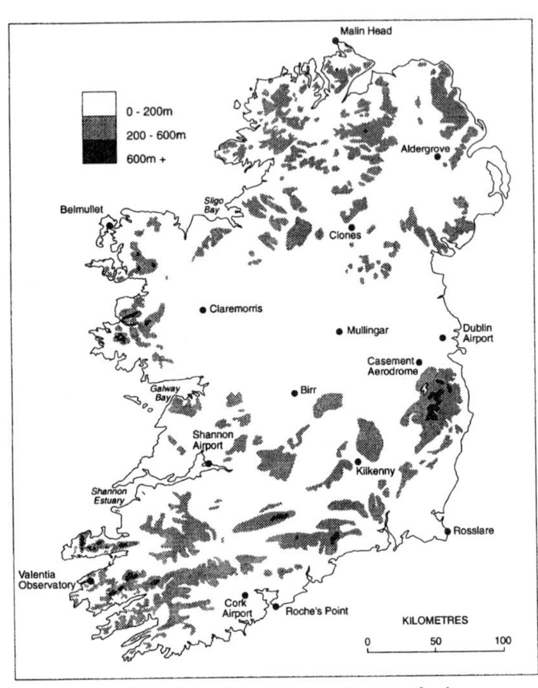

Figure 11.1 Location of weather stations and places referred to in the text

east the Antrim Basalt Plateau offers some protection from polar air masses to the lowlands around Belfast. Within this mountain perimeter the softer Carboniferous Limestones of the Central Plain are generally less than 150 m ASL and enjoy considerable shelter, often from several directions. Such a relief configuration inevitably produces a distinctive coastal-versus-interior set of climatic contrasts as well as producing distinctive geographies of most of the main weather elements associated with a particular airflow.

The Instrumental Record and Other Direct Sources of Climatic Information

Regular meteorological observations in Ireland began at Armagh Observatory in 1794 and have continued there without interruption since. During the early nineteenth century, observations commenced at a number of locations such as the Botanic Gardens, Dublin (1800), Markree Castle, County Sligo (1824), and Phoenix Park, Dublin (1829). Whilst these early records must be used with caution, from about 1880 onwards they are fairly reliable and are today supplemented by fourteen synoptic stations in the Republic of Ireland and one (Aldergrove) in Northern Ireland (Figure 11.1). These stations are manned by trained observers and provide hourly observations. Approximately seventy climatological stations and 650 rain-gauge sites (10 per cent of these are recording rain-gauges) exist, and further data from lighthouses, oil rigs, ships and buoys complete the network (Fitzgerald 1994). An overwhelming concentration on lowland locations is in evidence, though Betts (1982) notes the addition of sixteen stations above 300 m ASL since the 1960s in Northern Ireland. As with their counterparts in the UK Meteorological Office network, sites are chosen to minimise (though they cannot wholly exclude) site-specific influences on the recorded weather.

In common with countries such as Japan, Iceland and China where a long tradition of chronicling significant meteorological and other events exists, there is in Ireland also a wealth of documentary material which may be used for inferring climatic conditions before the era of observations got under way (Shields 1983; Tyrrell 1995). Early newspapers, estate records, weather diaries (some with instrumental observations such as those of William and Samuel Molyneux of Dublin (1684–1709), Thomas Neve of Ballyneilmore, County Derry (1711–1725) and Richard Kirwan of Cavendish Row, Dublin (1780–1808)) provide useful data back into the Little Ice Age times of the seventeenth and eighteenth centuries. Early Christian manuscripts are an as yet untapped source of weather information for the Medieval Warm Period around the end of the first millennium AD, whilst in even earlier times, reference to a storm on Lough Conn in the Annals of the Four Masters (Figure 11.2), allegedly in 2668 BC, is claimed to be the earliest documentary refer-

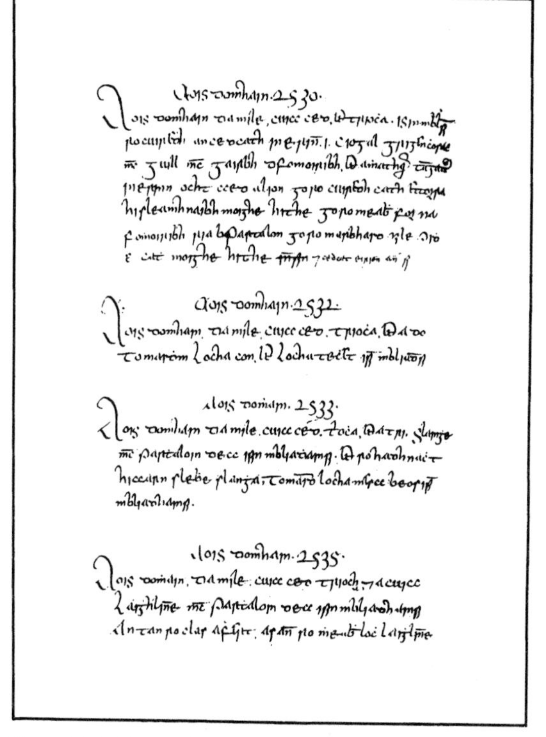

Figure 11.2 The earliest documentary reference to a meteorological event in Britain or Ireland: a storm on Lough Conn allegedly in 2668 BC – from the Irish Annals of the Four Masters

ence to a meteorological event in Ireland or Britain (Britton 1937).

TEMPERATURE

The equability of Irish climate is the characteristic which distinguishes it most from that of Britain, with the moderating influence of the North Atlantic Drift the primary cause. Flowing at 16–32 km/day, the surface water takes about eight months to reach the Kerry coast from Florida, by which time its temperature in January is about 10°C. This is some 3 to 4 degrees C warmer than the air over Ireland and has a dual effect in winter. First, it enables a transfer of sensible and latent heat to frontal systems, which are thus rendered more active at this season in western parts. Stable tropical maritime air masses which have cooled gradually on their journey from the Azores to the British Isles produce the stratus so prevalent in Irish weather. They may also be incorporated into rapidly developing wave depressions in mid-Atlantic and induced to release their water vapour load over western upland areas. Second, the winter warmth of the nearby ocean triggers convective motions in the vicinity of Ireland, particularly in unstable polar maritime air moving from the north-west in the rear of a depression. Such airflows may be warmed by up to 9 degrees C as they pass across the main axis of the North Atlantic Drift between Ireland and Iceland (Sweeney 1988). The convective cells generated tend to release the bulk of their precipitation over western Ireland, especially where their forced ascent over the mountains occurs.

The location of Ireland astride this oceanic conveyor, however, means that both the seasonal rhythm and year-to-year variations in annual average temperature are determined principally by changes in the thermal characteristics of the nearby sea. This is vividly demonstrated in Figure 11.3, where the Malin Head monthly temperature cycle and annual mean temperatures follow slavishly the offshore water temperatures at that location. The site's notably temperate regime is given statistical support in Table 11.1. The highest recorded sea temperatures at this location not surprisingly were during 1995, when the warmest summer on record occurred.

Thermally conservative marine influences also explain the relatively small year-to-year variations in annual temperatures apparent in Figure 11.3. Throughout Ireland, only one year in ten typically departs from the annual average by more than 0.5 degrees C. Mean annual temperatures also exhibit a south-west to north-east gradient as a consequence of differences in offshore sea temperatures, which range from 10°C off the south-west coast to less than 7°C off the north-east coast in late winter, and from 15°C to 13.5°C respectively in the same locations towards the end of summer. Accordingly, equability between summer and winter temperatures decreases towards the north-east as well as with distance from the coast. Mean minimum temperatures in the Belfast region (Aldergrove), for example, do not exceed 4°C until early May on average, whilst in the extreme southwest only briefly in February do mean minimum temperatures fall below this level (Table 11.1). In contrast, mean maximum temperatures at Aldergrove exceed those of Valentia from May until the beginning of September.

Mean January temperatures range from just over 7°C along the south-west coast to 3.5°C in the interior of Northern Ireland (Figure 11.4). The coldest month at coastal stations is typically February, as opposed to January at inland locations. The month of lowest average temperatures is also typically the month with the highest variability, since at this time of the year marked tendencies for prolonged spells of similar weather exist and either high- or low-index circulations frequently get entrenched for a few weeks at a time. In favourable years, winter can pass almost unnoticed in the extreme south-west; unfortunately, some summers have a similar tendency!

Even though July temperatures inland tend to be higher than on the coast, with daytime maxima above 19°C as opposed to 16°C, the contrast in summer mean temperatures is much less than in winter and the pattern of isotherms is not explicitly

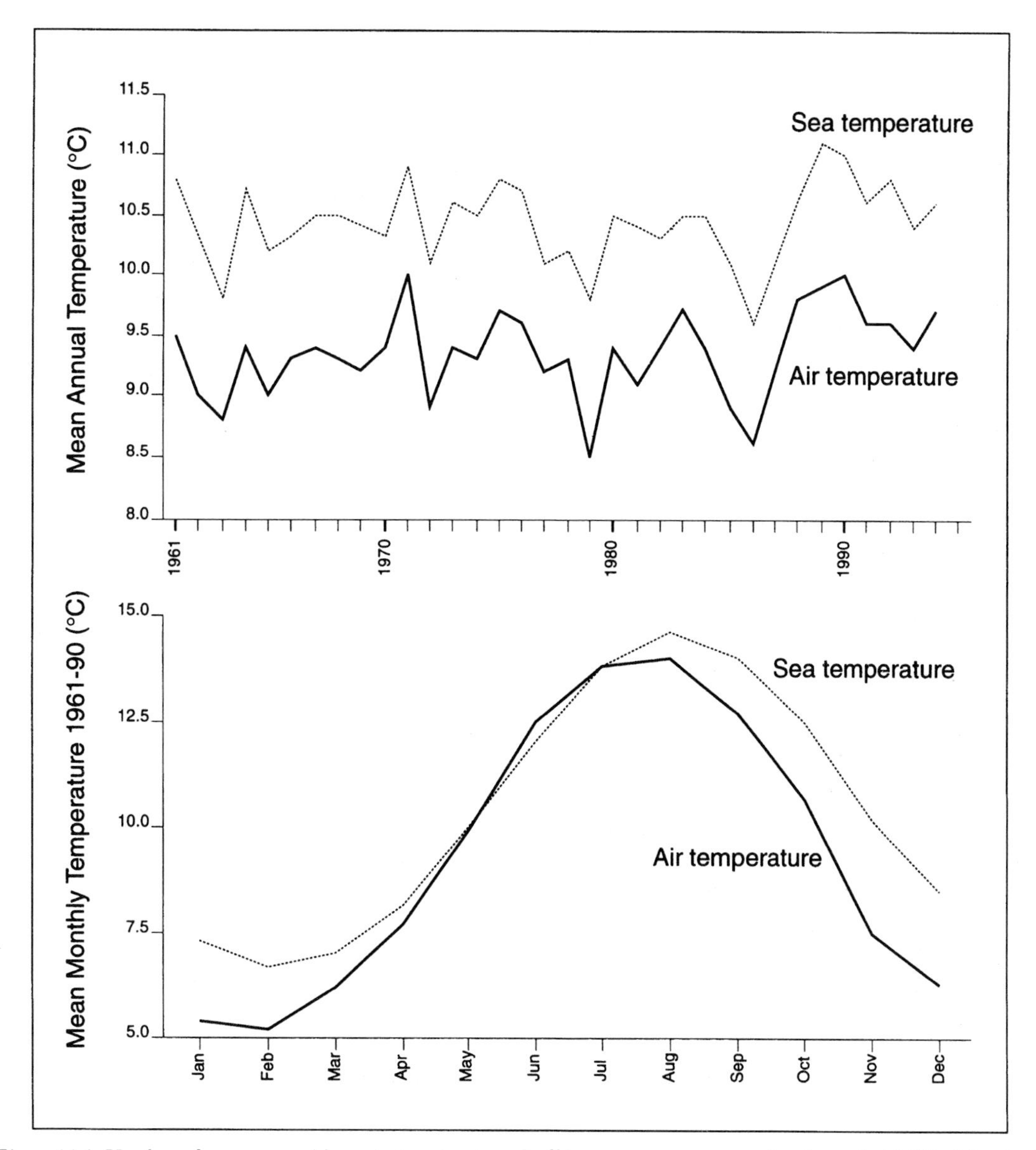

Figure 11.3 Yearly and mean monthly variations in air and offshore sea temperatures for the period 1961–94 at Malin Head

concentric, with more of a tendency to reflect radiation income differences. The latitudinal control is clearly detectable, with Cork at 16°C in August almost 2 degrees C warmer than Malin Head (Figure 11.4). Throughout the year, cities such as Dublin and Cork probably enjoy an enhancement of their temperatures by over 1 degree C due to their urban heat-islands (Sweeney, 1987; see also Box 11.1).

The oceanic influence also manifests itself in the absence of marked temperature extremes. Meskill (1995b) has described the summer of that year, which was one of the warmest in Ireland since at

Table 11.1 Mean monthly maximum and minimum temperatures (°C) for the period 1961–90

Location	*Alt. (m ASL)*	*Jan.*	*Feb.*	*Mar.*	*Apr.*	*May*	*June*	*July*	*Aug.*	*Sept.*	*Oct.*	*Nov.*	*Dec.*	*Year*
Dublin	81	7.4	7.6	9.6	11.8	14.7	17.8	19.4	19.0	16.8	13.7	9.7	8.2	13.0
(Casement)		1.8	1.7	2.6	3.6	6.1	8.9	10.8	10.6	9.0	6.9	3.4	2.5	5.7
Valentia	12	9.4	9.3	10.5	12.3	14.4	16.5	17.9	18.0	16.7	14.3	11.4	10.1	13.4
		4.2	3.9	4.7	5.8	7.8	10.3	12.0	12.0	10.7	8.8	6.1	5.1	7.6
Kilkenny	66	7.7	7.9	10.0	12.4	15.1	18.1	19.9	19.6	17.2	13.9	10.1	8.4	13.4
		1.4	1.6	2.3	3.4	5.6	8.4	10.4	9.9	7.9	6.1	2.8	2.1	5.1
Claremorris	71	7.2	7.6	9.6	12.0	14.5	17.0	18.4	18.2	16.1	13.2	9.5	7.9	12.6
		1.4	1.3	2.3	3.3	5.5	8.2	10.2	9.8	8.1	6.3	3.0	2.3	5.1
Malin Head	22	7.6	7.5	8.7	10.3	12.7	15.0	16.2	16.6	15.3	13.0	9.8	8.4	11.8
		3.2	2.9	3.7	5.0	7.1	9.6	11.4	11.4	10.1	8.3	5.2	4.2	6.8
Aldergrove	69	6.2	6.6	8.8	11.6	14.4	17.3	18.5	18.2	15.9	13.0	8.9	7.3	12.2
Airport		0.7	0.7	1.9	3.5	6.1	9.1	10.9	10.7	9.0	6.9	3.1	2.0	5.4

Note: ASL = above sea level

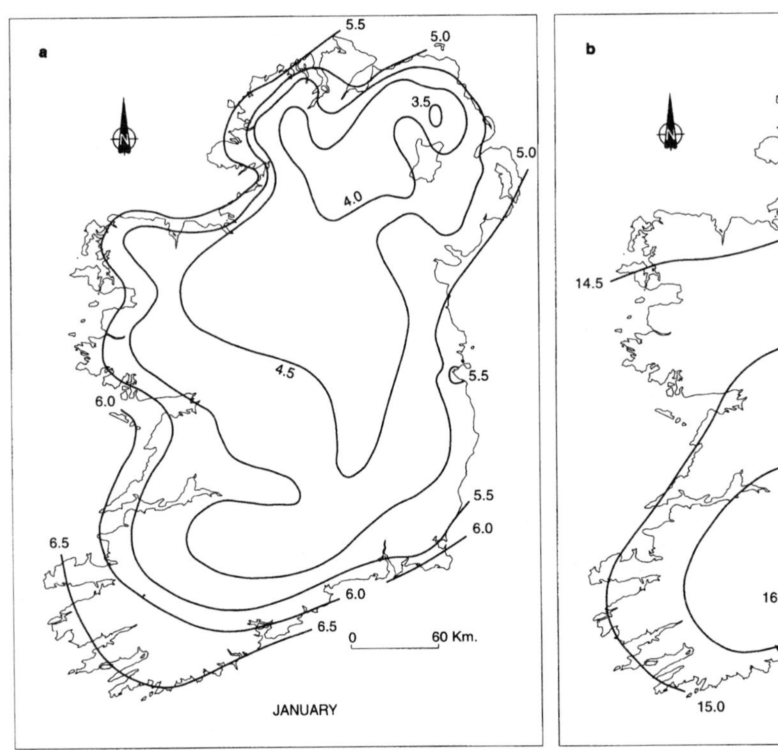

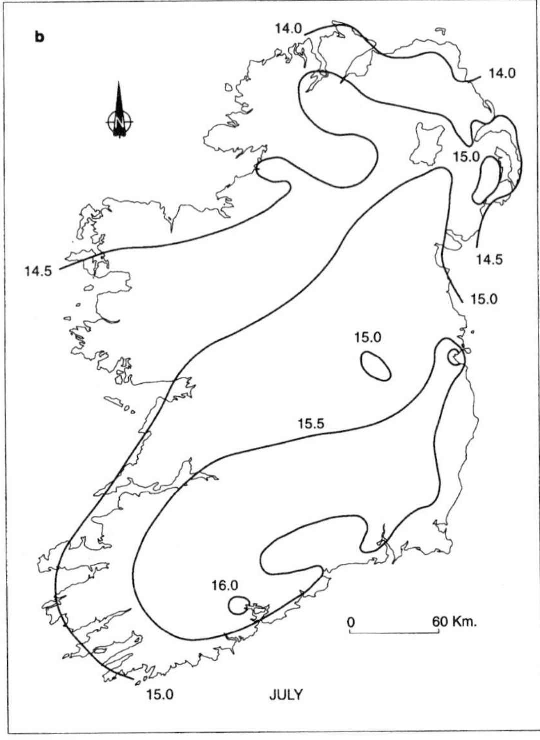

Figure 11.4 (a) Mean January temperatures (°C) for the period 1951–80
(b) mean July temperatures (°C) for the period 1951–80

Box 11.1

URBAN HEAT ISLAND OF DUBLIN CITY

The relationship between city size and the intensity of urban heat-islands in European conditions was suggested by Oke (1973) as capable of being expressed by the equation:

$$T_{(u-r)max} = 2.01 \log P - 4.06$$

in which T is temperature, P is population and the subscripts u and r refer to urban and rural respectively. This suggests a maximum intensity for a city the size of Dublin of 8 degrees C. A maximum urban–rural difference of 6 degrees C is shown in Figure 11.5 for the night of 22 November 1983, when light westerly winds of 5 knots and 3 oktas cloud cover (i.e. three-eighths of the sky had cloud cover) was observed at the airport just north of the city. The temperature 'cliff' close to the edge of the urban area and the peak warmth of the central business district are clearly apparent. The displacement towards the city of the isotherms on its south side reflects the influence of katabatic drainage of cold air downslope from the nearby Wicklow Mountains, while warm marine air is also in evidence on the low-lying tombolo on the north-east of the city. Like many cities in Britain (see Box 3.1) and Ireland, the urban heat-island is complicated by topographic and oceanic influences.

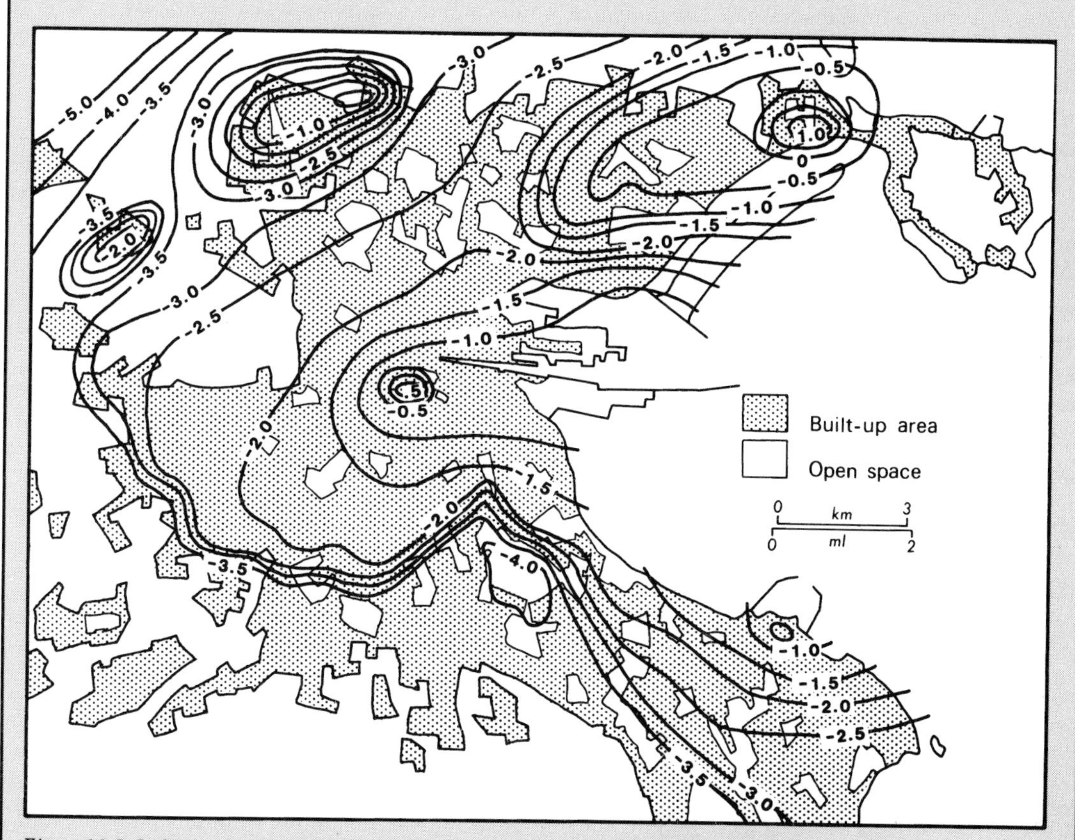

Figure 11.5 Isohyets showing Dublin's urban heat island (°C) on 22 November 1983

least, 1959. Westerly winds certainly subdued the maxima in exposed coastal areas but did not prevent Casement Aerodrome to the south-west of Dublin from recording 30.2°C on 29 June. In early August, in contrast, high pressure centred to the north encouraged an easterly airflow and it was the turn of western districts now to enjoy the warmest conditions, with maxima in excess of 30°C being widespread. Despite the record-breaking warmth of summer 1995, the highest temperature recorded in Ireland under standard conditions remains 33.3°C, measured at Kilkenny Castle on 26 June 1887 (Perry 1976). This is not a remarkably high value outside Ireland; temperatures 3 degrees C higher and more were observed in south-east England several times during the 1990s, and the comparable extreme maximum for Britain is 37.1°C (Burt 1992).

January 1987 provided an example of the synoptic conditions under which Ireland can anticipate some of its coldest weather. Betts (1987) described the conditions in Northern Ireland. On 10 January a strong anticyclone was situated over continental Europe while a depression moved south-eastwards to bring a cold polar continental airstream to much of the British Isles. This general situation persisted until 17 January, during which time Britain experienced some of its coldest weather of the century. Temperatures in Northern Ireland remained below freezing throughout the period 11–13 January. Aldergrove's maximum on the 12th was −3.2°C, whilst at Dungonnell to the north temperatures failed to rise above −4.5°C. The minima was also exceptionally low, falling to below −10°C at Newry (County Down) and at Loughall (County Armagh). Minima of −8°C were widespread, and on this occasion coastal districts were scarcely more than a degree or so warmer (Downpatrick, for example, noted −5.0°C) in strong easterly winds.

Generally, Ireland suffers less from very low minima than the rest of the British Isles. The absolute minima of −19.1°C (Markree Castle, County Sligo, 16 January 1881) and −19.4°C (Omagh, County Tyrone, 23 January 1881) are comparable to values reached in the end-of-year cold polar outbreak in central Scotland in 1995 and have been surpassed on several occasions by cold snaps even in southern England. Table 11.2 shows the overall contrast between coastal and inland locations for extreme temperature values, with maxima of over 30°C being more than a once in a century event at coastal locations though of less than a once in a half-century occurrence inland. Similarly, a value of approximately −7°C occurs every second year in the central plain, though only once in a hundred years at Valentia.

These spatial contrasts in temperature mean that frost susceptibilities, accumulated degree days, and therefore the length and efficiency of the thermal growing season are largely regulated by winter temperature variations.

Causes and Incidence of Frost

One of the situations most likely to give rise to frosts has been discussed in the preceding paragraphs. Generally, along the south, west and

Table 11.2 Extreme temperatures in Ireland and their return periods

Location	*Return period (years):* 2		5		10		50		100	
	Max.	Min.	Max.	Min.	Max.	Min.	Max.	Min.	Max.	Min.
Inland										
Birr	26.0	−7.3	27.8	−9.7	28.9	−11.4	30.7	−14.4	31.4	−15.5
Phoenix Park	25.5	−7.4	27.2	−9.8	28.3	−11.4	30.5	−14.4	31.3	−15.7
Coastal										
Roche's Point	23.0	−1.9	24.3	−2.9	25.0	−3.8	26.8	−5.2	27.4	−5.7
Valentia	24.9	−3.3	26.5	−4.6	27.4	−5.3	29.1	−6.7	29.5	−7.2

north-west coasts air temperature falls below freezing only on about ten nights during a typical winter. Frost may, however, be experienced on up to sixty-five nights in the midlands, especially where katabatic ponding of chilled air occurs. This may be common especially along the river valleys of the Barrow, Nore and Suir; the Shannon's multitude of lakes inhibits low nocturnal temperatures. The enclosed depressions in the karst limestone areas of north-west Clare make ideal frost hollows, and though extremely cold air is rare at this westerly location it does appear that in some circumstances substantial frost hollow effects can occur in such poljes. Reports from the cold winter of 1947 of cattle perishing in the Carran depression may be indicative of this (Haughton 1953). Sharp contrasts in the date of early and late frosts occur with distance from the coast. Along the south coast the last air frost of spring (2-year return period) is around 1 March; along the other coasts it is 1 April, and this recedes to 1 May throughout much of the rest of the island. Air frosts have been recorded in every month of the year at low elevations, though rarely from June to the end of September (Rohan 1975).

If an air temperature of 5.6°C is selected as corresponding to a soil temperature of 6°C, at which grass growth commences, the all-year-round growing season of the south coast gives way to 220 days of growth in the extreme north-east of the island. Of course altitude is the main determinant of growing season changes, with losses of about twenty days for every rise in altitude of 150 m. Frequently, it is wind and moisture excesses that limit cultivation at higher levels and not temperature alone, though in Northern Ireland isolated areas of crops may be found up to 300 m ASL (Betts 1982).

Isophene analysis shows that spring diffuses from Kerry north-eastwards over a period of about three weeks, an important economic consequence of Ireland's climate in providing a comparative advantage for the south-west in crop and cattle production. Such is the general mildness of winters in most parts of Ireland, though, that experiments comparing live-weight gains for outwintered and stall-fed cattle habitually show no significant differences between the two groups (Gleeson and Walsh 1967).

PRECIPITATION

In terms of its ability to convert oceanically derived water vapour into precipitation, Ireland possesses one of the world's most efficient climatic regimes. Located on an oceanic margin where an abundant supply of water vapour exists, and sitting astride the main depression tracks of the north-eastern Atlantic, Ireland has both the raw materials and the forcing mechanisms to ensure that precipitation will be a central feature of its climatology. The orographic effects of a relief configuration where all the land above 750 m lies within 56 km of the coast further emphasises these relationships.

Spatial and Seasonal Characteristics

The 1961–90 annual distribution confirms first the classic west-to-east gradient, with isolated mountain areas in the south and west receiving over 3,000 mm annually, compared with less than 750 mm for the Dublin area (Figure 11.6 and Table 11.3). The highest annual totals invariably occur in the west. In 1964, for example, 4,235 mm of rainfall was measured at Glenvickee, County Kerry (126 m ASL), though some insplashing from obstacles near the gauge may render this figure somewhat suspect (Rohan 1975). In contrast, some particularly dry years have been observed in eastern Ireland, such as 1887, when only 357 mm of rainfall was recorded at Dublin (Glasnevin). Second, the intimacy of the interaction between relief and receipt is clear, with significant rain shadow effects in the lee of the uplands, east of the Donegal and Wicklow Mountains and south east of the Cork and Kerry Mountains. Perhaps the strongest expression of the shelter effects of the surrounding topography is apparent in the upper parts of the Shannon estuary, where annual totals less than 1,000 mm are more typical of the English Midlands or the South

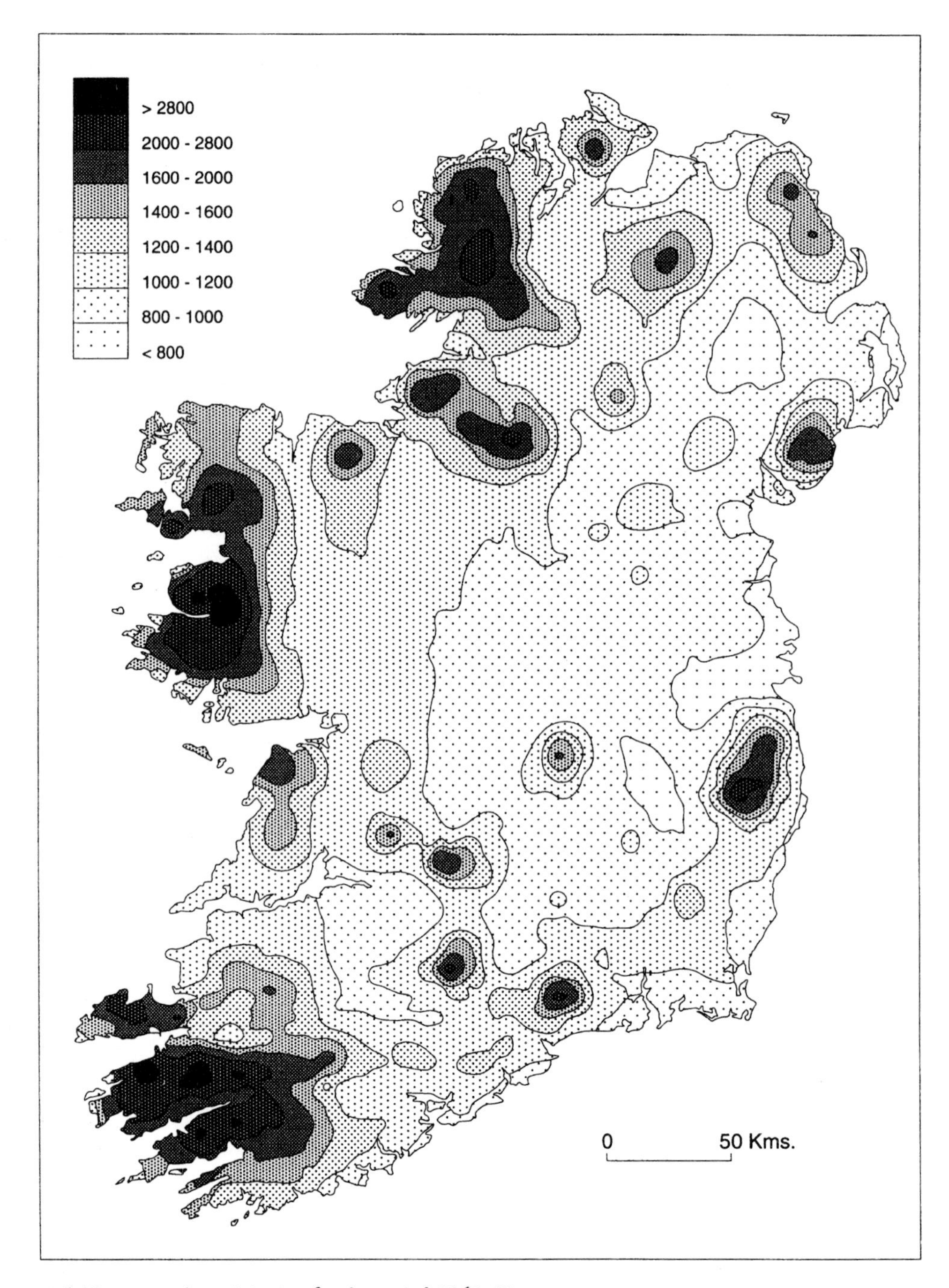

Figure 11.6 Mean annual precipitation for the period 1961–90

Table 11.3 Mean monthly and annual precipitation totals (mm) for the period 1961–90

Location	*Alt. (m ASL)*	*Jan.*	*Feb.*	*Mar.*	*Apr.*	*May*	*June*	*July*	*Aug.*	*Sept.*	*Oct.*	*Nov.*	*Dec.*	*Year*
Malin Head	25	114	77	87	58	59	65	72	92	102	119	115	103	1,061
Aldergrove	69	86	58	68	53	60	63	64	80	85	89	78	78	862
Belmullet	10	124	80	96	57	68	68	68	94	109	134	127	119	1,143
Dublin (Casement)	81	70	50	54	51	55	56	50	71	66	70	64	76	733
Shannon Airport	6	98	72	71	56	60	63	57	82	82	93	95	99	927
Valentia	9	167	123	122	77	89	80	73	111	125	157	147	159	1,430
Cork Airport	154	148	116	97	70	84	68	65	90	97	126	109	136	1,205
Mangerton, Co. Kerry	808	402	297	280	177	196	174	168	223	266	340	316	391	3,230

Note: ASL = above sea level

Downs than the west coast of Ireland. This area enjoys the protection of the mountains of Connemara and north Clare in northerly airstreams, and the mountains of Cork and Kerry in southerly airstreams. It is therefore largely dependent on precipitation borne only on westerly winds. This is the only location where the 1,000 mm isohyet reaches the west coast, which probably means that this is one of the most sensitive locations in Britain or Ireland to changes in the frequency of westerly-borne precipitation. Indeed, during the lull in westerliness in the 1960s and 1970s, the annual rainfalls at Shannon Airport, for example, have shown a significant downward trend (Sweeney 1985). Such contrasts are also overlain with other spatial gradients at a less significant scale, such as the north–south contrasts identified by Perry (1972). Most often these relate to preferred depression tracks, and the trend is most marked with systems passing to the north. Overall, however, the highest totals invariably occur in the west.

The spatial contrasts in annual receipt do not derive from contrasts in intensity, but rather from contrasts in duration. Precipitation is observed for about 6.5 per cent of the time in the Dublin area (comparable with much of lowland England), as compared with over 10 per cent of the time at sea level in the west. The number of days with rain changes from 150 to approximately 240 at these two locations and it is not uncommon for rain to be measured on all but three or four days during wet winter months along the western seaboard. Indeed, the recent increase in westerly circulation frequency (Mayes 1991) has caused severe hardship in some areas, especially in parts of counties Clare and Galway which are particularly exposed to airflows from this direction (see Box 11.2).

Some of the contrast between the east and the west is also due to a seasonal imbalance. For most parts of Ireland the February to July period is significantly drier than August to January (Logue 1978). This is especially so in western areas (Figure 11.8), and is undoubtedly related to the role of sea surface temperatures. During winter the warmth of the offshore areas means that convection is concentrated on the westerly areas; indeed, thunderstorm days reach a maximum in winter in western Ireland, in contrast to most places in Britain, where a marked summer maximum exists. Typically, thunder is observed on six or seven days per year: Valentia's annual mean is 7.1 days, Clones's 5.7 days and Cork's only 3.7 days. These frequencies are less than one-third of those from parts of south-east England. Summer is the most thundery of the seasons in most areas but in south-west Ireland at locations such as Valentia (Figure 11.9) about 50

Box 11.2

THE RETURN OF THE WESTERLIES AND WINTER WETNESS IN WESTERN IRELAND

After a long period of declining westerlies, signs of a resurgence in westerly circulation frequencies appeared in northern and western parts of Britain and Ireland during the 1980s and 1990s (Mayes 1991). This has meant increased winter rainfall in parts of western Ireland without the benefit of mountain shelter, in particular the coastal lowlands between Galway Bay and the Shannon estuary. Winter 1994/95 exemplified the problem, with many areas receiving double their normal winter complement of rainfall. Figure 11.7 illustrates the problems of these areas, many of which are karstic limestone areas with high winter water tables. This makes the flooding problems acute and long-lasting into the spring. At Gort, County Galway (155 m ASL), for example, it will be seen that measurable rain was recorded on every day except one between 1 December 1994 and 1 March 1995. Seventy-one wet-days (days with more than 1 mm of precipitation) were recorded during these eighty-nine days of winter 1994/95, with total amounts of 713 mm during this period (and in excess of 1,077 mm for the October to March period as a whole). It is hardly surprising that flooding of much farmland persisted well into the spring. In the longer term the picture is not promising since wetter winters in western Ireland seem to be likely consequences of CO_2-led global climate changes.

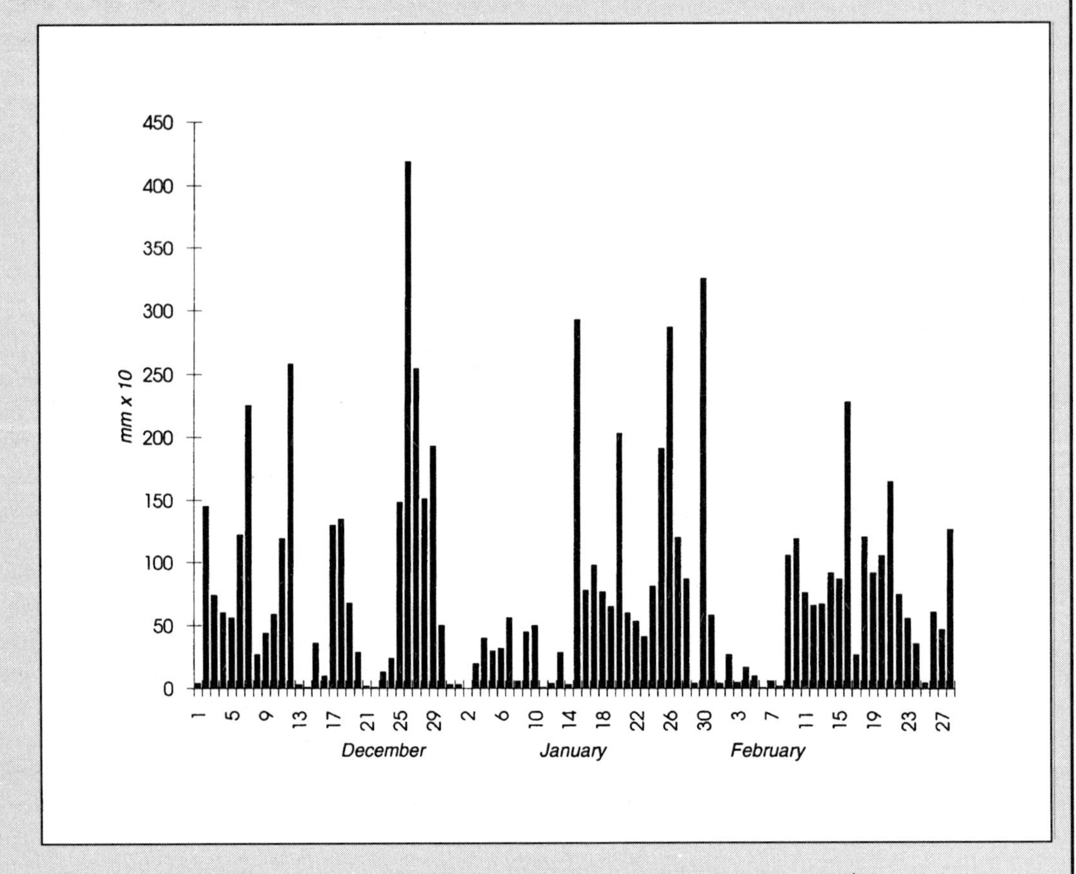

Figure 11.7 Daily rainfall (mm) at Gort (Derrybrien), Co. Galway, during the winter 1994–5

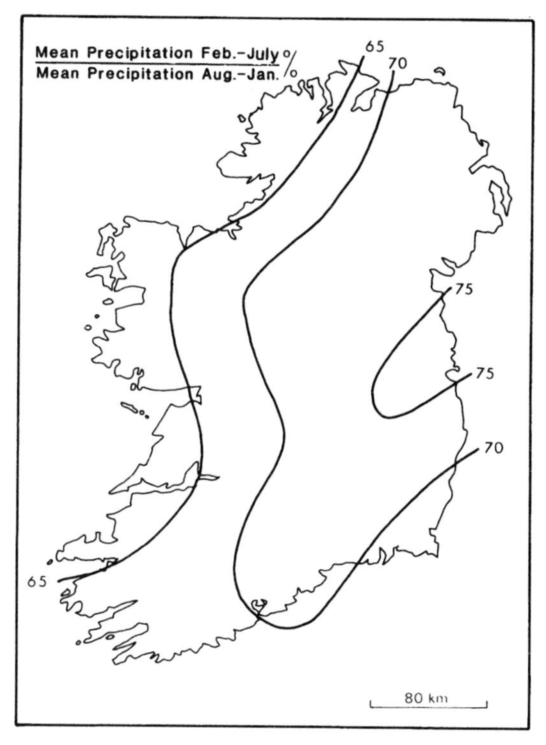

Figure 11.8 Seasonality in precipitation: February to July as a percentage of August to January totals

per cent of observations occur in the three winter months. The abiding warmth of the summer of 1995 has already been described. The frequently anticyclonic conditions also imposed stability on the atmosphere and limited the scope for thunderstorm activity despite the notably high surface temperatures. Nevertheless, occasional interruptions to this regime did occur, especially towards the close of June when southern Ireland experienced widespread thunder, hail and intense rainfall (Meskill 1995a). On 30 June a weak front over northern districts disturbed the atmosphere's stability by introducing cold air at high level. During the afternoon a convergence zone between south to south-east and north-easterly winds developed over the Galty Mountains to the north of Cork. Deep cumulus clouds formed in the rising air above the high ground where instability was sufficient to sustain several hours of intense thunderstorm activity. Rainfall, though typically localised, exceeded 30 mm in many places and hailstones of 20 mm diameter were recorded.

The oceanic warmth also assists the transfer of sensible and latent heat to frontal systems, which are at their most active at this season in western parts. During summer, by contrast, the main focus of convective activity is over land, especially the warmer areas of the east. This characteristic is seen in Figure 11.9, where the Clones and Aldergrove monthly thunder frequencies have a marked summer peak. Even stable westerly airflows may be rendered unstable by heating from below in their passage across the island, and cloud cover and rainfall amounts may increase from west to east on such occasions. The east thus gains a convective summer rainfall component which balances out its annual regime of precipitation. The wettest month over the century of records from the Phoenix Park in Dublin for example is August – rather unexpected for an island often heralded as exemplifying a maritime-controlled climate.

Daily extremes of rainfall tend not to be associated with frontal passages alone, of which there are about 170 in a typical year, but rather where orographic and/or convectional enhancement occurs, especially in summer. The two highest 24-hour totals exemplify this. First, the Mount Merrion thunderstorm of 11 June 1963 produced 184.2 mm in a 24-hour period in this south Dublin suburb. Some 75 mm of this amount fell in one hour (Morgan 1971). The annual average precipitation at this site is approximately 800 mm. In this case the urban heat-island was probably also an instrumental factor in intensifying convective influences. Even more exceptional, however, was the remnants of Hurricane Charley, which produced 24-hour falls of up to 280 mm in the Wicklow Mountains on 26 August 1986 (see Box 11.3 and Figure 11.10).

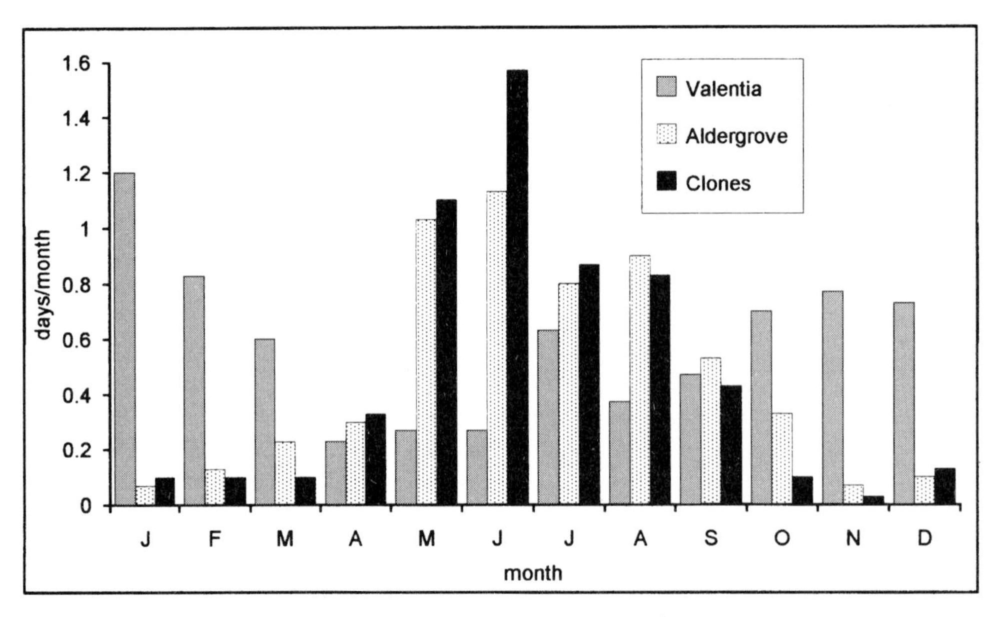

Figure 11.9 Mean monthly number of days with thunder for the period 1961–90

Box 11.3

HURRICANE CHARLEY – 25 AUGUST 1986

What is popularly known in Ireland as 'Hurricane' Charley can be traced back to a tropical storm which appeared off the coast of the Carolinas (USA) on 15 August 1986. Strengthening to formal hurricane status by the 17th, the disturbance tracked eastwards across the Atlantic and became incorporated into the mid-latitude circulation as a progressively weakening depression. On 24 August, when it was still over 500 km south-west of Kerry, it began to deepen rapidly. Over the next twenty-four hours it tracked along the south coast and into the Irish Sea, where it produced gusts in excess of 55 knots (force 10) and exceptionally large amounts of rainfall in the east and south of the country.

Although an estimated 280 mm of rain fell at Kippure (about double the average August total) this site is located at 750 m ASL and is not considered representative for official purposes. However, at Kilcoole, a lowland station just south of Dublin, 200 mm was measured for the day, thereby setting a new maximum daily record for Ireland. Over extensive areas of eastern Ireland record amounts of rainfall were recorded, enhanced by orographic lifting of this tropical airstream over the eastern slopes of the Wicklow Mountains (Figure 11.10).

Major geomorphological changes in the uplands were caused by the swollen mountain rivers and, inevitably, further downstream, severe problems arose, where the worst flooding for a century occurred in Dublin city. Both of the river Liffey's mountain tributaries, the Dargle and the Dodder, burst their banks during the height of the storm and flooded a total of 416 houses in the city, some to a depth of 2.5 m. Floating debris threatened to block the arches of city centre bridges dangerously close to being overtopped. Many thousands of trees were blown over, particularly fine old deciduous trees in full leaf and thus at a considerable aerodynamic disadvantage.

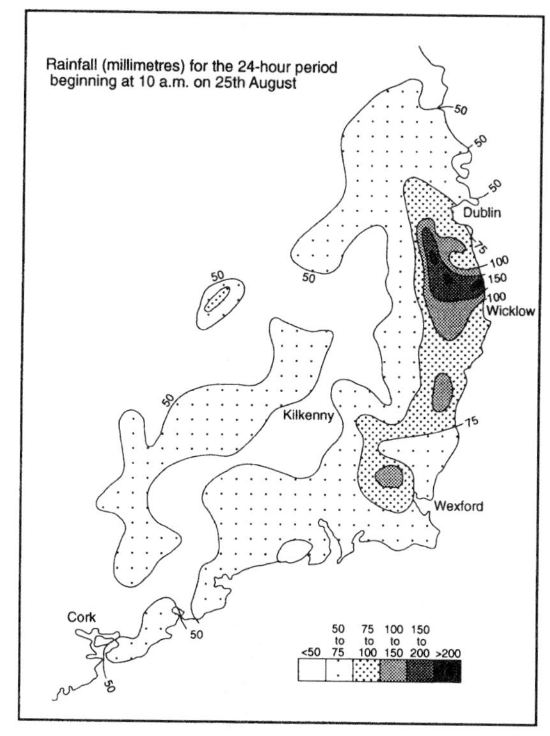

Figure 11.10 Rainfall in south-east Ireland associated with the passage of Hurricane Charley (for the 24 hours beginning 1000 GMT on 25 August 1986)

Snowfall

With January and February mean temperatures over 4°C, snowfall is not a major component of Ireland's precipitation regime. At locations near sea level, snow seldom persists for more than a day or so and is most likely to linger in the eastern and north-eastern interior areas where maritime influences are minimised. At all sites snow is observed to fall on far more days than it is seen to lie. Aldergrove has the highest number of days with snow or sleet observed to fall (averaging thirty-five days per year) but the cooler inland locations of Birr, Mullingar and Clones tend to record higher number of days with snow lying (the annual average at Clones is 11.3 days, one of the highest in the Republic). By way of a contrast, Valentia averages less than six days with snow or sleet falling, and in many southern parts no significant falls of snow occur in one year out of two. Valentia's average annual total of days of snow lying is only 0.9. Elsewhere the annual mean ranges mostly between two and seven days. The western mountains are remarkably free of heavy snowfalls, and westerly airstreams seldom cool sufficiently in their passage over these areas. Rather it is the mid-winter blast of continental polar air as the Siberian anticyclone grows westwards across Europe that ushers in bitterly cold air from the east. In its passage across the North Sea and then the Irish Sea enough moisture may be picked up to provide sometimes significant snowfalls in eastern coastal areas. Characteristically, though, it is the polar northerlies, often with small embedded polar lows, which brings the risk of heaviest falls, particularly if the boundary with warmer maritime air is stationary in the vicinity of Ireland. Snow rarely lies at low level beyond April anywhere in Ireland and is equally uncommon before November. Figure 11.12 shows that the snow 'season' is longer at more exposed locations such as Malin Head than at Valentia, which is protected from snow-bearing northerlies.

Synoptic Origins of Irish Precipitation

The sensitivity of precipitation receipt to airflow type can be demonstrated by mapping yields according to Lamb circulation types (Figure 11.13). This demonstrates that it is S, C and W airflow types that produce the heaviest falls of rain. Of these, the C type produces a relatively even distribution of precipitation across the island with only a slight reduction in the interior where oceanic water vapour may be less abundant. There appears to be a significant increase in rainfall with C airflows along the County Antrim coast, and this may well reflect the destabilising effect of the North Channel under such unstable conditions (Sweeney and O'Hare 1992), as well as convergence in low-level airflows as they squeeze between the Antrim Plateau and the Southern Uplands of Scotland.

West–east contrasts are, as expected, most marked with westerly circulations. Receipts in the extreme

Figure 11.11 Firemen monitoring the level of the river Dodder in central Dublin as it threatens to overtop the parapet of Ball's Bridge, 26 August 1986
Photo by courtesy of the Irish Times

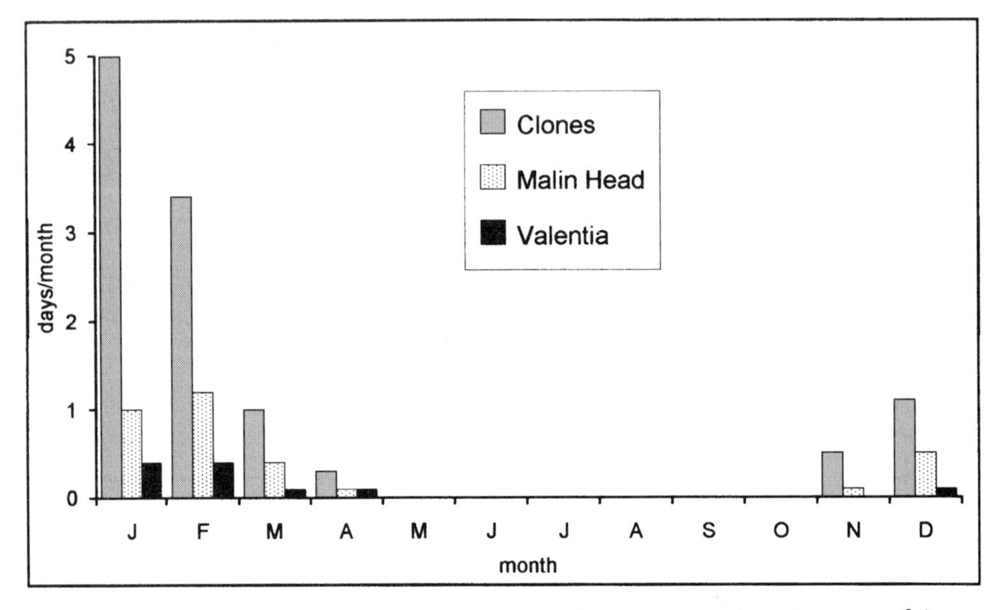

Figure 11.12 Mean monthly frequency of days with snow lying and covering more than 50 per cent of the ground at 0900 GMT for the period 1951–80, Malin Head, 1956–80

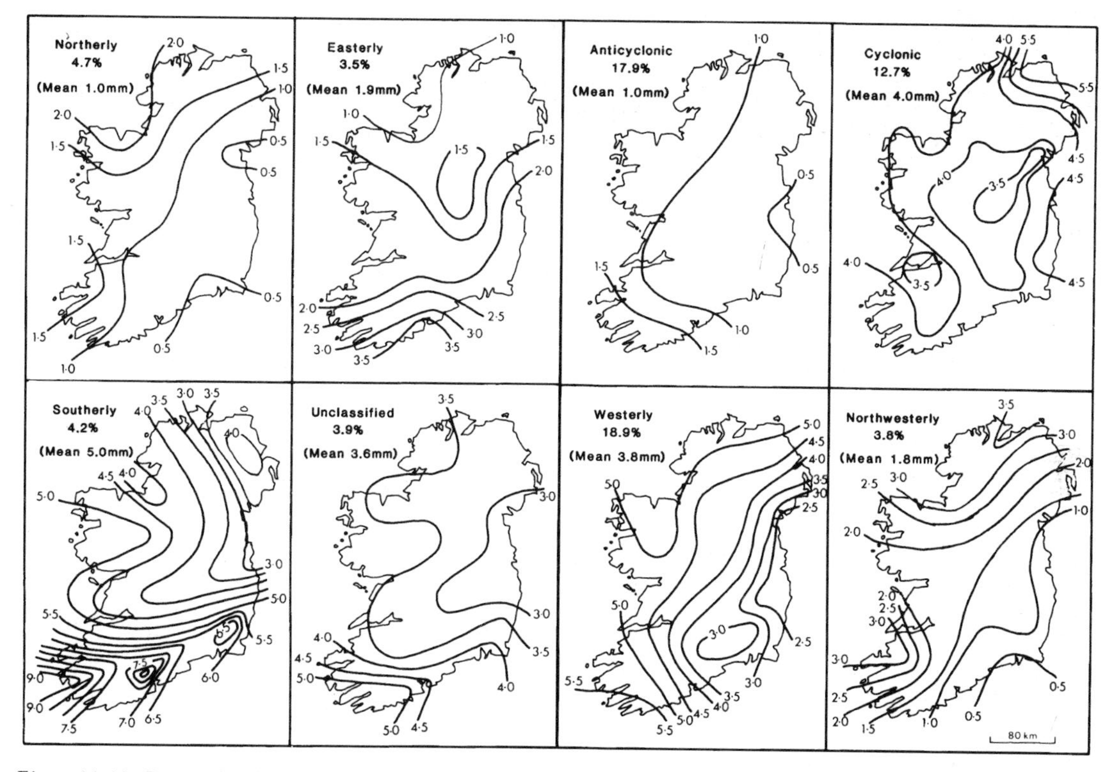

Figure 11.13 Geographical distribution of rainfall under different Lamb airflow types

west may be up to three times those along the east coast, especially in the shelter of the Wicklow Mountains. The dry winter of 1988/89 produced in Ireland a similar response to that over the British Isles as a whole, with an intensification of the west-to-east rainfall gradient as a consequence of the dominant westerly airflow (Betts 1990). On the other hand, May 1989 was generally dry in response to the pervasive anticyclonic character of the weather, the usual rainfall gradient being diminished at this time. The persistence of dry weather during much of 1988 and 1989 provoked problems of water shortages not usually associated with a country widely regarded as having an abundance of rainfall.

Different synoptic conditions can provide distinctive spatial patterns of rainfall. In easterly airflows marked rain shadows can be seen to the west of the Wicklow Mountains. Interestingly, Cork City experiences a daily average rainfall amount from westward-moving air across the Celtic Sea similar to that it receives from eastward-moving Atlantic airflows. Southerly circulations, especially in autumn when the sea offshore is warmest, are the highest rainfall-yielding airflows in Ireland, and this confirms the role of sea temperatures as the root cause of the autumn and winter maximum in most of the island.

Since cyclonic and westerly circulations account for about two-thirds of Irish rainfall, changes in the balance between these are of great potential significance for the national pattern. The make-up of the precipitation totals at four synoptic stations are shown in Figure 11.14. It is apparent that in eastern Ireland C airflows are the dominant provider, while in the west it is generally W airflows which are most important. Major reductions in the frequency of westerlies have occurred between the 1930s and the 1970s. The effect of these can be seen in the

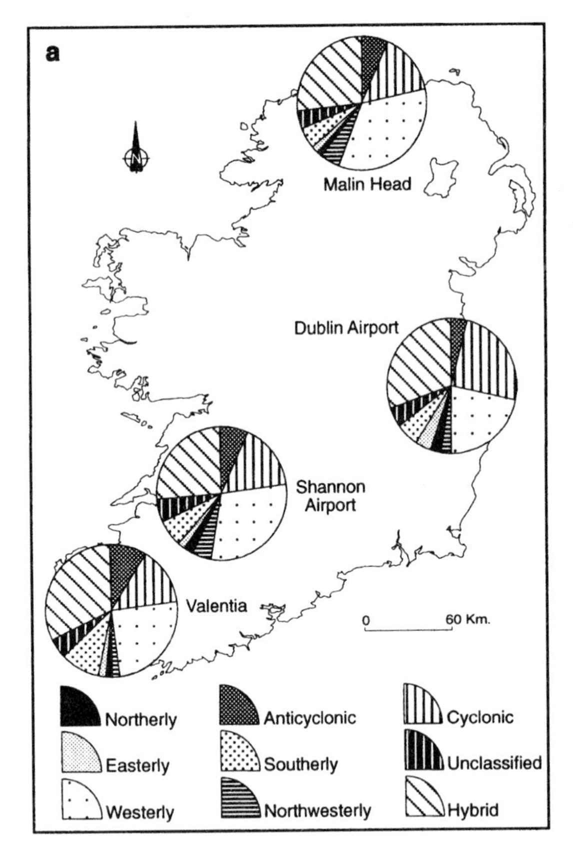

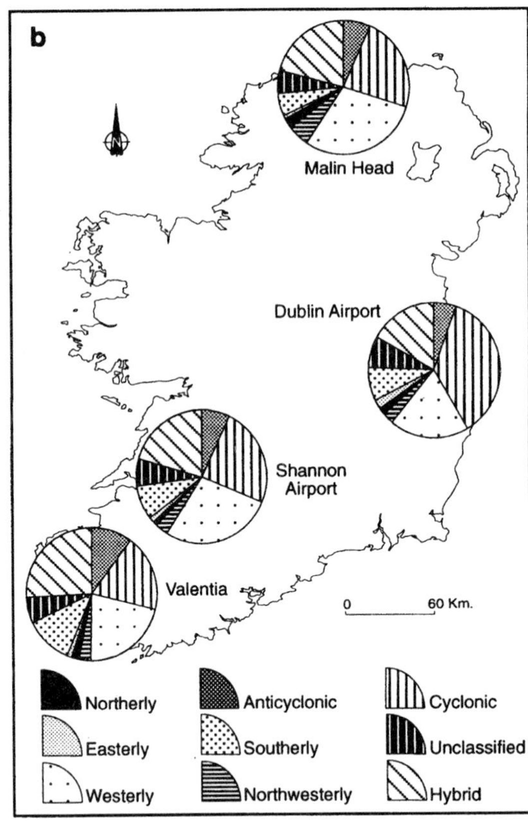

Figure 11.14 Contributions to annual totals at Malin, Shannon, Dublin and Valentia from Lamb circulation types for the periods (a) 1861–1961 and (b) 1961–90

precipitation make-up of the stations for the two periods also depicted in Figure 11.14.

Potential evapotranspiration averages 450–500 mm throughout the island, resulting in moisture surpluses at lowland locations ranging from 200 mm in the east to over 700 mm in the west. Potential evapotranspiration is highest during the best summer month, June (approximately 75 mm). From June on, therefore, soil moisture deficits may become pronounced in eastern parts. In dry summers these may become acute by September, making irrigation desirable on cultivated soils derived from sandy fluvioglacial deposits, particularly for vegetables and sugar beet. On average deficits in excess of 75 mm, measured at the end of 10-day periods, occur approximately four times per year in eastern parts, and water supply problems occur, especially in Northern Ireland (Betts 1990). During the exceptional summer of 1995 deficits in excess of 100 mm were recorded along much of the east coast of Leinster.

WIND

Wind is the most feared of the meteorological elements in Ireland, and severe storms are etched indelibly in the public consciousness. The Irish language has a rich variety of descriptors for wind hazards, perhaps inevitably in an island where the relatively frictionless surface of the sea is close by and where the North Atlantic depres-

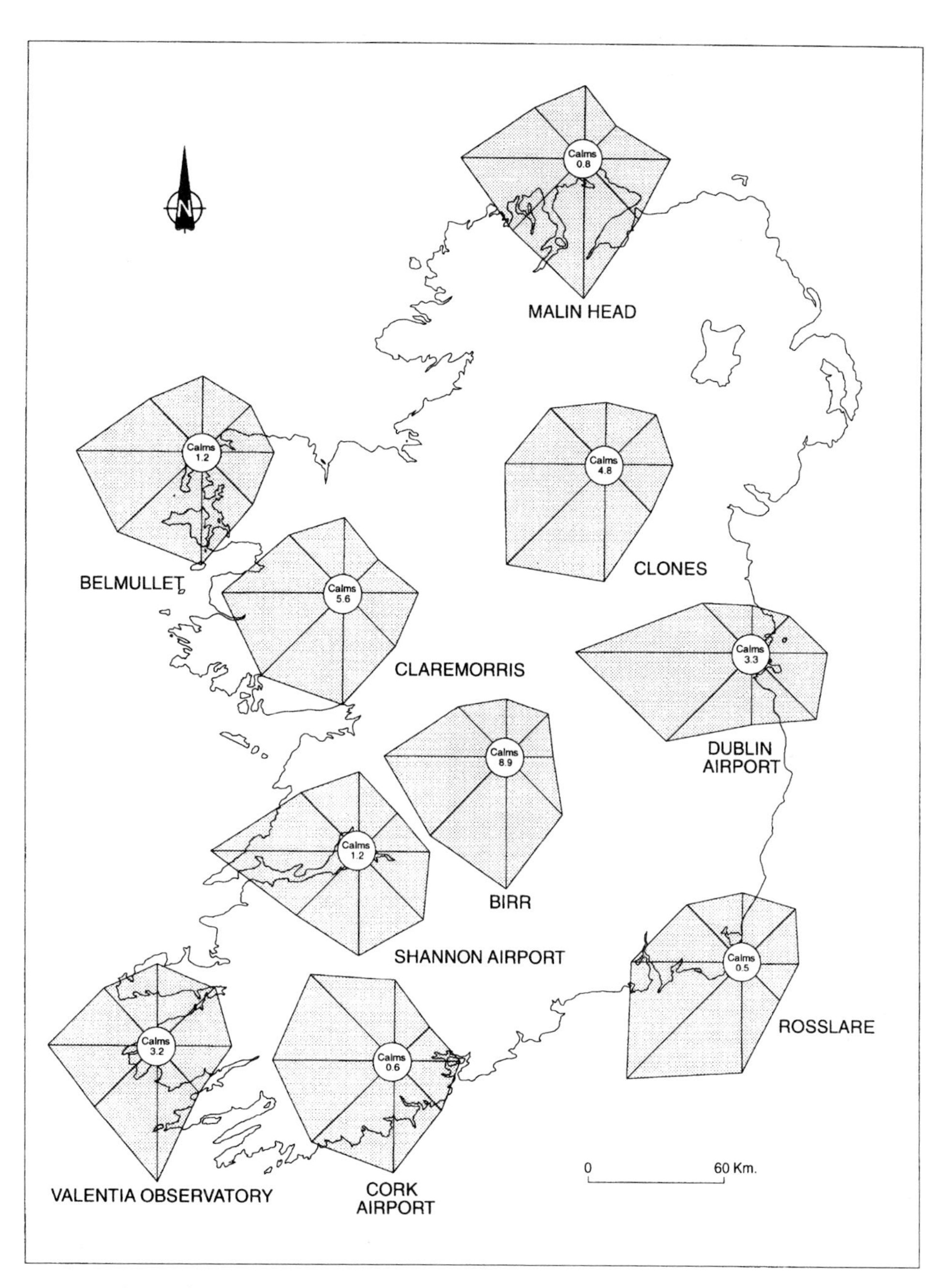

Figure 11.15 Wind roses for the period 1951–80

sions are often at their most mature state. For the fisherman of north-west Donegal, or the farmer from Clare, the ferocity of wind is a reality not often appreciated by the urban dweller of eastern Ireland, or the Brussels bureaucrat.

As with all regions of Ireland and Britain, the prevailing wind directions are from the southwest, though all quadrants except north and east are generally well represented in the annual wind rose (Figure 11.15). Directional frequencies are, however, modified considerably by the sheltering effect of nearby relief features. This is clearly seen in the reduction of southerly frequencies at Shannon and Dublin Airports, where funnelling of winds from the west and south-west along the river valleys is implied. Figure 11.16 shows a good example of the sheltering effect at Dublin Airport in July 1995. Such funnelling has also been noted in various locations in Northern Ireland, such as the Foyle valley (Betts 1982).

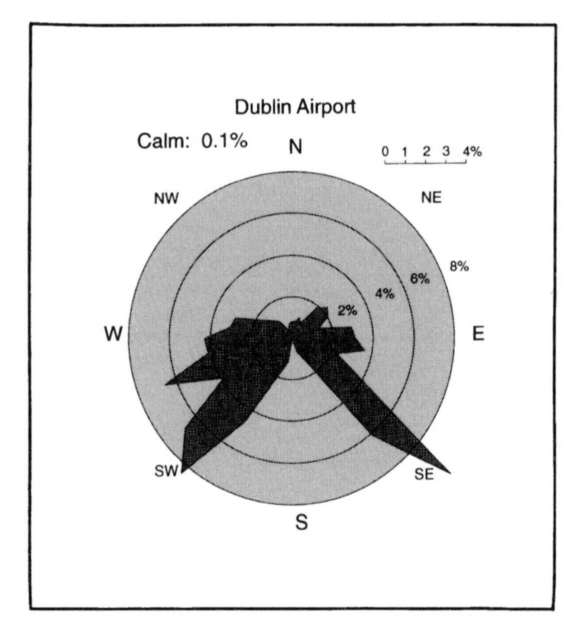

Figure 11.16 Wind rose for Dublin Airport for July 1995

In contrast with some other parts of the British Isles, westerlies tend to be more frequent during the summer months. The northward displacement of the Azores High tends to sharpen the south-to-north pressure gradient across Ireland at this time. Southerlies tend to peak in autumn and winter, while northerlies peak in April, contributing to delayed spring grass growth in some years. Approximately 20 per cent of the easterlies typically occur during February, associated with persistent blocking over Scandinavia which occasionally occurs at this time of the year.

A native of County Meath, Sir Francis Beaufort devised the scale of wind force which is now so widely used for relating wind speed to the movement of everyday objects such as leaves, trees, waves, etc. Force 8 (or gale force) on his scale is attained when the mean velocity over a period of not less than 10 minutes exceeds 34 knots, an event which occurs almost one day in five in north-western Ireland. It is perhaps inevitable that north and western Ireland should be one of Europe's windiest locations, close to the main depression tracks and with no sheltering landmasses upwind. Recent investment in wind farms has therefore been concentrated in the north-west. Yet the geographical variations in wind speed across Ireland are quite dramatic, illustrating well the frictional effect of terrain in removing the worst excesses of wind, even a short distance inland.

Average annual wind speeds range from over 16 knots in the extreme north-west to approximately half this amount in the south-central part of the island. Ireland thus exhibits a spatial variation in wind climate which is similar to the Western Isles–lowland England contrast in Britain. Table 11.4 highlights this by showing that the number of days with gales falls by over an order of magnitude, from over sixty days per year along the north-west coast to just over one day per year in the sheltered parts of the Central Plain. The popular perception of Ireland as an island of high winds and copious rainfall is thus once again demonstrably an image of the north-west coast. The 'two Irelands' are highlighted, with a much more quiescent climate prevailing in the interior of the island as opposed to a more robust variant on the western perimeter.

This is not to say that intense depressions do not result in high wind speeds on occasion at all loca-

Table 11.4 Average number of gale days per month for the period 1961–90 and highest gusts observed

Location	*Jan.*	*Feb.*	*Mar.*	*Apr.*	*May*	*June*	*July*	*Aug.*	*Sept.*	*Oct.*	*Nov.*	*Dec.*	*Year*	*Maximum gust (knots)*
Malin Head	11.2	8.6	8.0	3.4	2.3	1.3	0.8	1.5	3.8	6.7	8.7	9.7	66.0	98
Claremorris	1.2	0.9	1.0	0.1	0.1	0.1	0.0	0.0	0.2	0.4	0.5	0.7	5.2	96
Kilkenny	0.5	0.3	0.1	0.0	0.0	0.0	0.0	0.0	0.0	0.1	0.1	0.3	1.4	77
Cork Airport	3.2	2.3	1.8	0.7	0.4	0.0	0.1	0.2	0.7	1.2	1.8	2.6	15.1	94

Box 11.4

THE 'NIGHT OF THE BIG WIND' – 6/7 JANUARY 1839

This most notorious of all storms to affect Ireland was well recorded in early instrumental records and its effects dramatically reported in newspaper reports and first-hand accounts. An unusually deep depression travelling in a north-easterly direction to the north of Ireland seems to have been responsible. A pressure gradient of 37 mb seems to have existed over Ireland, producing gusts in excess of 100 knots in places (Shields and Fitzgerald 1989). Though loss of life was surprisingly low, damage to buildings, shipping and crops was severe. Between 20 and 25 per cent of the housing in Dublin was damaged and unusual occurrences were reported, such as salt deposition well inland and other 'freak' phenomena (Shields and Fitzgerald 1989), a sample of which are reproduced here.

> The damage which it has done is almost beyond calculation. Several hundreds of thousands of trees must have been levelled to the ground. More than half a century must elapse before Ireland, in this regard, presents the appearance she did last summer.
> (*Dublin Evening Post*, 12 January 1839)

> What appeared to be the most astonishing effect of the storm was the blowing of water out of the canal near this town. I visited it this morning, and it was nearly dry.
> (*Tuam Herald*, 19 January 1839)

> Trees, ten or twelve miles from the sea, were covered in salt brine – and in the very centre of the Island, forty or fifty miles inland, such vegetable matter as it occurred to individuals to test had universally a saline taste.
> (*Dublin Evening Post*, 12 January 1839)

> Comparing it with all similar visitations in these latitudes, of which there exists any record, we would say that, for the violence of the hurricane, and deplorable effects which followed, as well as for its extensive sweep, embracing as it did the whole island in its destructive career, it remains not only without a parallel, but leaves far away in the distance all that ever occurred in Ireland before. Ireland has been the chief victim of the hurricane – every part of Ireland, – every field, every town, every village in Ireland – have felt its dire effects, from Galway to Dublin – from the Giant's Causeway to Valentia. It has been, we repeat it, the most awful calamity with which a people were afflicted.
> (*Dublin Evening Post*, 12 January 1839)

tions. Table 11.4 shows also that most coastal locations experienced maximum gusts close to 100 knots during the 30-year standard observation period. It is likely that inland sites above 300 m have wind climatologies comparable to those of the exposed coastal locations. The highest wind speed at a low-level site was measured at Kilkeel in County Down during January 1974, when a gust of 108

Table 11.5 Mean monthly and annual sunshine totals (hours) for the period 1951–80

Location	*Alt. (m ASL)*	*Jan.*	*Feb.*	*Mar.*	*Apr.*	*May*	*June*	*July*	*Aug.*	*Sept.*	*Oct.*	*Nov.*	*Dec.*	*Year*
Aldergrove	68	45	66	101	158	185	177	141	142	108	81	57	37	1,298
Claremorris	69	50	68	98	140	172	154	122	130	98	78	55	38	1,203
Malin Head	25	41	64	106	158	200	182	136	145	110	75	47	29	1,293
Rosslare	25	63	77	122	179	219	212	194	183	144	109	75	57	1,634

knots was recorded. Not surprisingly, this was a storm which blew down many trees, damaged many buildings and interrupted the electricity supply for 150,000 consumers before going on to wreak further havoc in Britain. But not all storms causing damage in Britain pass over Ireland *en route*. Some of the most severe events to affect Britain, such as the October 1987 storm, passed Ireland by, as did the greatest British storm of recent centuries: Daniel Defoe's storm of 1703. Ireland was not so fortunate, however, on the night of 6/7 January 1839 when the 'Big Wind' struck large areas (see Box 11.4).

The so-called 'cyclonic bomb' whereby explosive deepening of an Atlantic depression occurs west of Ireland is the chief source of extremes of wind speed. Sometimes difficult to forecast, these are often associated with the rapid development of a secondary depression which may race across Ireland in a few hours. Even in an age of satellite observations, the value of the terrestrial Irish meteorological network in acting as an early-warning facility for areas further east is unquestionable in such circumstances.

SUNSHINE AND CLOUD

The seasonal and topographic controls on precipitation discussed above are instrumental also in explaining cloud cover and sunshine variations in Ireland. Table 11.5 shows that May and June are the sunniest months, with durations over seven hours per day in the extreme south-east of Ireland. This represents up to 46 per cent of the possible maximum figure and a sunshine regime which is very similar to that of the south-east coast of England. The north-west by contrast only manages between 5 and 6 hours during this the season of driest airflows across the country. It is striking how much of a reduction typically occurs in July and August when the European 'summer monsoon' sets in (see also Chapter 10).

December is the dullest month everywhere, with less than an hour per day of sunshine in the extreme north, representing only 13 per cent of the maximum possible even at this time of short days, and only 50 per cent of the corresponding figure for Rosslare in the extreme south east. These are values directly comparable with most of western Scotland at this time.

Sunshine data are reinforced by the cloud data, which, as expected, are almost a mirror image. Everywhere in Ireland cloud cover averages between 5.5 and 6 oktas, fairly typical of a maritime climate at these latitudes. On average skies are completely cloud covered about 33 per cent of the time and relatively clear (less than 2 oktas) about 16 per cent of the time. Cloud cover downwind of major mountain barriers may be noticeably less on occasion owing to föhn effects. This is noticeable, for example, in Dublin, which habitually enjoys higher sunshine on light southerly breezes because of subsidence after the air mass has traversed the Wicklow Mountains.

CONCLUSION

The climate of Ireland, like that of Britain, is the product of a struggle between tropical and polar air masses, a battleground invaded periodically by one or other variant of these air masses only to be reconquered by another. Thus whilst its oceanic location ensures that equability will be the major feature of its climatology, constant alternation in air mass dominance ensures that variability in weather types will be an equally striking characteristic.

The stage on which this struggle is acted out is not a featureless plain, but rather a distinctive topography of coastal upland and interior plain. This makes for a climatology of contrasts between a maritime fringe with all the oceanic influences for which Ireland is well known and an interior where shelter and continental influences produce conditions more representative of lowland England.

A winter climatic gradient from south-west to north-east, conditioned by the advection of heat energy from the Atlantic Ocean, and a summer one from south-east to north-west, reflecting a complex interplay between latitudinal, relief and oceanic influences, provides further layers of variation for the Irish climatic mosaic. This mosaic is essentially a compendium produced by the frequency of occurrence of specific airflow trajectories, and the extremes of weather embedded in them. The snapshot in time presented here will undoubtedly change in the future, as it has in the past, in response to the external forcing factors which provide the boundary conditions for the climate of Ireland.

REFERENCES

Betts, N. (1982) 'Climate', in J. Cruickshank and D. Wilcock (eds) *Northern Ireland: Environment and Resources*, Belfast: Queen's University of Belfast and University of Ulster.

—— (1987) 'A year of notable extremes: a review of Ulster weather 1986', *J. Meteorology (UK)*, 12: 159–162.

—— (1990) 'Ulster's recurring water supply problem: a portent for the 1990s', *J. Meteorology (UK)*, 9: 37–40.

Britton, C.E. (1937) 'A meteorological chronology to AD 1450', *Geophysical Memoir no. 70*, London: Meteorological Office.

Burt, S. (1992) 'The exceptional hot spell of early August 1990 in the United Kingdom', *Int. J. Climatol.*, 12: 547–567.

Fitzgerald, D. (1994) 'Data sets for climate change studies in the climatological archive of the Irish Meteorological Service', in J. Feehan (ed.) *Climate Variation and Climate Change in Ireland*, Dublin: UCD Environmental Institute.

Gleeson, P.A. and Walsh, M.J. (1967) 'Rearing dairy replacement heifers', *Research Report Animal Production*, 40–50.

Haughton, J.P. (1953) 'Land use in the Carran polje', *Irish Geography*, 2: 225–226.

Logue, J.J. (1978) 'The annual cycle of rainfall in Ireland', *Technical Note no. 43*, Dublin: Meteorological Service.

Mayes, J.C. (1991) 'Regional airflow patterns in the British Isles', *Int. J. Climatol.*, 11: 473–491.

Meskill, D. (1995a) 'The severe thunderstorm on 30 June 1995 in southern Ireland', *J. Meteorology (UK)*, 20: 304–308.

—— (1995b) 'The hot Irish summer of 1995', *J. Meteorology (UK)*, 20: 336–338.

Morgan, W.A. (1971) 'Rainfall in the Dublin area on 11th June 1963', *Internal Memorandum no. IM/72/71*, Dublin: Meteorological Service.

Oke, T. (1973) 'City size and the urban heat island', *Atmospheric Environment*, 7: 767–779.

Perry, A.H. (1972) 'Spatial and temporal characteristics of Irish precipitation', *Irish Geography*, 5: 428–442.

—— (1976) 'Temperature extremes in Ireland' *J. Meteorology (UK)*, 1: 148–149.

Rohan, P.K. (1975) *The Climate of Ireland*, Dublin: Stationery Office.

Shields, L. (1983) 'The beginnings of scientific weather observations in Ireland (1684–1708)', *Weather*, 38: 304–311.

Shields, L. and Fitzgerald, D. (1989) 'The "Night of the Big Wind" in Ireland 6–7 January 1839', *Irish Geography*, 22: 31–43.

Sweeney, J. (1985) 'The changing synoptic origins of Irish precipitation', *Trans. Inst. Br. Geogrs*, NS 10: 467–480.

—— (1987) 'The urban heat-island of Dublin City', *Irish Geography*, 20: 1–10.

—— (1988) 'Controls on the climate of Ireland', *Geographical Viewpoint*, 16: 60–72.

Sweeney, J. and O'Hare, G. (1992) 'Geographical variations on precipitation yields and circulation types in Britain and Ireland', *Trans. Inst. Br. Geogrs*, NS 17: 448–463.

Tyrrell, J.G. (1995) 'Paraclimatic statistics and the study of climate change: the case of the Cork region in the 1750s', *Climatic Change*, 29: 231–245.

PART 3

REGIONAL PERSPECTIVES ON CLIMATIC VARIABILITY AND CHANGE

12

REGIONAL PERSPECTIVES ON CLIMATIC VARIABILITY AND CHANGE

Julian Mayes and Dennis Wheeler

INTRODUCTION

Part One of this volume demonstrated how changes in the frequency of airflow types may influence regional weather over daily and monthly time scales. Part Two (Chapters 2 to 11) provided examples of these processes by illustrating the character of regional climates across the British Isles. Part Three presents a wider view of the regional effects of climatic variation and change over a range of time scales.

Climatic change can result from the changing frequency and character of airflow types. For example, it could be suggested that one reason for the warmth of the 1980s over the British Isles was an above-average frequency of southerly airstreams. This implies that regional and local warming can occur independently of hemispheric or global temperature changes. A separate component of climatic change may result from the 'forcing' factors that alter the intrinsic character of all airflow types. An obvious example of the latter is the enhanced greenhouse effect. Warming over the British Isles would then occur independently of any changes in the atmospheric circulations that would vary the frequency of those airflow types. The fact that such large-scale changes in the global climate system may provoke changes in the atmospheric circulation adds yet another dimension to regional climatic change.

Much attention has been given to greenhouse gas forcing of global temperatures. Of course, it should not be assumed that future warming is inevitable, but given this scenario, it is appropriate to consider these contrasts in a volume that seeks to focus on the regional scale. Recent public concern prompted by this issue should not, however, obscure the importance of other, extra-terrestrial or geological, factors that determine global and regional climates and these factors are reviewed in the following sections.

A further, contentious, issue is that of the significance of climatic change. The term can be taken to have two meanings. 'Statistical significance' refers to the numerical changes in a body of data (such as a change in precipitation or temperature) that have a possibly small but specifiable probability of occurring randomly and are indicative of climatic change. However, significance in a more general sense refers to whether a change or variation in a climatic variable was 'significant' to those who experience the event. An example could be an increase in the frequency of frost or drought that might be highly significant for the farming community.

The above comments lead to the question of timescales in climatic change. By convention and tradition, climate is usually represented in terms of 30-year averages, such as the 1961–90 averages cited in this volume. If two successive 30-year periods show a 'significant' change, does this represent climatic change or merely a shorter-term variation? The term 'climatic change' is usually reserved for variations that extend over periods of at least thirty years' duration. Though to some this may appear to be too short a time scale, it does have the merit of identifying changes that occur within an individual's lifetime. Finally, the issue of relevance raises the question of extreme, rather than average, events. This is especially applicable to the frequency of climatic hazards such as heavy rainfall and snowfall. Furthermore, description of climatic change by averages alone

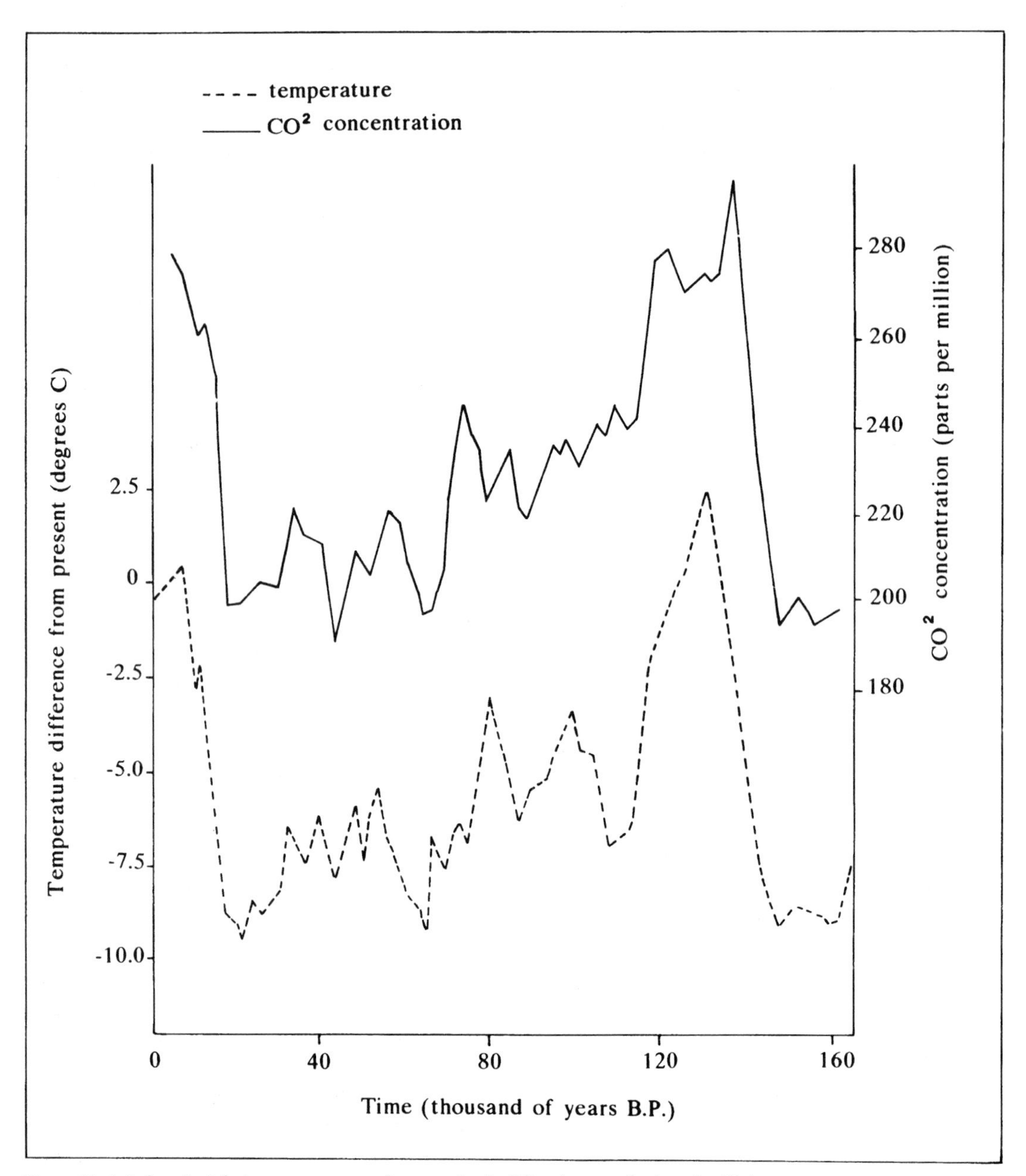

Figure 12.1 Inferred global temperature and atmospheric CO_2 changes during the Pleistocene

may obscure changes in the variability of weather and climate which themselves may have adverse effects on social and economic life.

Most of the averages cited in the preceding chapters are based on the 30–year observation period, but there is no doubt that climatic means differ even between such seemingly long intervals. Prompted largely by the 'global warming' debate, scientists are increasingly aware that past climates have been anything but constant. The magnitude of the changes that accompanied the great Pleistocene ice advances and retreats have been acknowledged for many decades but it is only more recently that the idea of some naturally constant or normal post-glacial climate has ceased to enjoy wide currency.

The variation of global temperatures over the past 200,000 years is summarised in Figure 12.1 and this chapter briefly reviews those factors that may explain these changes. While these factors are commonly viewed from a global perspective, as Figure 1.1 makes clear, their influence cascades down the climatic system with equally profound, though differing and perhaps unpredictable, consequences at smaller geographical and temporal scales.

EXTRA-TERRESTRIAL CONTROLS ON GLOBAL CLIMATE

The Sun's energy is the 'fuel' upon which the climate depends and changes in spatial or seasonal distribution are widely supposed to induce a climatic response. On the basis of our current understanding of the Sun's behaviour it is, however, difficult to distinguish such responses from other aspects of climatic variation.

The Sun and Variations in Solar Activity

The Sun radiates most of its energy in the visible light, infra-red and ultra-violet wavebands of the electro-magnetic spectrum and provides also a constant stream of charged particles known as the 'solar wind'. This energy is produced by the fusion of hydrogen to produce helium in the Sun's core, during which a small amount of matter is converted to energy. Fluid convection and radiation transmit this energy to the visible surface or photosphere where temperatures are 6,000°C. It has been estimated that the Sun loses 4 million tons of its mass each second by this process. The Earth's upper atmosphere intercepts the merest fraction of this massive output.

The Sun's behaviour is by no means constant, however, and its surface is marked by dark areas that come and go in rapid succession. These sunspots appear as dark areas on the photosphere because they are cooler than the general surface. They typically consist an umbra or core at about 4,000°C and an encircling penumbra which has a surface temperature of 5,000°C. Because of their lower temperatures, sunspots emit less radiant energy than the surrounding photosphere. They usually occur in groups and have a life span of a few hours to several months, and although they can be thousands of kilometres in diameter sunspots still represent only the minutest fraction of the solar surface. Their origin remains a mystery, but their most remarkable feature, which has prompted the search for corresponding atmospheric behaviour, is the regularity of sunspot activity that sees their numbers rise and fall over a cycle whose an average period is 11.2 years but can vary between 8 and 16 years.

The sunspot cycle is measured by the Wolf relative sunspot number, which is based on the number of groups of as well as the quantity of sunspots. The Wolf series begins in 1875 (Jones 1955), though sunspots have been carefully counted since 1750, and fragmentary information goes back to the seventeenth century.

Sunspots emit up to 75 per cent less radiation than the undisturbed photosphere, suggesting that total solar irradiance should be reduced during periods of maximum activity. In fact solar irradiance increases with sunspot activity – a fact explained by the presence of solar faculae. These are brighter areas, up to 2,000 degrees C hotter than the normal photosphere, with a cyclic incidence similar to that of sunspots. Their additional contribution to the Sun's radiated energy is believed to more than compensate for the reductions attributable to the sunspots. Recent satellite studies have shown that solar irradiance is indeed linked with sunspot number but varies only by 0.04 per cent. Such small fluctuations can be influential only if their consequences are exaggerated by positive feedback loops in the atmospheric system.

The magnetic properties of the sunspots are also important as they are foci for intense magnetic disturbances. The polarity of the sunspot fields is also cyclic, with a period of 22 years. This behaviour is described as the Hale cycle, and there are suggestions (Gribbin 1978; Burroughs 1992) that such fluctuations, perhaps through the agency of changes in the solar wind, have climatic consequences.

Despite the clarity of the sunspot and Hale cycles these subtle signals are soon lost in the noise of the climatic system. Thus while aspects of such geographically disparate phenomena as precipitation in South Africa (Tyson 1986), droughts in the western United States (Mitchell *et al.* 1978) and frequency of Chinese floods all show some evidence of 11- and 22-year cycles, the same periodicity is lacking in many other long periods of record. Mason's (1976) analysis of the three hundred years of Central England temperatures finds little evidence of an 11-year cycle, though a 23-year periodicity is well marked. Tabony's (1979) examination of two centuries of the precipitation record of England and Wales could find no evidence of either cycle.

For over a century, scientists at the Royal Greenwich Observatory have measured the umbra–penumbra area ratio that reveals secular changes which, though independent of the sunspot cycle, correlate with northern hemisphere temperatures (Hoyt 1979a, 1979b). Nordo (1955) has suggested a link between umbral : penumbral ratios and faculae, as both are indicative of convective activity within the Sun which possibly controls irradiance levels. A climatic link may be provided by the Baur index (Baur 1949), which measures the relative importance of solar faculae and sunspot areas. Lamb (1969) has found that over the past century the Baur cycle reveals a strong positive correlation with the dominance of westerly, zonal weather in the mid-latitudes – exactly the circumstances in which the rainfall gradients from west to east across the British Isles are most marked and when temperature regimes are at their most temperate.

Factors Influencing Energy Receipts at the Upper Atmosphere

Other factors may also change the energy received by the Earth's atmosphere, and theories have been offered based on periodic variations in the Earth's orbit about the Sun. These theories rely upon the Milankovitch cycles, named after the Yugoslavian astronomer Milutin Milankovitch who in the 1930s gave mathematical definition to the earlier suggestions of the Scottish astronomer James Croll. These cycles operate over a number of periods, but attention has focused on three of them, each connected with a different aspect of the Earth's orbital behaviour.

Eccentricity

The Earth's orbit describes an ellipse with a degree of eccentricity that varies from near circular (eccentricity 0.001) to a maximum of 0.054. At present the degree of eccentricity is 0.017 and is decreasing. These changes have an important cycle at about 100,000 years. In the near-circular phases the Earth's total energy receipts differ little between the seasons. At higher degrees of eccentricity, however, those receipts will vary more depending upon distance from the Sun, being greater at times of closer approach but less when the Earth is more distant.

Obliquity

Obliquity is the angle of the Earth's axis of rotation with respect to the orbital plane about the Sun. At present this angle is 23.4 degrees but varies on a cycle of 41,000 years from between 22.0 and 24.5 degrees. These changes influence the Earth's seasonal solar energy receipts, which are in most marked contrast when obliquity is at a maximum and the annual variation in daylight hours is at its greatest.

The precession index

The precession index determines the time of year when the Earth is closest to the Sun. There also exists a circular motion in the rotational axis which requires between 19,000 and 23,000 years to complete its cycle. At present, perihelion (the point at which the Earth is closest to the Sun) is in January. The precession index, however, determines that perihelion changes slightly each year, and in approximately 10,000 years' time it will occur in July. At present the northern hemisphere winter upper atmospheric energy receipts should be slightly higher than those in summer, whereas in 10,000 years' time the reverse will be true and northern hemisphere winter receipts will be less than those of summer.

It is tempting to view global changes exclusively in the light of total energy receipts, but variations in the balance of seasonal energy receipts can be equally important. For example, cool summers when snow accumulations might be able to survive until the following winter are an important element in initiating ice cover and of greater significance than cold winters.

Vernekar (1972) and Mitchell (1972) have shown that patterns of solar energy receipts reconstructed for the past 300,000 years on the basis of Milankovitch orbit variations are in phase with the periods of ice advance and retreat. Further support comes the polar ice cores, the annual accumulating layers of which trap minute air bubbles that preserve a record of contemporary climates which can be interpreted from the changing ratios in the isotopes of oxygen and hydrogen and of carbon dioxide concentrations, all of which are temperature dependent (Robin 1983). The data from the ice core at the Russian Vostok station in the Antarctic, for example, provide a record for the past 160,000 years and reveal clear evidence of a pattern of variability very close to those which might be anticipated from the Milankovitch cycles. Scientists often invoke CO_2 as the means of amplifying the weak astronomical effects (Genthon *et al.* 1986; Barnola *et al.* 1986; Jouzel *et al.* 1987). These findings support also the independent evidence from deep-sea sediment cores which are based on oxygen isotopes and foraminiferal remains to determine variations in sea surface temperatures over time scales of tens of thousands of years.

Cycles and Quasi-periodic Behaviour

Evidence of sunspot cycles and of periodicities in the Earth's orbit is well established but debate surrounds the character of the atmosphere's response to these changes. There are, however, two atmospheric phenomena which display clear evidence of periodic behaviour. The first of these is the quasi-biennial oscillation (QBO) of approximately 28 months in the equatorial stratospheric winds, which shift between easterly and westerly phases (Figure 12.2). Although there is no clear understanding of how this behaviour communicates itself with the tropospheric circulations, there is evidence to support a 2-year periodicity in some aspects of the world's climate. In Britain, for example, Tabony's

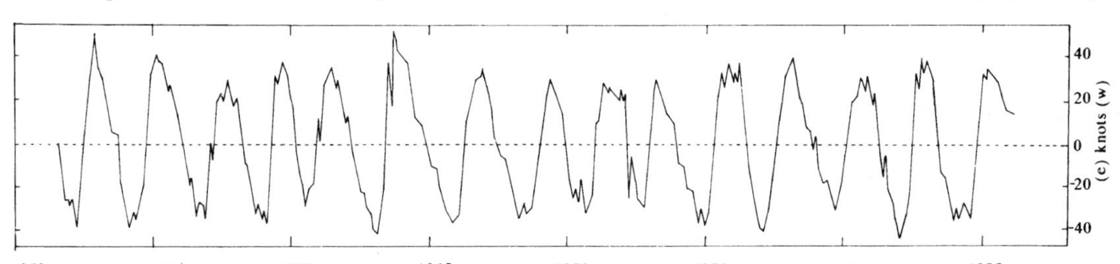

Figure 12.2 The quasi-biennial oscillation (QBO) shown by variability of equatorial stratospheric winds 1950–85 (after Burroughs 1992)

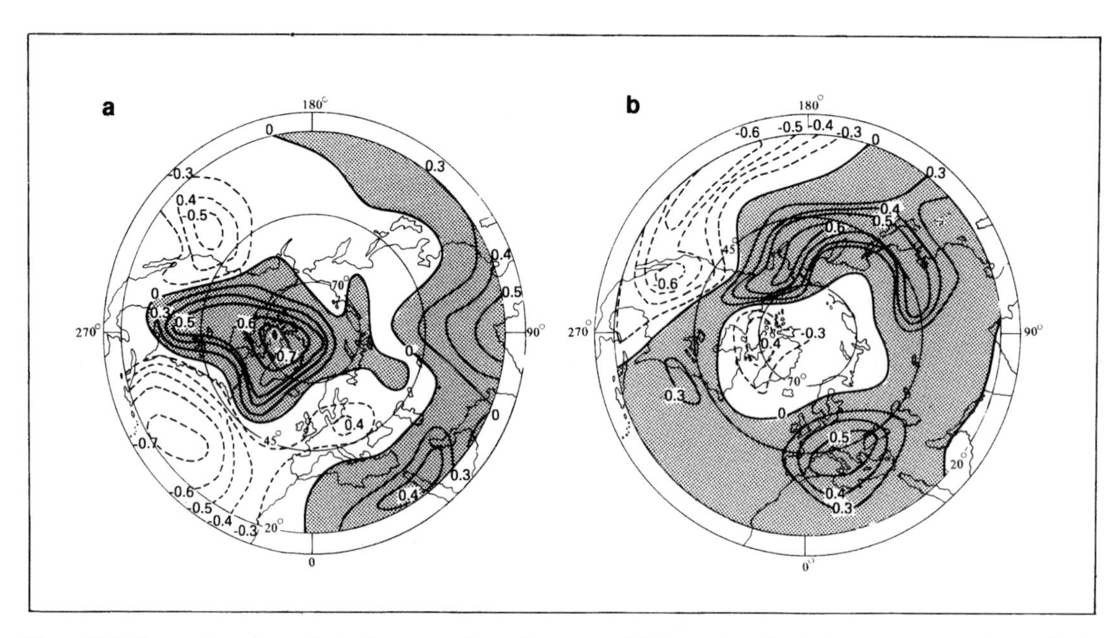

Figure 12.3 Lines of equal correlation between winter (January and February) sea level air pressure anomalies and solar activity when the stratospheric winds are in (a) westerly phase and (b) easterly phase (after Labitzke and van Loon 1988)

(1979) analysis of the precipitation record found strong support for 2- and 3-year cycles, and Perry (1977) has found evidence of this period in his study of the Oxford temperature record. Mason (1976) also found a 2-year signal in the three-century-long Central England Temperature series.

The studies summarised by Labitzke and van Loon (1988) established a link between the sunspot cycle and the QBO which depends on the correlations between polar stratospheric temperatures, the degree of solar activity and air pressure anomalies in the northern high latitudes. They suggested that surface pressure anomalies are positively correlated with solar activity in west-phase years of the QBO, and when the Sun is most active air pressure will be unusually high over the North American continent but lower than normal over the Pacific and Atlantic Oceans (Figure 12.3). During periods when the Sun is less active, air pressure is lower than normal over the North and South American continents, but higher over the Pacific and Atlantic. Such changes reflect variations in the character of the upper westerlies and their long wave forms. This determines in turn the approach made towards the British Isles by mid-latitude depressions with consequences for the regional distribution of rainfall, sunshine and the overall temperature levels.

One of the factors which further obscures the climatic response to external influences, and constitutes the second example of quasi-periodic behaviour, is the El-Niño–Southern Oscillation (ENSO). In the 1920s Sir Gilbert Walker observed that when air pressure is low across the Indian Ocean it tends to be high across the eastern Pacific and vice versa. This see-saw effect, the Southern Oscillation, brought in its train changes in the geography of sea surface temperatures and rainfall. These changes are closely associated with the El Niño events, which lead to the displacement of the focus of warm surface water and of the principal areas of precipitation eastwards across the Pacific towards the South American coastline. It normally takes about nine months for the movement from west to east across the Pacific to take place. However, feedback mechanisms within the complex Pacific oceanic circulations ensure that the system will eventually

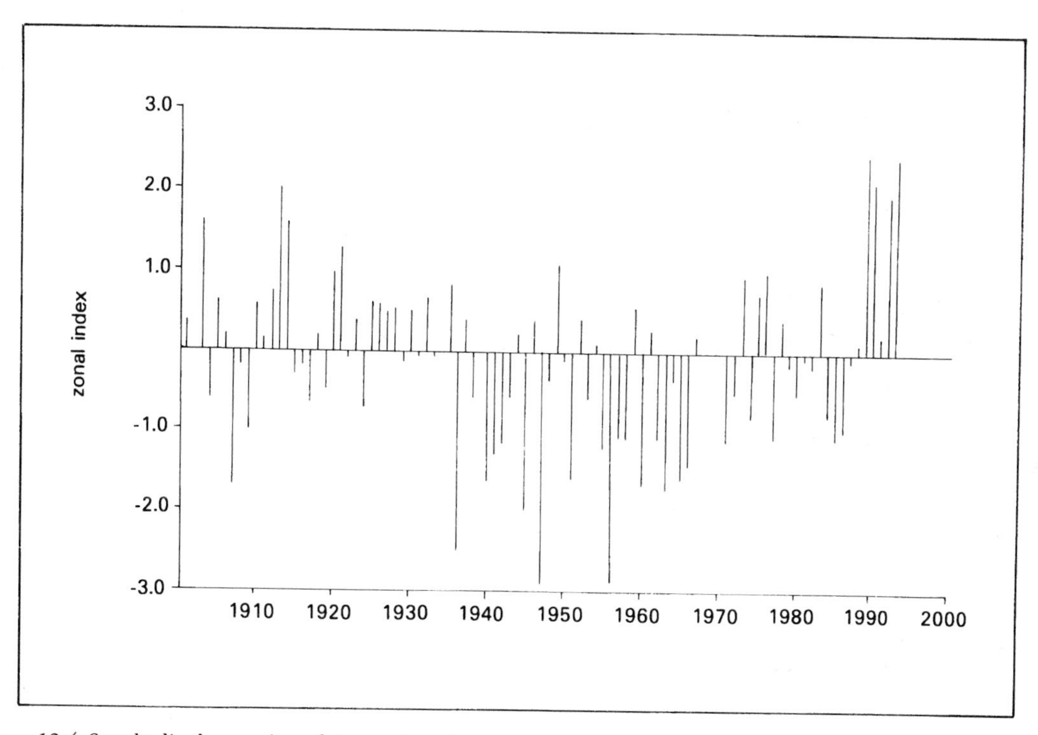

Figure 12.4 Standardised anomalies of the northern hemisphere winter zonal index between 35 and 65° N (after WMO 1995)

return to its 'normal' state, though the El Niño events may persist for a year or more. Bigg (1990, 1995) has shown that ENSO events have global repercussions for precipitation.

Similar oscillatory behaviour is now being studied in other oceanic areas. Of these the North Atlantic Oscillation (NAO) is of particular interest. The NAO is governed by the behaviour of the subtropical anticyclone and the subpolar low pressure zones. The pressure gradient between these two centres helps to determine the zonality of the mid-latitude westerly circulations. When air pressure is unusually high over the Azores but uncharacteristically low over Iceland the NAO is said to be in positive mode. When both centres are poorly developed, and the pressure gradient weakened, the mode is negative. In positive mode the westerly circulation strengthens, the zonality of upper westerlies increases and warm air and water are advected towards the British Isles. The mild winters of the late 1980s and early 1990s in Britain were a time when the NAO was positive to an unprecedented degree (Figure 12.4; WMO 1995). Negative phases, however, encourage low degrees of zonality with meridional flow in the upper westerlies, the Icelandic low tends to be displaced to the south-west and blocking becomes more frequent around the British Isles with the southwards penetration of polar anticyclones. The exceptionally cold winters of 1941/42 and 1962/63 fall into this category.

ATMOSPHERIC COMPOSITION AND CLIMATIC VARIABILITY

Vying with extra-terrestrial influences on global energy budgets are those 'internal' factors that determine the distribution of energy through the atmospheric system. In particular, anything that intercepts radiation on its passage downwards through the atmosphere can be expected to reduce surface energy receipts and lead to decreased temperatures. Clouds are particularly effective in this respect and globally reflect nearly one-quarter of all incoming radiation back into space.

Tropospheric cooling is most evident when dust occurs in the stratosphere. Major 'dusty' volcanic eruptions may have sufficient energy to inject millions of tons of pulverised rock and sulphurous material into the lower stratosphere, where it may remain for up to four years and exercise a notable influence on global climate. Lamb's (1970) dust veil index (DVI) measures the relative importances of these events and shows that veils, if initiated by low-latitude events, can spread across much of the respective hemisphere by the generally polewards drift of stratospheric air. All widespread veils produce immediate but temporary lowering of global temperatures, usually of the order of one-tenth of a degree to a degree Celsius or more, with a consequent diminution of global circulations and a tendency to more meridional-type airflows in the mid-latitudes. As time advances, however, and the veils clear first from the lower latitudes there may be a temporary steepening of the equator-to-pole temperature gradient, thereby reactivating the global circulations and supporting temporarily stronger zonal mid-latitude flows until the dust is wholly dispersed. Defant (1924) found that immediately following volcanic eruptions North Atlantic air pressure gradients fell by 18 per cent of its long-term mean. But within two years, as the dust veils cleared from lower latitudes, North Atlantic air pressure rose to 17 per cent above normal.

Through determining transparency to outgoing terrestrial infra-red radiation, greenhouse gases constitute another important aspect of the atmosphere's composition. Chapter 1 has already discussed the vital role that greenhouse gases have to play in determining global temperatures and their climatic consequences. Increased concentrations in greenhouse gases are now widely believed to be the prime mover in the 'global warming' problem (Houghton *et al.* 1990). That greenhouse gases are closely linked to climate variation is indisputable. Present-day general circulation models (GCMs) of the atmosphere attempt to predict future temperatures on the basis of increasing concentrations of greenhouse gases.

CLIMATIC CHANGES IN HISTORICAL TIME

Some indication has already been given of the great advances and retreats of the ice caps that characterised the Pleistocene period, particularly in respect of their association with the Milankovitch cycles. Ice core evidence is one of the most important sources of information from which inferences can be drawn on past climates. The Greenland Ice Core Project (GRIP) was launched in the early 1990s and has provided a section that reaches nearly 3,000 m to bedrock in the deepest part of the Greenland ice cap. In doing so, the core provides information extending over more than 250,000 years through the variations in the oxygen isotope ratios and concentrations of CO_2 and other greenhouse gases trapped at the time the snow was laid down (Figure 12.1). The cold and warm periods alternate relatively quickly, with long-term mean temperatures varying over a range of between 10 and 15 degrees C. Although closely linked with changes in the greenhouse gas composition of the atmosphere, the cold spells are also associated with those phases of the Milankovitch cycles when eccentricity was at its greatest and northern hemisphere summer energy receipts were at a minimum. This correlation of events with astronomical influences is noteworthy and should not be overlooked in the midst of the global warming debate. Attention in the following sections is, however, concentrated on the post-glacial climates and especially on the past millennium – a period for which there is a clearer regional picture and for which human responses to the demands of a changing climate can be more precisely gauged.

A substantial body of information exists regarding the climate of the British Isles over the past 11,000 years, and several phases have been identified. Stu-

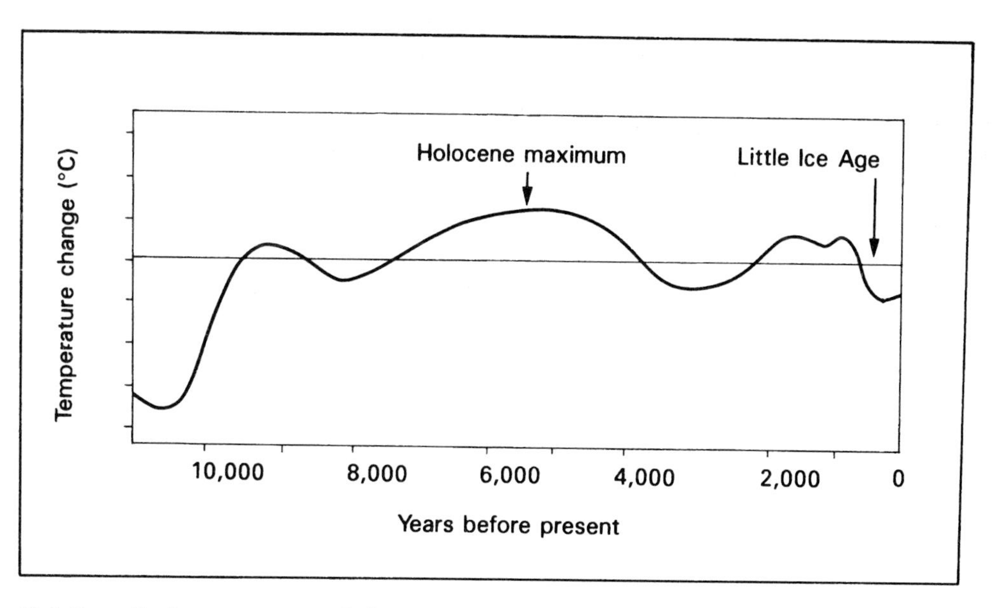

Figure 12.5 Generalised temperature trends during the Holocene compared with present day (horizontal line) (after Houghton *et al.* 1996)

dies, many of them using palynological (preserved pollen) evidence, allowed the Scandinavian botanists Blytt and Sernander to propose a bioclimatic zonation of the Holocene (Table 12.1) in which warm and cool, wet and dry phases alternate. The improved resolution of the GRIP core also provides evidence at this scale. The post-glacial Holocene period, although revealing global temperature variations of the order of only 2 degrees C following recovery from the last ice phase, has been important, as much of mankind's social and technological evolution has taken place during this time and, as Lamb (1995) has so cogently argued, the two are closely interwoven.

One important such phase in the Holocene that casts valuable light on the factors that possibly govern climate at scales below the global and hemispherical was the Younger Dryas event. This was a phase of sharp cooling that ended around 11,000 years ago and marked an interruption to the otherwise rapid warming that characterised the conclusion of the Pleistocene. It has been suggested that temperatures may have increased locally by as much as 7 degrees C within as little as about fifty years at the end of this oscillation (Dansgaard *et al.* 1989). This was, however, one of a number of climatic phases that seem not to be associated with changes in the greenhouse gas composition of the atmosphere. One mechanism for achieving such an abrupt temperature fluctuation in Western Europe is through changes in the North Atlantic conveyor. This is part of a global circulation of ocean currents that is driven by the sinking from the surface to deep layers of relatively dense saline waters in the North Atlantic (Pickering and Owen 1994). Increased amounts of fresh water, produced for example by the sudden melting of large bodies of ice, can discourage this mechanism by spreading cold, fresh water over the denser saline waters of the high latitudes and cutting off the northward flow of warm water into the North Atlantic. This disrupts the North Atlantic Drift (Rahmstorf 1995), driving it southwards and away from north-west Europe and the British Isles, where abrupt cooling may result.

Warm phases are also of importance because they may act as analogues for the present-day changes. The so-called Post-Glacial Climatic Optimum occurred during the Atlantic phase (Table 12.1) and marked a period when conditions became wetter

Table 12.1 The Blytt–Sernander zonation of the European Holocene

Period	*Zone number*	*Zone name*	*Climate*	*Approximate time span (years BP)*
Post-glacial	IX	Sub-Atlantic	Cool and wet	2450 to present
	VIII	Sub-Boreal	Warm and dry	4450–2450
	VII	Atlantic	Warm and wet	7450–4450
	VI	Late Boreal	Warming and dry	8450–7450
	V	Early Boreal	Warming and dry	9450–8450
	IV	Pre-Boreal	Sub-Arctic	10250–9450
Late glacial	III	Younger Dryas	Glacial readvance	11350–10250
	II	Allerød	Glacial retreat	12150–11350
	Ic	Older Dryas	Glacial readvance	12350–12150
	Ib	Bølling	Glacial retreat	12750–12350
	Ia	Older Dryas	Glacial	13500–12750

Sources: Goudie (1977); Pennington (1974)

but warmer in Britain, perhaps by as much as 2 degrees C above present-day level. No complete agreement exists for the precise timing and duration of this warm phase but it is widely thought to date from approximately 4,500 to 7,000 years BP. This period saw also the flourishing of the Bronze Age cultures with settlement moving on to higher ground. The subsequent clearance of land by these communities, together with higher rainfall, may have been instrumental in the significant changes to the upland flora, with the mixed woodland being replaced by blanket bog communities. Lamb *et al.* (1966) have suggested that this period was one of weak but more zonal circulations with a northwards displacement of depressions tracks and of the range of the subtropical anticyclones.

A fluctuating but gradual deterioration of climate took place from the end of the Post-Glacial Climatic Optimum. Temperature and precipitation regimes varied, but during the final millennium BC significant cooling and, more importantly, increases in precipitation were evident. Woodland, which had once covered Britain down to sea level, was in retreat from exposed north-western coastal areas, and between 2800 and 2500 BP many upland peat bogs from mid-Wales to north-west England grew more rapidly than at any time since. At the same time, the once flourishing Bronze Age communities of the Somerset Levels had to resort to constructing wooden trackways to maintain communications in the face of rising water levels (Godwin 1975). Archaeological evidence from eastern Britain suggests that conditions remained dry at this time, indicating a well-developed westerly airflow and, over eastern areas, rain shadow influence. Cooling of the Arctic from 2500 BP onwards lowered temperatures yet further but, more importantly, pushed depression tracks southwards. Lamb (1991) has noted the peculiar storminess of the North Sea at this time. More significantly, eastern Britain became much wetter – a good indication that depressions were now passing over or to the south of the British Isles, generating a more frequent easterly wind component and introducing through this change a new balance of regional climates.

The past millennium has not been immune to similar variations and has witnessed three major climatic phases (Figure 12.6): the Medieval Warm Period, the Little Ice Age and the recovery from the latter. The following sections review these phases, but a study of the evolving climate of the British Isles does more than convey an impression of the local conditions. The location of these islands places them at a focal point on the western oceanic margins of the planet's greatest landmass. Lamb *et al.* (1966, p. 174) noted that

> the changes of annual mean temperature observed in this country [Great Britain] in recent decades, and probably over the past 200 to 300 years, appear to be

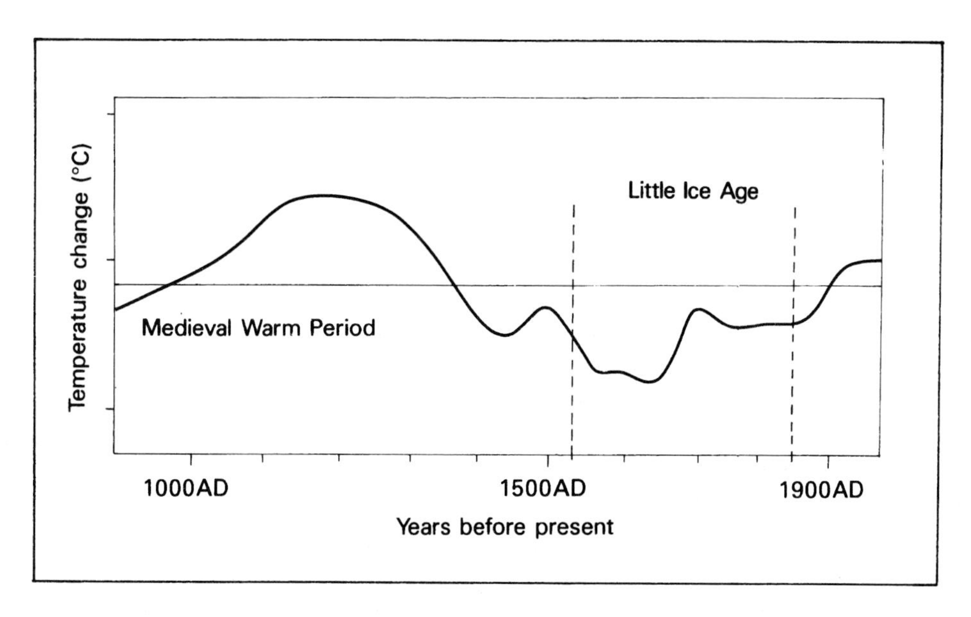

Figure 12.6 Generalised temperature trends since AD 800 compared with present day (horizontal line) (after Houghton *et al.* 1996)

similar in phase and magnitude to what can be established of changes affecting the whole of the Earth.

From this standpoint the events of the past 1,000 years in the British Isles may assume a greater significance.

The Medieval Warm Period

The Medieval Warm Period is the name given to the period between about AD 900 and AD 1200, which was characterised by warmth over much of the world. Results from techniques such as dendrochronology (tree ring analysis) and radiocarbon dating have indicated that there was a consistent pattern of warmth in mid- and high latitudes but with moister conditions at low latitudes (Hughes and Diaz 1994). In general, this was the last period when most parts of the world were warmer than in the late twentieth century, and, as such, can be interpreted as a possible analogue for any future warming.

Knowledge of the sequence of climatic events around the British Isles can be gained from analysis of written archives and other proxy sources (Ogilvie and Farmer, 1997). Lamb (1965, 1995) cites two separate features in the landscape of medieval Britain as clues to a warmer environment: ridge and furrow evidence for tillage on moorland, and records of vineyards across much of lowland England. Medieval colonisation of upland moorland is known to have reached heights of 320 m in Northumberland and 400 m on Dartmoor (see Box 12.1). Indeed, historical evidence for land reclamation and spreading settlement is commonplace (Beresford 1973).

Whilst the imprint of this activity on the landscape provides evidence to historians of population increase and more intensive human use of land, climatological inferences have to be made more cautiously. With increased evidence of a developing economy and increasing population pressure in the early medieval period – prior to the peak in warmth – it now seems that climatic amelioration may have been an enabling factor rather than a trigger for the colonisation of the uplands.

Records of medieval vineyards in what are now considered to be relatively frost-prone sites in southern England have been interpreted as evidence for warmer and drier summers (Lamb 1965) which were possibly 0.7 to 1.0 degrees C warmer than those of

Box 12.1

CLIMATIC CLUES FROM DARTMOOR

How can deserted settlements on Dartmoor add to our knowledge of earlier climates? Bronze Age settlements, such as those found at Grimspound and Broadun Ring, were established during the more clement conditions of the Post-Glacial Climatic Optimum. Farming as well as mining (for tin) drew people to the moor, where they remained, establishing some of the earliest recognisable dwelling houses in Britain. Here these groups remained until the onset of the Sub-Atlantic (Table 12.1), when wetter conditions made farming difficult in upland areas over much of the British Isles (Simmons 1964). As a result the land was abandoned, not to be repopulated until the Medieval Warm Period. There is also evidence (Keeley 1982), coincident with a change in climate, of widespread soil degradation brought about by clearance of the natural woodland which once flourished in the relatively genial climate of those times

The events at the close of the Medieval Warm Period also provide a clear indication of human response to climatic change. The remains of the farmstead at Hound Tor are known to have been occupied till the thirteenth century. The field beyond shows evidence of the ridge and furrow cultivation. By comparison with the environment today, it is possible to gain an impression of a gentler climate in medieval times. The aspect of the field is northerly, and the upper edge is very close to the top of the tor; the site is thus extremely exposed. This site gives an obvious impression not only of a warmer climate, but of a less wet and windy one too. It would take only a small northward displacement of depression tracks to recreate a much less hostile growing environment on this, Britain's most southerly, major upland area.

the mid-twentieth century. Evidence for wetter winters and drier summers can be inferred from the response of pressure systems to higher temperatures, which lead to a northward migration of their principal tracks that places northern Europe more securely under the influence of the Azores anticyclone in summer but in the path of a slightly invigorated westerly airstream in winter. The warm years of the late 1980s perhaps constitute an example of this synoptic pattern. More specific evidence of drier summers comes from the distribution of water-mills in Kent around 1275 (Manley 1952) and reports of summer drought in 1252/53 (Meaden 1976). Hoskins (1988) noted that a chronicle at Crowland Abbey alluded to the Fens drying up, suggesting that in eastern Britain at least, a drier phase had indeed commenced.

Human responses are often multi-causal, reflecting both economic and climatic stimuli. The use of human proxy evidence to identify the Medieval Warm Period and its subsequent termination reveals the dangers of inferring climatic changes from human records, from which it is often easy and convenient to exaggerate the effects of climatic change when human behaviour seemingly fits a climatic hypothesis. The study of physical features can often avoid this uncertainty. The analysis of peat bogs by Barber (1985), for example, shows an abrupt surface wetness peak around 1400 at Bolton Fell Moss near Carlisle. This latter example provides a contrast to the effect of the earlier dry period that caused serious subsidence at Carlisle Cathedral (Cormack 1994). At the wider geographical scale, the equally objective analytical techniques employed by Hughes and Diaz (1994) also reveal the reality of warming and climatic change across much of the world at this time.

Nonetheless, the scarcity of climatic information

for this period is unfortunate, but this could be a consequence of the fact that warming may cause less social disruption than cooling. By contrast, there is plentiful evidence of social stress associated with a climatic downturn after about 1300.

The Little Ice Age

The period between the end of the Medieval Warm Period and the early nineteenth century is generally known as the Little Ice Age and has been the subject of much academic interest (Grove 1988). It is now recognised that although the period included many severe winters, another of its principal characteristics was variability, with spells of warm and dry weather.

The onset of the Little Ice Age can be detected as early as 1200 in the marginal economies of the Norse Greenland colonies, where sea ice became a growing problem (McGovern 1981). In the British Isles the immediate consequences were seen in the increased number of catastrophic storms and sea floods. Lamb (1991) has carried out an exhaustive analysis of these events, which reached a peak in the thirteenth century. The North Sea area appears to have been particularly badly affected (Figure 12.7), and it was during this period that the old ports of Ravenspur (not far from Hull) and Dunwich (off the East Anglian coast) were destroyed. Harlech was choked by a spreading sand dune around the 1380s. It is tempting to attribute such changes to a southward displacement of depression tracks (in response to initial high-latitude cooling) leading to a more cyclonic period over the British Isles. As significant as the frequency of the depressions was the intensity of the gales which accompanied them. Douglas *et al.* (1978), using the evidence from documents from the Spanish Armada campaign of 1588, have deduced unprecedented intensification of the subpolar jet stream. This characteristic agrees with the suggestion of a steepening temperature gradient across the mid-latitudes encouraged by the cooling in the sub-Arctic regions.

Storms such as those that tormented the Spanish Armada were a feature of the British Isles throughout the Little Ice Age. The more southerly track of these depressions during this phase was tragically exemplified much later in its course by the Great Storm of 1703, so vividly described by Defoe

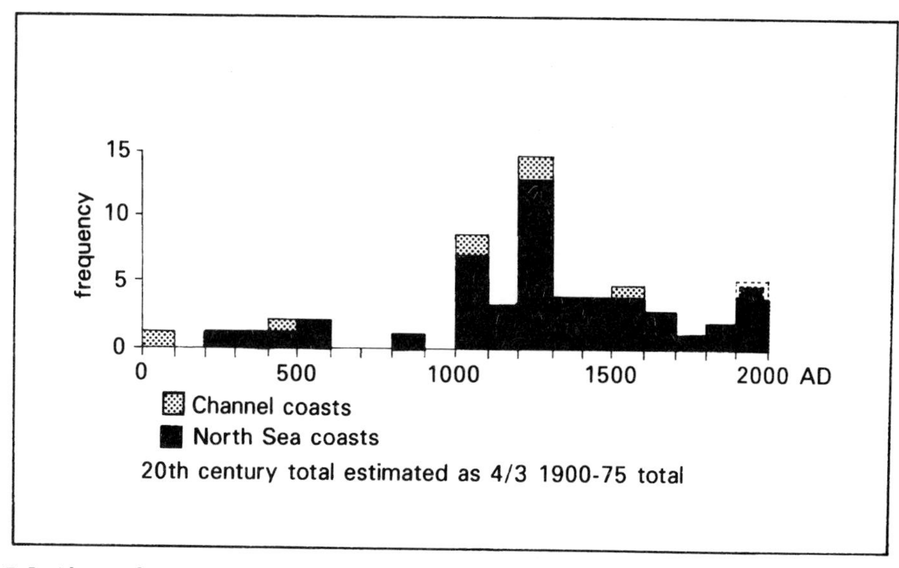

Figure 12.7 Incidence of severe storms on the coasts of southern and eastern Britain AD 0 to the present (after Lamb 1995)

(1704), which raged across southern counties, bringing a death toll of thousands and wreaking havoc unmatched in the region until the events of October 1987.

The first signs of changes to the temperature and precipitation regimes, however, became suddenly apparent in the summers between 1313 and 1317, when excessive rains led to crop failure and widespread famine across Europe. Matters were made worse by the winters of astonishing severity. The 1430s were especially bad, but the winters of the 1690s have scarcely a rival in the last 1,000 years. Consistent with an increased frequency of westerly weather, Barber (cited in Lamb 1995) has indicated that the conditions in north-west England were very wet indeed during the late fourteenth and early fifteenth centuries. The climate remained variable, however, and the later reversal of the west–east regional contrasts is suggested by Barber's archaeological observations of a drier phase in north-west England in the sixteenth century, while documentary sources confirm eastern England to have been plagued by floods, wet soils and overflowing ditches – problems not helped by lower evaporation and transpiration losses in the cool summers.

Generally cool and damp conditions held a firm grip from 1550 until about 1700 and mark the most profound depths of the Little Ice Age. It was a time of glacier advance over Europe, of interrupted communications across mountain passes and wide-scale abandonment of marginal land. Areas exposed to northerly weather must have been particularly at risk, and Parry (1975) has estimated that in the Lammermuir Hills of southern Scotland the upper limit of cultivation fell from 425 to 225 m between 1350 and 1600.

The privations and social disruption in Scotland have been summarised in Lamb (1995). Between 1693 and 1700 the rye harvest failed seven times. Evidence suggests that snow lay all year on many Scottish peaks. During the 1670s the surface of a loch in Strathglass failed to lose its ice cover in even the warmest summer. Such an occurrence was estimated by Lamb to require temperatures between 1.5 and 2 degrees C lower than those of the late twentieth century, a departure nearly three times as great as that registered in Manley's Central England Temperature series. This accentuation in south-to-north regional contrast is in keeping with the theory of a steeper temperature gradients across the mid-latitudes.

Variability remained a feature even of the most intense cold phases. The winter of 1683/84 was cold enough for frost fairs on the frozen Thames; the ground of south-west England froze to a depth of a metre whilst ice floes over 5 km wide could be seen in the Channel. Yet just two years later the winter of 1685/86 was the fourth warmest in Manley's Central England temperature series. Such extremes of behaviour are often associated with blocking situations in which different centres of activity can draw in air masses from contrasting sources.

The recovery from the Little Ice Age has been so gradual and faltering that it is difficult to put any reliable date to its conclusion. Even in the closing years of the nineteenth century, winters could display a level of cold that would excite much comment today. We may thus be confident that the Little Ice Age extended into the eighteenth century, and here for the first time our studies are aided by the appearance of a tolerably comprehensive, if not necessarily standardised, set of instrumental data.

Figure 12.8 shows how the Central England Temperatures have varied from the mid-seventeenth century onwards. The picture is one of warming from 1700 to the mid-twentieth century. Only then did temperatures become comparable with those of the Medieval Warm Period. Once again, yearly as well as decadal variability was very much a feature of the climate. Following one of the coldest winters on record (1740/41), the 1740s were years when southern England enjoyed a run of dry years with rainfall 5 to 10 per cent less than present, suggesting a stronger anticyclonic element. But from 1750 onwards the summers at least were wet, and accumulating surface waters led to well-documented peat bursts on Solway Moss in 1771 and 1772.

The 1780s were a notably poor decade. Crop failures occurred in Scotland in 1781 and 1783 while local diaries provide evidence of prolonged snow cover (Pearson 1973; Wheeler 1994). This decade was notable for its volcanic eruptions. July 1783 witnessed the major Icelandic eruptions of Eldeyjar, Laki and Skaptar Jökull. So great was the volume of dust produced by that outburst that the crops in Caithness were destroyed by the huge quantities that were deposited, and even in southern England the Sun was seen as if through a permanent mist. This phase of activity reached a peak with the Tambora eruption of 1815 in the East Indies, which caused 1816 to be widely described as 'the year without a summer'. Temperatures were low throughout the year and crop failures widespread. Wales seems to have been particularly wet, though Scotland reported drier and sunnier weather. This pattern suggests a strengthening and extension of the polar anticyclones, the depressions adopting more southerly tracks across Wales and central England.

Kington (1988) has been able to collate information from across Europe to produce daily weather maps for the period 1780–85. He found a notably low frequency in westerly days – only 66 per year compared with, for example, over 100 in the first half of the twentieth century. This contrast offers further evidence for an abundance of blocking situations during the Little Ice Age.

The question remains, what was the cause of the Little Ice Age? It was certainly a time of notable volcanic activity, and Lamb's (1970) list of major eruptions includes several notable events. However, the consequences of such activity are temporary, with recovery normally being made within four years. Positive feedback mechanisms operating through the medium of snow and ice cover extension may of course provide the means whereby these effects are magnified and made more long-lasting.

An alternative explanation lies in sunspot variations. There is no doubt that the period from 1645 until about 1715 was one in which scarcely a sunspot was seen (Eddy 1976). The coincidence of the so-called Maunder Minimum with the coldest period of the Little Ice Age may be of great importance. A problem is, however, posed by the duration of the Maunder Minimum: it does not explain the climatic deterioration, which began in the thirteenth century. Fortunately, other evidence exists, and although sunspots have been studied since 1700 naked-eye observations have been made in the Far East for much longer, and a record has been collated from Japanese, Korean and Chinese sources (Kanda 1933) that covers the period 28 BC to AD 1743. These show that the Maunder Minimum may not be unprecedented, and periods of inactivity were found between 1520 and 1604 (the Spörer Minimum), another before 1520 and another between AD 600 and AD 800. No sunspot records exist for Europe at this time, but aurora sightings were often noted, and these can be used as a surrogate measure as they are known to be associated with sunspot activity. Eddy (1976) found that the European aurora record supports the contemporary oriental evidence. During the Medieval Warm Period aurora and naked-eye sunspots were frequently sighted, suggesting this to have been a period of marked solar activity. On this basis it might be speculated that extra-terrestrial mechanisms were possibly important in the initiation of the Little Ice Age.

Further supporting evidence is found in the record of terrestrial ^{14}C which is assimilated into living vegetable tissue. This isotope is formed in the atmosphere by cosmic ray bombardment, but such activity is disrupted during periods of solar activity when galactic cosmic rays do not reach the Earth's atmosphere in such abundance, and consequently less ^{14}C is produced. The ^{14}C record in tree rings may therefore reveal the past production rates and corresponding levels of solar activity. A relatively inactive Sun should be indicated by higher ^{14}C quantities, and a more active Sun by lower concentrations. Research (Suess 1965) shows fluctuations broadly in line with those anticipated from the foregoing reconstructions. An important feature is

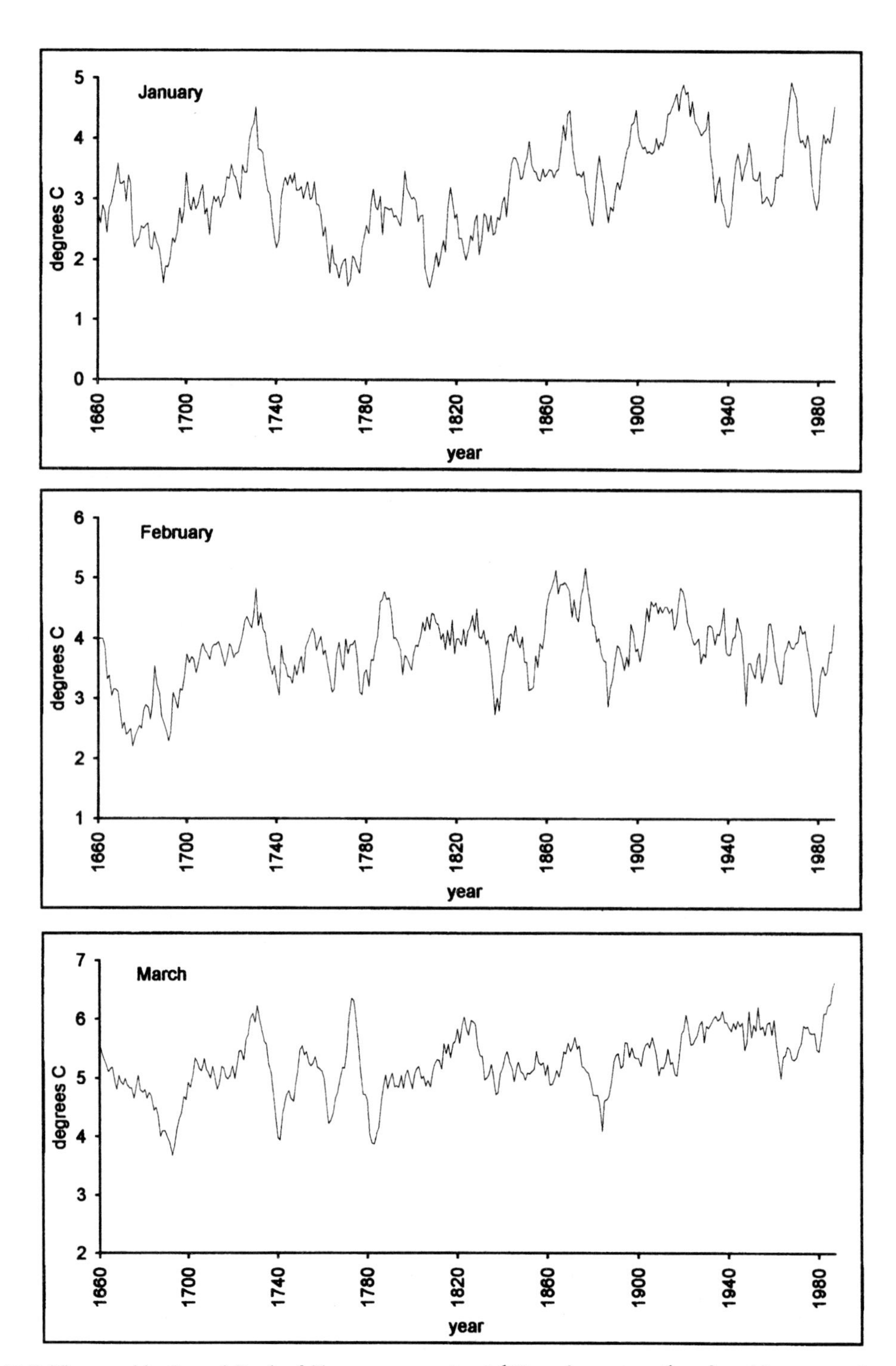

Figure 12.8 The monthly Central England Temperature series, 1659 to the present (based on 10-year running means)

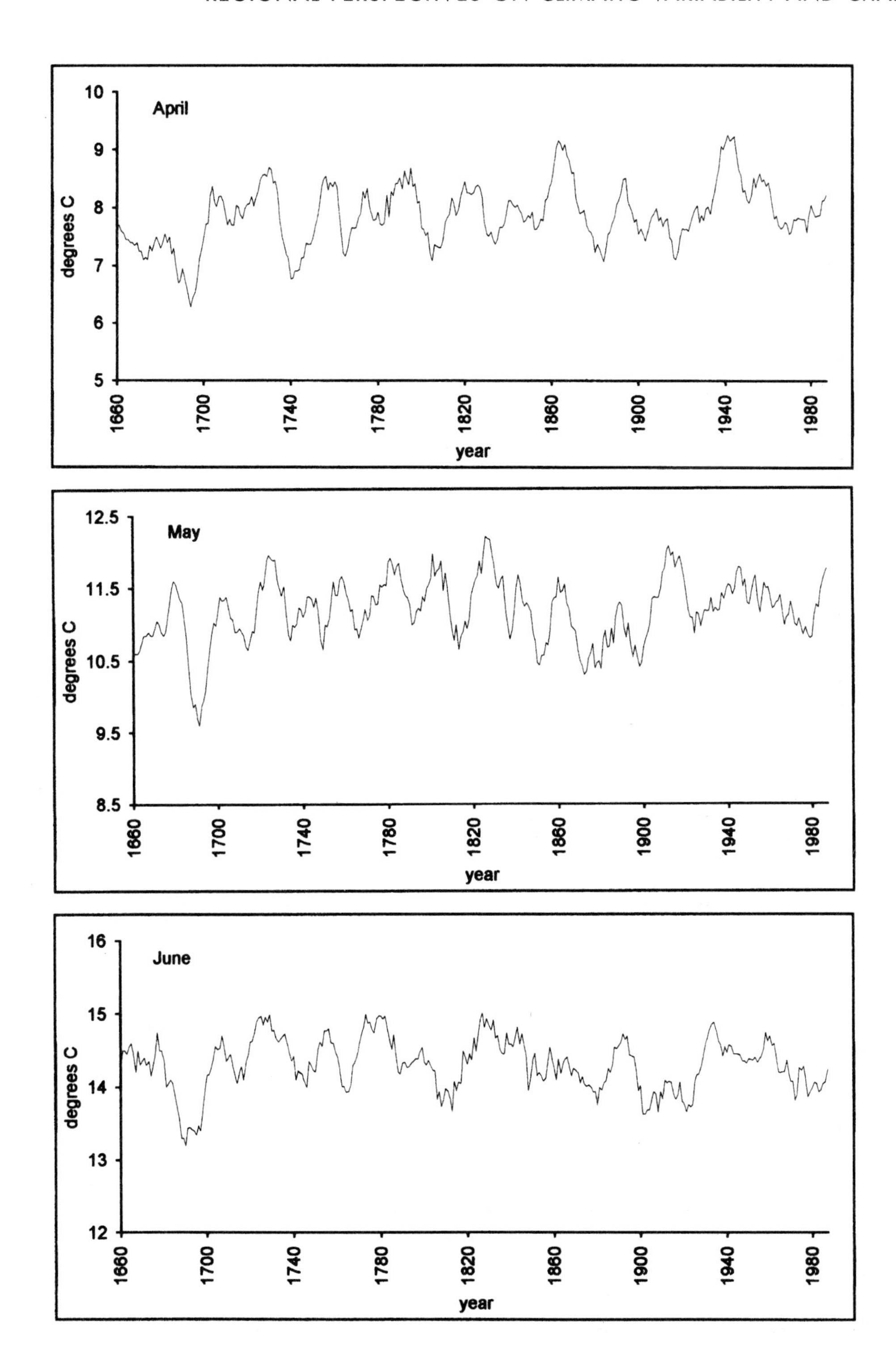

Figure 12.8 (Continued)

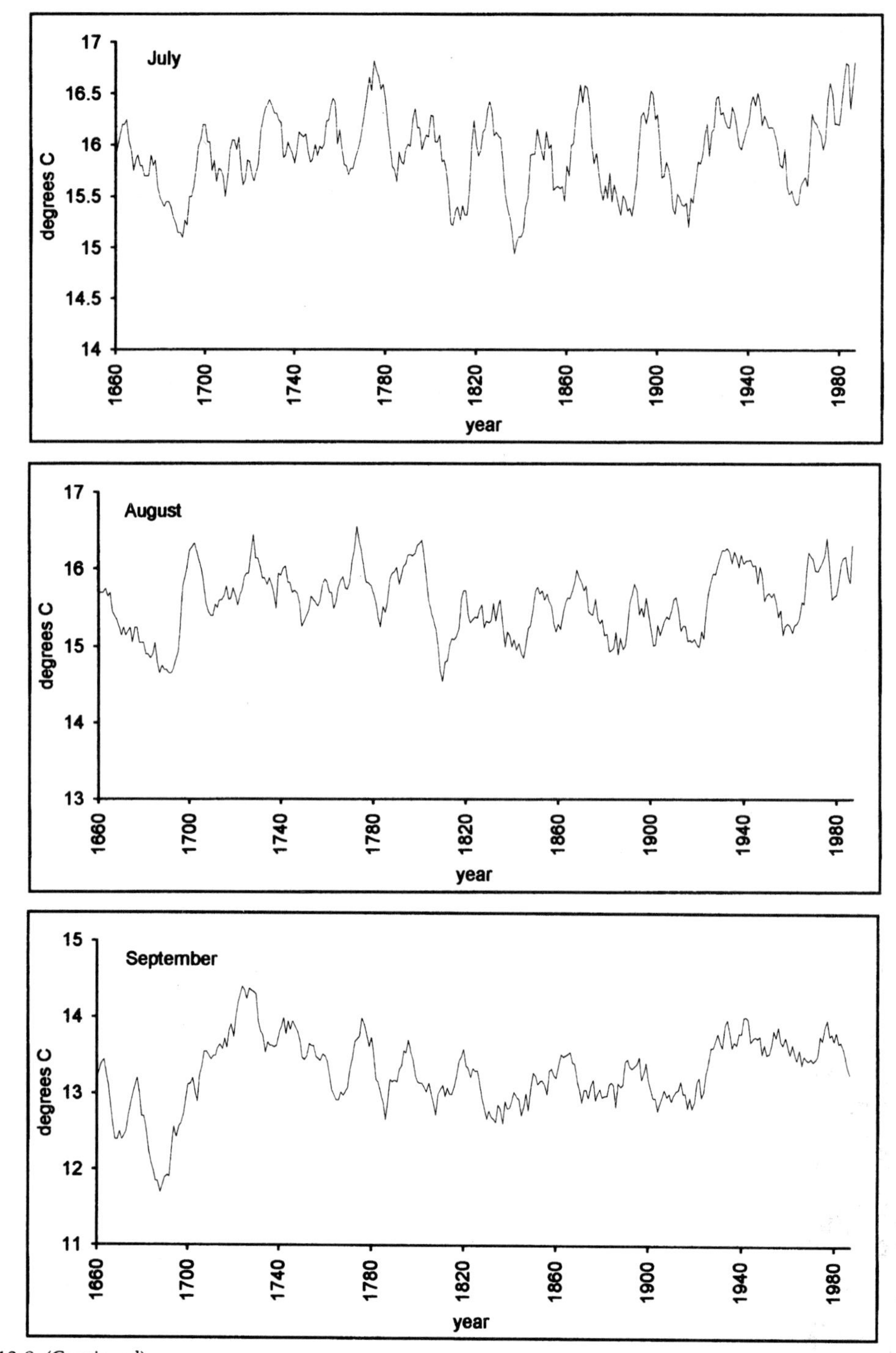

Figure 12.8 (Continued)

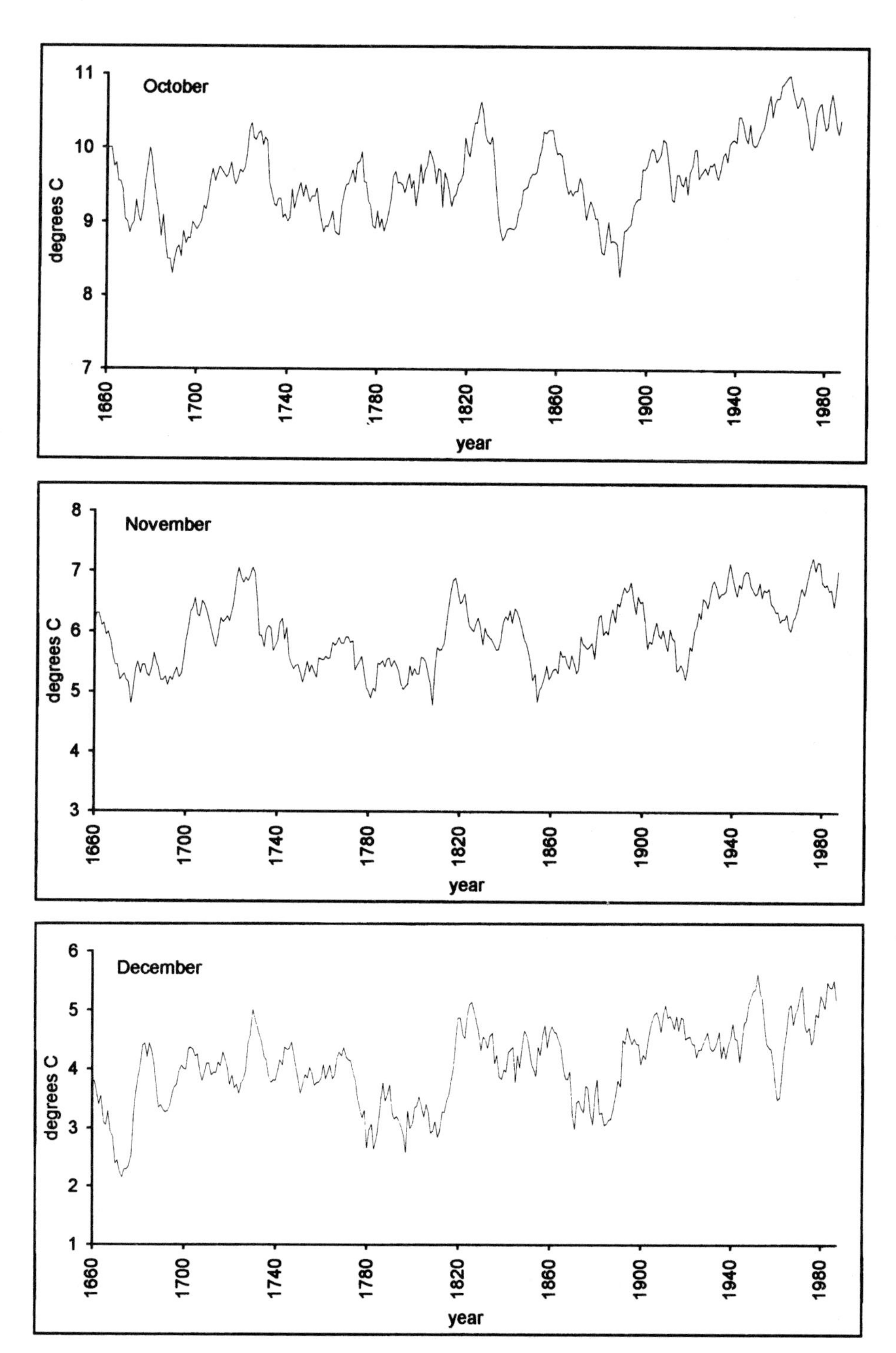

Figure 12.8 (Continued)

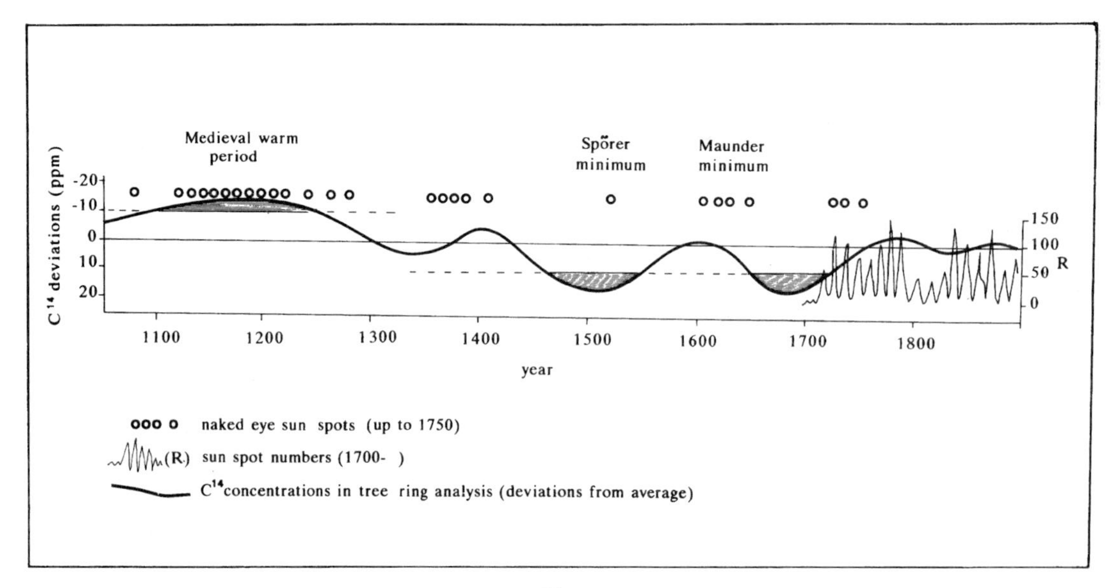

Figure 12.9 Changes in sunspot and auroral activity and ^{14}C in vegetation between AD 1100 and 1900. Deviations of ^{14}C concentrations in excess of 10 ppm are shown by shaded areas (after Eddy 1976)

the DeVries Fluctuation, a period when ^{14}C was unusually abundant, which coincides with the Maunder Minimum. Figure 12.9 shows a general agreement between the ^{14}C levels and the Medieval Warm Period.

These theories are not universally accepted. Stuiver and Braziunas (1992) have pointed to the complex relationship between ^{14}C and environmental factors. More recently, Frenzel *et al.* (1994) have suggested that the Europe-wide character of climate changes at the time of the Maunder Minimum, i.e. most marked cooling on the Atlantic margins but less pronounced in central Europe, is not consistent with solar forcing. This 'signature', he proposes, agrees more readily with changes in oceanic heat transport. Lamb's (1995) evidence of the changes in fishing productivity in the north-east Atlantic supports the latter view and presents a picture of a strong southwards movement of the cold waters of the East Greenland Current.

EVIDENCE FROM INSTRUMENTAL OBSERVATIONS AFTER THE 1850S

The mid-nineteenth century was a period of rapid development in instrumental observation and can be regarded as the period in which valuable networks of observing stations became established. This growth was stimulated by a sequence of loosely related events. Following the loss of the *Royal Charter* off Anglesey during a storm in October 1859 (Lamb 1991; Kington 1997), a storm warning service for shipping was established by the setting up of a department within the Board of Trade in 1860 that later evolved into the Meteorological Office. This coincided with an expansion of the pioneer network of weather stations. Collation of observations from these weather stations made it possible to construct basic daily weather maps. This in turn enabled H.H. Lamb to derive a catalogue of daily weather types back to January 1861 (Lamb 1972).

Knowledge of the geographical distribution of rainfall across the British Isles was greatly extended by the founding of the British Rainfall Organisation (BRO) in 1860 by George Symons (Folland and

Box 12.2

GORDON MANLEY AND THE CENTRAL ENGLAND TEMPERATURE SERIES

The Central England Temperature (CET) series, the longest of its kind in the world, was derived through painstaking study by the late Gordon Manley (Figure 12.10). For over thirty years Manley collated instrumental and non-instrumental data, and by careful cross-checking and calibration was able to synthesise monthly temperatures for a 'central England' area. A more complete description of the notable endeavours of Manley can be found in his original publications (Manley 1941, 1946, 1953, 1959, 1974). This unparalleled synthesis of data is presented in the form of monthly mean temperatures from 1659 to the present day (Figure 12.8). They are representative of conditions in a broad area from Lancashire to the East and West Midlands of England at an altitude of between 30 and 60 m above sea level. The sites chosen to provide the raw data should be of 'intermediate character', which is to say 'they should not for example be located in acute frost hollows, or on wide stretches of sandy soil, or exceptionally windswept ridges' (Manley 1974). Some of Manley's original work was, however, based on data and information from sites as far afield as Plymouth and Edinburgh, Lyndon (in the East Midlands) and Shrewsbury. He also relied heavily on his earlier studies of the climate of Lancashire (1946) and on the record of the Radcliffe Observatory in Oxford, where regular daily observations began in 1815. The Meteorological Office continues to produce the monthly CET, which is now derived from daily values and has been partially reworked (Parker *et al.* 1992). Though the sites chosen to represent this well-understood though loosely defined area have changed with time, it provides an incomparable yardstick by which scientists can gauge climatic behaviour. Thanks to Gordon Manley's efforts, scientists and the public alike can now judge more objectively the peculiarities of present-day weather.

Figure 12.10 Portrait of the late Professor Gordon Manley
Photo by courtesy of University of Durham

A full account of Manley's life and a bibliography of his writings and publications has been prepared by Tooley and Sheail (1985).

Wales-Smith 1977; Eden 1995). Rainfall observation acquired the status of a favoured pastime of the (mostly) leisured classes of Victorian Britain. An original network of 500 gauges grew to 3,500 by 1900 and 5,000 by 1918. More comprehensive monitoring of weather was initiated by the formation of a 'Meteorological Committee' in 1866, set up in response to a call for more observations by Edward Sabine, President of the Royal Society (Burton 1994). As a result, meteorological observatories were set up at Falmouth, Armagh, Glasgow, Aberdeen, Valentia (south-west Ireland) and Stonyhurst (Lancashire), with their observations being supervised from the existing King's Observatory at Kew. These records were soon supplemented by significant numbers of climatological stations from the latter decades of the century, especially at coastal resorts, where 'Health Resort' stations were established.

A Climatic Downturn in the Late Nineteenth Century

Evidence from written archives and early observations indicate that an emphatic recovery from the Little Ice Age had occurred by the mid-nineteenth century. By the 1860s a widespread recession of European glaciers had been noted, and in the summer of 1868 a temperature of 100.6°F (38.1°C) was recorded in a Glaisher stand at Tonbridge in Kent on 22 July. This is 1 degree C higher than any temperature observed in a Stevenson screen, but is not accepted as the highest temperature recorded in the British Isles because of known differences between the two types of screen (Burt 1992).

By the late 1870s a sharp deterioration commenced that contributed to a decline in British agriculture that was to last fifty years. The year 1879 was unusually inclement and comparable to the worst years of the Little Ice Age; it was the coldest year in London since detailed records were first kept in 1841 (Brazell 1968). After a particularly wet summer, the wheat harvest was reduced to about half of average and crops remained unharvested till the end of the year, even in East Anglia (Lamb 1995). The collapse in agriculture was most acute in the English lowlands, where much arable land was converted to grass pasture. However, this resulted from a conjunction of climatic and economic factors: the free trade policy of the time allowed imports of cheap wheat to reach Great Britain from North America and refrigeration soon gave a new opportunity to import meat. These events were compounded by a run of severe winters in the late nineteenth century. In addition to the famous blizzards of 1881 and 1890 (Chapter 2), the winter of 1885/86 had notable snowfalls from October to May and was the 'longest' winter between the 1850s and the turn of the century (Bonacina 1957). The years 1894/95 continued the run of severe winters around the 95th year of each century that was such a curious feature of the Little Ice Age. On the morning of 11 February, Braemar recorded the lowest air temperature observed in the British Isles of −27.2°C. It was said to be the best winter of the nineteenth century for skating on Lake Windermere and Loch Lomond, and was also the last really severe winter in London (Figure 12.11), with a fair being held on 30 cm of ice on the river at Kingston upon Thames (Currie 1995; Pike 1995). The mean monthly temperature at Kew was below the 1931–60 average in all but one month in each year from 1885 to 1888, for all but two months in 1889 and 1890, and for each month in 1891 and 1892 (Brazell 1968).

Airflow Type Variations in the Early Twentieth Century

This cooling phase was not sustained, and the first significant change in climate in the twentieth century was a general warming over the first three decades. This was most conspicuous over high latitudes, possibly because of a northward retreat of the polar ice boundary. The warming was accompanied by an increase in the frequency of westerly winds in the mid-latitudes (Figure 12.12). This amplified the warming tendency over Western Europe by assisting the advection of mild, maritime air into the area. Both Manley (1973) and Lamb (1995) concur that

Figure 12.11 Ice on the Thames at Kingston Bridge, January 1895
Copyright Kingston Heritage Centre, by courtesy of Tim Emerson

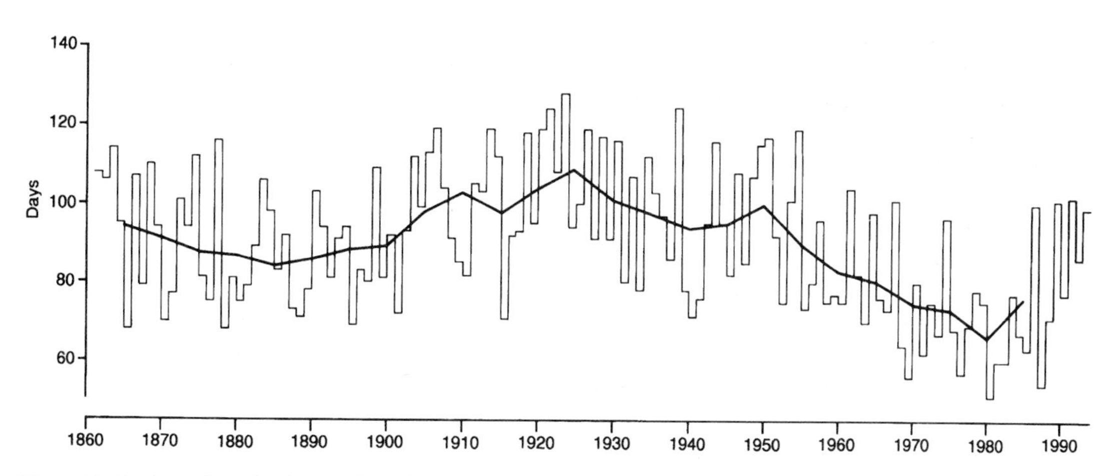

Figure 12.12 Annual totals of westerly airflow type in Lamb classification (1861–1993) with 10-year running mean plotted at 5-year intervals (after Lamb 1972, updated from data originally published in *Climate Monitor* and *Climate News*)

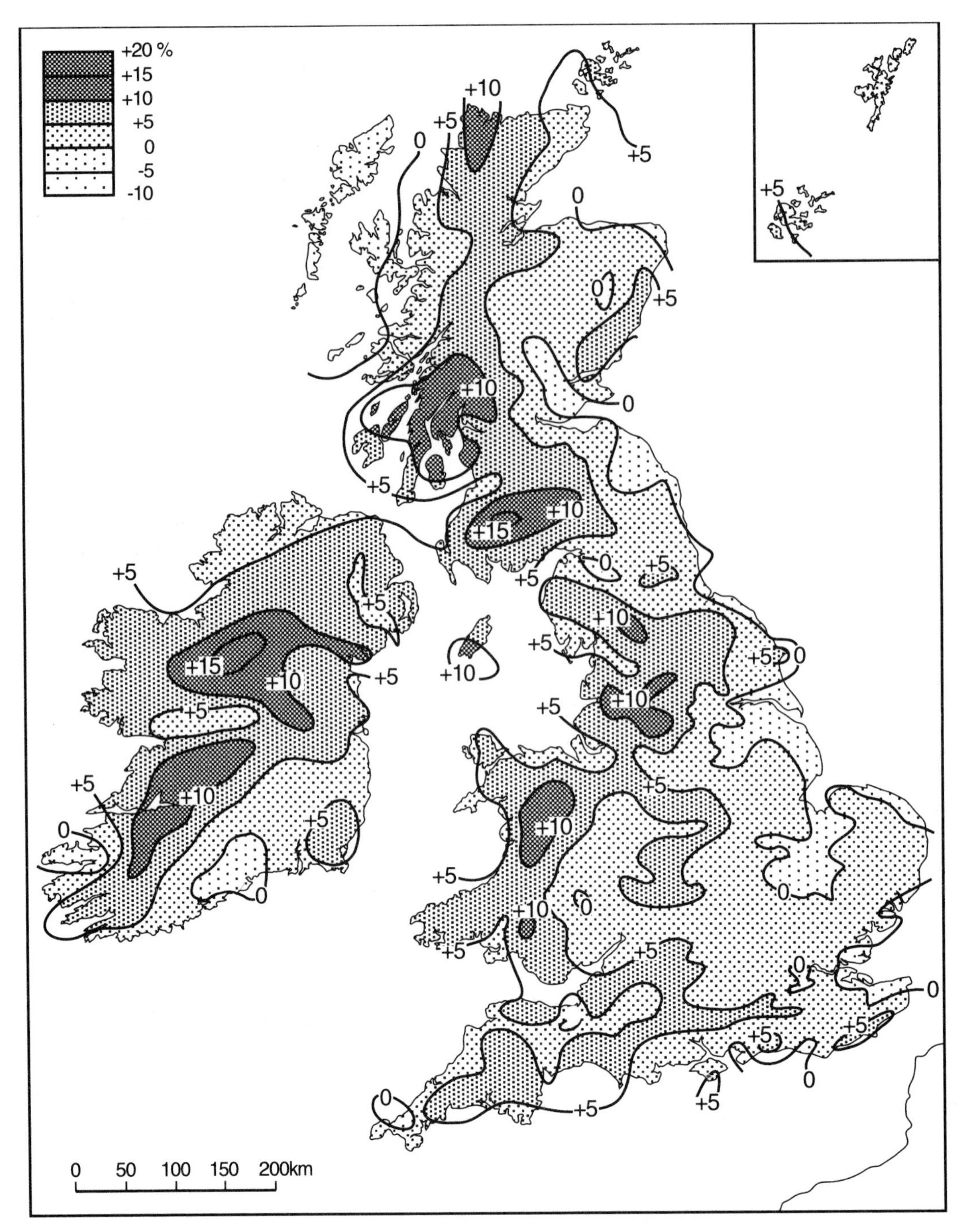

Figure 12.13 Average annual precipitation 1916–50 expressed as a percentage of 1881–1915 average (after Mayes 1995)

the only severe winters of the century up to 1939/40 were those of 1916/17 and 1928/29.

Another emphatic consequence of the upturn in westerly winds was an increase in rainfall over much of the British Isles. Nowhere was this change more conspicuous in both absolute and percentage terms than over the west-facing mountains of Scotland, northern England and Wales. Figure 12.13 shows that rainfall over these wettest parts of the country increased by up to 10 per cent in places whilst the driest eastern coastal localities – especially those lying in a rain shadow in westerly winds – became marginally drier (Gregory, 1956). In contrast, Howells (1985) has shown that the frequency of heavy daily rainfalls over south-east England dropped at around this time. Mapping of the annual rainfall anomalies by the British Rainfall Organisation showed the clear influence of exposure to the west, which led to the conclusion that the wetter conditions here were 'undoubtedly a reflection of the increasing vigour of the westerly circulation' (Salter and Glasspoole 1923).

Renewed Blocking and Cooling after the 1940s

The decade of the 1940s is often cited as representing a peak of warmth which was concluded by a cooling from the 1950s to the 1970s. For example, in the early 1970s Manley wrote: 'the signs of a recession from what appears to be a peak of warmth about 1950 must be considered very seriously. The economic load if our climate becomes cooler, will be quite considerable' (Manley 1973, p. 21). The fine summers of 1940 and 1947 were followed by remarkable warmth in 1949, when the Central England temperature reached 10.6°C, a value that was equalled in 1990. However, in addition to the famous winter of 1947 the decade included severe weather in the winters of 1939/40 and 1944/45, and the deep snow of January 1945 is described by Stirling (1982). The winter of 1950/51 was the snowiest of the century at upland sites, where snow patches remained well into summer 1951. Such seasonal extremes could be interpreted as a facet of greater continentality in the climate; Figure 12.12 shows that westerly activity dipped temporarily around both 1940 and 1947, a precursor of changes to follow over the coming two decades.

Lamb (1977) has presented a case for a heightened frequency of climatic extremes between the 1940s and the 1970s. This coincides with, and is generally attributed to, an increase in blocking activity in the global circulation which in turn is associated with a downturn in northern hemisphere temperatures. By the 1960s a greater frequency of gales was being noted in the North Sea which could be related to both a change in depression tracks and also a greater frequency of north-westerly and northerly winds (Lamb 1995).

The weather experienced in any one location can result from either the changing frequency of airflow types or the changing character of a particular type itself (a 'within-type' change). Whilst the frequency of airflow types was changing, so the cooling to the north of the British Isles was altering the character of north-westerly and northerly winds. This was due to a southward movement of the edge of the sea ice; by spring 1968 and again in 1969 Iceland was half surrounded by ice (Lamb 1995). This caused a corresponding movement of the 'polar front' – the boundary between the ocean currents of the Arctic and the North Atlantic Drift and the preferred location for cyclogenesis. It has also been suggested that the extra ice could have been caused by a decrease in the salinity of the oceans' surface layers resulting from greater runoff of fresh water from northern Canadian rivers.

The Changing Character of the Seasons from the 1940s

Cooling was observed in all four seasons after the 1940s but the timing, size and implications of the fluctuation differed quite considerably.

Spring

Cooling was most marked in spring, and the 10-year running mean temperature on the Central England

Temperature series dropped by 1 degree C between the 1940s and the 1970s. This is partly a reflection of the fact that springs in the 1940s were the warmest for any decade in the CET series. Whilst fourteen of the sixteen springs between 1938 and 1953 had temperatures above the 1920–60 average, only one such 'warm' spring was noted between 1962 and 1980 (Lamb 1995). A deterioration of spring weather between the 1950s and the 1970s was reported by Shone (1980), who found that most of the driest, sunniest and warmest springs of the period were in the 1950s and that the 'cool and wet spring of 1979 in Britain was a bitter epitaph to a wretched winter'.

One of the most pervasive changes in the weather of an individual spring month since the mid-twentieth century has been an increase in March rainfall. Between the periods 1941–70 and 1961–90, March rainfall increased by more than 50 per cent in the central Scottish Highlands. This is an impressive change, especially in view of the overlap between the reference periods, and calls into question the use of 30-year periods as a baseline for the analysis of climatic anomalies (Mayes 1996).

Autumn

Following an abrupt increase from the 1920s, autumn temperatures remained close to the peak attained during the 1940s for several decades. This is over a degree Celsius warmer than the typical autumn temperature over the previous two hundred years. Figure 12.8 shows that one of the changes in monthly temperature in the mid-twentieth century has been a warming in October (see also Clark 1979; Lyall 1976). This change has been linked to a change in the pressure distribution involving a south-westward shift in the mean position of the Icelandic low and a strengthening of high pressure over continental Europe, factors that have encouraged a higher frequency of south or south-westerly airstreams across the British Isles (Lyall 1976). More recently, a number of colder Octobers have been recorded, with a higher frequency of northerly winds (contrary to changes over the year as a whole), including those of 1974, 1981 and 1993.

Winter

Winter temperatures on the western seaboard of the European continent are sensitive to the vigour of the westerly circulation. It is therefore not surprising to find that a reduction in winter temperature mirrors the downturn in westerly airflows after the 1920s. The snowstorms of early 1947 were regarded as the worst since 1814 (Manley 1952; Bonacina 1957), and February 1947 was the coldest in the Central England record with a mean temperature of −1.9°C. The winter of 1962/63 was notable for the intensity and persistence of the severe cold, although extreme temperature records were not broken (Shellard 1968). Moor House in the northern Pennines had thirty-four consecutive days of subzero temperatures and Kew, west London, had nine such days (Booth 1968). Although mean temperatures were generally a little lower than those of 1947 – and the duration of the severe weather more dominant through the winter – conditions in 1963 were probably easier for much of the population and the amount of snow deposited was rather less, except in south-west England (Shellard 1968).

Burroughs (1980) has noted that whilst cold winters have usually tended to be dry (and mild winters wet), a number of winters in the 1970s were either dry and mild (1975/76) or wet and cold (1976/77 and 1978/79). Since then, outbreaks of severe weather have tended to occur in individual months rather than persist for a whole season. After a cold, snowy December in 1981, the first part of January 1982 was intensely cold, and Braemar equalled its UK minimum temperature record of −27.2°C, set in 1895. However, most of the latter part of the winter was mild. February 1986 and January 1987 both had mean temperatures below freezing but neither spell of cold weather dominated either winter.

A more persistent feature since the 1970s has been a tendency for winters to claim an increasing proportion of annual rainfall (Figure 12.14). The

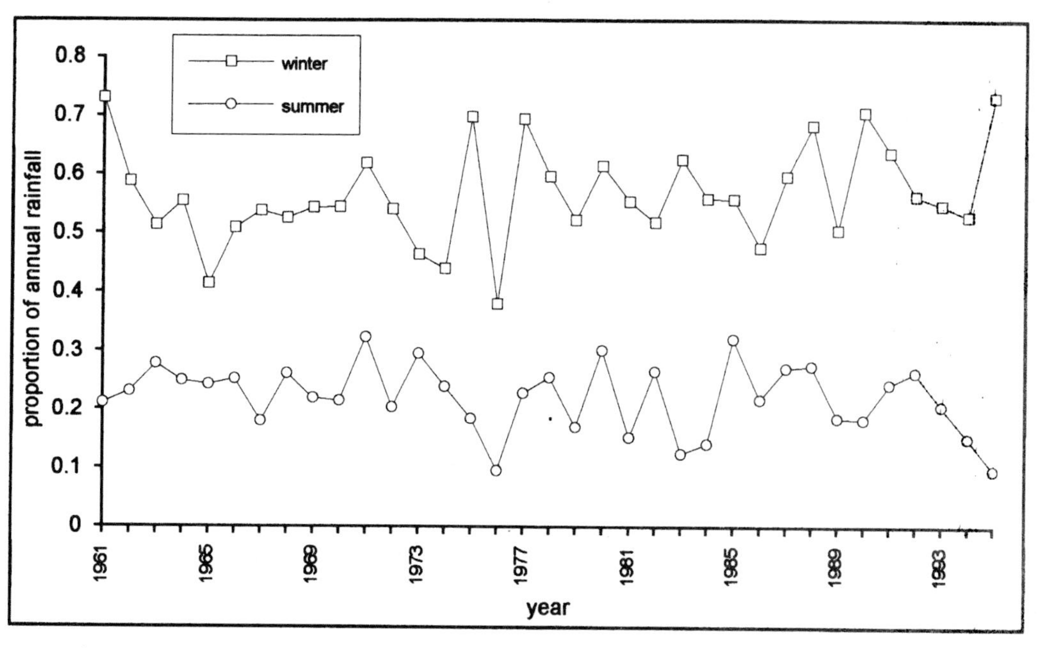

Figure 12.14 Proportion of annual rain falling in the winter half-year (October–March) and summer (June, July and August) in England and Wales for the period 1961–95

ratio of winter half-year to summer half-year rainfall increased to 1.34 for the period 1979/80 to 1988/89 (Marsh *et al.* 1994).

Summer

One of the clearest climatic fluctuations has been a decrease in summer rainfall between the 1960s and the 1980s. Four of the eleven driest summers between 1766 and 1984 in England and Wales occurred after 1976 (Folland *et al.* 1984). This decrease was initially widespread in the 1970s, but several very wet summers have been experienced since 1985, especially in the north and west of the British Isles. Summers in the 1980s were therefore wetter than average in north-western areas but significantly drier than average further south-east (Figure 12.15). The return of some wet summers in the late 1980s has also contributed to an increase in the variability of summer rainfall over recent decades.

A more holistic view of summer weather can be gained from the calculation of indices that integrate the temperature, sunshine and rainfall into a single value that could be said to represent how 'fine' a summer has been. That of N.E. Davis (1968) has been widely accepted, combining weighted expressions of maximum temperature, total rainfall and sunshine as follows:

$$I = 18T + 20S - 0.276R + 320$$

where T = mean daily maximum temperature (°C), S = mean daily sunshine duration (hours) and R = total precipitation (mm).

Whilst there is a case for referring instead to mean temperature, rain days and a measure of wind speed, the index provides a means of comparing the general character of summer weather. Summer 1968 was notable for several periods of wet weather in southern England caused by a southward displacement of depression tracks in a blocked airflow pattern (Hughes 1979). A relatively high frequency of easterly winds resulted in highest scores in western Britain and Ireland; the value at Glasgow

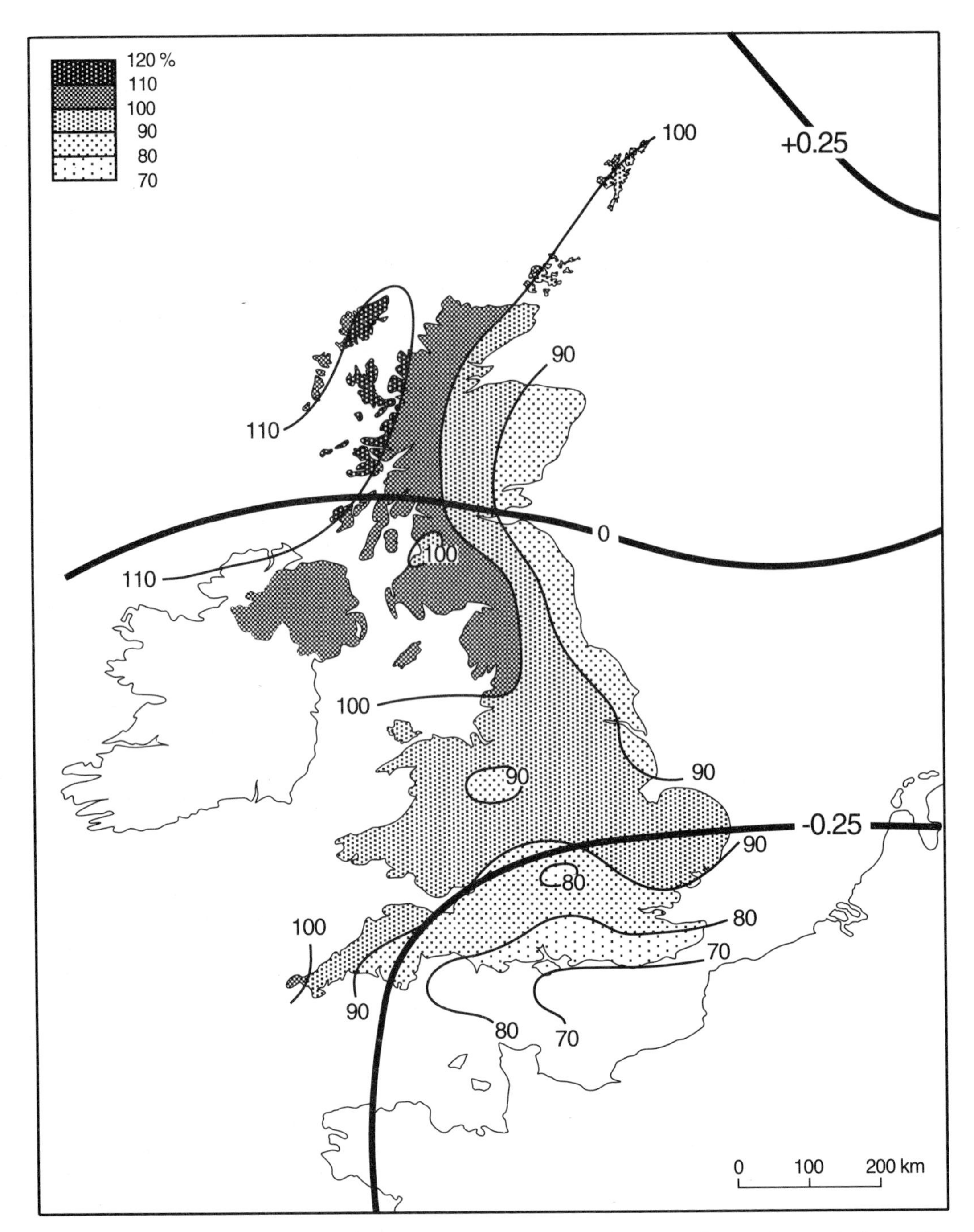

Figure 12.15 Summer precipitation 1981–90 expressed as a percentage of 1951–80 figures. A scenario for possible future changes by the early to mid-twenty-first century is denoted by bold lines showing the deviation in millimetres per day averaged over the years 31 to 70 of the Hadley Centre transient climate model (after Rowntree *et al.* 1993)

exceeded that achieved across East Anglia and east Kent.

The regional 'character' of 1968 was succinctly expressed by Manley, who obviously savoured the benefits of the lack of westerly winds in north-western Britain: 'Around London, the very poor weather of the 1968 summer holiday season is fresh in the memory, but at Manchester it was agreeably dry and sunny, and in W. Scotland superb' (Manley 1973, p. 18).

It is interesting to note that one of the largest totals of 'wet-days' (days with rainfall of more than 1.0 mm) in summer 1968 was forty-five at Cambridge, whilst the total in Glasgow was only seventeen. According to the Davis index, the highest scoring summer in the twentieth century (so far) was 1976, although summer 1995 surpassed 1976 over parts of Ireland and north-west Britain.

A Changing Geography of Rainfall Distribution

The geographical distribution of changes in average annual rainfall between 1941–70 and 1961–90 is illustrated in Figure 12.16. After widespread increases between 1881–1915 and 1916–50, especially in western areas, it might be thought that this latter comparison would show decreases in western or north-western areas in response to a decline in frequency of westerly airflow types from an annual average of 100.9 days in 1916–50 to only 76.3 days in 1961–90. Figure 12.16 shows that the reverse has been observed, with increases over many north-western areas, especially over upland areas.

Whilst western Scotland recorded increases of more than 5 per cent, parts of the east coast experienced decreases of a similar percentage magnitude, changes that point to an enhanced rain shadow effect on the eastern side of the Highlands. These changes are partly in response to the relative dryness of 1941–70 in some of these north-western areas, but more importantly reflect an abrupt increase in rainfall after the late 1970s. Rainfall over western Scotland was above average every year from 1979 to 1994 (Figure 12.17). This sustained elevation of rainfall amounts has had important consequences for flood frequency (Marsh *et al.* 1994; Black and Bennett 1995) and is unprecedented across Scotland as a whole since at least the 1860s, despite the recent dryness of eastern Scotland (Smith 1995).

There is considerable evidence that rainfall during the summer half-year varies quite independently from that of the winter half. Comparison of the 1941–70 and 1961–90 periods has revealed that rainfall in the winter half-year (October to March) increased far more in north-western than in south-eastern areas, with changes ranging from increases of 20 per cent in the western Scottish Highlands to decreases of up to 5 per cent on the coasts of East Anglia and Kent (Mayes 1996). This pattern is indicative of an enhanced role for moist westerly airstreams and of orographic enhancement of precipitation over western upland areas. In the summer half-year decreases were far more uniform across both Britain and Ireland and widely exceeded 10 per cent.

Regional Airflow Patterns and Rainfall Anomalies

The Lamb catalogue has recorded an increase in the W-type frequency only since the mid-1980s (Figure 12.12) and does not appear to offer a full explanation for the rainfall changes noted above. It is possible that Scotland itself has experienced a greater increase in westerly airflow types which has not extended across the whole of the British Isles and therefore has not been 'captured' in the Lamb classification (Smith 1995). There is some evidence of an increase in westerly activity over Scotland as early as the 1970s (Figure 12.18a). This is identified in the regional airflow classification (Mayes 1991), which is in marked contrast to the muted and much later increase suggested by Lamb's work. This is consistent with an anticyclonic anomaly over southern parts of the British Isles and is confirmed by the decline in cyclonicity over south-east England during the 1980s (Figure 12.18b). Regional airflow types can thus be used as a means of relating the

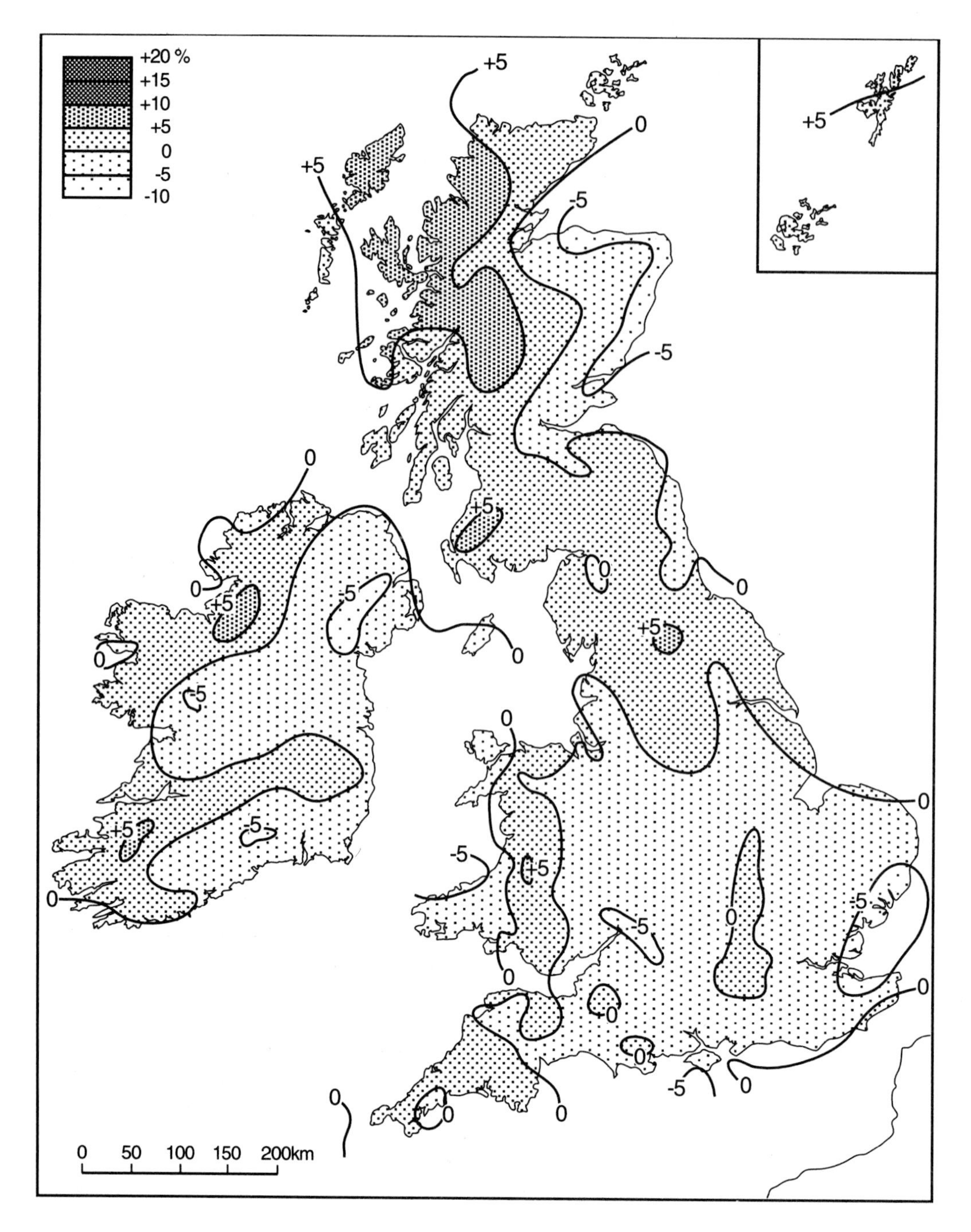

Figure 12.16 Average annual precipitation 1961–90 expressed as a percentage of 1941–70 figures
By courtesy of the Journal of the Chartered Institution of Water and Environmental Management

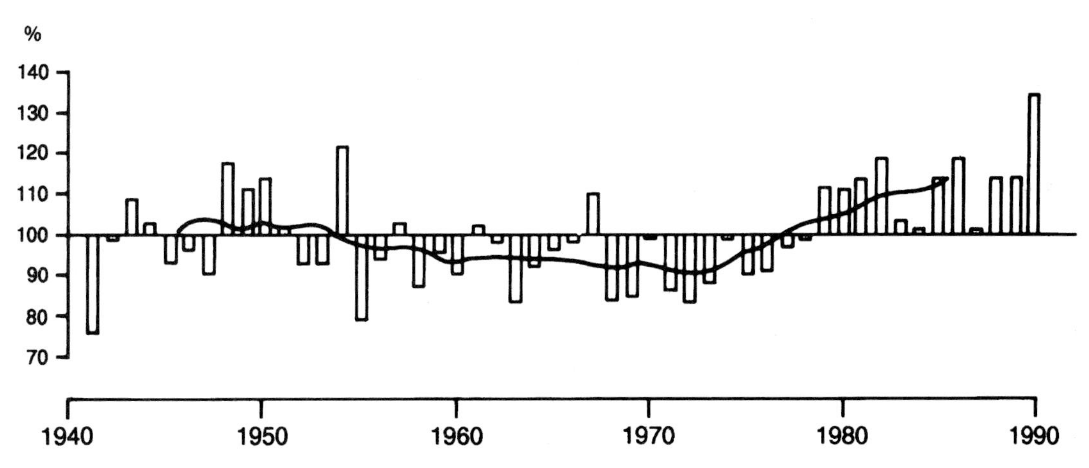

Figure 12.17 Annual precipitation totals in western Scotland 1941–90 expressed as percentages of 1941–90 average with 10-year running mean (after Mayes 1995)
By courtesy of the Journal of the Chartered Institution of Water and Environmental Management

broad pattern of atmospheric circulation around the British Isles to the regional pattern of weather and climatic anomalies.

Changing Patterns of Snowfall

Snowfall frequency can be highly sensitive to winter temperature in places such as the British Isles where the mean winter temperature is close to the point at which snow turns to rain. Manley (1971) has eloquently reviewed the status of snow as a climatic feature of the British Isles, and in *Climate and the British Scene* he observed that 'the more impressive extremes of our winter weather occur with a somewhat dangerous rarity' (Manley 1952, p. 197).

The importance of snowfall as a climatic hazard originates partly from Britain's unpreparedness for such events, which occur only on an irregular basis. A vivid example of this variability was the occurrence of severe snowstorms in south-west England and south Wales in February 1978 and January 1982 (see Chapters 2 and 6), a cluster that helps to explain why the interpretation of return periods for snowfall is such a risky enterprise. Table 12.2 shows how the monthly frequency of lying snow at two sites has varied through the twentieth century. The mild Januaries of the 1920s and 1930s, the relatively snow-free Decembers of the 1940s and 1950s and the general mildness of the winters of the 1970s are all clearly seen.

REGIONAL WEATHER PATTERNS AND THE ATMOSPHERIC CIRCULATION

In order to gain a fuller understanding of the causes of regional weather anomalies, it is helpful to express variations of temperature, sunshine and precipitation in an integrated single index of the climatic character of different locations. One such index that expresses the climatic characteristics of the whole year was devised by Hatch (1973) and is based on temperature and sunshine (each having a 30 per cent weight), total rainfall, rain days and wind speed (10 per cent each) together with rain duration and humidity (5 per cent each). Average values are shown in Figure 12.19. The south-eastward gradient reflects increasing distance from the average track of Atlantic depressions.

The purpose of this section, however, is to investigate how this geographical pattern of the 'average' climate of the British Isles responds to variations in the behaviour of the atmospheric circulation. The first consideration is the latitude of depression tracks. Eastward-moving depressions typically track

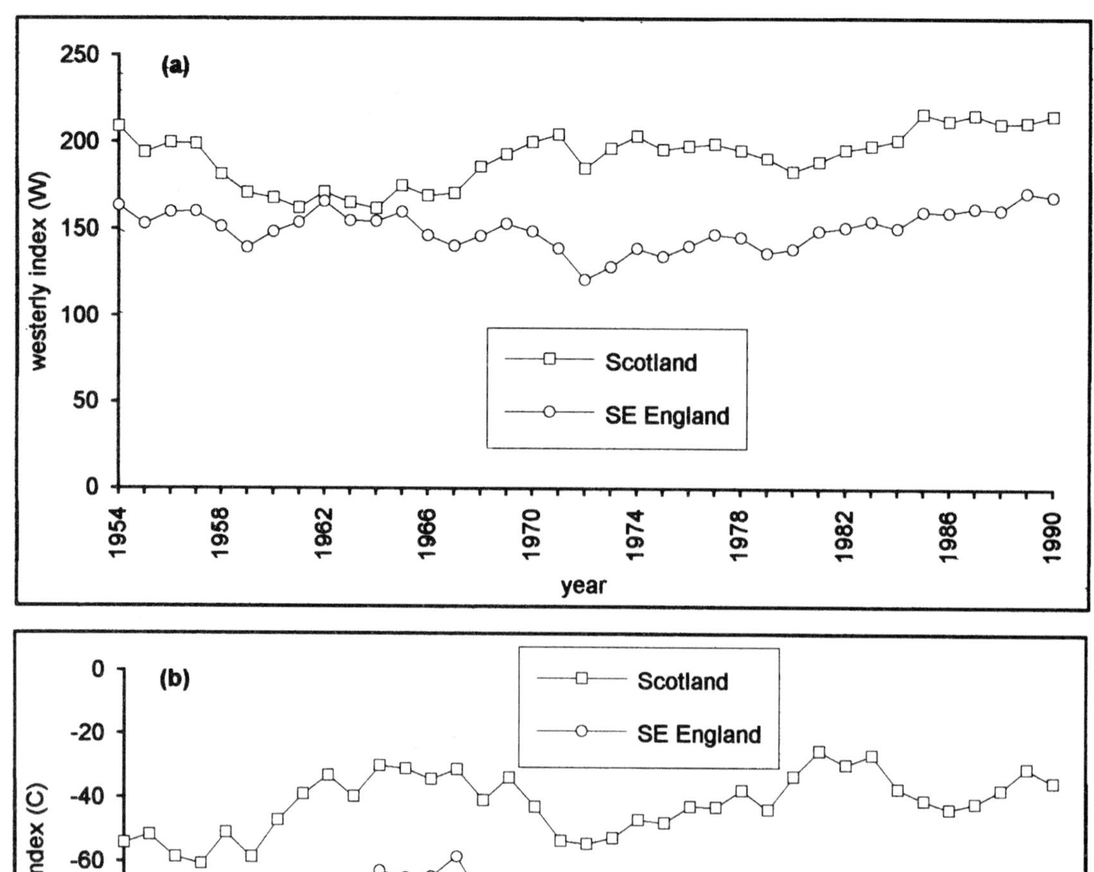

Figure 12.18 Ten-year moving average values of (a) the W (westerly) index and (b) the C (cyclonicity) index for Scotland and south-east England from the Mayes regional airflow classification, 1950–95

between Scotland and Iceland. Northward shifts in the tracks place the British Isles more frequently on the southern side of depressions where westerly winds predominate (in contrast to easterly components on the northern side) and increase the significance of the Azores subtropical anticyclone. Northern Britain may then be covered by a westerly airflow whilst southern parts have either an

Table 12.2 Days with snow lying (more than 50 per cent cover at 0900 GMT): decadal means of monthly averages for Balmoral and Eskdalemuir

	Jan.	*Feb.*	*Mar.*	*Apr.*	*May*	*Oct.*	*Nov.*	*Dec.*	*Year*
Balmoral (282 m)									
1911–20	13.8	13.8	13.6	4.3	0.4	0.6	5.6	13.7	65.8
1921–30	13.6	9.7	7.9	3.2	0.4	1.1	7.9	10.4	54.2
1931–40	12.6	14.4	10.5	4.3	0.4	1.7	2.4	11.5	57.8
1941–50	16.5	14.4	11.8	1.9	0.2	0.5	2.6	7.2	55.1
1951–60	17.8	17.1	9.3	1.0	0.0	0.1	1.7	9.1	56.1
1961–70	14.8	17.1	13.0	3.2	0.2	0.3	7.4	15.8	71.8
1971–80	12.9	12.6	6.8	2.5	0.3	0.6	6.0	9.5	51.2
1981–90	18.3	15.5	8.7	2.4	0.0	0.0	3.0	8.1	56.0
Eskdalemuir (242 m)									
1911–20	7.9	5.4	6.3	2.9	0.0	0.5	2.9	4.9	30.8
1921–30	6.7	4.2	3.1	0.5	0.0	0.2	2.3	5.4	22.4
1931–40	5.3	4.5	3.5	1.0	0.1	0.2	0.4	5.6	20.6
1941–50	10.1	8.6	5.7	0.2	0.1	0.0	0.8	4.7	30.2
1951–60	12.7	10.6	4.7	0.7	0.1	0.1	1.3	3.6	33.8
1961–70	8.5	12.1	4.8	0.6	0.0	0.0	2.7	6.3	35.0
1971–80	8.2	6.2	4.7	0.8	0.0	0.0	1.8	5.3	27.0
1981–90	12.4	9.0	4.1	1.3	0.0	0.0	1.2	5.7	33.7

Source: Derived and updated from data for 1911–20 to 1961–70 published in Manley (1971)

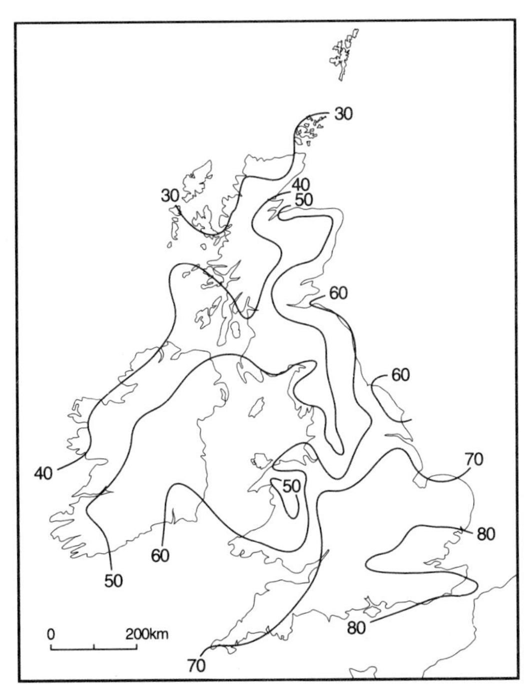

Figure 12.19 Average annual values of the Hatch climatic index

anticyclonic westerly or anticyclonic airflow type. Western Scotland may be very wet owing to the combination of orographic and frontal rainfall whilst drier conditions are likely in eastern Scotland, owing to the operation of the rain shadow, and over south-east England, owing to proximity of the high pressure. The following examples show how this synoptic type has produced notable regional weather anomalies most evident in precipitation deficiencies.

1921

The drought of 1921 was one of the longest of the twentieth century, lasting for almost all of the calendar year. Annual rainfall over most of south-east England was less than 60 per cent of average (Figure 12.20a). In east Kent barely half the average fell, whilst in western Scotland it was unusually wet year. The year 1921 began with a mild 'westerly' winter and a cold snap in April. But July was to be the seventh consecutive month with above average temperatures and the fourth consecutive dry month. The largest negative anomaly of rainfall was in

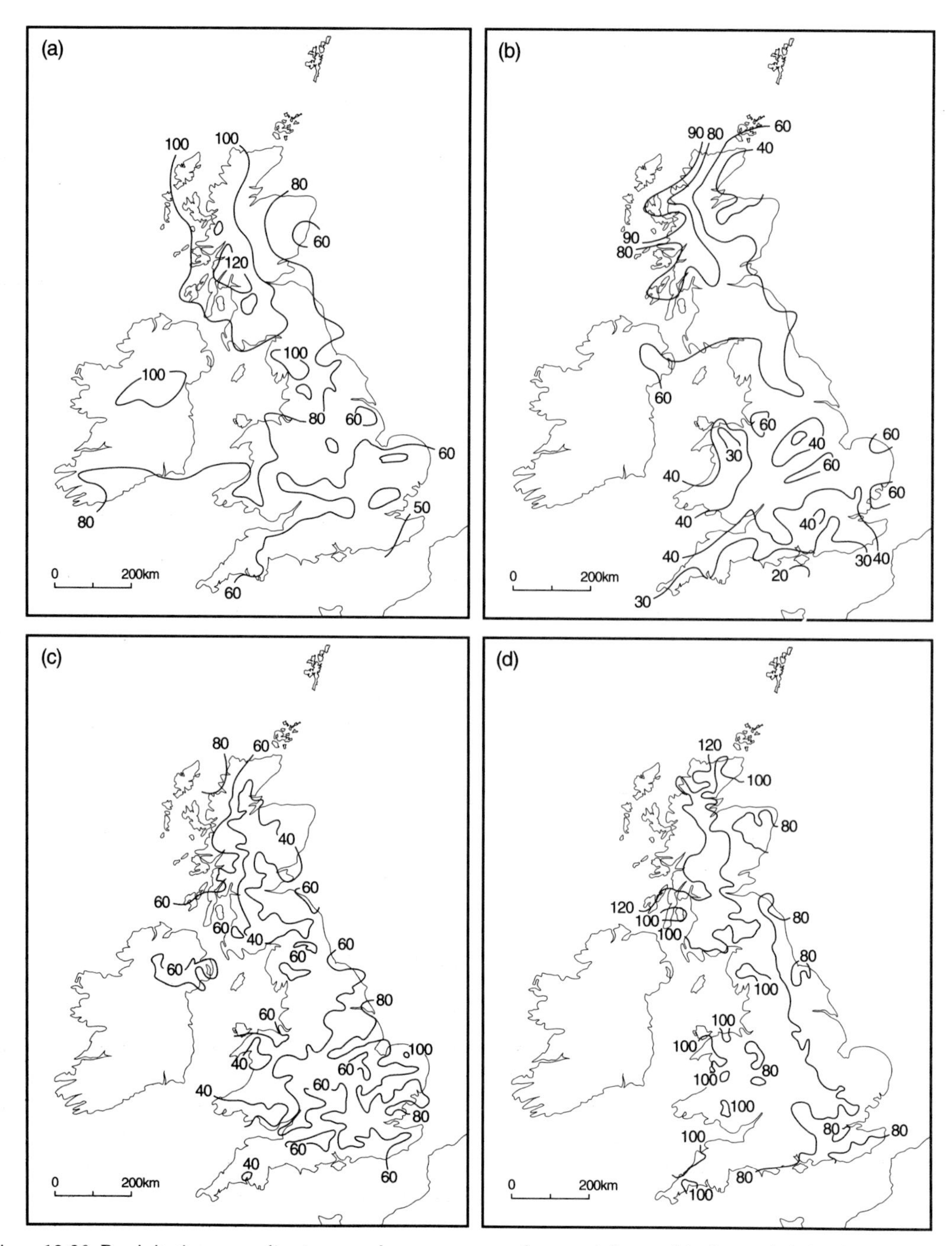

Figure 12.20 Precipitation anomalies (expressed as percentage of average) for notable dry periods in the twentieth century: (a) January–December 1921; (b) April–August 1976; (c) April–August 1984; and (d) August 1988–February 1992

central southern and south east England, where the total was as low as 360 mm. (349 mm below average).

1975/76

The famous drought of 1975/76 was memorable for its severity over most of the British Isles and also for its exceptional persistence (Figure 12.20b). It has attracted an array of superlatives in the meteorological literature – the proximity of two very fine summers in 1975 and 1976 making it all the more remarkable. It produced the highest values on a drought index for south-east England in three hundred years (Wigley and Atkinson 1977). Not since 1749/50 had a period from one summer to the following spring been so dry in southern Britain. At Oxford, every month from May 1975 to August 1976 had below average rainfall with the sole exception of September 1975. It was therefore not surprising that this was the driest 16-month period on record for England and Wales (Marsh and Lees 1985). The severity of the drought was heightened by the acute hydrological impact of an exceptionally dry winter being sandwiched between two hot, dry summers. Winter 1975/76 had only 61 per cent of average rainfall over England and Wales. With very limited recharge of aquifers and reservoirs, water supply problems were experienced quite early in the spring of 1976. The drought was most severe in south-eastern England but was felt widely across England and Wales, and the most stringent water supply restrictions were experienced in South Wales, where water was cut off for up to seventeen hours a day to domestic consumers. North-west England and much of western Scotland escaped the attentions of this notable drought and were more frequently subject to the passage of fronts associated with cyclonic systems displaced northwards by the high pressure over southern England.

1984

This was a relatively intense spring and summer drought (Figure 12.20c), bracketed by wet winters. For England and Wales as a whole, the period from February to August was the second driest of the century (after 1976). However, with a tendency for high pressure to be centred further west than in 1976, the drought was this time most noteworthy in northern and western parts of Britain and Ireland. This is clearly illustrated by the distribution of rainfall percentage anomalies. Scotland had the driest April to August period since the Scottish rainfall series began in 1869. As Marsh and Lees (1985) have pointed out, the highly variable rainfall of 1984 contributed to a developing tendency for wetter winters and drier summers during the 1980s.

1988–92

This was an exceptionally persistent event that was punctuated by short-lived but often abrupt phases of wet weather. In terms of origin it showed some similarity with the 1975/76 drought. However, Marsh *et al.* (1994) contend that it was probably unprecedented this century for the persistence of the rainfall deficit. The dry conditions were immediately evident in declining runoff rates, but a succession of dry winters which offered limited scope for aquifer recharge reduced the water tables in even the most reliable groundwater areas such as the chalk of south-east and eastern England. The winter of 1991/92 provided only one-third of the normal recharge volumes and during the following summer many boreholes recorded their lowest known levels; Figure 12.21 shows one such case.

The origin of the dry weather lay in a series of high-pressure areas centred over or to the south of the British Isles. Blocking anticyclones rarely managed to displace a vigorous westerly airflow over northernmost parts of the British Isles. This resulted in an exceptional dichotomy between the wet weather that often held sway in north-western areas and simultaneous dryness over south-eastern parts (see Figure 12.20d).

The persistence of south-west and west winds led to eastern Scotland often falling within a rain shadow region downwind of the Scottish Highlands. The westerly flow was especially active in

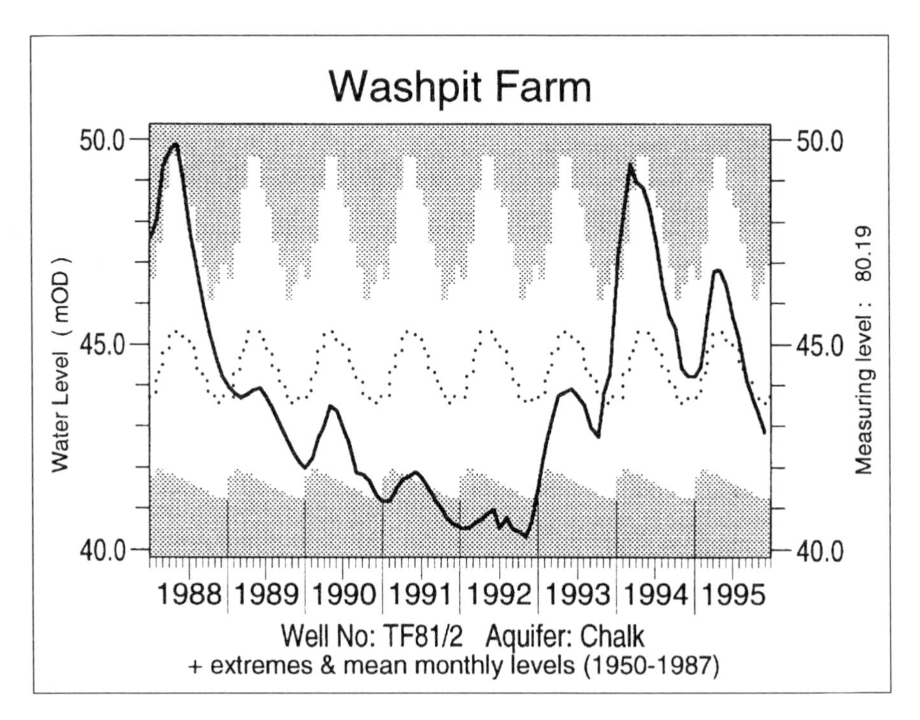

Figure 12.21 Groundwater levels (solid line) for the Washpit Farm borehole (Norfolk) for the period 1988–95. The figure shows also the average monthly levels (dotted line) and the highest and lowest recorded monthly levels (shaded areas)
By courtesy of National Water Archive

the early months of 1990; whilst the Western Highlands attained monthly totals in excess of 1,000 mm in January, Fyvie Castle, in the rain shadow of the Grampian Mountains close to Aberdeen, had a mere 4 mm. In February Fort William had a total of 538 mm, making it the wettest February in a hundred years of records. In the same month, Aberdeen enjoyed the sunniest February of the century.

As in 1921 and 1975/76, the airflow patterns influenced the distribution of both sunshine and temperature. The largest positive anomalies of both elements were concentrated in the south-eastern half of England (Mayes 1991). The main features of the drought are shown in Table 12.3. The incidence of the highest officially accepted air temperature in the British Isles (37.1°C), measured on 3 August 1990 at Cheltenham (Burt 1992; Eden 1995; and Chapter 5 of this volume), emphasises the anticyclonic nature of southern Britain at that time.

Southward Shifts in Depression Tracks

When large areas of the British Isles are located to the north of a depression track, the geographical distribution of weather is reversed owing to the prevalence of easterly winds. The markedly different weather experienced in easterly winds is a reflection not just of the more continental source area for these winds but of the reversal in the geographical distribution of exposure and shelter. Rain-shadow areas shift to the now sheltered western coasts, and orographic enhancement of precipitation occurs in the vicinity of east-facing hills and mountains, such as the eastern Cairngorms (Weston and Roy 1994). For example, 1968 was remarkable not just for the heavy rainfalls over southern England but for the lengthy

Table 12.3 The main features of the drought of 1988–92 and their timing

The main features of the drought as a whole can be summarised as follows:

- The four-year period from spring 1988 was the wettest over Scotland since records began in 1869.
- Over the five years from 1988 the English lowlands had accumulated a rainfall deficit of 20 per cent which was amplified to a runoff and recharge deficit of 50 per cent.
- The years 1988–1992 formed the warmest 5-year period on the Central England Temperature record.
- Over England and Wales, November 1988 to January 1989 was the driest such period since 1879.
- Winter 1989/90 was the wettest in England and Wales on record but spring 1990 was the driest since 1893.
- The year 1990 had the highest Central England temperature (10.8°C) on record.

spells of fine weather over northern Scotland, where the largest positive air pressure anomalies existed. In addition to the fine summer weather noted elsewhere in this chapter, settled and bright weather characterised most of the second half of the year on Shetland, and at Lerwick wind speeds were below average in each month from June to December (Irvine 1969).

Two other years in the twentieth century illustrate the effects of a southward shift in depression tracks combined with a reduced pressure gradient between northern and southern districts. In 1937 only eighty-six W days, in the Lamb classification, were recorded. With a larger than usual number of depressions passing to the south, north-west Scotland recorded more southerly than westerly winds. This meant that the coast of the mainland north of the Isle of Skye was often sheltered – Ullapool recorded only 60 per cent of average rainfall. However, in a remarkable contrast, south-east England was notably wet under the heightened influence of depressions, and Clacton (Essex) and Boston (Lincolnshire) recorded 153 and 152 per cent of their respective averages.

A similar reversal of the north-west to south-east gradient of rainfall was observed in 1960, which had a similar pattern of southerly-tracking depressions. After the wettest winter since that of 1915/16, England and Wales had the second wettest year on record, largely because of persistent wet weather in the second half of the year which gave rise to sustained flooding, especially in the West Country. Regional weather anomalies were similar to those of 1937; at Cape Wrath in the far north-west of Scotland, annual rainfall amounted to 820 mm (68 per cent of average). In contrast, rainfall percentages of average reached 150 per cent in such normally dry eastern locations as Skegness and Shoeburyness, with 160 per cent being recorded at Brighton. Ventnor on the Isle of Wight recorded 1,356 mm, over one and a half times the total at Cape Wrath. This can be compared with 1990, a year when the gradient of rainfall anomalies was quite the opposite, giving totals of as much as 1,813 mm at Cape Wrath but only 622 mm at Ventnor. At Fort Augustus, the total of 1,984 mm represented 181 per cent of average.

THE GEOGRAPHICAL DISTRIBUTION OF REGIONAL WEATHER ANOMALIES

An earlier section showed how the regional distribution of temperature, rainfall and sunshine may fluctuate according to changes in the character of the atmospheric circulation. Since geographical variations in each of these basic elements are influenced by the same changes in circulation patterns, it is logical to link them in a single climatic anomaly index in order to gain a more holistic impression of regional weather conditions. A TRS (temperature, rainfall and sunshine) index has been compiled that quantifies the annual standardised anomalies of each element in relation to long-term averages. It thus

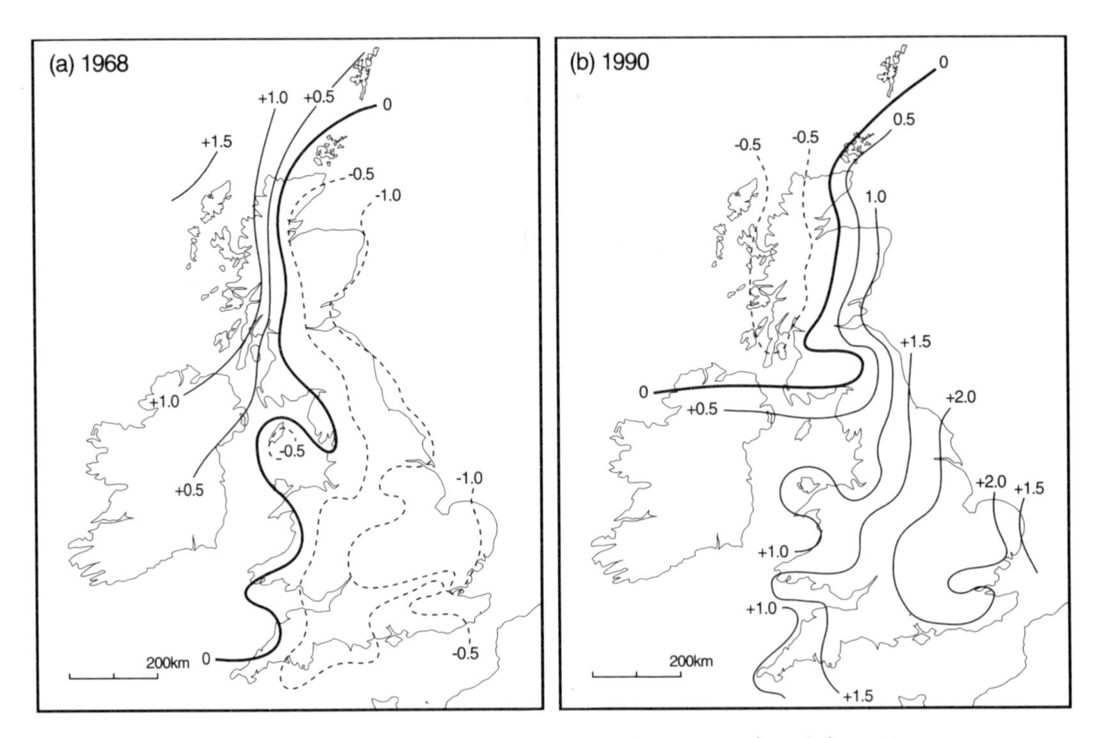

Figure 12.22 Distribution of scores on the annual weather (TRS) index in (a) 1968 and (b) 1990

represents anomalies from Hatch's climatic gradient. The index is calculated as follows:

$$TRS_i = \frac{T_i - T_x}{SD_x} + \frac{S_i - S_x}{SD_x} - \frac{R_i - R_x}{SD_x}$$

where i is the year, T_i, S_i and R_i are respectively the annual value of temperature, rainfall or sunshine, T_x, S_x and R_x are the 1961–90 average value and SD_x is the 1961–90 value of the standard deviation of each variable. Thus a year that is warmer, drier and sunnier than the respective 1961–90 averages has three positive terms in the index. The values for each element are expressed in terms of standard deviations from the mean, and values of the index itself approximate to the normal distribution and represents a standardised series. The working of the index can be illustrated by 1968 and 1990 (Figure 12.22). A 'reverse gradient' weather pattern prevailed in 1968 arising from cyclonic blocking over southern Britain whilst 1990 shows an enhanced gradient.

Of the years between 1961 and 1990, 1968 was by far the driest and sunniest at Stornoway, and the resulting TRS score was also highest on record. At Heathrow it was the wettest year, and the TRS score was also the lowest recorded, whilst 1990 was the warmest year at Stornoway and also the wettest by a wide margin. The year 1990 was also dull, and the resulting TRS score was below average. At Heathrow temperature and sunshine were both more than two standard deviations above average and it was the equal warmest (with 1989), the sunniest (1989 was the second sunniest) and the driest year on record. Time series plots of the TRS index from 1961 to 1990 at four contrasting locations (Figure 12.23) show a preponderance of negative scores at Stornoway after the late 1970s. This picture is in marked contrast to the outstandingly high scores of 1989 and 1990 in England and Wales.

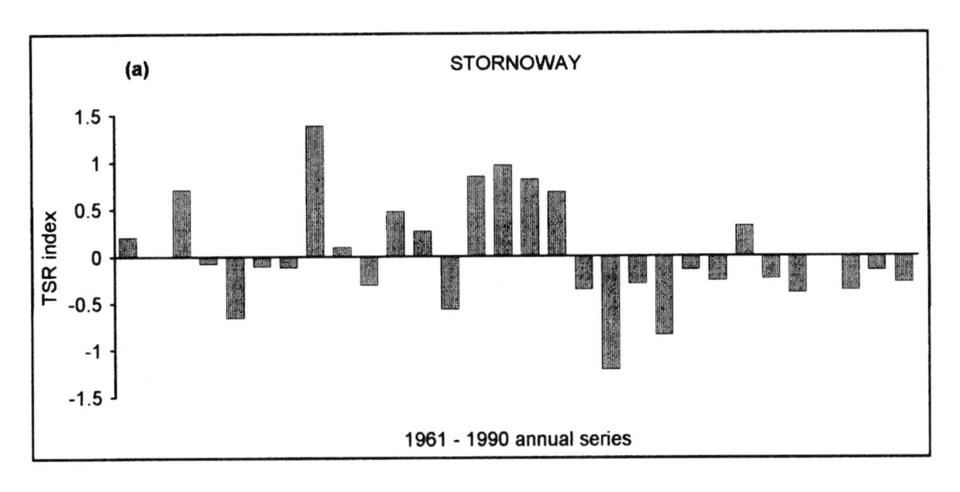

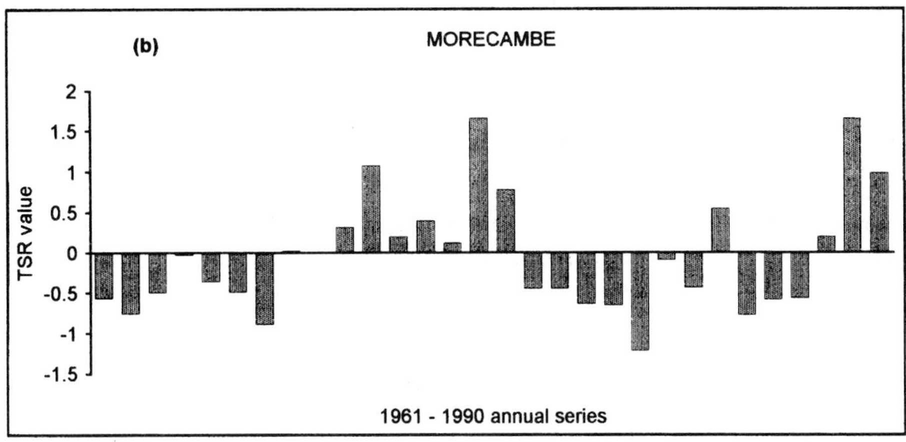

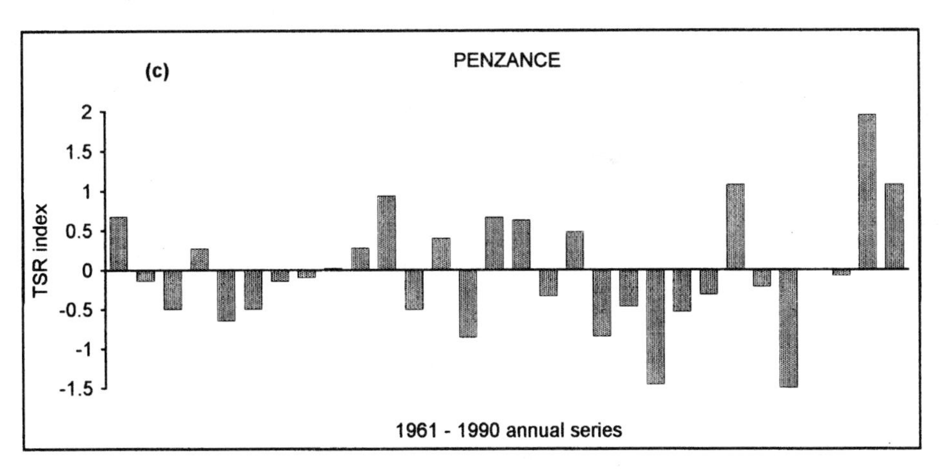

Figure 12.23 Time series of the annual weather (TRS) index, 1961–90, for (a) Stornoway, (b) Morecambe, (c) Penzance and (d) Wye (Kent)

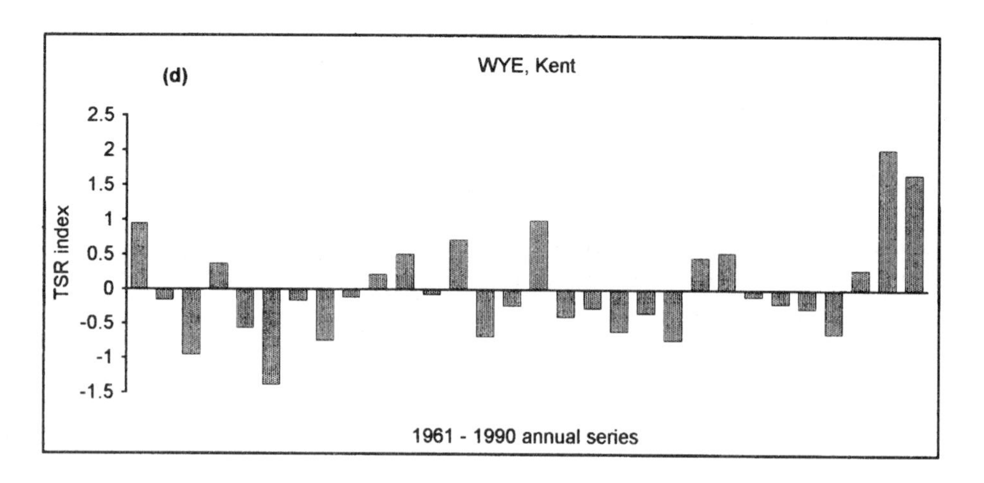

Figure 12.23 (Continued)

CLIMATE AND CLIMATIC REGIONS IN THE FUTURE

Earlier sections in this chapter have shown that some periodicity exists in the climatic record. However, this is insufficiently strong or well understood to allow a simple projection of past patterns into the future. One of the problems in this area is the role that increasing concentrations of greenhouse gases may have on the world's climate. To establish at least a range of scenarios – they are not, strictly speaking, forecasts – computer programs known as General Circulation Models, or GCMs, are used. These models treat the atmosphere as a series of layers, perhaps ten or more, with the surface subdivided by a grid of squares of side between 300 and 1,000 km. Each of these 'atmospheric boxes' can be set to have a certain temperature, moisture level, wind field and air pressure. The models also run on the basis of prescribed levels of greenhouse gas concentrations. Taking these specified conditions as the starting point, the models are run until they settle into an equilibrium condition with no further changes to the simulated behaviour of the atmosphere. The Intergovernmental Panel on Climate Change (IPCC) has presented a number of detailed reports (Houghton *et al.* 1990, 1992, 1995 and 1996) which examine the results of these computational exercises and other important issues surrounding GCMs and their applications.

These 'equilibrium' models have produced tolerably good simulations of current climates using present-day conditions as their starting point. They are, however, notoriously poor at predicting climate at the regional scale, and the British Isles might be described by as few as two grid squares. More importantly, the models indicate that temperatures globally should have risen by 1.0 to 1.8 degrees C since the nineteenth century in response to greater concentrations of greenhouse gases. In fact, the response has been far more muted, and is about 0.5 degrees C. The IPCC has attributed this discrepancy to the cooling effect of sulphate particles in the atmosphere, which have had an influence on northern hemisphere daytime temperatures.

These results have not dissuaded scientists from rerunning the models using possible future concentrations of greenhouse gases to provide an overall impression of global temperatures changes and their spatial variations. This usage implies a sudden increase in greenhouse gas concentrations. In reality, the increases are continually changing, with temperatures evolving over time. Importantly, there is also a time lag, largely as a result of the thermal inertia of the oceans, between atmospheric gas composition and temperature changes. Equilibrium

Box 12.3

A SCENARIO OF THE CLIMATE OF THE BRITISH ISLES IN THE TWENTY-FIRST CENTURY

Scenarios of the British Isles' climate response to global warming have undergone a revision in the early 1990s. Those based upon 'equilibrium' models took no account of the slow response of oceans to thermal changes. Time-dependent models allow a more refined view of the rate of climatic change.

The Climatic Change Impacts Review Group (CCIRG) was established by the Department of the Environment to assess the impact of climatic change using the latest climate models. The 1991 report (CCIRG 1991) was based on equilibrium models whereas its 1996 report (CCIRG 1996) uses the Hadley Centre time-dependent coupled ocean–atmosphere model (Hulme 1996). Scenarios of the climate of the British Isles for the 2020s and 2050s assume that some changes to global energy production (a major source of greenhouse gases) will be made over the coming decades. A warming rate of +0.2 degrees C per decade is predicted. However, the transient model distinguishes between different regions of Britain and Ireland according to the strength of the maritime influence.

The main findings of this work (Hulme 1996) are as follows:

- Annual rainfall to increase by 5 per cent by the 2020s and by 10 per cent by the 2050s.
- Winter temperatures to increase by the 2050s by between 0.8 degrees C in the north-west and 2 degrees C in the south-east.
- Summer temperatures to increase by the 2050s by between 1.2 degrees C in the north-west and 1.8 degrees C in the south-east. Assuming an unchanging variability of summer temperature the probability of a summer being as warm as that of 1995 increases from 0.013 to 0.33 by the 2050s.
- Whilst summer rainfall is expected to increase in northern areas, a decrease in summer rainfall in south-east England of about 10 per cent by the 2050s (Figure 12.24) corresponds to an increase in potential evapotranspiration (PE) of as much as 40 per cent. By contrast, PE decreases in north-west Scotland in both summer and winter.
- Winter rainfall is expected to increase in all areas, but the increase may be largest in southern England.
- The frequency of frost is expected to drop by about 50 per cent by the 2050s, and the frequency of days with temperatures exceeding 25°C could double.

models can tell us little about rates of climate change or the progression towards equilibrium. In this sense there is also an important distinction to be drawn between temperature changes that have been realised and those that will inevitably take place (even if emission of further greenhouse gases ceases) because of the time lag effect. The latter are 'committed' temperature increases.

One way around this problem is to use time-dependent or transient models in which atmospheric gas composition changes are progressive and not sudden. These models are more complex and take into account the differing response times of the atmosphere and the oceans. This linking, or coupling, of the oceans and the atmosphere is a key feature of transient models and has led them to produce predictions different from those derived from equilibrium GCMs. This is especially the case in the North Atlantic area (Rahmstorf 1995), where oceanic influences are particularly important.

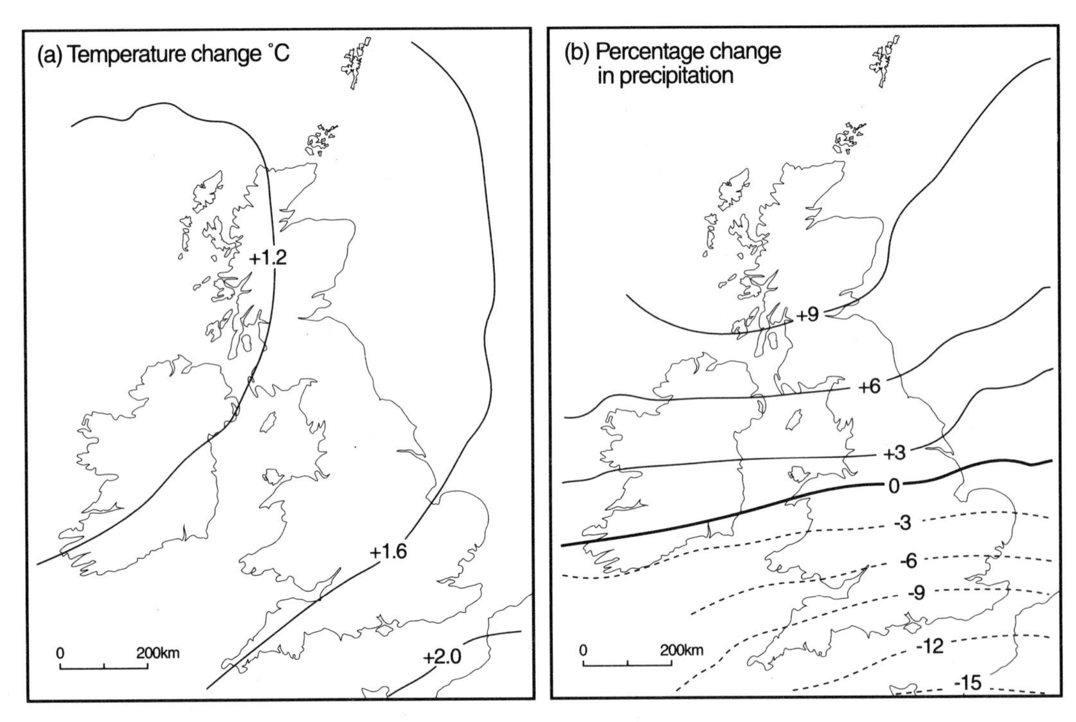

Figure 12.24 Scenarios of (a) summer (June–August) temperature, and (b) summer precipitation for the 2050s, expressed as differences from 1961–90 averages (after Hulme 1996)

Another means of examining the present-day warm, and future possibly yet warmer, conditions is to seek analogues in the past such as the Post-Glacial Climatic Optimum, but a significant problem is posed by the mid-Holocene and other times in the past because the Earth's orbital parameters were then different from today's. The planet was closer to the Sun in the northern hemisphere summer. As a result, the northern hemisphere received approximately 5 per cent more summer radiation than at present. It is significant that warming then was a summer phenomenon, whereas GCMs propose that at present it is winters which drive the annual figures upwards.

Modelled Changes in Seasonal Rainfall over North-west Europe and the British Isles

Regional-scale predictions should be interpreted with particular care. It is nonetheless useful to consider possible regional scenarios because they provide a basis for 'downscaling' a scenario to the British Isles regional level. Rowntree *et al.* (1993) have presented scenarios of precipitation changes for winter and summer based upon the Hadley Centre (Meteorological Office) transient climate model. The model is run for a simulated period of 75 years assuming an increase in CO_2 of 1 per cent per annum, and the scenarios illustrated are based on averaging of the results between years 31 and 70. The results indicate that increases in mean daily precipitation in winter might extend eastwards from the British Isles across much of Northern Europe whilst decreases will be found over much of the Mediterranean basin. The pattern of anomalies is transferred northwards in summer with the area of decreases extending across most of central Europe, including south-east Britain (Figure 12.24). Clearly, any decline in summer precipitation over Central and Northern Europe could be viewed as more important than a similar change in the Mediterranean since summer rainfall is already insignificant in that area. In contrast, parts of Scandinavia (and, to a lesser extent, Scotland) remain in a zone of increasing rainfall in summer. These changes result from an intensification and northward displacement of depression tracks. As discussed in Chapter 1, an enhanced north–south temperature gradient provides additional energy to the westerly circulation and to the low-pressure systems within it. This has

Box 12.4

A SCENARIO OF GLOBAL CLIMATE CHANGE IN THE TWENTY-FIRST CENTURY

The Intergovernmental Panel on Climate Change (IPCC) was established jointly in 1988 by the World Meteorological Organization and the United Nations Environment Programme, and acts as the main forum for global climate change research. A range of scenarios for greenhouse gas emissions and climatic change were published in 1990 (Houghton *et al.* 1990). Following a revision in 1992, a new report was published in 1996. The following scenario of climatic change is based on new assumptions about the rate of greenhouse gas emissions and of the sensitivity of the climate system to this perturbation. This is the IS92a scenario, which is based on a limited adoption of renewable energy sources, and acknowledges a steadying of the concentrations of CFCs, a powerful group of greenhouse gases, following the phasing out of their production in response to ozone depletion. The adoption of this scenario is an important departure from previous work which has used a 'Business-as-Usual' energy use scenario. The main findings are as follows:

- The 'best estimate' warming figure for 2100 is +2.0 degrees C, lower than the value given in the 1990 report owing to the role of sulphate aerosol cooling and slightly reduced emissions of greenhouse gases. Nevertheless, this still represents the greatest rate of warming in the past 10,000 years. Only 50 to 90 per cent of the eventual (equilibrium) warming will have actually been observed by 2100, because of the thermal inertia of the world's oceans; the world will be committed to further warming even if greenhouse gas concentrations remain steady from that date.
- Mean sea level is expected to rise by 50 cm by 2100, though the range of possible values is large from +15 to +95 cm, depending upon which emissions and sensitivity scenarios are used.
- Maximum warming is expected to occur in the northern hemisphere high latitudes in winter.
- Largest increases in precipitation are also expected in northern hemisphere high latitudes, with decreases at lower latitudes, especially during summer. Both trends have been observed over recent decades.

It should be remembered that natural variability in the global climate will offset or accelerate many of these trends. Even if progress to a warmer world occurs at the rates suggested, this change is not likely to be smooth. However, the fact that such anomalous seasons or years may occur as a result of natural variation in the climate does not necessarily contradict the longer-term development of the global warming scenarios listed above.

Source: IPCC Working Group I 1995 Summary for Policy makers, United Nations Environment Programme Internet site [http://www.unep.ch/ipcc]; Houghton *et al.* (1996)

been confirmed by Hall *et al.* (1994), who note the effects of below average sea temperatures around the south of Greenland in energising the circulation in the transient model.

A cautionary note to be added at this point is that these scenarios represent possibilities, not certainties. They are limited by our knowledge of the climate system and by our ability to incorporate this knowledge into climate models. They assume not just that warming will continue, but that the climate system will respond in the way and at the rate which we expect. They are, however, plausible, not just in relation to current knowledge, but in relation to the pattern of weather anomalies in recent warm periods in the British Isles, such as 1975/76 and 1988–92. Even if the rate of global and mid-latitude warming is uncertain, it is the possibility of distinctive geographical patterns of changes that should arouse our concern today, particularly in relation to the planning of water resources in the twenty-first century. Local temperature changes will be influenced by other factors as well. Coastal districts, for example, may be expected to experience slower temperature changes if onshore winds predominate. This may lead to a less rapid warming on exposed northern and western coasts of the British Isles, where oceanic influences are generally more powerful. Conversely, if onshore winds are less frequent then the warming rate may increase significantly.

THE RESPONSE OF ATMOSPHERIC CIRCULATION AND RAINFALL TO GLOBAL WARMING AND POSSIBLE ANALOGUES WITH THE RECENT PAST

The preceding chapters have provided evidence of the sensitivity of weather and climate to changes in the westerly circulation. Stronger and more persistent westerlies are a well-known cause of mild, wet weather in winter and of cool, changeable weather in summer. In addition, we have seen how intensified westerly activity can also enhance the north-north-west to south-south-east climatic gradient as the driest, sunniest and warmest weather (in relation to local averages) may become concentrated in the more sheltered south and south-east. The purpose of this section is to examine a scenario of atmospheric circulation and rainfall for the early twenty-first century based on analogues rather than on modelled scenarios.

Comparison with Circulation Patterns in the Twentieth Century

Circulation and rainfall patterns in relatively warm periods in the twentieth century could be interpreted as a potential analogue for future warming. Lough *et al.* (1983) found that much of Ireland, Scotland and northern England were wetter in the relatively warm period 1934–53 than in the cooler period 1901–20, whilst much of southern Britain was drier. It is noteworthy that this situation agrees with some GCM scenarios. This approach was developed by Palutikof (1987), who showed that these regional rainfall variations were reflected in river runoff changes, with decreases in the warmer period in 'eastern' rivers.

Figure 12.12 shows that the decrease in the westerliness in the mid-twentieth century was reversed in the 1980s. It is possible to verify this latest increase against the frequency of west and south-westerly winds. Unlike the W airflow type, such winds rose from a relatively low frequency in the 1950s, and the rise was followed by an increase as early as about 1970 at Stornoway and Lerwick, and rather later at Kew. The increased frequency of westerly winds across northern Scotland is thus consistent with the greater incidence of the W airflow type in the Mayes regional airflow classification (Figure 12.18a). The average north–south pressure gradient has also increased, and the highest annual value in a series extending back to 1911 was in 1990 (Figure 12.25). There is also a consistent trend towards increasing southerliness that commenced in the early 1950s (Murray 1993; Figure 12.26). This may result from a change in prevailing winds from west-south-west to south-west, possibly associated with depressions periodically acquiring a north-easterly rather than an easterly track. It

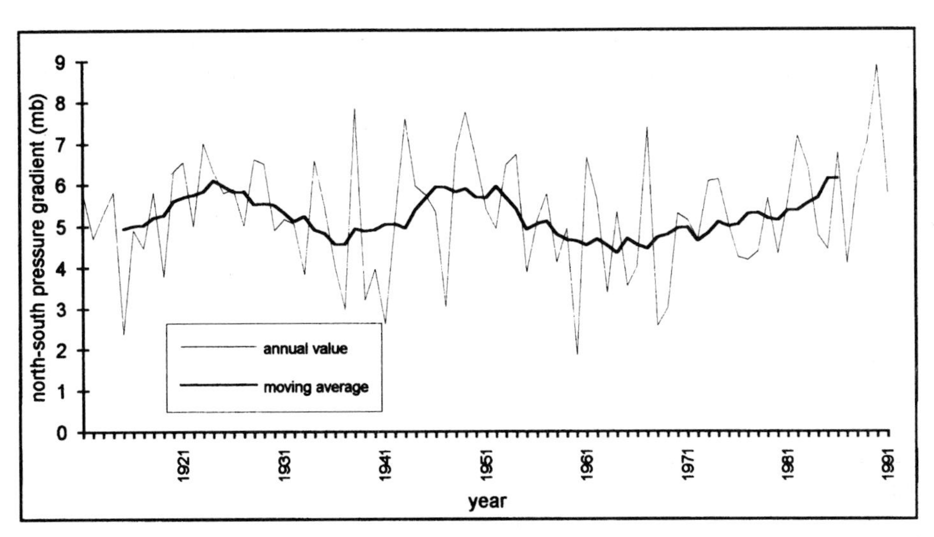

Figure 12.25 Annual and 10-year moving average values of latitudinal pressure gradient across the British Isles 1911–91 (calculated as the difference between the mean of Plymouth, London and Gorleston minus the mean of Stornoway and Lerwick)

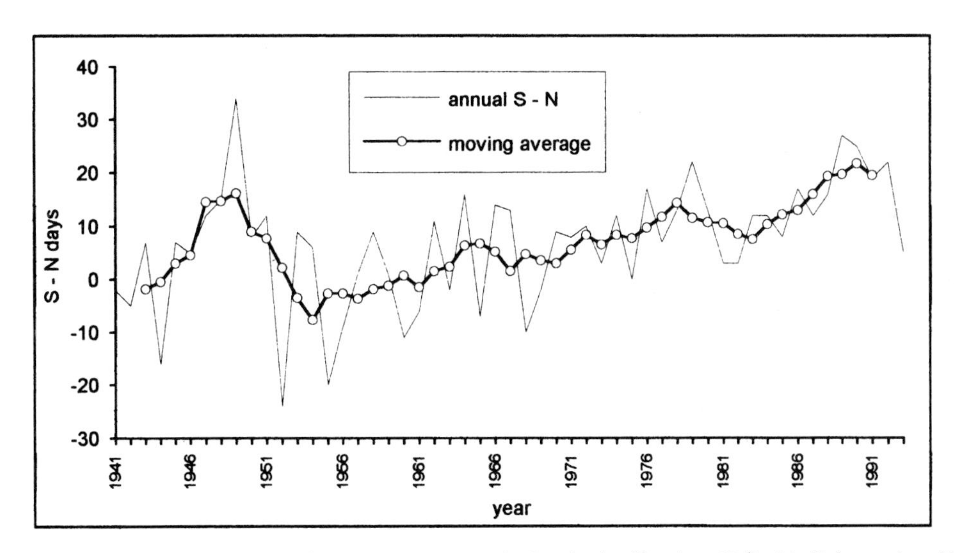

Figure 12.26 The increasing frequency of S weather types in the Lamb classification, 1941–93 (S days minus N days shown as 5-year moving averages)

should also be noted that unsettled 'progressive' weather types sometimes appear as more southerly than westerly, leading to misunderstandings of what constitutes meridional and blocked patterns.

The Role of the North Atlantic Oscillation

The North Atlantic Oscillation (NAO) is associated with many aspects of the westerly circulation. In positive mode the westerlies are strengthened. This condition, which is most clearly marked in winter, prevailed over the first three decades of the twentieth century and has done so again since about 1980 (Figure 12.4), the increase in pressure gradients being confirmed by Trenberth (1991). The corresponding negative anomalies of sea surface temperature north of 50° N have been remarkably persistent in the 1970s and 1980s (see, for example, Folland *et al.* 1992). Increased temperature gradients have also been noted in larger-scale hemispheric analyses of air temperature in the troposphere by Weber (1990).

The consequences of changes in the NAO have included such varied phenomena as a strengthening of the westerly circulation across Northern Europe, the development of unusually deep depressions (Namias 1987; Lamb 1991), and a reduction in the temperature of polar maritime air reaching Scotland as a result of cooling of the high-latitude North Atlantic (Parker and Folland 1988). This latter aspect points once again to the importance of the oceanic setting of the British Isles.

The effect of global warming on the frequency of gales around the British Isles has aroused considerable interest, partly because of the existence of conflicting scenarios. Whilst there is no evidence for an overall increase in average wind speeds (Palutikof *et al.* 1997), there was a cluster of strong winds associated with deep winter depressions in the late 1980s and early 1990s. If winter temperature gradients over the North Atlantic increase in the future then such storms may become a more persistent feature.

REGIONAL CLIMATES AND LANDSCAPE

Manley's *Climate and the British Scene* (1952) rightly places great emphasis on the manner in which climate contributes to the landscape of the British Isles. It is appropriate therefore to conclude by considering how possible regional climate changes might find more general expression in the future. The two ultimately critical factors are temperature and precipitation, which together determine natural as well as cultivated vegetation. The balance between these two elements, which operates through potential evapotranspiration, also determines the quantities of water in our rivers, its seasonal distribution and even, in part, its chemical character. It also determines water resource potential.

Land Use and Agriculture

In just the same way as the Medieval Warm Period witnessed successful and widespread vine production in England, so too might future warmer climates encourage the adoption of crops currently regarded as exotic; it would also exclude, possibly through warmth and summer dryness, other crops with which we are currently familiar. Because of the economic implications of such changes this is an important area of research (Parry and Duncan 1995).

The most obvious consequence of possible global warming would be the northwards shift in the limits of many crops. It is widely argued (Parry 1990) that crop limits would move northwards by between 200 and 330 km for every degree Celsius increase in temperature. Although there are important local variations to this picture, some scenarios indicate that by the year 2050 large parts of England, Wales and Ireland could support, for example, grain maize production. The climate of southern England could also permit sunflowers to be grown on a large scale (Figure 12.27). On the other hand, a combination of warmer and drier summers would reduce the potential in south-east England for crops such as winter wheat, which could be advantageously located in previously marginal areas of northern and western Britain. Drier summer soils would also adversely influence the production of crops such as main-crop potatoes in northern and western areas where it currently flourishes. It is nevertheless wrong to argue on the basis of a simple analogy with those southern European areas where these crops now flourish. The boundary conditions are again different, especially with regard to the more marked seasonality of the northerly latitudes and the maritime character of the north-west European climate.

Parry (1990) has suggested also that a 1 degree C

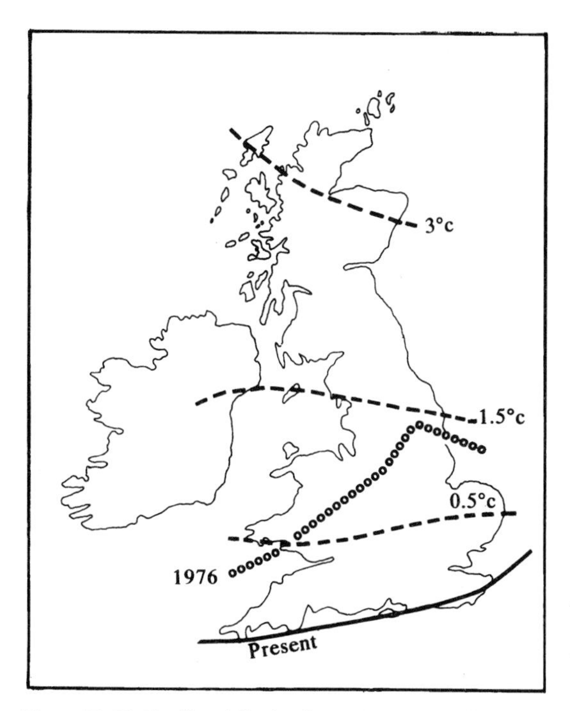

Figure 12.27 Predicted limits for grain-maize cultivation under different warming scenarios and as it would have been for the conditions during the summer of 1976 (after Parry 1990)

increase in temperature, which might occur by the year 2030, would lead to an upwards altitudinal shift in cultivation limits of 140 m. In theory this would result in one-third of unimproved moorland currently submarginal for cereal production becoming viable for hardier varieties such as rye. In practice, other limiting factors, such as steep and difficult terrain, would remain, and this would reduce the proportion of land that could be used within the new climatic limits. It is, however, just such 'marginal' areas that are most likely to show profound responses to climate change, as they did at the conclusion of the Medieval Warm Period. This 'new' land might therefore be little more than a reclaiming of ground lost with the onset of the Little Ice Age.

Down-scaling of coarse resolution scenarios is always problematical, and in this respect the work of Pepin (1995) is particularly interesting. He has taken the case of England's most extensive area of marginal land: the Pennines and Lake District uplands. Using existing relationships between altitude and temperature and the manner in which they vary according to the airflow type, Pepin was able to show that downscaled responses interpreted from GCM scenarios of global warming depend critically upon airflow. He shows, for example, that continental influences, through for example south-easterly airflows, might be highly beneficial in terms of bringing land into successful cultivation. Westerly airflows, however, generally having steeper climatic gradients, would be far less beneficial to upland areas, which would be exposed to a relatively severe climate under this regime. At present it is only through such detailed studies, taking present-day climatic relationships and expressing them within possible scenarios, that a detailed subregional picture can be established. Assumptions have to be made that present climatic associations between lapse rates and air circulation types will persist under future regimes. This, however, is a problem common to many 'downscaling' exercises carried out using a combination of GCM predictions and empirical relationships to define smaller-scale climatic behaviour.

Water Resources and River Systems

If precipitation forecasts are hedged with uncertainty, then predictions on water resource availability and river flow must be regarded as speculative at best. There need, for example, be no decrease in summer rainfall for resources to be diminished, and higher temperatures alone will encourage evaporation and transpiration, thereby reducing river flow. Arnell and Reynard (1993) have proposed that potential evapotranspiration may increase by between 5 and 29 per cent depending upon the scenario of future climate that is selected. Unfortunately, hydrological models based on climatic scenarios, whilst producing regional patterns of possible runoff change, tend to differ as much between themselves as between the regions. Nevertheless, the consensus view is that there will be greater runoff reduction in the south and east of Britain (where evapotranspiration already accounts for more than 50 per cent of precipitation) and a more muted response in the north and west. Given also the prediction of greater winter precipitation, it is not difficult to concur with any suggestion of higher runoff during that season. The assumption is made, of course, that the longer growing season and the higher temperatures of winter do not outweigh the gains made from anticipated higher precipitation. The summer-to-winter contrast in river regimes is likely therefore to be more marked in the future, and recent evidence (Marsh *et al.* 1994) has found that the period of warming in the 1980s and 1990s has indeed been accompanied by an increase in the winter : summer precipitation ratios. At the same time, those areas, such as north-east England and eastern Scotland, where precipitation is often in the form of snow may demonstrate another change. Higher temperatures may lead to less snowfall, and consequently less spring snow melt as water is not delayed in snow packs before entering the river system.

In other areas, especially the south-east of England, where groundwater discharge is so important, the picture is made yet more difficult to discern because of the geological control on river regimes. Where groundwater discharge is dominant, summer river flow may be maintained or may increase because of the time lag between winter recharge and summer runoff. This is one case in which wetter winters may compensate for drier summers. Another important factor, though it goes far beyond the scope of the present text, is the pattern of changing water demand. Higher temperatures alone have been estimated to account for a 5 per cent increase in domestic water demand within the next fifty years. Drier and warmer conditions, especially in south-east England, will also encourage increased demand for water from potentially competing uses such as irrigation and domestic water. However, as indicated above, the south-east of England will be one of those areas least able to sustain this level of demand. Any increase in the responsiveness of domestic demand to warm, dry weather as a result of increased use of washing machines, garden sprinklers and other items of household technology could have serious implications for water management, particularly in relation to competing uses such as irrigation of crops and the use of water in the leisure industry – the watering of golf courses being a good example. On the other hand, there are possible beneficial effects. The warm and dry summer of 1995 in much of Britain showed only too clearly that many coastal and inland resorts will reap handsome economic returns from an invigorated tourist trade in the future if such conditions become more commonplace.

Changing climates have consequences far beyond water supply. Other, undesirable, outcomes, especially from drier conditions, would be the subsidence in clay-rich soils, which shrink when suffering moisture loss (Doornkamp 1993). Much depends on the degree of moisture loss and the character of the soil, but insurance companies in particular are alive to this potentially expensive consequence as building foundations fracture.

Other threats result from sea level rise. This is a consequence of the warming and expansion of the surface waters of the oceans, as well an increase in ocean water mass due to ice cap and glacier melting. IPCC scenarios suggest that sea levels may rise between 32 and 65 cm by 2100. If this proves to be the case, then several parts of low-lying England – the Fenland and the Thames valley in particular – will run the risk of more frequent inundation (Whyte 1995; Parry and Duncan 1995)

Droughts and Their Effects

Droughts remain the most obvious consequence, but do not occur with uniform effect across the British Isles. Their foci depend upon the location of the blocking highs that so often bring them into being. For this reason, any use of well-documented droughts as an analogue for future conditions must be adopted with caution. Nonetheless, the events of 1988–92, which were characterised by a run of hot summers and mild winters, provide at least an indication of some possible consequences of climate

Box 12.5

DROUGHTS AND THE ECONOMIC LANDSCAPE: THE DRY SUMMER OF 1995

Droughts and dry spells not only bring stress to the natural system, but also impose burdens on society. As water demands continue to rise, it is increasingly difficult to disentangle the purely hydrological and meteorological aspects of drought from the socio-economic dimension. This observation notwithstanding, the events of the summer of 1995 provoked a public response far more critical of the water supply industry than did those of the comparable summers of 1975 and 1976. It demonstrated also some of the features frequently predicted for future climates with wetter winters and drier summers, the higher temperatures of the latter provoking further water loss through heightened evaporation, transpiration and water demand (Marsh and Turton 1996).

The winter of 1994/95 was generally wet with frequent and widespread flooding, and at its conclusion reservoir stocks were high with no indication of the difficulties yet to come. March produced close to average rainfall but April was dry with high pressure active over or to the west of the British Isles for much of the month, and river flows began to decline. May and June provided no respite, and hot, dry and sunny weather exaggerated river flow recessions and increased soil moisture deficits. July and August continued the drought, and at this point the water supply problem became serious in regions as disparate as Yorkshire, north-west England and Cornwall. If the rapid onset of the drought was similar to that of 1984, August in particular offered a scene reminiscent of the drought of 1976. The month was characterised by a well-developed ridge from the Azores, causing large parts of southern England to have less than 10 per cent of the usual rainfall. In common with much of the summer, temperatures were also exceptionally warm, this being indeed the warmest August in the Central England Temperature series (Horton *et al.* 1996). As in 1976, the exceptional weather provided a welcome stimulus for tourism, highlighting at least one postive economic consequence of warm, dry summers.

September provided a respite, as it had done in 1976, but by this stage the matter had become a topic of public debate and of government concern. October 1995 was one of the warmest in the CET record and a return to dry conditions did little to alleviate concerns or to provide an optimistic outlook for the approaching winter. If the events of the year were a harbinger for some of those of the twenty-first century, the resource, economic and political implications may be profound, and any gains from tourism and recreation will have to be balanced against the water resource implications.

change, and their ecosystem impacts have been reported in Cannell and Pitcairn (1993). Aquatic and terrestrial flora and fauna all revealed measurable responses to the changed winter and summer conditions. If these two years do indeed provide a reliable analogue for the future then significant environmental change can be anticipated in natural as well as agricultural systems.

CONCLUSION

It is difficult to predict with confidence the degree to which future trends will follow those established in the 1980s and 1990s. Climate variation and its regional expressions respond to a wide range of stimuli which interact in a complex manner. Whilst temperature might be modelled with reasonable confidence, the 'downscaling' of these results in the setting of the British Isles is complicated by the contrasting thermal responses of the north-east Atlantic and the near Continent. The future behaviour of atmospheric circulations will determine patterns of exposure and shelter, and will thereby help to influence not only precipitation but also regional and local temperatures. Regional-scale variations can otherwise only be discussed in a more speculative manner. But scenarios of future climates must acknowledge the intrinsic variability of the atmospheric circulation over a variety of time scales. Even if global warming unfolds as computer models predict, cold winters will still be recorded and their existence does not contradict any longer-term processes of climatic change.

Social, economic or physical consequences of climatic changes cannot be overlooked and will highlight the significance of regional weather and climate both as a hazard and as a resource. But whatever climatic pattern might present itself in the future it will continue to contribute to the diversity of the landscapes of the British Isles. The essence of our regional climates will also continue to be derived from the way in which the atmospheric circulation generates weather phenomena of all kinds. Despite the existence of climate models that are capable of producing plausible scenarios of future conditions, perhaps a more certain consequence of the unfolding pattern of climatic change will be the excitement of experiencing the unexpected. We may well recall, though in the setting of the British Isles rather than North America, the observation made by Mark Twain:

> There is a sumptuous variety about the weather, that compels the stranger's admiration – and regret. The weather is always doing something there; always attending strictly to business; always getting up new designs and trying them on people to see how they will go.

Might it always be so.

REFERENCES

Arnell, N.W. and Reynard, N.S. (1993) *Impact of Climate Change on River Flow Regimes in the United Kingdom*, Wallingford: Institute of Hydrology.

Barber, K.E. (1985) 'Peat stratigraphy and climatic change: some speculations', in M.J. Tooley and G.M. Sheail (eds) *The Climatic Scene*, London: Allen & Unwin.

Barnola, J.M. *et al.* (1986) 'Vostok ice core provides 160,000-year record of atmospheric CO_2', *Nature*, 329: 408–413.

Baur, F. (1949) 'Zurückführung des Grossweters auf rolare Erscheinungen', *Archiv. F. Met. Geophys. und Biokl.*, A: 358–374.

Beresford, M.W. (1973) 'Founded towns and deserted villages of the Middle Ages', in A.R.H. Baker and J.B. Harley (eds) *Man Made the Land*, Newton Abbot: David & Charles.

Bigg, G.R. (1990) 'El Niño and the Southern Oscillation', *Weather*, 45: 2–8.

—— (1995) 'The El Niño event of 1991–94', *Weather*, 50: 117–124.

Black, A.R. and Bennett, A.M. (1995) 'Regional flooding in Strathclyde December 1994', *Hydrological Data UK: 1994 Yearbook*, Wallingford: Institute of Hydrology/British Geological Survey.

Bonacina, L.C.W. (1957) 'Snowfall in Great Britain during the decade 1946–1955', *British Rainfall*, 95: 219–230.

Booth, R.E. (1968) 'The severe winter of 1963 compared with other cold winters, particularly that of 1947', *Weather*, 23: 477–479.

Brazell, J.H. (1968) *London Weather*, London: HMSO.

Burroughs, W.J. (1980) 'Average temperature and rainfall figures in British winters', *Weather*, 35: 75–79.

—— (1992) *Weather Cycles: Real or Imaginary?*, Cambridge: Cambridge University Press.

Burt, S. (1992) 'The exceptional hot spell of early August 1990 in the United Kingdom', *Int. Jnl Climatol.*, 12: 547–567.

Burton, J. (1994) 'The Meteorological Office observatory programme 1867 to 1883', in B.D. Giles and J.M. Kenworthy (eds) *Observatories and Climatological Research*, Occasional Publication 29, Durham: University of Durham Press.

Cannell, M.G.R. and Pitcairn, C.E.R. (1993) *Impacts of the Mild Winters and Hot Summers in the United Kingdom in 1988–1990*, London: HMSO.

CCIRG (1991) *The Potential Effects of Climate Change in the United Kingdom*. London: HMSO.

—— (1996) *The Potential Impacts of Climate Change and Adaptation in the United Kingdom*, London: HMSO.

Clark, J.B. (1979) 'An investigation into recent October temperature trends over Central England', *Weather*, 33: 101–109.

Cormack, P. (1994) *English Cathedrals*, London: Artus Books.

Currie, I.J.M. (1995) 'The great frost: the winter of 1894/95', *Weather*, 50: 66–73.

Dansgaard, W., White, J.W.C. and Johnsen, S.J. (1989) 'The abrupt termination of the Younger Dryas climate event', *Nature*, 339: 532–537.

Davis, N.E. (1968) 'An optimum summer weather index', *Weather*, 23: 305–317.

Defant, A. (1924) 'Die Schwankungen der atmosphärischen Zurkulation über dem nordatlantischen Ozean in 25-jähringen 1881–1905', *Geogr. Ann.*, 6, 13–41.

Defoe, D. (1704) *The Storm*, London.

Doornkamp, J.C. (1993) 'Clay shrinkage induced subsidence', *Geographical Journal*, 152: 196–202.

Douglas, K.S., Lamb, H.H. and Loader, C. (1978) 'Weather observations and a tentative meteorological analysis of the

period May to July 1588', *Climatic Research Unit Publ. CRU RP 6a*, Norwich: University of East Anglia.
Eddy, J.A. (1976) 'The Maunder Minimum', *Science*, 192: 1189–1202.
Eden, G.P. (1995) *Weatherwise*, London: Macmillan.
Folland, C.K. and Wales-Smith, B.G. (1977) 'Richard Towneley and 300 years of regular rainfall measurement', *Weather*, 32: 438–445.
Folland, C.K., Parker, D.E. and Newman, M. (1984) *Worldwide Marine Temperature Variations on the Season to Century Timescale*, Corvallis: 9th Climate Diagnostics Workshop.
Folland, C.K., Karl, T.R., Nicholls, N., Nyenzi, B.S., Parker, D.E. and Vinnikov, K.Ya. (1992) 'Observed climate variability and change', in J.T. Houghton, B.A. Callander and S.K. Varney (eds) *Climate Change 1992: The Supplementary Report to the IPCC Scientific Assessment*, Cambridge: Cambridge University Press.
Frenzel, B., Pfister, C. and Gläser, B. (1994) (eds) *Climatic Trends and Anomalies in Europe 1675–1715*, Stuttgart: Gustav Fischer Verlag.
Genthon, C. *et al.* (1986) 'Vostok ice core: climatic response to CO_2 and orbital forcing changes over the last climatic cycle', *Nature*, 329: 414–418.
Godwin, H. (1975) *History of the British Flora*, 2nd edn, Cambridge: Cambridge University Press.
Goudie, A.S. (1977) *Environmental Change*, Oxford: Oxford University Press.
Gregory, S. (1956) 'Regional variations in the annual rainfall over the British Isles', *Geographical Jnl*, 122: 346–353.
Gribbin, J. (1978) 'Astronomical influences', in J. Gribbin (ed.) *Climatic Change*, Cambridge: Cambridge University Press.
Grove, J.M. (1988) *The Little Ice Age*, London: Methuen.
Hall, N.M.J., Hoskins, B.J., Valdes, P.J. and Senior, C.A. (1994) 'Storm tracks in a high-resolution GCM with doubled carbon dioxide', *Q. J. R. Meteorol. Soc.*, 120: 519, 1209–1230.
Hatch, D.J. (1973) 'British climate in generalised maps', *Weather*, 28: 509–516.
Horton, E.B., Cullum, D.P.N. and Folland, C.K. (1996) 'Global and regional climate in 1995', *Weather*, 51: 202–210.
Hoskins, W.G. (1988) *The Making of the English Landscape*, London: Hodder & Stoughton.
Houghton, J.T., Jenkins, G.J. and Ephraums, J.J. (1990) (eds) *Climate Change: The IPCC Scientific Assessment*, Cambridge, Cambridge University Press.
Houghton, J.T., Callander, B.A. and Varney, S.K. (eds) (1992) *Climate Change: The Supplementary Report to the IPCC Scientific Assessment*, Cambridge: Cambridge University Press.
Houghton, J.T., Meira Filho, L.G., Bruce, J., Hoesung Lee, Callander, B.A., Haites, E., Harris, N. and Maskell, K. (1995) *Climate Change 1994: Radiative Forcing of Climate Change*, Cambridge: Cambridge University Press.
Houghton, J.T., Meira Filho, M.G., Callander, B.A., Harris, N., Kattenberg, A. and Maskell, K. (eds) (1996) *Climate Change 1995: The Science of Climate Change*, Cambridge: Cambridge University Press.
Howells, K.A. (1985) 'Changes in the magnitude–frequency of heavy daily rainfalls in the British Isles', unpublished PhD thesis, University of Wales.
Hoyt, D.V. (1979a) 'An empirical heating of the Earth by the carbon dioxide greenhouse effect', *Nature*, 282: 388–390.
—— (1979b) 'Variations in sunspot structure and climate', *Climatic Change*, 2: 79–92.
Hughes, G.H. (1979) 'The summers of 1968 and 1977: some similarities and contrasts', *Weather*, 34: 319–325.
Hughes, M.K. and Diaz, H.F. (1994) *The Medieval Warm Period*, Dordrecht: Kluwer.
Hulme, M. (1996) *The 1996 CCIRG Scenario of Changing Climate and Sea-Level for the UK*, Technical Note no. 7, Norwich: Climatic Research Unit.
Irvine, S.G. (1969) '1968 – Shetland's record breaking weather year', *Weather*, 24: 234–235.
Jones, H.S. (1955) *Sunspots and Geomagnetic-Storm Data Derived from Greenwich Observations, 1874–1954*, London: HMSO.
Jouzel, J., Lorius, C., Petit, J.R., Genthon, C., Barkov, N.I., Kotlyakov, V.M. and Petrov, V.M. (1987) 'Vostok ice core: a continuous isotope temperature record over the last climatic cycle (160,000 years)', *Nature*, 329: 403–407.
Kanda, S. (1933) 'Ancient records of sunspots and auroras in the far east and the variation of the period of solar activity', *Proc. Imp. Academy (Tokyo)*, 9: 293–296.
Keeley, H.C.M. (1982) 'Pedogenesis during the later prehistoric period in Britain', in A.F. Harding (ed.) *Climatic Change and Later Prehistory*, Edinburgh: Edinburgh University Press.
Kington, J. (1988) *The Weather of the 1780s over Europe*, Cambridge: Cambridge University Press.
—— (1997) 'A history of observing the weather', in M. Hulme and E. Barrow (eds) *The Climate of the British Isles: Present, Past and Future*, London: Routledge.
Labitzke, K. and van Loon, H. (1988) 'Association between the 11-year solar cycle, the QBO and the atmosphere, Part I: The troposphere and the stratosphere of the Northern Hemisphere winter', *J. Atmos. Terres. Phys.*, 50: 197–206.
Lamb, H.H. (1965) 'The early medieval warm epoch and its sequel', *Palaeogeography, Palaeoclimatology, Palaeoecology*, 1: 13–37.
—— (1969) 'The new look of climatology', *Nature*, 223: 1209–1215.
—— (1970) 'Volcanic dust in the atmosphere: with a chronology and assessment of its meteorological significance', *Phil. Trans. R. Soc. London, series A*, 266: 425–533.
—— (1972) *British Isles Weather Types and a Register of the Daily Sequence of Circulation Patterns, 1861–1971*, Geophys. Mem. no. 116, London: HMSO.
—— (1977) *Climate: Present, Past and Future*, vol. 2: *Climatic History and the Future*, London: Methuen.
—— (1991) *Historic Storms of the North Sea, British Isles and Northwest Europe*, Cambridge: Cambridge University Press.
—— (1995) *Climate, History and the Modern World*, London: Routledge.
Lamb, H.H., Lewis, R.P.W. and Woodroffe, A. (1966) 'Atmospheric circulation and the main climatic variables between 8000 and 0 BC: meteorological evidence', in J.S. Sawyer (ed.) *World Climate from 8000 to 0 BC*, London: Royal Meteorological Society.
Lough, J.M., Wigley, T.M.L. and Palutikof, J.P. (1983) 'Climate and climate impact scenarios for Europe in a warmer world', *J. Clim. Appl. Meteor.*, 22: 1673–1684.
Lyall, I. (1976) 'Recent trends in British October weather', *Weather*, 31: 322–327.
McGovern, T.H. (1981) 'The economics of extinction in Norse Greenland', in T.M. Wigley, M.J. Ingram and G. Farmer (eds)

Climate and History: Studies in Past Climates and Their Impact on Man, Cambridge: Cambridge University Press.

Manley, G. (1941) 'The Durham meteorological record', *Q.J.R. Meteorol. Soc.*, 67: 363–380.

—— (1946) 'Temperature trends in Lancashire, 1753–1945', *Q.J.R. Meteorol. Soc.*, 72: 1–31.

—— (1952) *Climate and the British Scene*, London: Collins.

—— (1953) 'The mean temperature of Central England, 1698 to 1952', *Q.J.R. Meteorol. Soc.*, 79: 242–261.

—— (1959) 'Temperature trends in England, 1698–1957', *Archiv für Met. Geophys. u. Bioklim.*, 9: 413–433.

—— (1971) 'The mountain snows of Britain', *Weather*, 26: 192–200.

—— (1973) 'Climate in Britain over 10,000 years', in A.R.H. Baker and J.B. Harley (eds) *Man Made the Land*, Newton Abbot: David & Charles.

—— (1974) 'Central England temperatures: monthly means 1659–1973', *Q.J.R. Meteorol. Soc.*, 100: 389–405.

Marsh, T.J. and Lees, M. (1985) *The 1984 Drought*, Wallingford: Institute of Hydrology, British Geological Survey.

Marsh, T.J., Monkhouse, R.A., Arnell, N.W., Lees, M.L. and Reynard, N.S. (1994) *The 1988–92 Drought*, Wallingford: Institute of Hydrology, British Geological Survey.

Marsh, T.J. and Turton, P.S. (1996) 'The 1995 drought: a water resources perspective', *Weather*, 51: 46–53.

Mason, B.J. (1976) 'Towards the understanding and prediction of climatic variations', *Q. J. R. Meteorol. Soc.*, 102: 473–498.

Mayes, J.C. (1991) 'Recent trends in summer rainfall', *Weather*, 46: 190–196.

—— (1995) 'Changes in the distribution of annual rainfall in the British Isles', *Water and Environmental Management*, 9: 531–539.

—— (1996) 'Spatial and temporal fluctuations of monthly rainfall in the British Isles and variations in the mid-latitude westerly circulation', *Int. J. Climatol.*, 16: 585–596.

Meaden, G.T. (1976) 'North-west Europe's great drought: the worst in Britain since 1252–1253?', *J. Meteorology (UK)*, 1: 379–383.

Mitchell, J.M. (1972) 'The natural breakdown of the present interglacial and its possible intervention by human activities', *Quaternary Res.*, 2: 436–445.

Mitchell, J.M., Stockton, C. and Meko, D.M. (1978) 'Evidence of a 22-year rhythm of drought in the western United States related to the Hale solar cycle since the 17th century', paper presented at Solar–Terrestrial Influences on Weather and Climate, Columbus, Ohio, 24–28 July 1978.

Murray, R. (1993) 'Bias in southerly synoptic types in decade 1981–90 over the British Isles', *Weather*, 48: 152–154.

Namias, J. (1987) 'Factors relating to the explosive North Atlantic cyclone of December 1986', *Weather*, 42: 322–325.

Nordo, J. (1955) 'A comparison of secular changes in terrestrial climate and sunspot activity', *Videnskaps-Akademiets Institut for Vaer- ag Klimatorskning*, Oslo, Rept. no. 5.

Ogilvie, A.E.J. and Farmer, G. (1997) 'Documentary evidence for changes in climate in England and the North Atlantic region during the medieval period', in Hulme, M. and Barrow, E. (eds) *The Climate of the British Isles: Present, Past and Future*, London: Routledge.

Palutikof, J.P. (1987) 'Some possible impacts of greenhouse gas induced climatic change on water resources in England and Wales', in *The Influence of Climate Change Variability on the Hydrologic Regime and Water Resources* (Proceedings of the Vancouver Symposium; IAHS Publ. no. 168), Vancouver: IAHS.

Palutikof, J., Holt, T. and Skellern, A. (1997) 'Wind: resource and hazard', in M. Hulme and E. Barrow (eds) *The Climate of the British Isles: Present, Past and Future*, London: Routledge.

Parker, D.E. and Folland, C.K. (1988) ' The nature of climate variability', *Meteorol. Mag.*, 117: 201–210.

Parker, D.E., Legg, T.P. and Folland, C.K. (1992) 'A new daily Central England Temperature series, 1772–1991', *Int. J. Climatol.*, 12: 317–342.

Parry, M.L. (1975) 'Secular climatic change and marginal land', *Trans. Inst. Br. Geogrs*, 64: 1–13.

—— (1990) *Climate Change and World Agriculture*, London: Earthscan.

Parry, M.L. and Duncan, R. (1995) *The Economic Implications of Climate Change in Britain*, London: Earthscan

Pearson, M.G. (1973) 'Snowstorms in Scotland, 1782–1786', *Weather*, 28: 195–201.

Pennington, W. (1974) *The History of British Vegetation*, 2nd edn, London: English Universities Press.

Pepin, N. (1995) 'The use of GCM scenario output to model effects of future climatic change on the thermal climate of marginal maritime uplands', *Geogr. Anal.*, 77A: 167–184.

Perry, J.D. (1977) 'The relationship between the strength of the quasi-biennial oscillation and the equatorial stratosphere and the mean anomaly of the monthly mean maximum screen temperature at Oxford', *Meteorol. Mag.*, 106: 212–219.

Pickering, K.T. and Owen, L.A. (1994) *An Introduction to Global Environmental Issues*, London: Routledge.

Pike, W.S. (1995) 'Rivalry on ice: skating and the memorable winter 1894/95 winter', *Weather*, 50: 48–54.

Rahmstorf, S. (1995) 'Bifurcations of the Atlantic thermohaline circulation in response to changes in the hydrological cycle', *Nature*, 378: 145–149.

Robin, G. de Q. (1983) *The Climate Record in Polar Ice Sheets*, Cambridge: Cambridge University Press.

Rowntree, P.R., Murphy, J.M. and Mitchell, J.F.B. (1993) 'Climate change and future rainfall predictions', *J. Inst. Water and Envir. Man.*, 7: 464–470.

Salter, M. de C.S. and Glasspoole, J. (1923) 'The fluctuations of annual rainfall in the British Isles considered cartographically', *Q. J. R. Meteorol. Soc.*, 49: 207–229.

Shellard, H.C. (1968) 'The severe winter of 1962–63 in the United Kingdom: a climatological survey', *Meteorol. Mag.*, 97: 129–141.

Shone, K.B. (1980) 'A survey of British spring weather 1950 to 1979', *Weather*, 35: 68–75.

Simmons, I.G. (1964) 'Pollen diagrams from Dartmoor', *New Phytologist*, 63: 165–180.

Smith, K. (1995) 'Precipitation over Scotland, 1757–1992: some aspects of temporal variability', *Int. J. Climatol.*, 15: 543–556.

Stirling, R. (1982) *The Weather of Britain*, London: Faber & Faber.

Stuiver, M. and Braziunus, T.F. (1992) 'Evidence of solar activity variations', in R.S. Bradley and P.D. Jones (eds) *Climate since A.D. 1500*, London: Routledge.

Suess, H. (1965) 'Secular variations of the cosmic-ray-produced carbon 14 in the atmosphere and their interpretations', *Jr. Geophys. Res.*, 70: 5937–5952.

Tabony, R.C. (1979) 'A spectral filter analysis of long period records in England and Wales', *Meteorol. Mag.*, 108: 102–112.

Tarling, D.H. (1978) 'The geological–geophysical framework of ice ages', in J. Gribbin (ed.) *Climatic Change*, Cambridge: Cambridge University Press.

Tooley, M. and Sheail, G.M. (1985) 'The life and work of Gordon Manley', in M. Tooley and G.M. Sheail (eds) *The Climatic Scene*, London: Allen & Unwin.

Trenberth, K.E. (1991) 'Recent climatic changes in the Northern Hemisphere', in M.E. Schlesinger (ed.) *Greenhouse-Gas-Induced Climatic Change: A Critical Appraisal of Simulations and Observations*, Amsterdam: Elsevier.

Tyson, P.D. (1986) *Climate Change and Variability in Southern Africa*, Oxford: Oxford University Press.

Vernekar, A. (1972) 'Long-period global variations in incoming solar radiation', *Meteorological Monographs*, 12: no. 34.

Weber, G.R. (1990) 'Tropospheric temperature anomalies in the northern hemisphere 1977–1986, *Int. J. Climatol.*, 10: 3–19.

Weston, K.J. and Roy, M.G. (1994) 'The directional-dependence of the enhancement of rainfall over complex orography', *Meteorological. Applications*, 1: 267–275.

Wheeler, D. (1994) 'The weather diary of Margaret MacKenzie of Delvine (Perthshire): 1780–1805', *Scot. Geogr. Mag.*, 110: 177–182.

Whyte, I.D. (1995) *Climatic Change and Human Society*, London: Arnold.

Wigley, T.M.L. and Atkinson, B.W. (1977) 'Dry years in S. E. England since 1698', *Nature*, 265: 431–434.

WMO (1995) *The Global Climate System Review: climate system monitoring June 1991–November 1993*, Geneva: World Meteorological Organization.

GLOSSARY OF TERMS

The following glossary of terms lists those items italicised in the text which may require definition. Whilst every endeavour has been made to limit the use of technical terms to those necessary for a relatively straightforward discussion of climatic issues their use cannot be altogether avoided. In recognition of this inevitability, and to avoid any distracting elaboration in the text, the following definitions should act as a comprehensive guide. For a full meteorological glossary the reader is referred to *The Meteorological Glossary* prepared by the UK Meteorological Office and published by HMSO.

adiabatic: refers to a process in which heat neither enters nor leaves a system but in which temperature changes nonetheless occur. In atmospheric examples it results from the expansion and contraction of 'parcels' of air as they rise (or descend). If no condensation of water vapour takes place, cooling (heating) occurs at the dry adiabatic rate of 0.98 degrees C/100 m of altitude gained (lost). When condensation occurs, **latent heat** is released and the cooling takes place at a slower saturated adiabatic rate which is variable and depends upon the rate and quantity of condensation.

air mass: a body of air, usually of the order of hundreds or thousands of kilometres across, within which there is little horizontal differentiation of temperature and humidity.

albedo: a measure of the reflectivity of any surface in response to direct solar radiation. Measured on a scale of zero (no reflectivity) to 1.0 (perfect reflectivity).

anabatic wind: a local wind which develops over heated hillsides initiating the upslope motion of warmer, buoyant, air. *See also* **katabatic wind**.

blocking: the interruption of the eastwards progress of mid-latitude weather systems. Usually the result of a stationary anticyclone with a low **zonal index** in the upper westerlies.

cold front: the movement of a boundary between warm and cold air in such a way that the latter displaces the former. *See also* **warm front**.

conveyor belt: the inwards and ascending motion of tropical maritime air within a mid-latitude depression.

Coriolis force: an apparent force acting on the direction of a free-moving body on a rotating system. Air moving over the Earth's rotating surface is subject to this force, causing it to deflect to the right of its direction of motion in the northern hemisphere.

cut-off low: a closed circulation of polar air isolated to the equatorwards side of the normal passage of depressions. It occurs under conditions of a low **zonal index**.

cyclogenesis: the development or strengthening of an air circulation around a low pressure centre.

dew point: the temperature at which a body of air reaches saturation point (100 per cent relative humidity). Air can be cooled to its dew point without the need for further moisture to be added.

environmental lapse rate: the prevailing rate of temperature change with altitude. The rate is variable according to the air mass, season and local effects; it may be positive or negative but is usually taken to have a long-term mean of −0.65 degrees C/100 m. *See also* **adiabatic** changes.

Ferrel cell: the mid-latitude circulation immediately polewards of the **Hadley cell**.

frost hollow: a valley floor area into which cold air will drain by the motion of **katabatic winds**. A location for pronounced and unseasonal frosts.

geostrophic wind: the wind which results from the balance between the pressure gradient and the **Coriolis** forces. In detail it should be applied only to air moving in a straight line.

greenhouse gases: those gases that are largely transparent to short-wave (solar) radiation but opaque with regard to long-wave (terrestrial) radiation.

Hadley cell: a thermal circulation of rising warm air over the equatorial latitudes and descending at approximately 30° N (and 30° S) where it sustains the subtropical anticyclones.

jet stream: a narrow band of strong wind (often in excess of 100 km/h), usually found close to the tropopause and often several hundreds of kilometres in length but only a few kilometres in depth and width.

katabatic wind: a local wind which develops over cold surfaces. Heat loss from the air in contact with the cooling land surface causes it to become denser and to be displaced downslope, usually into a **frost hollow**. *See also* **anabatic wind**.

latent heat: heat that brings about a change of state but not of temperature in the receiving body. *See also* **sensible heat**.

meridional airflow: northwards or southwards motion of the upper westerly circulations indicative of a low **zonal index** with a high probability of **blocking**.

polar low: a non-frontal low-pressure system which usually forms in **unstable** polar airstreams. Polar lows are commonly smaller than fully developed frontal lows.

radiation fog: when air cools, especially on calm nights, its temperature may fall to its **dew point**, leading to condensation of water vapour and to fog.

radiation window: those wavebands in the electromagnetic spectrum within which greenhouse gases are transparent to radiation. The term is commonly applied within the long-wave (terrestrial) radiation ranges.

Rossby wave: a wave or major deformation in the upper westerly circulation causing airstreams to meander across the mid-latitudes. *See also* **zonal index**.

secondary low: a depression that forms within the circulation of an already existing system. Secondary lows most commonly occur on trailing cold fronts of parent systems.

sensible heat: heat gain or loss that brings about a change in temperature of the receiving body. *See also* **latent heat**.

specific heat: the quantity of heat energy required to raise the temperature of 1 gram of any substance by 1 degree C.

stable: the atmosphere is said to be stable when it has no tendency to mix vertically. It is found to prevail in anticyclonic systems or when air masses pass over relatively cool land or sea surfaces. *See also* **unstable**.

stratosphere: the layer in the atmosphere immediately overlying the **troposphere** and extending from approximately 10 to 50 km altitude.

temperature inversion: the point in the atmosphere at which temperature decreases with altitude give way to a layer in which the temperature increases. The **tropopause** is a quasi-permanent temperature inversion, but others may result from low-level radiation cooling, subsidence or the presence of a front.

tropopause: the boundary between the **troposphere** and **stratosphere**. Its height varies from approximately 15 km in equatorial latitudes to 8 km in the polar regions.

troposphere: the lowest layer of the atmosphere below the **tropopause**. Also known as the 'weather layer'.

unstable: the atmosphere is said to be unstable when it has a tendency to mix vertically. This commonly occurs in relatively cool air masses whose lower layers have been heated by passage over a relatively warm land or sea surface. *See also* **stable**.

warm front: the movement of a boundary between warm and cold air in such a way that the former displaces the latter. *See also* **cold front**.

zonal index: a measure of the strength of the westerly circulation over the mid-latitudes (35°–60° m/s). *See also* **meridional airflow**.

GENERAL INDEX

INDEX OF PLACES

Entries in bold indicate tabulated statistics for the cited locations.